ANNUAL REVIEW OF NUCLEAR AND PARTICLE SCIENCE

ANNUAL REVIEW OF NUCLEAR AND PARTICLE SCIENCE

VOLUME 39, 1989

BIP-94

ANNUAL REVIEWS INC   4139 EL CAMINO WAY   P.O. BOX 10139   PALO ALTO, CALIFORNIA 94303-0897

International Standard Serial Number: 0163-8998
International Standard Book Number: 0-8243-1539-1
Library of Congress Catalog Card Number: 53-995

TYPESET BY AUP TYPESETTERS (GLASGOW) LTD., SCOTLAND
PRINTED AND BOUND IN THE UNITED STATES OF AMERICA

PREFACE

Volume 39 begins with reminiscences about 1939. We are honored to have a special prefatory chapter by John A. Wheeler concerning those heady and momentous days fifty years ago when the scientific world was just beginning to comprehend the phenomenon of nuclear fission.

The regular contents of the volume are as catholic in coverage as ever. The energy range spanned is more than 20 decades, from the milli-electron volts important in muon-catalyzed fusion (Breunlich, Kammel, Cohen & Leon) to 10^{20} eV for the highest energy cosmic rays (Wdowczyk & Wolfendale). The unusual is the focus for nuclear physics, with reviews on heavy-particle radioactivity (Price), exotic light nuclei (Détraz & Vieira), and nuclear systems with strangeness (Chrien & Dover). Astrophysics and nuclear/particle physics meet in the chapters on supernovae and the hadronic equation of state (Kahana) and on solar neutrinos (Davis, Mann & Wolfenstein).

The evolution in our concepts in the past 50 years is nowhere more evident than in the particle physics reviews. Quarks are at the heart of all and are nowhere explicitly manifest, as we learn from the survey of searches for fractionally charged particles (Smith). The subjects of D mesons (Morrison & Witherell) and especially deep inelastic lepton-hadron scattering (Mishra & Sciulli) have as their bases descriptions in terms of quarks. Perturbative QCD (Altarelli) is a theoretical description of energetic hadronic interactions in terms of quarks and gluons; its tests give credence to that picture in a very detailed way. Even purple cows have masses and those masses have physical manifestations; Donoghue reviews the evidence on "current" quark masses and chiral symmetry.

The chapter on technical issues in arms control (Lynch, Meunier & Ritson) is of more than ordinary interest as the great powers begin to talk and even act on reduction of nuclear and conventional armaments.

This volume is the last to benefit from the expert touch of Associate Editor Roy F. Schwitters, who relinquishes his stewardship because of the burden of other duties. Roy's place is now taken by Vera G. Lüth of the Stanford Linear Accelerator Center. Harry Gove and I shall miss Roy's congenial and effective contributions to our common work, but Vera is a worthy successor. We salute Roy in his new endeavors and welcome Vera to the panel of editors.

We note with particular sadness the death of former Editor Emilio Segrè on 22 April 1989 at the age of 84. Segrè's scientific achievements are well known and are in any event sketched in his autobiographical prefatory chapter in Volume 31 of this series. A member of the Rome school under

(*continued*) v

Fermi in the early 1930s, Segrè went on to carve out an independent and most distinguished career at Palermo, Berkeley, Los Alamos, and at Berkeley again since 1945. The Nobel Prize (shared with Owen Chamberlain) is just one of his many honors. His services to the physics community are many, perhaps best exemplified by his association with the *Annual Review of Nuclear Science*, as it was then called. He was a member of the first Editorial Committee (listed opposite the title page of Volume 2), serving for six years in that role before becoming Editor in January of 1958. He was Editor for nineteen years, retiring officially at the end of December 1976, but serving nearly a year beyond that date to see Volume 27 through to publication. He thus was associated with this series for 25 years and responsible editor of 20 volumes.

At the start of Segrè's tenure as its Editor, this series stood almost alone as an authoritative review of nuclear science. Its usefulness and importance was never doubted. By the end of his service, the whole structure and mechanisms of scientific communication had changed. Preprints, conference talks, and telephone calls replaced the refereed journals of record as first sources of information; many other review journals and books had come into existence. It was characteristic of Segrè's pragmatic attitude that in his last year as Editor he raised seriously with the Editorial Committee the question of whether to close down the *Annual Review of Nuclear Science* in the light of these changes. I know he was fully prepared to implement the termination of the series if that had been recommended. In the event, continuation was urged. Now, twelve years later, the present Editors endeavor to preserve his high standards. As Editor of more than half the volumes, Emilio Segrè *is*, more than anyone else, the *Annual Review of Nuclear and Particle Science*. The success of the series over the years is a fitting tribute to his memory. To him we dedicate this volume.

J. D. JACKSON
EDITOR

This volume is dedicated
to the memory of
EMILIO SEGRÈ
1905–1989

Annual Review of Nuclear and Particle Science
Volume 39, 1989

CONTENTS

ANNUAL REVIEWS INC. is a nonprofit scientific publisher established to promote the advancement of the sciences. Beginning in 1932 with the *Annual Review of Biochemistry*, the Company has pursued as its principal function the publication of high quality, reasonably priced *Annual Review* volumes. The volumes are organized by Editors and Editorial Committees who invite qualified authors to contribute critical articles reviewing significant developments within each major discipline. The Editor-in-Chief invites those interested in serving as future Editorial Committee members to communicate directly with him. Annual Reviews Inc. is administered by a Board of Directors, whose members serve without compensation.

SOME GENERAL INTEREST OR RELATED ARTICLES IN OTHER *ANNUAL REVIEWS*

From the *Annual Review of Astronomy and Astrophysics*, Volume 27 (1989):

Dreams, Stars, and Electrons, Lyman Spitzer, Jr.

X Rays from Normal Galaxies, G. Fabbiano

Supernova 1987A, W. David Arnett, John N. Bahcall, Robert P. Kirshner, and Stanford E. Woosley

From the *Annual Review of Materials Science*, Volume 19 (1989):

Fractal Phenomena in Disordered Systems, R. Orbach

From the *Annual Review of Physical Chemistry*, Volume 40 (1989):

Atomic-Resolution Surface Spectroscopy with the Scanning Tunneling Microscope, R. J. Hamers

John A. Wheeler

Annu. Rev. Nucl. Part. Sci. 1989. 39: xiii–xxviii

FISSION IN 1939: The Puzzle and the Promise

John Archibald Wheeler

Physics Department, Princeton University, Princeton, New Jersey 08544; and Physics Department, University of Texas, Austin, Texas 78712

KEY WORDS: compound nucleus, channel count, fissility parameter, saddle point, barrier penetration.

Never more actively, responsibly, and productively than today do historians of science study the evolution of modern physics. Their enterprise founds itself on written records, historical training, and scholarship. As they build, however, the living lamps of memory one by one go out. Therefore, the present account—by one early participant in fission physics—will perhaps be more useful if it is conceived, not as history, but as memories, impressions, atmosphere.

I stood in the winter cold at Pier 97, North River, New York on Monday, January 16, 1939 to welcome Niels Bohr (Figure 1), about to debark from the Swedish-American liner *Drottningholm*. In my wait Enrico and Laura Fermi (Figure 2) joined me. Fermi had been at Columbia University for less than a month since his Rome-to-Stockholm trip. As Bohr cleared customs we greeted him, his son Erik, and his long-time colleague Léon Rosenfeld. Their upcoming three-month stay at the Institute for Advanced Study in Princeton had for Bohr an overriding purpose: Clarify the quantum. To that end pursue the long-continued dialogue (1) with Einstein (Figure 3). Do everything possible, man to man, to reach agreement with him.

January 3rd, however, had opened to Bohr a second vista. That day, just before the *Drottningholm* sailed, Otto Robert Frisch—a friend of mine since my 1934–1935 year in Copenhagen—fresh back from Sweden, reported to him the conclusions to which he and his aunt, Lise Meitner, had been forced (2) by the not yet published findings of Otto Hahn and Fritz Strassmann (3). "I had hardly begun to tell him," Frisch writes (4), "when he struck his forehead with his hand and exclaimed, 'Oh what

xiii

0163–8998/89/1201–0xiii$02.00

idiots we have all been! Oh but this is wonderful! This is just as it must be.' "

Show that fission as then known really does proceed just as it must: This goal held itself out ever more invitingly to Bohr with each new day of pacing up and down the deck with the one shipboard companion or the other. Nothing of this new goal or of fission itself did I or anyone in America know when I greeted Bohr at the pier. How could I have anticipated that he would invite me to join him in this enterprise? Or that we would extend the work (5): **predict** as yet undiscovered features of fission?

Bohr felt obligated not to let out word of fission until Frisch with an ionization chamber or otherwise could demonstrate splitting and send in his findings for publication (6). Niels and Erik went into Manhattan with Enrico and Laura to visit old friends, father and son spending a night or two there before coming down to Princeton. Rosenfeld and I, however, took the next train. He was unaware of Bohr's self-imposed commitment to silence on fission. He revealed the exciting news on the one-hour trip. I got him to report the great discovery that very evening at the regular Monday 7–9 p.m. Journal Club (Figure 4).

Bohr arrived a day or two later, discovered that the cat was out of the bag, told me more and we got to work. The aim was straightforward. The burgeoning world of experimental findings: bring them into order within the framework of Bohr's compound-nucleus model of nuclear reactions. This model he had first enunciated in 1935, during my time in Copenhagen.

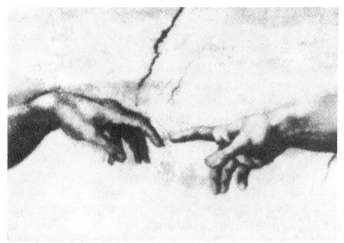

Figure 1   The Old World reached across to the New World—at Pier 97 North River, New York, Monday January 16, 1939—with the word of fission [Detail from Michelangelo's *The Creation of Adam*, Sistine Chapel.]

Figure 2 The Fermis shortly after their arrival in the United States. Courtesy of Wide World Photos, Inc.

Figure 3 Niels Bohr and Albert Einstein in dialogue in the Leyden home of Paul Ehrenfest in 1933. The restoration of the negative of Ehrenfest's photograph and the production of the print were done by William R. Whipple. Courtesy of the American Institute of Physics Niels Bohr Library.

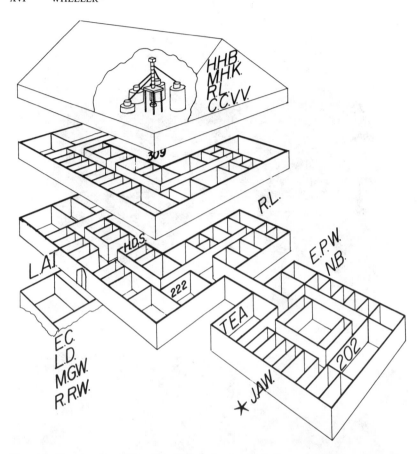

Figure 4   Star: location of the Fine Hall (today Jones Hall) ground floor seminar room where Rosenfeld made known the discovery of fission. Subsequent Princeton work on fission was divided between the main floor of Fine Hall (Eugene P. Wigner, Bohr, and the writer) and the adjacent Palmer Physical Laboratory. Cyclotron: Edward Creutz, Luis A. Delsasso, Milton G. White, and Robert R. Wilson. Main floor: Louis A. Turner, Henry D. Smyth, and Rudolf Ladenburg. Attic: Van de Graaff generator of neutrons: Henry H. Barschall, Morton H. Kanner, Ladenburg, and Cletus C. Van Voorhis. TEA: "where we explain to each other what we don't understand." Room 222: graduate courses in physics. By January 1939 Einstein had vacated his former Fine Hall office (E.P.W.) for new quarters at the Institute for Advanced Study, then abuilding a mile away, but nearby, over the fireplace in the Professors' Room (# 202), are engraved his famous words, "Raffiniert ist der Herr Gott, aber boshaft ist Er nicht." Of physics colloquia in Room 309 he attended occasional ones, including the one on the mechanism of fission; and there he gave his last talk (7) before his death.

Since then he and I had been working on the development and application of this model (Figure 5).

As we proceeded with our work, we found we had to introduce concepts new to nuclear theory: **fissility parameter**, nuclear **potential-energy** surface, **saddle-point energy** as threshold for fission, **channel** open for over-the-barrier fission, **channel count** as determiner of the contribution of fission to level width, and **spontaneous fission** via barrier penetration. The analysis culminated in a 25-page paper, "The Mechanism of Nuclear Fission." As if omen of a new world of weapons, it appeared in the issue of the *Physical Review* dated the first of September, 1939, the very day World War II began. What was the background for the collaboration of the American junior partner in this work?

No better symbol do I know than Michelangelo's great Sistine Chapel painting for the lightning stroke of January 16th, 1939 that brought the word of fission from Europe to America. No fitter image, either, can I offer for the electricity of learning that had flowed from the outstretched

Figure 5 A few of the participants in the Third Washington Conference on Theoretical Physics of February 18, 1937, where the compound-nucleus model of nuclear reactions received considerable attention. Niels Bohr, front; I. I. Rabi and George Gamow, second row; Fritz Kalckar and John Wheeler, third row; and Gregory Breit, directly behind Wheeler.

finger of Europe to the outheld finger of America for many decades before 1939.

In 1876 the new Johns Hopkins University in Baltimore, under the leadership of Daniel Coit Gilman, became the first great institution of higher learning in America explicitly to dedicate itself to the Europe-inspired research ideal. Henry A. Rowland brought to it preeminence in physics as other men brought to it a like spirit of exploration and discovery in other fields. What we would today call graduate-level training for research took first place in those days. Relative to it, any undergraduate education provided was regarded as preparatory, secondary, gap filling. Education was not a teacher facing a class. Education was colleagues, young and old, facing an issue.

Seminar courses provided the most stimulating source of education (8) even for one who had entered Hopkins so late as 1927, especially seminars focused on such books as Born and Jordan's *Quantenmechanik*, conducted by Gerhard Dieke, Maria Mayer, and Karl Herzfeld—relatively new arrivals from The Netherlands, Germany, and Austria; Compton and Allison's *X-Rays in Theory and Experiment*, guided by Joyce Bearden; Wintner's *Spektraltheorie der unendlichen Matrizen—Einführung in den analytischen Apparat der Quantenmechanik*, inspired by Wintner himself—from Leipzig; Ames and Murnaghan's *Theoretical Mechanics*, converted into tools for the future by Murnaghan; assorted Rayleigh works on optics, seminars given coherence by A. H. Pfund; and the then hot Rutherford, Chadwick, and Ellis *Radiations from Radioactive Substances*, illuminated by Norman Feather, in his early twenties, whom R. W. Wood had just recruited from Rutherford's own Cavendish Laboratory group. Nuclear physics was clearly the wave of the future. So, in the traditional Hopkins spirit of hands-on involvement in experimental research, after some small work in x rays with Bearden and in spectroscopy with Dieke, I learned from Feather how to determine for myself the 3.5-day half-life of radon. The most important requirement was simple. Sit in the dark for half an hour. Then the fully dilated pupil easily picks up the flash that the alpha particle makes when it hits the zinc sulfide screen. Come back every few hours and repeat the measurement of counting rate. Voilà—the decay curve.

How prophetic it was of the future that I should be asked to give a seminar report on the 1930 paper (9) of W. Bothe and H. Becker on the artificial excitation of nuclear "gamma radiation." How did the excitation get from A to B? A mystery, a puzzle, an enigma! This paper was a doorway to the discovery of the neutron as the neutron was a doorway to the discovery of fission.

A penetrating radiation with mysterious properties! What a challenge to try to understand it. We interested students, especially Robert T. K. Murray and I, followed the subsequent efforts to unravel the mystery, not least among them the attempt (10), the failed 1932 attempt, of Irene Curie and Frederic Joliot, "Émission de protons de grande vitesse par les substances hydrogénées sous l'influence des rayons γ très pénétrants." How brief the time was from the finding of Bothe and Becker to Chadwick's discovery (11) of the neutron! Minds were better prepared for the new particle at the Cavendish Laboratory than anywhere in the world because Rutherford—"Ce jeune homme dévine tout" in the words of Becquerel—had been arguing as early as 1920 that such a particle should exist.[1]

We graduate students raised with each other question after question about the neutron, and followed with excitement each week's new findings. Can neutrons be bottled? Are free neutrons present in the atmosphere? What will a neutron do to a nucleus?

Clearly nuclear physics was becoming wide open territory. So my first year of postdoctoral research found me in New York in the fall of 1933 with Gregory Breit, working on questions of nuclear barrier penetration, resonant and nonresonant nuclear reactions, and the production of electron-positron pairs out of the emptiness of the vacuum.

By the spring of 1935, with Breit's support, I was applying to go to Copenhagen for a second year of a National Research Council fellowship, to work with Niels Bohr, because—I wrote—"he sees further ahead than any man alive."

No one starting a year with Bohr could have had a more marvelous introduction to the great men and the great open issues of physics than the International Conference on Physics held at London and Cambridge in the fall of 1934: J. J. Thomson, seventy-seven, frail, and white haired, host at a reception at Trinity College. Ernest Rutherford, head thrown back, in impressive discourse, with a circle of delegates, a lordly presence at an evening reception in the rooms of the Royal Society. Max Born, deprived of his position in Göttingen and newly arrived in the United Kingdom, writing in huge letters on the blackboard, "NUCLEAR PHYSICS," then with eraser and chalk—to laughter—altering the title to read "UNCLEAR PHYSICS." Thirty-three-year-old Enrico Fermi reporting radioactivities produced in many an element by neutron irradiation. Less than a month later (11 a.m., October 27, 1934) came the Rome group's great discovery that hydrogenous substances moderate neutrons and this moderation of the neutrons "increases the activation intensity by a factor which, depending on the geometry used, ranges from a few tens to a few hundreds."

[1] Rutherford in his 1920 Bakerian Lecture to the Royal Society of London had predicted the neutron's likely properties.

There was not one of the many papers (8) in solid-state physics or nuclear physics which I did not find truly significant. One, by Gray and Tarrant (13)—on the anomalous back-scattering of gamma rays—incited me to prove the effect to be, not back-scattering at all, but a minishower. This work later in the Copenhagen year brought me into meeting with Lise Meitner (Figure 6), initiator of experiments of this kind (14) and a guiding spirit of the Hahn-Strassmann uranium work.

Bohr's institute, smaller than many a house, his small group, and the man (Figure 7) and his way of work I have described elsewhere (8, 15). Formulator of complementarity, he was also the personal embodiment of complementarity. Who could switch so totally—as occasion warranted—from one mode of operation to another? From boldness to caution? From breadth to concentration? From one who never makes an advance except when in solitary thought to one who never makes an advance except in discussion with another?

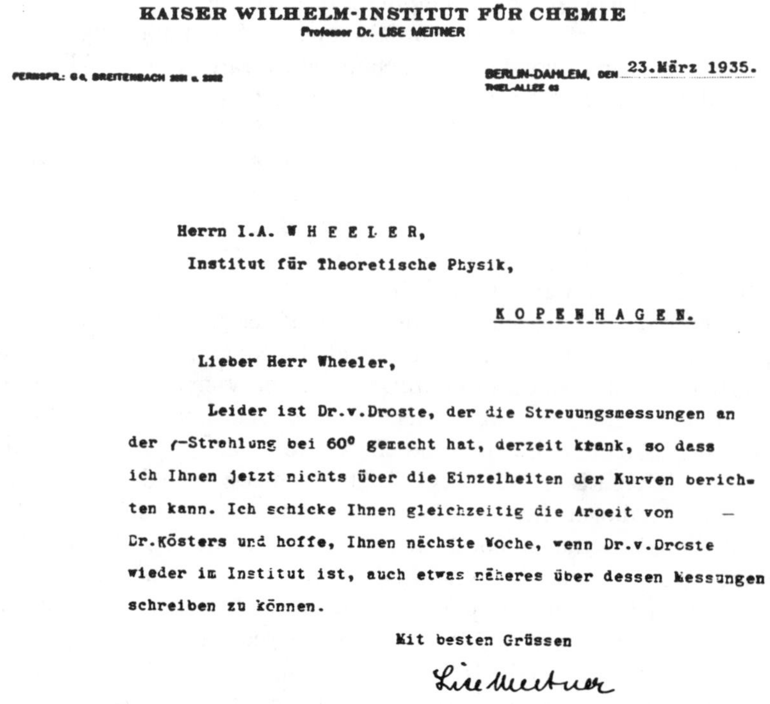

KAISER WILHELM-INSTITUT FÜR CHEMIE
Professor Dr. LISE MEITNER

FERNSPR.: 64, BREITENBACH 2001 u. 2002   BERLIN-DAHLEM, den 23.März 1935.
THIEL-ALLEE 63

Herrn I.A. W H E E L E R,
Institut für Theoretische Physik,

                    K O P E N H A G E N.

    Lieber Herr Wheeler,

        Leider ist Dr.v.Droste, der die Streuungsmessungen an
der ʅ-Strahlung bei 60° gemacht hat, derzeit krank, so dass
ich Ihnen jetzt nichts über die Einzelheiten der Kurven berich-
ten kann. Ich schicke Ihnen gleichzeitig die Aroeit von   —
Dr.Kösters und hoffe, Ihnen nächste Woche, wenn Dr.v.Droste
wieder im Institut ist, auch etwas näheres über dessen Messungen
schreiben zu können.

                    Mit besten Grüssen

                    Lise Meitner

Figure 6   Meitner letter about gamma radiation scattered nearly backward by heavy nuclei.

Figure 7   Niels and Margrethe Bohr on motorcycle (Courtesy of Aage Bohr).

Eastertime 1935, Christian Møller returned from a visit to Fermi's group in Rome. Bohr called a seminar to hear and discuss the new findings, most impressive among them being the high cross section of many nuclei for the interception of a neutron. Møller had not gotten half an hour down the road when Bohr interrupted him and took his place. Head lowered, pacing back and forth, he murmured over and over, "Now it comes. Now it comes. Now it comes...". Suddenly it did come. Then and there Bohr sketched out the compound-nucleus model of nuclear reactions.[2] It stood totally at variance with the earlier conception of the nucleus as an open planetary system. The incoming particle, in the new view, has in nuclear matter a mean free path that, compared to nuclear dimensions, is not long but short. The new nucleus, the compound nucleus, retains no memory of how it was formed. How—and how quickly—it breaks up or deexcites depends only on its energy and its angular momentum, not on its history.

[2] Reference (16) contains a carefully documented chronology of the birth and elaboration of the compound-nucleus model. I was not present in Copenhagen for the developments that Bohr, and Bohr and Kalckar, made in this model subsequent to June 1, 1935, but I remember vividly the Eastertime seminar at which the idea arose, as well as the Bohr-Kalckar manuscript of some months later and their report on this work in Washington in February of 1937 (Figure 5).

By the time of the Third Washington Conference on Theoretical Physics of February 18, 1937, Bohr with Fritz Kalckar was well on the way to developing a comprehensive account of nuclear energy levels and nuclear reaction probabilities on the foundation of this compound-nucleus model. To think of the mean free path of nucleons inside the nucleus as short compared to the nuclear diameter was to have justification, they pointed out (17), for adopting a liquid-drop model for the nucleus. This droplet model had been advanced by George Gamow some years earlier [(18); after Bohr's compound-nucleus model (18a)]. However, no one had ever really pushed it until the compound-nucleus model converted it from vision to tool.

The liquid-drop model gave an approximate way to estimate the quantities that totally define the compound-nucleus model of nuclear reactions: (a) the energy levels of the nucleus, and (b) the probabilities—per second—that the state in question will send out this, that, or the other particle, or a photon, with this, that, or the other energy.

A less global approach to nuclear reaction processes I had imbibed from my 1933–1934 year with Breit: Let scattering be the key to knowledge! Deduce law of force from phase shifts in scattering. Deduce these phase shifts from the variation of scattering cross section with angle (19). Already while I was with him I had started working on the scattering of alpha particles in helium. I was inspired to go on in my Copenhagen time and at Chapel Hill (1935–1938) by new results on alpha-alpha scattering reported at the October 1934 London-Cambridge conference. It was obvious that neither helium nucleus is a simple particle. Therefore new methods had to be developed to treat the interaction between nuclei (20): the scattering matrix to give a precise description of the phenomenology of scattering, and the method of resonating group structure to define an effective interaction potential.

Despite the rich subsequent development of this clustering model by many workers, I—like others—soon found it quicker to make progress in understanding the broad features of nuclear reactions by applying the compound-nucleus model. That model I described and extended in work with graduate students at Chapel Hill and in lectures I gave at Princeton over a period of some weeks in 1937, as well as in the fall of 1938 after I moved to Princeton. Thus somehow fate had put me with the right ideas in the right place at the right time and with the right man when Bohr arrived on January 16, 1939 with the news of fission.

Once at work together, Bohr and I undertook a detailed analysis of fission regarded as an exciting new application of a compound-nucleus-plus-liquid-drop model. This work took not only the three months of Bohr's stay in Princeton but two additional months of finishing up until I could send it in for publication (June 28, 1939). The topics that had to be taken up are seen in this quotation from the paper (21):

[We] estimate quantitatively in Section I by means of the available evidence the energy which can be released by the division of a heavy nucleus in various ways, and in particular examine not only the energy released in the fission process itself, but also the energy required for subsequent neutron escape from the fragments and the energy available for beta-ray emission from these fragments.

In Section II the problem of the nuclear deformation is studied more closely from the point of view of the comparison between the nucleus and a liquid droplet in order to make an estimate of the energy required for different nuclei to realize the critical deformation necessary for fission.

In Section III the statistical mechanics of the fission process is considered in more detail, and an approximate estimate made of the fission probability. This is compared with the probability of radiation and of neutron escape. A discussion is then given on the basis of the theory for the variation with energy of the fission cross section.

In Section IV the preceding considerations are applied to an analysis of the observations of the cross sections for the fission of uranium and thorium by neutrons of various velocities. In particular it is shown how the comparison with the theory developed in Section III leads to values for the critical energies of fission for thorium and the various isotopes of uranium which are in good accord with the considerations of Section II.

In Section V the problem of the statistical distribution in size of the nuclear fragments arising from fission is considered, and also the questions of the excitation of these fragments and the origin of the secondary neutrons.

Finally, we consider in Section VI the fission effects to be expected for other elements than thorium and uranium at sufficiently high neutron velocities as well as the effect to be anticipated in thorium and uranium under deuteron and proton impact and radiative excitation.

A new feature of capillarity entered in the case of fission, the concept of fission barrier. The very idea was new and strange. More than one distinguished colleague objected that no such quantity could even make sense, let alone be defined. According to the liquid-drop picture, is not an ideal fluid infinitely subdivisible? And therefore cannot the activation energy required to go from the original configuration to a pair of fragments be made as small as one pleases? We obtained guidance on this question from the theory of the calculus of variations in the large, maxima and minima, and critical points. This subject I had absorbed over the years by osmosis from the Princeton environment, so thoroughly charged by the ideas and results of Marston Morse (22). It became clear that we could find a configuration space to describe the deformation of the nucleus. In this deformation space we could find a variety of paths leading from the normal, nearly spherical configuration over a barrier to a separated configuration. On each path the energy of deformation reaches a highest value. This peak value differs from one path to another. Among all these maxima, the minimum measures the height of the saddle point or fission threshold or the activation energy for fission. The fission barrier *was* a well-defined quantity!

Bohr knew from his own student research on water jets that a work of Lord Rayleigh would have something to say about the capillary oscillations of a liquid drop. We rushed up to the library on the next floor of Fine Hall and looked it up in the *Scientific Papers* of Rayleigh. This work furnished a starting point for our analysis. However, we had to go to terms of higher order than Rayleigh's favorite second-order calculations to pass beyond the purely parabolic part of the nuclear potential, that is, the part of the potential that increases quadratically with deformation. We determined—as soon also did Feenberg, von Weizsäcker, Frenkel, and others—the third-order terms to see the turning down of the potential. These terms enabled us to evaluate the height of the barrier, or at least the height of the barrier for a nucleus whose charge was sufficiently close to the critical limit for immediate breakup.

We found that we could reduce the whole problem to finding a function f of a single dimensionless variable x. This "fissility parameter" measures the ratio of the square of the charge to the nuclear mass. This parameter has the value 1 for a nucleus that is already unstable against fission in its spherical form. For values of x close to 1, by a power series development we could estimate the height of the barrier and actually give quite a detailed calculation of the first two terms in the power series for barrier height, or f, in powers of $(1-x)$. The opposite limiting case also lent itself to analysis. In this limit the nucleus has such a small charge that the barrier is governed almost entirely by surface tension. The Coulomb forces give almost negligible assistance in pushing the material apart.

Between this case (the power series about $x = 0$) and the other case (the power series about $x = 1$) there was an enormous gap. We saw that it would take a great amount of work to calculate the properties of the fission barrier at points in between. Consequently we limited ourselves to interpolation between these points. In the decades since that time many workers (among them Wladyslaw J. Swiatecki at Berkeley, Vladimir Strutinski and his colleagues in the USSR, and Ray Nix and his colleagues at Los Alamos) have revealed many previously unsuspected features of the fission barrier. This is not the place to go into the deeper theoretical considerations on prompt neutrons, delayed neutrons, the physics of fission product decay, and many another topic that came up, nor to detail the many impressive experimental results that were obtained on these and other topics week by week.

Two other issues of comparable or greater challenge came up in doing our paper: (*a*) figuring the rate of decay associated with spontaneous fission, and (*b*) determining the probability for fission of a nucleus with excitation in excess of the barrier summit. The first of these forced us to introduce a measure for the inertia associated with a deformation. The second led us to the concept of channels for fission associated with the transition state. Both introduced lines of thought still under active exploration and development today.

For our immediate needs, however, our simple "poor man's" interpolation was adequate. With it, knowing—or estimating from observation—the fission barrier for one nucleus, we could estimate the fission barrier for all the other heavy nuclei, among them plutonium 239. Thanks to the questioning of Louis A. Turner (Figure 8), soon to write his great and timely review of nuclear fission (23), we came to recognize that this substance, which up to then one had never seen except through its radioactivity [McMillan & Abelson (24), June 15, 1940[3]] would be fissile. This conclusion was soon to lead to a preposterous dream: by means of a neutron reactor such as never before existed, one could manufacture kilograms of an element never before seen on Earth. By the fall of 1944 the E. I. duPont de Nemours Company had converted this dream into the reality of a double alchemy. In total silence neutrons flow from place to place, splitting some uranium nuclei, converting others (the U^{238}) to U^{239}. Then nature itself took hold to complete the alchemy. Over the ensuing hours and days it let the U^{239} spontaneously transform—through two beta transformations—to Pu^{239}. By this double miracle, duPont at Hanford, Washington, was able, week by week, to supply ponderable masses of a strange and totally new element, plutonium, to its Los Alamos, New Mexico customer.

The barrier height of a compound nucleus against fission was not the only factor relevant for fission. Equally important in governing the probability of this process was the excitation, or "heat of condensation," delivered by the uptake of a neutron to form the compound nucleus in the first place. On this point an important development occurred on a snowy morning when I was occupied with classes and not with Bohr. He, having breakfast at the Nassau Club with Rosenfeld and with an arrival of the night before, George Placzek, faced Placzek's continuing skepticism about the very existence of fission. Placzek asked, how can it possibly make sense that slow neutrons and fast neutrons cause uranium to split but not neutrons of intermediate energy? Bohr stopped but said not a word, left with Rosenfeld, crossed the campus to Fine Hall still without a word and there, when Placzek and I joined them, explained the great idea (25) that had just come to him: that the slow neutron fission takes place in the rare

---

[3] According to Irving (24a), J. Schintlmeister and F. Hernegger identified neptunium and plutonium in June 1940 but reported their findings only at the end of 1940.

Figure 8   Louis A. Turner, who pointed the way to the manufacture of plutonium. Courtesy of Argonne National Laboratory Photographic Archives.

isotope U^{235} and the fast neutron fission in the abundant isotope U^{238}. Thus an incoming neutron delivers up a high heat of condensation when it enters into a nucleus with 143 neutrons, because it can form a new neutron pair. This excitation puts the compound nucleus U^{236} over the barrier summit. Therefore, the U^{236} must split when it is formed by slow neutron capture. Moreover, the cross section for fission of the rare U^{235}, like the cross section for the "fission" of boron,

$$n^1 + B^{10} \rightarrow He^4 + Li^7,$$

must exceed by far the geometrical cross section of the nucleus for sufficiently slow neutrons. That circumstance makes it understandable why an isotope present to only one part in 139 imparts to natural uranium the observed substantial fission cross section. However, the cross section must fall off inversely as the velocity of the neutron for U^{235} as for B^{10}; hence the negligible fission cross section of natural uranium for neutrons of intermediate energy. In contrast, when a slow neutron enters U^{238} to form the compound nucleus U^{239}, no new neutron pair is formed. The heat of condensation delivered up is not enough to exceed the fission barrier. Only neutrons of a substantial kinetic energy striking U^{238} can produce U^{239} with enough energy to surmount the barrier. Hence, the existence of fast neutron fission in natural uranium.

Was it reasonable to expect so great a difference between U^{235} and U^{238} from the estimated odd-even difference in neutron binding? Could not the fission barrier differ equally drastically from the one nucleus to the other? Might not this difference be the dominant factor? How could one be sure that the proposed attribution of slow fission to U^{235} and fast fission to U^{238} really made sense until one was clear about these energies? Fortunately Bohr and I had just been through the systematics of nuclear energies in the course of calculating the release of energy in various actual and potential fission processes. Therefore, we could estimate the difference between the excitation developed by neutron capture in the two uranium isotopes as almost a million volts, in favor of fission of U^{235}. From our interpolation for fission barriers we estimated on the other hand a barrier almost 1 MeV lower for U^{235} than for U^{238}. Thus we concluded there was about a 2-MeV margin in favor of the fission of the rare isotope. In later years, after the development of the collective model it became clear that individual particle effects can modify significantly barrier heights and barrier shapes from the predictions of the simple liquid-drop model. However, the qualitative conclusions are not affected; U^{235} is the fissile nucleus.

Placzek, wonderful person that he was, a man of the highest integrity, often a thoroughgoing skeptic about new ideas, said to me over and over in those early spring days of 1939 that he could not believe that the small amount of U^{235} could be the cause of the slow neutron effects in natural uranium. I therefore bet him a proton to an electron, $18.36 to a penny, that Bohr's diagnosis was correct. A year later Alfred Nier at Minnesota had separated enough U^{238} to make possible a test and sent it to John Dunning at Columbia to measure its fission cross section (26). On April 16, 1940, I received a Western Union money order telegram for one cent with the one-word message "Congratulations!" signed Placzek (27).

Literature Cited

1. Wheeler, J. A., Zurek, W. H., eds., in *Quantum Theory and Measurement.* Princeton Univ. Press, NJ (1983), pp. 3–49
2. Meitner, L., Frisch, O. R., *Nature* 143: 239–40 (1939)
3. Hahn, O., Strassmann, F., *Naturwissenschaften* 27: 11–15 (1939)
4. Frisch, O. R., *What Little I Remember,* Cambridge Univ. Press (1979), p. 116
5. Bohr, N., Wheeler, J. A., *Phys. Rev.* 56: 426–50 (1939)
6. Frisch, O. R., *Nature* 143: 276 (1939)
7. Wheeler, J. A., et al., "Mercer Street and Other Memories," in *Albert Einstein: His Influence on Physics, Philosophy and Politics,* ed. P. C. Aichelburg, U. R. Sexl. Braunschweig: Vieweg & Sohn (1979), pp. 201–11
8. Wheeler, J. A., "Some men and moments in the history of nuclear physics: The interplay of colleagues and motivations," in *Symp. on the History of Nuclear Physics, Univ. Minn., 1977,* Minneapolis: Univ. Minn. (1979), pp. 217–322
9. Bothe, W., Becker, H., *Z. Phys.* 66: 289–306 (1930)
10. Curie, I., Joliot, F., *Comptes Rendus* 194: 273–75 (1932)
11. Chadwick, J., *Nature* 129: 312 (1932); *Proc. Roy. Soc. London Ser. A* 136: 692–708 (1932)
12. Deleted in proof
13. Gray, L. H., Tarrant, G. T. P., *Proc. Roy. Soc. London Ser. A* 136: 662–91 (1932); 143: 681–706 (1934); 143: 706–24 (1934)
14. Brown, L. M., Moyer, D. F., *Am. J. Phys.* 52: 130–36 (1984)
15. Wheeler, J. A., "Niels Bohr: The Man and His Legacy," in *The Lesson of Quantum Theory: Niels Bohr Centenary Symposium October 3–7, 1985,* ed. J. de Boer, E. Dal, O. Ulfbeck. Amsterdam: North-Holland (1986), pp. 355–67

16. Rüdinger, E., gen. ed., Peierls, R., vol. ed., *Niels Bohr Collected Works: Volume 9 Nuclear Physics (1929–1952).* Amsterdam: North-Holland (1986), pp. 14–27
17. Bohr, N., Kalckar, F., *Mat.-Fys. Medd. Dan. Vidensk. Selsk.* Vol. 14, No. 10 (1937)
18. Gamow, G., *Constitution of Atomic Nuclei and Radioactivity,* Oxford: Clarendon (1931), pp. 18–19
18a. Gamow, G., *Structure of Atomic Nuclei and Nuclear Transformations,* Oxford: Clarendon, 2nd ed. (1937), pp. 4, 26–38
19. Wheeler, J. A., *Phys. Rev.* 45: 746 (1934); *Phys. Rev.* 59: 16–26 (1941); *Phys. Rev.* 59: 27–36 (1934)
20. Wheeler, J. A., *Phys. Rev.* 52: 1107–22 (1937)
21. Bohr, N., Wheeler, J. A., *Phys. Rev.* 56: 427–28 (1939)
22. Morse, M., *The Calculus of Variations in the Large.* New York: Am. Math. Soc. (1934); *Functional Topology and Abstract Variational Theory.* Paris: Gauthier-Villars (1938); Morse, M., Cairns, S., *Critical Point Theory in Global Analysis and Differential Topology: an Introduction.* New York: Academic (1969)
23. Turner, L., *Rev. Mod. Phys.* 12: 1–29 (1940)
24. McMillan, E., Abelson, P. H., *Phys. Rev.* 57: 1185–86 (letter) (1940)
24a. Irving, D., *The German Atomic Bomb: The History of Research in Nazi Germany.* New York: Simon & Schuster (1967)
25. Bohr, N., *Phys. Rev.* 55: 418–19 (1939) (letter)
26. Nier, A., Booth, E., Dunning, J., Grosse, A., *Phys. Rev.* 57: 546 (1940) (letter)
27. Placzek, G., April 16, 1940, telegram, on deposit at the National Museum of Science and Technology, Washington, DC

Annu. Rev. Nucl. Part. Sci. 1989. 39: 1–17

LIGHT QUARK MASSES AND CHIRAL SYMMETRY

John F. Donoghue

Department of Physics and Astronomy, University of Massachusetts, Amherst, Massachusetts 01003

KEY WORDS: strange quark, sigma term, contents of proton, πN scattering, quantum chromodynamics.

CONTENTS

1. INTRODUCTION

The Lagrangian of quantum chromodynamics (QCD) is

$$\mathscr{L}_{\text{QCD}} = -\frac{1}{4} F_{\mu\nu}^A F^{A\mu\nu} + \sum_j \bar{\psi}_j (i\slashed{D} - m_j)\psi_j, \qquad 1.$$

where the summation runs over all of the quarks, and

$$F_{\mu\nu}^A = \partial_\mu A_\nu^A - \partial_\nu A_\mu^A + igf^{ABC} A_\mu^B A_\nu^C$$

$$D_\mu = \partial_\mu - ig \frac{\lambda^A}{2} A_\mu^A. \qquad 2.$$

The purpose of this article is to review critically what is known about the masses of light quarks, i.e. m_u, m_d, and m_s, which are parameters in this

1

Lagrangian. The prime tool in this discussion is the theory of chiral symmetry, although some description of less rigorous phenomenological estimates is also given. In chiral symmetric techniques, the quark mass is used as a parameter in an expansion of physical quantities in a power series in the energy. In this way, it is very similar to a coupling constant. The expansion to first order in the quark mass is very well defined. However, I argue that a full treatment to second order has not yet been given, and that the results at second order could differ appreciably from those at lowest order.

The attitude taken in this review is somewhat conservative. This is because the standard analysis of symmetry breaking leads us into conflict with the conventional assumptions when the πN sigma term is included. This forces us to question the validity of some assumptions. The use of symmetry gives us a well-defined approximation scheme. In contrast, the use of models or dynamical assumptions, such as the quark model or the Zweig rule, are on a less well-defined footing. I try here to keep the two separate for as long as possible. One possible resolution of the sigma term problem could be that the u, d quark masses are somewhat larger than is conventionally thought. There does not appear to be any firm evidence against this possibility.

The dimensional parameters of QCD (1) are the quark masses, m_i, and the scale parameter Λ, which is needed to specify the coupling constant in the quantum theory. The masses of the heavy quarks (c, b, t?) play little role in the analysis of the light quark masses and are thus ignored throughout the paper. That leaves (Λ, m_u, m_d, m_s) as the fundamental parameters of low energy QCD. Because quarks are confined, one cannot observe the masses directly as poles in propagators or thresholds for q$\bar{\text{q}}$ production. Rather we must learn about the quark masses from their influence on observables, e.g. the pion mass is a function of the quark masses. As always is the case in field theory, the functional dependence on quark masses must be renormalized, and the choice of renormalization prescription is crucial. When renormalized in perturbation theory, the masses depend on the scale chosen in the renormalization prescription. As this scale is varied, the masses "run" in much the same way as the coupling constant does (2). From this observation it becomes clear that there is no universal definition of the quark mass.

The discussion of the masses is intimately linked to the matrix elements of the operator $\bar{\psi}\psi$. This is because the masses enter the theory only in the combination

$$\mathscr{H}_m = \bar{\psi}m\psi. \qquad\qquad 3.$$

The mass and $\bar{\psi}\psi$ do not really exist separately. Only by describing how the operator $\bar{\psi}\psi$ is renormalized can one give meaning to the quark mass. At low energies there is no rigorous method for calculating the hadronic matrix elements of $\bar{\psi}\psi$ (at least not until lattice techniques overcome some of their current difficulties). This precludes a rigorous calculation of m_q at present. Phenomenological models may of course be applied, and these are discussed briefly below. The ratios of light quark masses are somewhat better defined. This is because the dependence on $\bar{\psi}\psi$ may totally or partially cancel in the ratio. For this reason, most of this article is devoted to quark mass ratios.

The most important qualitative property of the light quark mass parameters in $\mathscr{L}_{QCD}$ is that they are "small." This we have learned from the theory of chiral symmetry (3–6). If the masses were exactly zero, QCD would have an exact symmetry:

$$\psi_L \rightarrow \exp(i\boldsymbol{\alpha} \cdot \lambda)\psi_L$$

$$\psi_R = \exp(i\boldsymbol{\beta} \cdot \lambda)\psi_R$$

$$\psi_{\substack{L \\ R}} = \tfrac{1}{2}(1 \pm \gamma_5)\psi$$

$$\psi = \begin{pmatrix} u \\ d \\ s \end{pmatrix}, \qquad\qquad 4.$$

where λ^A are the Gell-Mann matrices of SU(3). This is called $SU(3)_L \times SU(3)_R$ chiral symmetry, in that the worlds of left-handed and right-handed quarks have separate SU(3) invariances. We expect that this zero-mass world would have the $SU(3)_L \times SU(3)_R$ symmetry "dynamically broken" to $SU(3)_V$. "Broken" is a common but misleading expression in this phrase ("hidden" would be better). The axial (L-R) symmetry remains exact, but the vacuum is not a singlet eigenstate of the symmetry. This realization requires the existence of exactly massless Goldstone bosons, which would be the π, K, η_8 pseudoscalar mesons. Conventional symmetries, such as isospin, relate the properties of particles that fall in multiplets, such as the neutron and proton, and require that they have the same mass. With dynamically broken symmetries, the relations are between a state and that state with the addition of a Goldstone boson. For example the vacuum and one-pion states would be related to each other by symmetry transformations, as would the proton and proton-plus-pion systems. For zero-mass, zero-energy Goldstone bosons these states have the same energy. The nature of the "broken" or "hidden" part of the chiral symmetry is then manifest in low energy relations between processes involving different numbers of Goldstone bosons.

In the real world, of course, the quark masses are not zero, and the masses connect the left-handed and right-handed sectors of the theory,

$$\mathscr{H}_m = \bar{\psi} m \psi = \bar{\psi}_L m \psi_R + \bar{\psi}_R m \psi_L$$

$$m = \begin{pmatrix} m_u & 0 & 0 \\ 0 & m_d & 0 \\ 0 & 0 & m_s \end{pmatrix}. \qquad \qquad 5.$$

This provides a real, explicit breaking of the chiral symmetry. If the masses were equal, one would have a residual vector SU(3) symmetry, but in the real world, this and the SU(2) of isospin are both broken because the quark masses are all different. However, it is here that the "smallness" of the masses enters. The relations of chiral symmetry, slightly modified by nonzero masses of π and K, appear to work fairly well. We are not far from the limit of massless quarks, and the π, K, η are "almost" Goldstone bosons. Similarly for isospin and vectorial SU(3): the mass differences are not too large and the isospin and SU(3) relations work fairly well.

This gives us our best handle on quark masses. If masses and mass differences are small, then physical quantities may be expanded in a power series in the mass or mass difference about the symmetry limit. Deviations of physical quantities from their symmetry values are probes of the masses. The connection is the cleanest if one is able or content to work to first order in the masses. The perturbation is $\bar{\psi} m \psi$, and, as always in first-order perturbation theory, the matrix elements are evaluated with the zeroth-order wavefunctions (i.e. in the symmetry limit). This means that deviations are linear in the masses, with well-defined symmetry properties. When one forms quark mass ratios at first order, the matrix element of $\bar{\psi} \psi$ cancels exactly in the ratio. It is also possible to work to second order in the quark masses, although, as discussed below, the treatment is more subtle.

Even though most readers will be aware of this, it is important to repeat the standard caution. The mass parameters of the Lagrangian are not the same as the effective inertial masses that are introduced in phenomenological quark models. The latter are of order $m_u \approx m_d \approx 330$ MeV, $m_s \approx 550$ MeV. These represent more the energy of a quark in a confined hadron; they do not follow directly from QCD but rather are parameters in some effective model.

The review starts with the standard first-order treatment of meson masses in Section 2. Section 3 provides the same treatment in baryons and also introduces the sigma term. The problem posed by the experimental value of the sigma term is discussed here. The present status of the analysis of σ is given in Section 4. In Section 5 we describe the formulations of the meson mass to second order in the energy expansion. Some comments on phenomenological models are given in Section 6, while the last section is a brief summary.

2. LOWEST-ORDER TREATMENT OF MASSES IN MESONS

This section describes the standard treatment of quark masses (7) to first order in the expansion in the mass. This is the analysis that most of the particle physics community views as well established. The results are characterized by

$$\frac{\hat{m}}{m_s} = \frac{m_\pi^2}{2m_K^2 - m_\pi^2} = \frac{1}{25} \qquad\qquad\qquad 6a.$$

$$\frac{m_d - m_u}{m_d + m_u} = 0.29, \qquad\qquad\qquad 6b.$$

with $\hat{m} = (m_u + m_d)/2$. These relations form the benchmark to which later analysis is compared.

The pseudoscalar mesons π, K, η_8 would be massless Goldstone bosons in the limit that m_u, m_d, $m_s \to 0$. To first order in the quark mass, their masses are then governed by the matrix element of

$$\mathcal{H}_m = \bar{\psi}m\psi = m_u\bar{u}u + m_d\bar{d}d + m_s\bar{s}s. \qquad\qquad 7.$$

In particular for the pion

$$\begin{aligned} m_\pi^2 &= \langle\pi|m_u\bar{u}u + m_d\bar{d}d + m_s\bar{s}s|\pi\rangle \\ &= \langle\pi|m_u\bar{u}u + m_d\bar{d}d|\pi\rangle \\ &= (m_u + m_d)B_0, \qquad\qquad\qquad 8. \end{aligned}$$

where

$$B_0 = \langle\pi|\bar{u}u|\pi\rangle = \langle\pi|\bar{d}d|\pi\rangle. \qquad\qquad 9.$$

The pion is massless for $m_u = m_d = 0$, for any value of m_s, because of the $SU(2)_L \times SU(2)_R$ symmetry in that limit. This implies that m_s cannot influence the pion mass in first order. The constant B_0 is here evaluated in the chiral limit ($m_q = 0$) and hence the equality of the $\bar{u}u$ and $\bar{d}d$ matrix elements. The constant B_0 is sometimes evaluated via soft pion theorems as

$$B_0 = -\frac{1}{F_\pi^2}\langle 0|\bar{q}q|0\rangle, \qquad\qquad 10.$$

where $q = u$, d, or s and $F_\pi = 94$ MeV. Similar treatment for the remaining particles yields

$$m_{K^+}^2 = (m_s + m_u)B_0$$
$$m_{K^0}^2 = (m_s + m_d)B_0$$
$$m_{\eta_8}^2 = \tfrac{1}{3}(4m_s + m_u + m_d)B_0. \qquad\qquad 11.$$

The latter relation is equivalent to the Gell-Mann Okubo formula

$$m_{\eta_8}^2 = \tfrac{4}{3}m_K^2 - \tfrac{1}{3}m_\pi^2. \qquad\qquad 12.$$

If we neglect isospin breaking, this easily yields

$$\frac{m_u + m_d}{m_s + m_d} = \frac{m_\pi^2}{m_K^2}, \qquad\qquad 13.$$

or equivalently Equation 6a.

One might try to determine $\hat{m}/m_s$ using the η_8 and kaon masses. At first order the η_8 can be identified with the $\eta(549)$, and one obtains

$$\frac{\hat{m}}{m_s} = \frac{4m_K^2 - 3m_\eta^2}{2m_K^2 - m_\pi^2} = 0.15. \qquad\qquad 14.$$

The mixing of η_8-η_0, for example, only modifies the mass at second order in $(m_s - \hat{m})$. While formally valid, this is not a good method to determine $\hat{m}/m_s$. It involves a cancellation between two large numbers $4m_K^2$ and $3m_\eta^2$, to extract a very small difference. Small changes in these values are amplified. Both m_K^2 and m_η^2 receive corrections of order m_s^2, and the difference $4m_K^2 - 3m_\eta^2$ is sensitive to these. The estimate in Equation 6a is more reliable.

In order to extract the isospin-violating mass difference $m_d - m_u$ one needs to be aware that electromagnetic corrections also break isospin invariance. These need to be accounted for if we are to extract the quark masses. The crucial ingredient is Dashen's theorem (8), which says that to lowest order in the quark masses the electromagnetic shifts in the pions and kaons are equal,

$$(m_{K^0}^2 - m_{K^+}^2)_{EM} = (m_{\pi^0}^2 - m_{\pi^+}^2)_{EM} + O(\alpha m_q^2). \qquad\qquad 15.$$

For the pions, the electromagnetic difference is the actual difference, to lowest order. This leaves the quark mass portion of the kaon difference as

$$(m_{K^0}^2 - m_{K^+}^2)_{\text{quark mass}} = (m_{K^0}^2 - m_{K^+}^2 - m_{\pi^0}^2 + m_{\pi^+}^2) \qquad \text{16.}$$

or, with the help of Equations 8 and 11,

$$\frac{m_d - m_u}{m_s - \hat{m}} = \frac{m_{K^0}^2 - m_{K^+}^2 - m_{\pi^0}^2 + m_{\pi^+}^2}{m_K^2 - m_\pi^2} = 0.023 = \frac{1}{43}, \qquad \text{17.}$$

which is equivalent to Equation 6b when combined with Equation 6a.

Another first-order symmetry determination of isospin breaking in the quark masses comes from the transitions $\psi' \to \psi\pi^0$ and $\psi' \to \psi\eta$. Here the first reaction is forbidden in the limit of exact isospin invariance, and the second is forbidden in the SU(3) limit. Hence the amplitudes are proportional to $m_d - m_u$ and $m_s - \hat{m}$ respectively. A simple application of degenerate perturbation theory then leads to the prediction

$$\frac{\Gamma(\psi' \to \psi\pi^0)}{\Gamma(\psi' \to \psi\eta^0)} = \frac{27}{16}\left(\frac{m_d - m_u}{m_s - \hat{m}}\right)^2 \left(\frac{p_\pi}{p_\eta}\right)^3, \qquad \text{18.}$$

where the momentum factors account for p-wave phase space. The current experimental value is

$$\frac{\Gamma(\psi' \to \psi\pi^0)}{\Gamma(\psi' \to \psi\eta^0)} = 0.036 \pm 0.009, \qquad \text{19.}$$

which implies a quark mass ratio

$$\frac{m_d - m_u}{m_s - \hat{m}} = 0.033 \pm 0.004, \qquad \text{20.}$$

somewhat ($\sim 40\%$) above the estimate using meson masses. This difference is perhaps an indication of the reliability of first-order estimates.

Although it is incidental to the above determination of the ratio, it is worth noting that there exists a lovely theoretical analysis (9) that can calculate the $\psi' \to \psi\eta$ and $\psi\pi^0$ rates. It uses the heavy quark multipole expansion for the $\psi' \to \psi + 2$ gluon transition, with the anomaly analysis providing the appropriate hadronic matrix element. Estimates of possible electromagnetic corrections indicate that they are quite small (10). The origin of this feature is easy to understand. When considering electromagnetic corrections to the two-gluon exchange in the multipole expansion, the one gluon plus one photon state is forbidden by the color singlet nature of the pion and the two gluon plus one photon state is forbidden by C invariance. The lowest-order correction involves three gluons plus a photon. This is highly suppressed by heavy quark mass factors in the multipole expansion.

3. LOWEST-ORDER TREATMENT IN BARYONS AND THE SIGMA TERM PROBLEM

There are two main sources of information on quark masses in baryons: (*a*) the baryon mass differences and (*b*) the sigma term in πN scattering. The baryon mass differences will, in an obvious way, yield information on quark mass differences. Less obvious is the sigma term, which will tell us about the magnitude of u, d masses. The sigma term is not a familiar object; it is introduced in this section and described in more detail in Section 4.

Should the quark mass ratio in baryons be the same as that in mesons? To the extent that the masses are treated as phenomenological parameters, the answer is in general negative. However, in principle one should be able to connect the ratios, as in each case there is a relation of the phenomenological parameters to the parameters of the QCD Lagrangian. To lowest order this relation is simple, and the quark mass ratios are meant to be the same when treated in first order.

The mass splittings of baryons are due to the quark masses, which we can write in the following form:

$$\mathcal{H}_{\text{mass}} = \tfrac{1}{3}(m_u + m_d + m_s)\,(\bar{u}u + \bar{d}d + \bar{s}s)$$
$$- \tfrac{1}{3}(m_s - \hat{m})\,(\bar{u}u + \bar{d}d - 2\bar{s}s)$$
$$- \tfrac{1}{2}(m_d - m_u)\,(\bar{u}u + \bar{d}d). \qquad 21.$$

The second operator is that which governs SU(3)-violating splitting, while the third governs isospin violation. The matrix elements can be parameterized in terms of two reduced matrix elements for the octet operators, and one for the SU(3) singlet:

$$\langle P | \mathcal{H}_{\text{mass}} | P \rangle = m_0 - \tfrac{1}{3}(m_s - \hat{m})\,(3F - D) - \tfrac{1}{2}(m_d - m_u)\,(D + F)$$

$$\langle N | \mathcal{H}_{\text{mass}} | N \rangle = m_0 - \tfrac{1}{3}(m_s - \hat{m})\,(3F - D) + \tfrac{1}{2}(m_d - m_u)\,(D + F)$$

$$\langle \Sigma^+ | \mathcal{H}_{\text{mass}} | \Sigma^+ \rangle = m_0 - \tfrac{1}{3}(m_s - \hat{m})2D - (m_d - m_u)F$$

$$\langle \Sigma^0 | \mathcal{H}_{\text{mass}} | \Sigma^0 \rangle = m_0 - \tfrac{1}{3}(m_s - \hat{m})2D$$

$$\langle \Sigma^- | \mathcal{H}_{\text{mass}} | \Sigma^- \rangle = m_0 - \tfrac{1}{3}(m_s - \hat{m})2D + (m_d - m_u)F$$

$$\langle \Lambda^0 | \mathcal{H}_{\text{mass}} | \Lambda^0 \rangle = m_0 + \tfrac{1}{3}(m_s - \hat{m})2D$$

$$\langle \Xi^- | \mathcal{H}_{\text{mass}} | \Xi^- \rangle = m_0 + \tfrac{1}{3}(m_s - \hat{m}) \cdot (3F + D) + \tfrac{1}{2}(m_d - m_u)\,(D - F)$$

$$\langle \Xi^- | \mathcal{H}_{\text{mass}} | \Xi^- \rangle = m_0 + \tfrac{1}{3}(m_s - \hat{m}) \cdot (3F + D) - \tfrac{1}{2}(m_d - m_u)\,(D - F). \qquad 22.$$

Note that this parameterization involves only vectorial SU(3) and not the chiral symmetry. From the larger SU(3) mass splittings we extract

$$(m_s - \hat{m})F = \tfrac{1}{2}(m_\Xi - m_N) = 188 \text{ MeV}$$

$$(m_s - \hat{m}) = \tfrac{3}{4}(m_\Lambda - m_\Sigma) = -58 \text{ MeV}, \qquad\qquad 23.$$

as well as the baryon Gell-Mann Okubo formula (11)

$$m_\Xi + m_N = \tfrac{3}{2}m_\Lambda + \tfrac{1}{2}m_\Sigma. \qquad\qquad 24.$$

One might hope to isolate the ratio $(m_d - m_u)/(m_s - \hat{m})$ from the isospin-breaking effects. However, this cannot be done using only symmetry techniques. The isovector octet portion of the electromagnetic interactions has the same SU(3) properties as the quark mass difference, and hence the two cannot be separately identified. Symmetry considerations can give the following results: (a) The combination $m_{\Sigma^+} + m_{\Sigma^-} - 2m_{\Sigma^0} = 1.7$ MeV is purely electromagnetic in origin, and (b) the masses satisfy a sum rule [the Coleman-Glashow relation (12)]:

$$m_{\Sigma^+} - m_{\Sigma^-} + m_N - m_p + m_\Xi - m_{\Xi^0} = 0 \qquad (=0.4 \pm 0.6 \text{ MeV expt.}) \quad 25.$$

valid for both the electromagnetic and quark mass contributions. Relation (a) follows from the $I = 2$ character of that combination of masses, while (b) follows from Equations 22 and the U spin singlet character of the electromagnetic interactions.

The sigma term is a matrix element that arises in $\pi N \to \pi N$ scattering. (See Equation 34 below.) In the chiral limit, the scattering amplitude would vanish for zero four-momentum pions (Adler zeros). The zero momentum limit then provides a measure of symmetry breaking, i.e. quark masses. In practice, zero four-momentum cannot be reached for massive pions. However, Cheng & Dashen (13) showed how to extrapolate physical amplitudes to a particular unphysical configuration (the Cheng-Dashen point) where, up to calculable corrections,

$$\Sigma \approx \sigma \equiv \langle P | m_u \bar{u}u + m_d \bar{d}d | P \rangle. \qquad\qquad 26.$$

The extrapolation procedure is discussed below. For now let us consider the consequences of the apparent experimental result (14, 15)

$$\sigma \approx 60 \pm 8 \text{ MeV}. \qquad\qquad 27.$$

The matrix element is not sensitive to isospin violation and can be parameterized as

$$\sigma = 3\hat{m}Z \approx 60 \text{ MeV}$$

$$Z = \tfrac{1}{3}\langle P | \bar{u}u + \bar{d}d | P \rangle. \qquad\qquad 28.$$

This is to be compared with the result of the SU(3) analysis, which can be written as

$$\delta_s \equiv \langle P|(m_s - \hat{m})(\bar{u}u + \bar{d}d - 2\bar{s}s)|P\rangle = (m_s - \hat{m})(3F - D)$$

$$= 3(m_\Xi - m_\Lambda) \approx 600 \text{ MeV}. \quad 29.$$

If the matrix element of $\bar{s}s$ vanished, then the matrix elements would be equal, i.e. $(3F - D)$ would be equal to $3Z$. We then would have

$$\frac{\hat{m}}{m_s - \hat{m}} = \frac{\sigma}{\delta_s}\left(\frac{3F - D}{3Z}\right) \to \frac{\sigma}{\delta_s} = \frac{1}{10}. \quad 30.$$

This ratio is considerably larger than the ratio extracted from the mesons. However, the question of whether or not the $\bar{s}s$ matrix element is small is controversial at present (15–17). Certainly it need not vanish, and models with a sizeable matrix element do exist (16). More generally, the relation between σ and the quark masses is

$$\frac{\sigma}{m_\Xi - m_\Lambda} = \frac{3\hat{m}}{m_s - \hat{m}}\frac{1}{1 - y}$$

$$y \equiv \frac{2\langle P|\bar{s}s|P\rangle}{\langle P|\bar{u}u + \bar{d}d|P\rangle}. \quad 31.$$

Numerically this works out as

$$\frac{\hat{m}}{m_s - \hat{m}} = \frac{1}{10}(1 - y)\left(\frac{\sigma}{60 \text{ MeV}}\right). \quad 32.$$

To agree with Equation 6a would require $y \approx 1/2$. The fact that we cannot yet confirm nor rule out this large value of y points to our true ignorance of much of hadron dynamics.

A sizeable value of y does not conform to standard intuitions. However, this should not stop us from considering it, as our intuition can be readjusted if we are convinced that it must be. Perhaps the least appealing consequence of this picture is what it predicts for the proton mass in the limit $m_s \to 0$. Again using first-order perturbation theory one finds, by using Equation 29,

$$m_p = m_p^{(0)} + \langle P|m_s\bar{s}s|P\rangle + \langle P|m_u\bar{u}u + m_d\bar{d}d|P\rangle$$

$$\approx m_p^{(0)} + 310 \text{ MeV}\frac{y}{1 - y} + \sigma, \quad 33.$$

which indicates that the proton mass would only be ~ 560 MeV if the strange quark were massless. It is possible that such a situation could occur if the dimensional parameters derived from QCD, such as the string tension or vacuum energy, shifted by 40% as a result of the strange quark mass. However, this consequence is not particularly pleasing.

What we have seen here is a conflict among

1. The experimental value of the σ term,
2. The first-order SU(3) analysis,
3. The standard quark mass ratio, and
4. Standard intuition on hadron physics, built into many of the QCD-inspired models.

The discrepancy is about a factor of two, so that minor adjustments will not resolve the issue. This drives home a procedural point. Models cannot be used to support the standard quark masses if they agree with items 2 and 4 above, as this combination is in conflict with experiment. This caveat excludes most determinations of quark masses using dynamical models. It also encourages attempts to use symmetry methods to second order, as discussed in Section 5.

4. PHENOMENOLOGICAL STATUS OF THE SIGMA TERM

The sigma term comes from the investigation of $\pi^{\pm}p \rightarrow \pi^{\pm}p$ scattering (18). Specifically one takes the isospin-even scattering amplitude, with the nucleon pole term subtracted, $\bar{D}^{+}(v, t)$ where $v = (s-u)/4m_p$ and s, t, u are the Mandelstam variables. The amplitude Σ is defined by

$$\Sigma = F_\pi^2 \bar{D}^{+}(v = 0, t = 2m_\pi^2). \qquad\qquad 34.$$

To lowest order in the quark mass $\Sigma = \sigma$, where σ is defined in Equation 26, but more generally

$$\Sigma = \sigma + \Delta. \qquad\qquad 35.$$

The correction Δ is small and calculable in chiral perturbation theory. The most recent theoretical progress is the full evaluation of Σ up to one loop order (a difficult calculation) by Gasser and his collaborators (15). Their result is

$$\Delta = 4.5 \pm 1.2 \text{ MeV}. \qquad\qquad 36.$$

The kinematic location $v = 0$, $t = 2m_\pi^2$ is called the Cheng-Dashen point. Unfortunately it is not in the physical region for πN scattering. This means that the scattering amplitudes must be extrapolated off shell using analyticity properties and dispersion theory. These features are always built into the partial wave analysis of πN scattering, so that these fits to low energy data determine Σ. Early analyses yielded values that varied wildly, but recently the determinations have been more uniform. The Karlsruhe group's value is generally the standard (14), with

$$\Sigma = 64 \pm 8 \text{ MeV}. \qquad 37.$$

It is also possible to obtain Σ directly from the physical region. In this procedure the scattering lengths are parameterized by Σ using chiral perturbation theory. A specific combination of scattering lengths then reveals Σ. Gasser et al (15) used this to obtain

$$\Sigma = 61 \text{ MeV} \qquad 38.$$

from the πN scattering lengths. A measurement of one of the scattering lengths in pionic atoms (19) disagrees with values obtained in πN experiments. This disagreement has to be resolved; plans are underway for further measurements at low energy.

It is fair to say that most theorists' favorite solution to the problem of the sigma term would be to have the experimental value shift by a factor of two. At present there is little indication that this is about to happen. Similarly, with the work of Gasser et al (15) the theoretical foundations seem to be reasonably solid.

5.   HIGHER-ORDER TREATMENT OF MASSES

This section is somewhat more technical than the preceding ones, as the method of effective Lagrangians needs to be introduced in order to treat the quark mass beyond leading order. The formalism used here is that of Gasser & Leutwyler (5), who imagine writing the Lagrangian for QCD coupled to an external c-number source χ

$$\mathscr{L}_{\text{QCD}}^{(\chi)} = -\tfrac{1}{4}F_{\mu\nu}^{\text{A}}F^{\text{a}\mu\nu} + \bar{\psi}i\not{D}\psi - \bar{\psi}_{\text{L}}\chi\psi_{\text{R}} - \bar{\psi}_{\text{R}}\chi^{+}\psi_{\text{L}}. \qquad 39.$$

QCD is recovered by the choice $\chi = m$, with m being the quark mass matrix. The modified theory has a chiral symmetry if

$$\chi \rightarrow \exp{(i\boldsymbol{\alpha} \cdot \boldsymbol{\lambda})}\chi \exp{(-i\boldsymbol{\beta} \cdot \boldsymbol{\lambda})} \qquad 40.$$

in the notation of Equation 4. The object is then to construct the most general effective Lagrangian, containing the Goldstone bosons ϕ^{A} and the field χ, that is chirally invariant. This is organized in a power series in the energy. The result to order E^4 is

$$L = L_2 + L_4$$

$$L_2 = \frac{F_\pi^2}{4}\text{Tr}(\partial_\mu U \partial^\mu U^+) - \frac{F_\pi^2}{4} B_0 \text{Tr}(\chi^+ U + U^+ \chi)$$

$$L_4 = L_1[\text{Tr}(\partial_\mu U \partial^\mu U^+)]^2 + L_2 \text{Tr}(\partial_\mu U \partial_\nu U^+)\text{Tr}(\partial^\mu U \partial^\nu U^+)$$

$$+ L_3 \text{Tr}(\partial_\mu U \partial^\mu U^+ \partial_\nu U \partial^\nu U^+) + L_4 B_0 \text{Tr}(\partial_\mu U \partial^\mu U^+)\text{Tr}(\chi^+ U + U^+ \chi)$$

$$+ L_5 B_0 \text{Tr}[\partial_\mu U \partial^\mu U^+ (U\chi^+ + \chi U^+)] + L_6 B_0^2[\text{Tr}(\chi^+ U + U^+ \chi)]^2$$

$$+ L_7 B_0^2[\text{Tr}(\chi^+ U - U^+ \chi)]^2 + L_8 B_0^2 \text{Tr}(\chi^+ U \chi^+ U + U^+ \chi U^+ \chi)$$

$$U = \exp\left(i\frac{\lambda^A \phi^A}{F_\pi}\right). \qquad\qquad 41.$$

This expansion is used in the following way. The Lagrangians with $\chi = m$ can be expanded algebraically in terms of the meson fields and then matrix elements taken, in order to obtain the tree level vertices. In addition, loop diagrams that utilize vertices coming from L_2 contribute to matrix elements at the same power of energy as do tree vertices from L_4. These then need to be included; indeed, the loops renormalize the coefficients of L_4. This program is called chiral perturbation theory, and is too extensive to be reviewed fully here.

The crucial point is that deviations from the lowest-order behavior are governed by the coefficients $L_1, \ldots, L_9$. For example Gasser & Leutwyler (5) find

$$\frac{m_\pi^2}{2m_K^2} = \frac{\hat{m}}{m_s + \hat{m}}(1 - \Delta_m), \qquad\qquad 42.$$

where

$$\Delta_m = \frac{8}{F_\pi^2}(m_s - \hat{m})B_0(2L_8^r - L_5^r) + \mu_\eta - \mu_\pi$$

$$= \frac{8}{F_\pi^2}(m_K^2 - m_\pi^2)(2L_8^r - L_5^r) + \mu_\eta - \mu_\pi$$

$$\mu_i = \frac{m_i^2}{32\pi^2 F_\pi^2}\ln m_i^2/\mu^2 \qquad\qquad 43.$$

and L_i^r are the coefficients renormalized at scale μ. The coefficient L_5^r can be determined from the pseudoscalar decay constants

$$\frac{F_K}{K_\pi} = 1 + \Delta_F$$

$$\Delta_F = \frac{4}{F_\pi^2}(m_K^2 - m_\pi^2)L_5^r + \frac{5}{4}\mu_\pi - \frac{1}{2}\mu_K - \frac{3}{4}\mu_\eta \qquad\qquad 44.$$

such that

$$L_5^r(\mu = m_\eta) = 2.2 \times 10^{-3}. \qquad\qquad 45.$$

Unfortunately L_8^r only appears in the analysis of meson masses, and is not known from independent physics. In order to obtain an estimate of the sensitivity of the mass ratio to this quantity, let us consider a reasonable range, i.e. $L_8^r = 0$ to L_5^r. This results in

$$\Delta_m = -0.46 \to +0.46$$

$$\frac{\hat{m}}{m_s} = \frac{1}{37} \to \frac{1}{14}. \qquad\qquad 46.$$

There is a strong sensitivity.

The isospin-violating mass ratio is governed by the same parameter Δ_m

$$\frac{m_d - m_u}{m_s - \hat{m}}(1 + \Delta_m) = \frac{(m_{K^0}^2 - m_{K^+}^2)_{\text{quark mass}}}{m_K^2 - m_\pi^2}. \qquad\qquad 47.$$

The problem with applying this relation is that the electromagnetic contribution to $m_{K^0}^2 - m_{K^+}^2$ (and hence the right-hand side of the above equation) has not been determined to next order in the quark masses. This would involve the calculation of mass effects in the electromagnetic self-energies, and would likely involve further low energy parameters. Similarly it is not consistent to use values of $(m_d - m_u)/m_s$ obtained from other systems (such as $\psi' \to \psi\pi^0$) unless they have also been evaluated to second order in the quark masses.

Equations 42, 46, and 47 summarize the present status of the analysis that relies only on symmetry methods. Unfortunately, without an independent measure of the low energy parameter L_8 it is not possible to determine the quark mass ratio to second order in the energy expansion. The reasonable range of values for the ratio spans roughly a factor of two, and includes a value of up to $\hat{m}/m_s \sim 1/14$, large enough to remove the problem with the sigma term if it were to be correct.

6. MODELS

Many of the attempts to deal with the dynamical features of QCD at low energy also contain the light quark masses as parameters. These can then be applied to yield information about the masses from a wider range of applications than we have discussed here. Unfortunately, the models do not yield uniform answers. There are models that support the lowest-order mass ratio and then have problems with the σ term (20). Other models fit a somewhat larger mass ratio and then agree with the sigma term (21). Still others have nonlinear SU(3) breaking so that neither the lowest-order nor second-order analyses have much validity (17). Some have large values of y, while others do not. It may well be that in the long run we can sort out these issues by studying the different models in detail. However, at the least, it is not the time for a review article on these attempts. What I do in this section is discuss two simple but general models relevant for quark masses.

The first model is an attempt to estimate the sign and possible magnitude of the low energy constants needed to complete the second-order analysis of the previous section. The issue addressed is the relation of the meson masses to the quark masses. The expansion for small masses is of the form

$$m_M^2 = 2m_q B_0 + (2m_q)^2 C_0 + \ldots , \qquad\qquad 48.$$

or equivalently the relation may be inverted

$$2m_q B_0 = m_M^2 - m_M^4 C_0 / B_0^2 + \ldots . \qquad\qquad 49.$$

For the other extreme, large masses, the magnitude of the quark mass is better defined and the relationship between mesons and quarks is linear:

$$m_M = 2m_q + \text{constant} + \ldots \qquad (m_q \text{ large}). \qquad\qquad 50.$$

The sign of the deviation at low masses is apparently fixed by the requirement that the general form interpolate between these two limits. One simple guess that reproduces both limits is

$$m_M^2 = 2m_q B_0 \left(1 + \frac{2m_q}{B_0} \right), \qquad\qquad 51.$$

i.e. $C_0 = 1$. Since we know from quark models that the D meson mass is in the linear region, we would expect in this formula $B_0 \lesssim 1$ GeV, perhaps being as small as the $600 \to 700$ MeV scales that often appear in the chiral expansions. The inverse would give

$$2m_q B_0 = m_M^2 [1 - (m_M/B_0)^2 + \ldots]$$

$$\frac{\hat{m}}{m_s} = \frac{m_\pi^2}{2m_K^2} \left(1 + \frac{m_K^2 - m_\pi^2}{B_0^2} + \ldots \right). \qquad\qquad 52.$$

For $B_0 \lesssim 1$ GeV the trend is to make the mass ratio larger, perhaps by as much as 50%. This is consistent with the range estimated in the last section, but with a clear sign.

The second model concerns the meaning of the magnitudes of masses in quark models. In particular it is often assumed that $m_s - \hat{m} \approx 150$ MeV because the proton and lambda, differing by one strange quark, have masses $m_\Lambda - m_p \approx 150$ MeV. However, explicit models with light quarks tend to find a larger value for $m_s - \hat{m}$ for the same Λ-P mass splitting. The reason for this is instructive. In first-order perturbation theory, the mass splitting is due to the operator $m_s \bar{s}s$, i.e.

$$m_\Lambda - m_p \approx \langle \Lambda | m_s \bar{s}s | \Lambda \rangle - \langle P | m_s \bar{s}s | P \rangle$$

$$\equiv m_s Z'. \qquad\qquad 53.$$

The factor Z' when calculated in a quark model generally turns out to be less than one (22). If the quark's Dirac wavefunction has upper and lower components u and l,

$$Z' = \frac{\int d^3 x(u^2 - l^2)}{\int d^3 x(u^2 + l^2)}, \qquad\qquad 54.$$

with the denominator being the normalization integral. In momentum space

$$Z' = \left\langle 1 - \frac{p^2}{2E(E + m_q)} \right\rangle. \qquad\qquad 55.$$

Explicit calculations in the bag model give $Z' = 0.48$, while if one estimates the momentum via the uncertainty principle $[p = (p_x^2 + p_y^2 + p_z^2)^{1/2} \approx \sqrt{3}/R$ with $R \approx 1$ fm], one obtains $Z' \approx 0.7$. These values then produce estimates of m_s up to a factor of two larger, which supports even $m_s \approx 300$ MeV. The main point is that hadron mass splittings are not the same as quark mass splittings for light quarks even within simple quark models. While *any* use of the quark mass magnitudes is subject to the caveats described in the introduction, the indiscriminate use of $m_s \approx 150$ MeV is particularly inappropriate.

7. SUMMARY

We have surveyed here the information on quark masses, with an emphasis on the results of symmetry methods. The problem associated with the sigma term has led us to question the standard first-order quark mass ratios. Unfortunately we do not find firm evidence beyond leading order, but significant ($\approx 50\%$) departures from the conventional values are not ruled out.

Much recent work has focused on the sigma term. As described in Section 3, the experimental value forces us to abandon at least one of the standard assumptions about low energy physics. A larger value of $(m_u + m_d)/m_s$ would help greatly in the understanding of the sigma term, and this apparently remains one of the possible solutions.

Literature Cited

1. Marciano, W., Pagels, H., *Phys. Rep.* 36: 276 (1978); Collins, J. C., Soper, D. E., *Annu. Rev. Nucl. Part. Sci.* 37: 383 (1987)
2. Georgi, H., Politzer, H. D., *Phys. Rev.* D14: 1829 (1976)
3. Pagels, H., *Phys. Rev.* 16C: 219 (1975); Adler, S., Dashen, R., *Current Algebra.* New York: Benjamin (1968)
4. Weinberg, S., *Physica* 96A: 327 (1979)
5. Gasser, J., Leutwyler, H., *Ann. Phys.* 158: 142 (1984); *Nucl. Phys.* B350: 465 (1985)
6. Gasser, J., Leutwyler, H., *Phys. Rep.* 87C: 77 (1982)
7. Gell-Mann, M., Oakes, R. J., Renner, B., *Phys. Rev.* 175: 2195 (1968)
8. Dashen, R., *Phys. Rev.* 183: 1245 (1969)
9. Voloshin, M., Zakharov, V., *Phys. Rev. Lett.* 45: 688 (1980)
10. Donoghue, J. F., Tuan, S. F., *Phys. Lett.* 164B: 401 (1985)
11. Gell-Mann, M., *Cal. Inst. Technol. Rep.* CTSL-20 (1961); Okubo, S., *Prog. Theor. Phys.* 27: 949 (1962)
12. Coleman, S., Glashow, S., *Phys. Rev. Lett.* 6: 423 (1961)
13. Cheng, T. P., Dashen, R., *Phys. Rev. Lett.* 26: 594 (1971); Brown, L. S., Pardee, W. J., Reccei, R. D., *Phys. Rev.* D4: 2801 (1971)
14. Höhler, G., In *Methods and Results of Phenomenological Analysis, Landolt-Börnstein,* ed. K.-H. Hellwege, Vol. 9b2. Berlin: Springer (1983); Koch, R., Z. *Phys.* 15C: 161 (1982)
15. Gasser, J., In *Proc. 2nd Int. Conf. on πN Physics,* ed. W. Gibbs, B. Nefkins. Los Alamos Natl. Lab. (1987), p. 266; Gasser, J., Sainio, M., Svarc, A. To be published (1989)
16. Donoghue, J. F., Nappi, C., *Phys. Lett.* 168B: 105 (1986)
17. Jaffe, R. L., *Phys. Rev.* D21: 3215 (1980)
18. Reya, E., *Rev. Mod. Phys.* 46: 545 (1974); Cheng, T. P., *Phys. Rev.* D13: 2161 (1976); Dominguez, C. A., Lanqacker, P., *Phys. Rev.* D24: 1905 (1981); Gasser, J., *Ann. Phys.* 136: 62 (1981); Jaffe, R. L., Korpa, C. L., *Comments Nucl. Part. Phys.* 17: 105 (1986)
19. Bovet, E., et al, *Phys. Lett.* 153B: 231 (1985)
20. Dominguez, C. A., de Rafael, E., *Ann. Phys.* 174: 372 (1987); Gasser, J., Leutwyler, H., *Nucl. Phys.* B94: 269 (1975); Socolow, R., *Phys. Rev.* 137: 1221 (1965)
21. Donoghue, J. F., Johnson, K., *Phys. Rev.* D21: 1975 (1980); Donoghue, J. F., Golowich, E., Holstein, B. R., *Phys. Rev.* D12: 2875 (1975)
22. Donoghue, J. F., Golowich, E., Holstein, B. R., *Phys. Rep.* 131: 319 (1986)

Annu. Rev. Nucl. Part. Sci. 1989. 39: 19–42

HEAVY-PARTICLE RADIOACTIVITY ($A > 4$)

P. Buford Price

Department of Physics, University of California, Berkeley, California 94720

KEY WORDS: cluster emission, superasymmetric fission, alpha decay.

CONTENTS

1.   INTRODUCTION

All nuclei with $Z > 40$ are metastable with respect to radioactive decay into two nuclear fragments with masses M_1 and M_2 for which $Q \equiv M(A, Z) - M_1(A_1, Z_1) - M_2(A_2, Z_2) > 0$. The case $A_2 = 4$ corresponds to alpha decay; cases in which $A_1 \approx A_2$ correspond to spontaneous fission; and the inter-

19

0163–8998/89/1201–0019$02.00

mediate cases, in which $4 < A_2 < A_1$, we call heavy-particle radioactivities. Among the heavy nuclei there are thousands of possible two-body decays with positive Q. For most such combinations of A_1, Z_1 and A_2, Z_2 the decay rates are undetectably low, but for certain combinations corresponding to fragments with nearly closed shells the high Coulomb barrier (proportional to Z_1Z_2) is partly compensated by a high Q, and the decay may be detectable.

A rule of thumb for guessing which decay modes might have high enough Q to occur at detectable rates is to let the heavier fragment be close to ^{208}Pb, which has a large mass defect as a result of its closed neutron and proton shells, and to let the lighter fragment be a tightly bound, even-even, neutron-rich nuclide such that the parent is stable enough to be amenable to study. An example is ^{232}U $\rightarrow$ ^{208}Pb $+$ ^{24}Ne, which has a branching ratio of $\sim 10^{-12}$ relative to alpha decay. Few cases of heavy-fragment emission have a higher branching ratio than this, which explains why heavy-particle radioactivity was not discovered until many decades after alpha radioactivity and spontaneous fission.

In order to choose favorable cases for experimental study, one needs a quantitative model for decay rates, and we discuss several models below. All of them involve tunneling through a potential barrier. They fall into two classes. Cluster models depict a cluster of nucleons as forming inside the parent nucleus with a probability that decreases with cluster size and then tunneling out without change of size or shape, as shown in Figure 1a. Unified models treat alpha decay, heavy-particle radioactivity, and cold (spontaneous) fission as equivalent processes, differing only in the degree of mass asymmetry, in which the parent nucleus deforms into two fragments, usually keeping constant the volumes of the portions that distort into separate fragments (Figure 1b).

Figure 2 shows examples of the dependence of potential energy on separation of the centers of the two fragments for several types of nuclear potential, calculated for the decay mode ^{232}U $\rightarrow$ ^{24}Ne $+$ ^{208}Pb. The ordinate is displaced by the amount Q, so that in this figure the total energy of the two fragments is zero. The abscissa is the separation, r, of the centers of the two fragments. Because of the short range of the strong force, when the two fragments are farther apart than the "touching distance" $R_t \equiv R_1 + R_2$, their radii have the well-defined values $R_1(A_1, Z_1)$ and $R_2(A_2, Z_2)$, and the potential energy is dominated by the repulsive Coulomb potential, which is $V_C \equiv Z_1Z_2e^2/r$ if both nuclei are spherical. At separations $r < R_t$ the potential energy drops dramatically because of the influence of the attractive strong force and the weakening of the Coulomb repulsion.

According to classical mechanics, decay cannot occur unless Q is greater than the height of the barrier, which is the case only for fission of nuclei

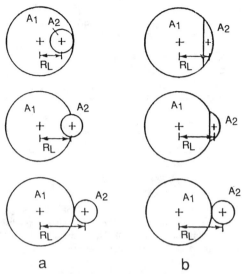

a   b

Figure 1   Heavy-particle decay viewed (*a*) as emission of a cluster at constant radius A_2 and (*b*) as distortion into two compact shapes at constant volume V_2.

with $Z^2/A > 49$, according to the liquid drop model. In the quantum mechanical picture, developed in 1928 by Gamow and by Condon & Gurney, the decay rate is proportional to the product of a frequency factor and a penetrability given in the WKB approximation by $P \approx \exp(-K)$, with

$$K = \frac{2}{\hbar} \int_{R_a}^{R_b} \sqrt{2m(r)(V-Q)} \, dr. \qquad\qquad 1.$$

The integration is between the two turning points; the mass parameter $m(r)$ is equal to the reduced mass μ at distances $r > R_t$ and depends on the specific model for distances $r < R_t$; $V(r)$ is the energy barrier, which includes the nuclear potential, the Coulomb potential, and possibly a centrifugal potential. In the cluster models the decay rate includes a factor S that expresses the probability of preformation of a particular cluster of nucleons inside the parent. In the fission models, the deforming parent nucleus follows a valley in a configuration space of several variables.

It is instructive to consider the simplest model of a pure Coulomb potential for spherical fragments at $r > R_t$ and a square-well potential at $r < R_t$, for which the quantity K in the exponent of the penetrability P is

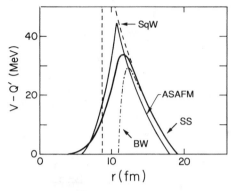

Figure 2   Potential energy barriers for $^{232}U \rightarrow {}^{24}Ne + {}^{208}Pb$ in the analytic superasymmetric fission model (ASAFM) (5); proximity-potential model (SS) (27; also unpublished additional calculations); Blendowske-Walliser model (BW) (28; also unpublished additional calculations); and square-well model (SqW) (13). The ordinate is displaced by an amount Q', which equals $Q + E_v$ for the ASAFM and equals Q for the others.

$$K = \frac{2}{\hbar} \sqrt{\frac{2\mu}{Q}} \, Z_1 Z_2 e^2 \, (\arccos \sqrt{x} - \sqrt{x - x^2}), \qquad 2.$$

where $x = Q/B$ and where $B = Z_1 Z_2 e^2 / R_t$ is the height of the barrier at $r = R_t$. Until about ten years ago everyone who looked at this expression concluded that only for alpha decay is the value of K small enough to give measurable decay rates. The product $Z_1 Z_2$ is, of course, smallest for $Z_2 \ll Z_1$. For proton radioactivity there are only a few proton-rich nuclides for which Q is positive and the decay rate competes with the beta-decay rate. For a large number of nuclides with $Z > 60$, Q is relatively large for alpha decay because of the large mass defect of ^{4}He. It was not until 1980 that Sandulescu et al (1) did the systematic calculations necessary to discover that a number of nuclides heavier than ^{4}He would have a high enough Q to be emitted at a detectable rate when the heavy daughter is close to doubly magic ^{208}Pb.

Poenaru, Ivascu, Sandulescu, and Greiner deserve credit for calling attention to the rich variety of possible heavy-particle radioactivities, including emission of ^{5}He (unbound but possibly detectable via correlated n + ^{4}He) (2), ^{8}Be (detectable via correlated ^{4}He + ^{4}He), ^{14}C, ^{18}O, ^{24}Ne, ^{30}Mg, and many others (3–7). Their recent review of the various theoretical approaches (8), including discussion of alpha decay and fission models, makes interesting reading. This chapter, which concentrates on the experimental status of heavy-particle radioactivities and excludes alpha decay and fission, complements their approach.

Though the 1980 paper by Sandulescu et al did not make quantitative predictions of decay rates, it should have stimulated experimenters by pointing out specific decay modes that might be detectable. Unfortunately the paper went unnoticed except in the Soviet Union, where Aleksandrov et al (9) mounted a search for ^{14}C emission from ^{223}Ra, ^{224}Ra, and ^{226}Ra. Although they did observe ^{14}C decay of ^{223}Ra, they were beaten by Rose & Jones (10), who observed this decay mode several months earlier, using a semiconductor $E + \Delta E$ counter telescope to detect 11 events with $Z = 6$ and kinetic energy ~ 30 MeV during a run of 189 days. Using a square well + Coulomb potential, Rose & Jones had calculated penetrabilities for heavy-fragment emission from a number of available radioactive sources and had concluded that the rate of emission of ^{14}C from ^{223}Ra should be most favorable.

Rose & Jones measured a branching ratio $B(C/\alpha) = 8.5 \pm 2.5 \times 10^{-10}$. They were not able to identify the isotope but concluded from the relative penetrabilities that it was almost certainly ^{14}C. Radiation damage and alpha-particle pulse pileups in their detectors deterred them from detecting any other cases of heavy-particle radioactivity, and all further results have been obtained with techniques capable of discriminating against intense backgrounds of alpha particles or fission fragments.

2. EXPERIMENTAL METHODS

Using their Analytical Superasymmetric Fission Model (ASAFM), Poenaru and coworkers (6, 7) calculated Q values, partial half-lives, and branching ratios relative to alpha decay for some 10^5 combinations of 2200 parent nuclei and 140 fragments with $2 \leq Z \leq 26$. They prepared tables containing this information for fragments with partial half-lives shorter than 10^{50} s emitted from nuclides with $47 \leq Z \leq 106$ and half-lives longer than 10^{-6} s. The most recent table (7) is based on a version of the ASAFM containing parameters adjusted to fit heavy-particle decay rates that had been measured as of 1987. When scanning tables such as theirs, looking for decay modes to investigate, one needs to keep in mind four practical constraints, which reduce the number of experimentally detectable examples of heavy-particle radioactivity to a few dozen at best:

1. The kinetic energy of the light fragment is typically 2 to 2.5 MeV per nucleon. Its range, no more than ~ 15 mg cm^{-2} in an element such as uranium, is comparable to that of alpha particles emitted by the same parent. To identify heavy-particle radioactivity by direct detection of the emitted fragment, the source must be thinner than ~ 1 or 2 mg cm^{-2}.

2. The decay rate should be high enough to observe at least a few decays

per year in a sample of practical dimensions. The lowest rate likely to be detectable without an extraordinary effort is $\sim 10^{-29}$ yr^{-1}.

3. The branching ratio relative to alpha decay and spontaneous fission should not be so low that the signal is swamped by these backgrounds. The lowest branching ratio detected to date is $\sim 10^{-16}$ relative to alpha decay (11). A decrease to 10^{-18} seems conceivable with improvements in technique.

4. The parent nuclide must be available in large enough quantities, with a long enough half-life, and free enough of interfering radioactive impurities to make the experiment feasible. These constraints rule out most nuclides with $Z > 94$ and pose great difficulties for the study of some of the otherwise attractive short-lived nuclides that would decay into the vicinity of doubly magic ^{132}Sn, ^{100}Sn, or ^{208}Pb.

2.1 Sources

Rose & Jones used a source of ^{227}Ac (half-life 21.7 yr) with which ^{223}Ra (half-life 11.4 d) and other members of the ^{235}U series were in secular equilibrium. Price & Barwick (12) used a source of natural ^{232}Th in a negative search for ^{26}Ne emission. One has to be aware, in using naturally occurring members of the ^{235}U, ^{238}U, and ^{232}Th decay chains, of the possibility of ambiguity as to which nuclide in the chain is the emitter of heavy particles.

Price et al (13) and Barwick et al (14) installed track-recording polycarbonate films at the collector inside the vacuum chamber of the ISOLDE on-line isotope separator at CERN, where beams of various short-lived nuclides were produced in spallation reactions by bombardment of a thick ThC$_2$ target with 600-MeV protons. Radium and francium isotopes were selectively ionized in a high-temperature surface-ionization source, and the beam extracted from the ion source was separated into its constituent atomic masses by the ISOLDE electromagnetic isotope separator. Each of the 60-keV beams of masses 221 to 225 was implanted in a separate collecting cup surrounded by polycarbonate film. For $A = 221$ both Fr and Ra are alpha emitters; for $A = 222$, 223, and 224 all the francium undergoes beta decay to an alpha-emitting Ra isotope; for $A = 225$ both Fr and Ra end up via beta decay as alpha-emitting ^{225}Ac. With the ISOLDE-II machine it was possible to produce sources with $Z \leq 89$ of up to a few 10^{13} atoms with half-lives down to a fraction of a second. The higher intensity of ISOLDE-III may make it possible to create adequate quantities of a few sources that are predicted to decay by heavy-particle radioactivity into the Sn region.

The other sources used were prepared either by chemical separation from natural uranium and thorium or by bombardment of heavy nuclides in a high-flux reactor. Isotopic analysis of radioactive impurities and careful purification are essential when studying actinides with low branching ratios for heavy-particle radioactivity.

2.2 Ionization Detectors

To search for heavy particles at branching ratios below 10^{-10} relative to alpha decay, the problem of discriminating against the enormous alpha-particle flux has been approached in two ways. The first is to install a $\Delta E + E$ particle identification telescope at the focal plane of a magnetic spectrometer set to transmit only monoenergetic fragments with the desired kinetic energy and charge state. Figure 3 shows schematically the trajectories of $^{14}C^{+6}$ ions and of singly and doubly charged alpha particles from ^{223}Ra in the superconducting solenoidal spectrometer used at Orsay (15, 16). Figure 4 shows the distribution of carbon isotopes, their energies and charge states, collected from ^{223}Ra with a gas ionization detector that measures position, ΔE, and E at the focal plane of an Enge split-pole spectrograph at Argonne National Laboratory (17). At both facilities it was possible to show that the carbon emitted by ^{223}Ra had $A = 14$. The main background in such a spectrometer, which proved not to be a serious problem, is from alpha decays of radon gas atoms throughout the system. The small solid angle (5 to 60 msr) and limited availability of magnetic spectrometers seem to restrict their use to the study of a few radionuclides with branching ratio greater than $\sim 10^{-11}$.

The second technique, used by our Berkeley group and by researchers at Dubna (USSR), is to record latent tracks of heavy particles in a plastic or glass detector that is insensitive to particles with ionization rate below that of the particles to be detected. After exposure, the tracks are developed by chemical etching and are located by either manual or automated scanning, after which their dimensions are measured in a microscope. With a detector of optimized sensitivity, calibrated with heavy ions accompanied by alpha particles at the dose expected in an actual experiment, it is possible to measure charge and range at branching ratios at least as low as 10^{-16}. Since all but two of the measurements in Table 1 were made with this technique, a brief description is given below. For a fuller discussion see (12, 18, 19).

Figure 5 shows the geometry of an etched track of a normally incident particle. To identify the particle one relates two quantities, s and R, to the charge Z and energy E. We define $s = v_T/v_G = 1/\sin \phi$ as the ratio of etch rate along the track to the general etch rate (ϕ is the half angle of the conical etchpit); calibrations with ions at a heavy-ion accelerator establish the relation between s and dE/dx for a given detector, etching reagent,

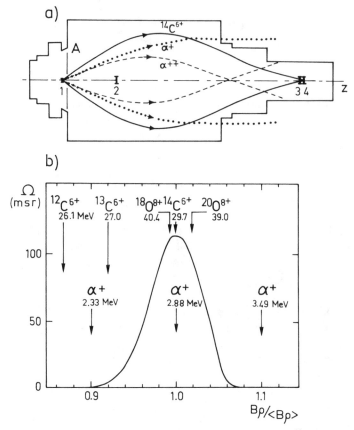

Figure 3 The Orsay spectrometer (15): (*a*) trajectories for 29.7-MeV ^{14}C and singly and doubly ionized alpha particles from ^{223}Ra; (*b*) transmission curve for ^{14}C as a function of $B\rho$.

and background alpha-particle fluence. For a track etched to the end of its range, the quantities D_1, D_2, and R, together with a knowledge of G, the amount of bulk material etched away, provide two sets of values of s_i, R_i that give two determinations of the charge and energy of the particle. The average value of s_1 is a function of etchpit mouth diameter, D_1, and zenith angle, θ:

$$\bar{s}_1 = [1+r^2]\sec\theta/[1-r^2],\qquad\qquad 3.$$

where $r \equiv D_1/2G$. A least-time analysis shows (18) that the value of D_1 is determined by etchant traveling at a rate v_T for a distance h given by

$$h = G\bar{s}_1/(1+\bar{s}_1\cos\theta)\qquad\qquad 4.$$

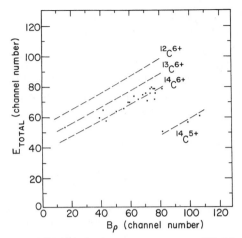

Figure 4 Identification of ^{14}C with the Argonne spectrometer (17). Mass lines were determined from a heavy-ion calibration.

followed by etching at a rate v_G along the direction indicated in Figure 5 until the mouth is reached. The average value of the range, $\bar{R}_1$, corresponding to $\bar{s}_1$ is given by $\bar{R}_1 = R - h/2$. Measurement of the diameter of the rounded tip, D_2, gives an independent identification. The larger the value of s, the sooner the etchant will reach the tip and the larger its diameter will become. Once the etchant reaches the tip, its diameter grows at a rate v_G. The average rate $\bar{s}_2$ at the average range $\bar{R}_2 = R/2$ is given by

$$\bar{s}_2 = R/(G - D_2/2). \qquad\qquad 5.$$

Polycarbonate film such as Rodyne (13) records tracks of carbon nuclei at the energies of interest, up to 2.5 MeV per nucleon. Although alpha particles do not themselves produce tracks, at alpha fluences above $\sim 3 \times 10^{10}$ cm^{-2} the short recoil tracks of backscattered C and O atoms reach such a high density as to make ^{14}C tracks difficult to detect. BP-1 phosphate glass (20) (Schott Glass Technologies) records ^{14}C tracks with

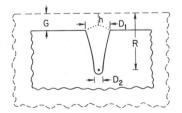

Figure 5 Etchpit of a normally incident particle of range R after removing a layer of thickness G (18).

better discrimination against alpha-produced elastic recoils and may make possible the study of carbon emission by a number of nuclides in the Ba to Ra region with branching ratios below 10^{-11}.

Two types of track detectors have been used to search for O and Ne decays. Polyethylene terephthalate plastic film (12, 19) is insensitive to nuclei up to carbon and can be used in the presence of alpha-particle fluences up to $\sim 10^{12}$ cm^{-2}. PSK-50 phosphate glass (Schott) is insensitive to nuclei up to oxygen and has good charge resolution for Ne and Mg nuclei (21, 25) at alpha fluences up to $\sim 10^{13}$ cm^{-2}. LG-750 phosphate glass (Schott) is ideal for the detection of Mg and Si nuclei and can withstand alpha fluences as high as 10^{14} cm^{-2} when properly calibrated (11, 22).

Etched tracks at approximately normal incidence on glass detectors appear as circular black spots on a light background and can be located automatically with a commercial image-processing system (11), which ensures a high and constant detection efficiency even at very low branching ratios.

2.3   Reduction of Fission Fragment Background

Even-even transuranic nuclides tend to have spontaneous fission rates orders of magnitude higher than for heavy-particle emission. Because the range of fission fragments is only about half as great as the range of particles emitted in heavy-particle radioactivity, they can usually be eliminated by covering the source with an Al film of uniform thickness, ~ 12 μm. The residual range of Ne, Mg, and Si nuclei after penetrating the Al film is typically 9 to 10 μm, which is sufficient for their charge and energy to be determined with phosphate glass detectors (11).

2.4   Gamma-Ray Spectroscopy

Certain emitted heavy particles decay into radioactive nuclides that can be identified by their characteristic gamma-ray spectra. The case of ^{233}U, which is predicted to decay by emission of both ^{24}Ne and ^{25}Ne with comparable branching ratios, is a favorable one to study by this method. ^{24}Ne decays into ^{24}Na (half-life 15 hr), which emits a 2754-keV gamma ray followed by a 1369-keV gamma ray and which can be sought by a delayed coincidence technique. ^{25}Ne decays into ^{25}Na (half-life 1 min), which decays away before the sodium can be chemically separated from the uranium and other radioactive elements. In a search for ^{24}Na gamma-ray lines in 10 g of ^{233}U, Balysh et al (23) obtained a null result, from which they concluded that the branching ratio for ^{24}Ne emission is less than 9.5×10^{-13} at the 68% confidence level. When compared with the branching ratio of $7.5 \pm 2.5 \times 10^{-13}$ measured for emission of all isotopes of Ne with track detectors (24), the gamma-ray limit is not stringent enough to rule out any combination of ^{24}Ne and ^{25}Ne. An improvement by a factor ten in the sensitivity of the gamma-ray method, which could be achieved in this case by increasing the source size, would make it possible to determine the relative yields of ^{24}Ne and ^{25}Ne.

The method could also be applied to ^{234}U, which is predicted to emit both ^{24}Ne and ^{26}Ne at comparable rates. The ^{24}Na has a much greater half-life than ^{26}Na, and the branching ratio inferred from its gamma-ray line intensities could be compared with the branching ratio for all Ne, obtained with track detectors (21, 25).

3. MEASUREMENTS OF HALF-LIVES FOR HEAVY-PARTICLE EMISSION

Table 1 is a compilation of measured partial half-lives for ions from ^{14}C to ^{34}Si, compared with values predicted by four unified models and three cluster models. Other models have also been developed (39–46) but are not included here, either because they made too few predictions or because some of their predictions disagreed with data by several orders of magnitude.

3.1 Carbon Emission

Soon after ^{14}C emission from ^{223}Ra was reported by Rose & Jones (10) and confirmed by Aleksandrov (9) three other groups reproduced the result, with similar branching ratios relative to alpha decay (13, 15, 17). Two groups used electronic detectors at the focal plane of a magnetic spectrometer (15, 17). Our own group, in collaboration with Ravn at CERN, used polycarbonate track-recording film in conjunction with beams of Fr and Ra isotopes at the ISOLDE isotope separator to confirm carbon emission from ^{223}Ra and to discover carbon emission from ^{222}Ra and ^{224}Ra with branching ratios of $3.7 \pm 0.6 \times 10^{-10}$ and $4.3 \pm 1.2 \times 10^{-11}$ respectively (13). We were able to infer that the ions were C with mass 14 by showing that their ranges were 10 to 15% greater than the ranges expected for emission of ^{12}C and ^{13}C and in good agreement with ranges calculated for ^{14}C ions with kinetic energy equal to $QA_1/(A_1 + A_2)$, the value expected for ground-state to ground-state decay.

With their magnetic spectrometer, Hourani et al (16) confirmed the carbon emission of ^{222}Ra, obtained a branching ratio consistent with the one we obtained, and verified that the isotope was ^{14}C. In addition, they observed the ^{14}C decay of ^{226}Ra, with a branching ratio of $3.2 \pm 1.6 \times 10^{-11}$. In followup experiments, our group, in collaboration with Ravn at CERN and Hourani and Hussonnois at Orsay, used polycarbonate detectors to obtain better statistics for ^{14}C emission by ^{226}Ra (branching ratio =

Table 1 Nuclei with measured partial half-lives for heavy-particle radioactivity

Decay	E_k (MeV)	Poe. (7)	P-P (35)	SqW (13)	SS (27)	BM (36)	BW (28)	SK (37)	log T (s)	−log B	Ref.
				Theoretical predictions of log T (Refs.)					Measured		
221Fr (14C)	29.28	14.4	16.5	15.2	16.0	14.0	15.5	14.5	>15.77	>13.3	13;14
221Ra (14C)	30.34	14.3	14.7	14.1	14.8	>12.4	14.2	13.3	>14.35	>12.9	13;14
222Ra (14C)	30.97	11.2	13.3	11.2	11.6	11.2	11.8	11.8	11.02±0.06	9.43±0.06	13;16
223Ra (14C)	29.85	15.2	15.6	15.0	15.7	15.3	15.1	14.2	15.2±0.05	9.21±0.05	10;9;15;13;17
224Ra (14C)	28.63	15.9	18.0	16.0	16.8	16.1	16.2	17.0	15.9±0.12	10.37±0.12	13
225Ac (14C)	28.57	17.8	19.0	18.7	19.7	18.8	18.6	18.3	>18.34	>12.4	14
226Ra (14C)	26.46	21.0	22.9	21.0	22.2	21.2	21.1	22.5	21.33±0.2	10.6±0.2	16;14
231Pa (23F)	46.68	25.9	>23.4	26.0	25.5	----	----	----	>24.61	>12.74	30
230Th (24Ne)	51.75	25.3	26.1	24.8	24.9	24.4	24.8	25.2	24.64±0.07	12.25±0.07	29
232Th (26Ne)	49.70	28.8	29.6	29.1	28.4	28.7	27.9	29.4	>27.94	>10.3	12
231Pa (24Ne)	54.14	23.4	23.4	23.7	23.5	21.6	23.4	22.3	23.38±0.08	11.37±0.08	30
232U (24Ne)	55.86	20.8	21.8	20.7	20.0	20.2	20.8	20.6	21.06±0.1	11.7±0.1	31
233U (24Ne)	54.27	24.8	24.4	24.9	24.8	23.7	25.4	23.6	24.82±0.15	12.12±0.15	24;32
233U (25Ne)	54.32	25.0	>24.4	25.1	24.4	----	----	23.6			
234U (24Ne)	52.81	26.3	26.4	25.8	25.7	25.5	25.6	26.4	25.25±0.05	12.36±0.05	21;25
234U (26Ne)	52.87	26.5	>26.4	26.2	25.0	25.9	26.4	26.5			
234U (28Mg)	65.26	25.8	>26.4	25.4	25.7	25.7	25.4	25.8	25.75±0.06	12.86±0.06	21;25
237Np (30Mg)	65.52	27.5	27.7	28.3	27.7	>27.3	29.9	26.8	>27.27	>13.4	29
238Pu (30Mg)	67.00	25.7	>26.4	25.9	24.6	25.6	25.8	25.7	25.7±0.25	16.25±0.25	11
238Pu (28Mg)	67.32	26.0	>26.4	25.5	----	25.7	26.9	25.4			
238Pu (32Si)	78.95	25.1	26.4	25.7	----	26.0	25.7	25.5	25.3±0.16	15.86±0.16	11
241Am (34Si)	80.60	24.5	25.5	26.5	26.2	25.3	28.8	23.8	>25.3; >24.2	>15.1; >14.1	29;22;34
$\sigma_{\log T}$		0.45	1.33	0.28	0.67	0.71	0.35	0.82			

$2.9 \pm 1.0 \times 10^{-11}$) and to set very stringent upper limits on the branching ratios for ^{14}C decay of the nuclides ^{221}Fr, ^{221}Ra, and ^{225}Ac produced at ISOLDE (14).

3.2 Neon Emission

According to the rule of thumb by which the heavy fragment is close to doubly magic ^{208}Pb, neon fragments should be emitted mainly from isotopes of uranium and neighboring elements. Two groups have used track detectors to search for neon emission in long-lived radioactive sources. Using polyethylene terephthalate films, Sandulescu, Tretyakova, and coworkers at Dubna reported Ne-decay branching ratios of $5.6 \pm 1.0 \times 10^{-13}$ for ^{230}Th (29), of $4.3 \pm 0.9 \times 10^{-12}$ for ^{231}Pa (30), and of $7.5 \pm 2.5 \times 10^{-13}$ for ^{233}U (24). Their targets were 0.33 to 0.56 mg cm^{-2} thick, which broadened the range distributions of the Ne ions too much for their isotopic distribution to be determined. In the case of ^{231}Pa, they were able to obtain an upper limit (at 90% confidence) of 1.8×10^{-13} on the branching ratio for ^{23}F decay, which is predicted to occur at a rate lower than, but within a factor of $\sim 10^2$ of, the rate for ^{24}Ne emission.

Using Cronar polyester film, our group detected neon decay of ^{232}U with a branching ratio of $2.0 \pm 0.5 \times 10^{-12}$ (31). The branching ratio is predicted to be several orders of magnitude greater for ^{24}Ne than for any other Ne isotope. This is a consequence of the much higher Q for ^{24}Ne than for the other isotopes. The calculated range for ^{24}Ne, with both fragments in the ground state, ranges from 10 to 30% greater than for the other Ne isotopes. Because the ^{232}U source was only 5 μg cm^{-2} thick, it was thus possible to show experimentally that all of the Ne ions were ^{24}Ne. Barwick (32), using Cronar detectors, confirmed the observation by the Dubna group (29) of Ne emission by ^{233}U. His branching ratio, $5.3 \pm 2.3 \times 10^{-13}$, was consistent with theirs.

^{232}Th is expected to emit ^{24}Ne, ^{26}Ne, and ^{28}Mg at branching ratios of $\sim 10^{-12}$ to $\sim 10^{-11}$, but because of its long lifetime, $\sim 10^{10}$ yr, the event rate is probably extremely low. Price & Barwick (12) placed a thick foil, of area 77 cm^2, of natural Th for 10 months in contact with Cronar film but found no tracks due to Ne or Mg. They set an upper limit (90% confidence level) of 5×10^{-11} on the branching ratio for emission of ions with $Z \geq 10$, which is not in conflict with predictions of the models. With automated scanning of an area a factor larger, exposed for a year, one would probably see a few tracks of ^{24}Ne.

3.3 ^{234}U—A Species with Four Hadronic Decay Modes

Figure 6 shows the partial half-lives calculated by Poenaru & Ivascu (33) for emission of heavy particles with A_2 from 4 to 117 for ^{234}U, a nuclide

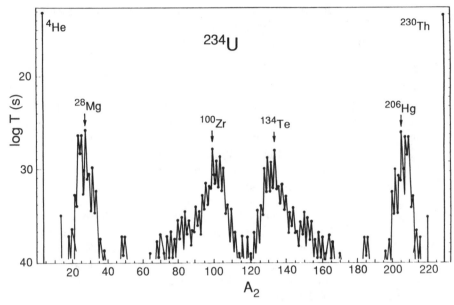

Figure 6   Partial half-lives for alpha decay, heavy-particle radioactivity, and cold fission of ^{234}U (33).

that is predicted to undergo alpha decay, cold fission, and emission of ^{28}Mg, ^{24}Ne, and ^{26}Ne at detectable rates. Our group, in collaboration with Moody and Hulet (Lawrence Livermore National Laboratory), used phosphate glass detectors sensitive only to ions with $Z > 8$ to see if we could detect Ne and Mg decay as well as spontaneous fission of this nuclide. Here is an example of a source for which it was necessary to know the exact isotopic composition, for in this case, even with the ^{232}U level reduced to 2×10^{-3}%, nearly 20% of the Ne events were from ^{232}U instead of ^{234}U. Two exposures and analyses were made, in each case with calibrated glass plates. In the first (21), 14 Ne decays and 3 Mg decays were detected after three-month exposures to ~ 1 mg cm^{-2} sources of total mass 123 mg. In the second (25), by increasing the source size and exposure time, we were able to identify 108 Ne decays and 36 Mg decays. Figure 7 shows the raw data for the second experiment. The existence of a Ne and a Mg population is quite clear. The partial half-lives, $1.8 \pm 0.2 \times 10^{25}$ s for Ne emission and $5.6 \pm 0.9 \times 10^{25}$ s for Mg emission, are fairly close to the values predicted by the models in Table 1 but are far from the values predicted by some of the other models. It is interesting that a single method,

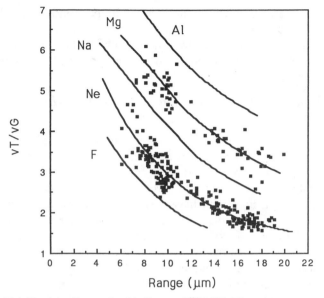

Figure 7 Identification of ions emitted in decays of ^{234}U (25). Three times as many Ne ions were emitted as Mg ions.

ASAFM, accounts reasonably well for the measured rates of alpha decay, heavy-particle radioactivity, and spontaneous fission.

3.4   *Mg Decay, Si Decay, Alpha Decay, and Fission of* ^{238}Pu

^{238}Pu provides a useful test of models because the predictions of partial half-lives for Mg and Si decay differ by a greater factor than in the case of heavy-particle emission by lighter nuclei. The experiment proved to be difficult because the rates of Mg decay and Si decay are lower by a factor of $\sim 10^{-8}$ than the rate of spontaneous fission and lower by a factor of $\sim 10^{-16}$ than the alpha decay rate. We and our Livermore colleagues used a highly purified source of ^{238}Pu and a hemispherical array of phosphate glass detectors sensitive only to ions with $Z \geq 10$ to detect Mg and Si events (11). The glass plates were covered with an Al foil to absorb fission fragments. To avoid absorbing the Mg and Si ions before they reached the detectors, we kept the detectors plus source at < 10 torr. Glass, in contrast to plastic track detectors, retains its full sensitivity while in a vacuum. The total area of glass was made large enough to reach a branching ratio as low as 10^{-16} while limiting the alpha-particle fluence to 1×10^{14} cm^2. In a 172-d exposure to the 10.5-mg ^{238}Pu source, we collected five Si events and two Mg events, corresponding to branching ratios of $\sim 10^{-16}$ and $\sim 5 \times 10^{-17}$ (11). Only four of the models predicted branching ratios in reasonable accord with those values.

3.5   Enhancement in Emission Rates of Even-Even vs Odd-A Nuclides

Neutron and proton pairing affect nuclear masses and decay rates. For alpha decay, the hindrance factor for decay from an odd-A nuclide compared with an even-even nuclide is ~ 5 to 10. For spontaneous fission, hindrance factors are $\sim 10^4$ to 10^7. From the suite of measurements of ^{14}C emission from nuclides with masses from $A = 221$ to 226, a hindrance factor of $\sim 10^2$, intermediate between the factors for alpha decay and for spontaneous fission, was inferred for ^{14}C decay (14). This fact was incorporated in an empirical way into several of the models (26–28; also unpublished additional calculations), which thereafter gave more reliable predictions of branching ratios.

Referring to Table 1, one sees that nine of the twelve measurements of partial half-lives were made on even-even nuclides and three on odd-A nuclides. Using the square-well model, the author has done a least-squares analysis, varying three parameters—a radius factor r_0, a frequency factor v_e for even-even nuclides, and a frequency factor v_0 for odd-A nuclides— with a preformation factor of unity. Values of log T calculated with $r_0 = 0.928$ fm, $v_e = 4.3 \times 10^{26}$, and $v_0 = 1.1 \times 10^{25}$ are given in the column of the table labeled SqW. The standard deviation in log T given in the last row of the table is smaller for this model than for any of the others. None of the seven null results in the table is inconsistent with the half-lives calculated with these parameters. The hindrance factor, defined as v_e/v_0, is ~ 40. Only three odd-A nuclides have been observed to emit heavy particles, too small a number to permit one to determine whether the hindrance factor depends on A_2.

3.6   Other Searches for Emission of Ions with $Z \geq 12$

Three groups (22, 29, 34) have used track detectors to search for Si decay of ^{241}Am. Their upper limits (2σ) on the branching ratio for Si decay relative to alpha decay were 4.2×10^{-13} (34), 5×10^{-15} (29), and 7.4×10^{-16} (22). No filter was used to absorb the fission fragments. In the latter search, $\sim 45,000$ fission tracks were recorded. According to the predictions of the more successful models, an increase of at least an order of magnitude in sensitivity would be necessary in order to detect heavy-particle emission.

Barwick (32) made an unsuccessful search for emission of ^{34}Si from ^{240}Pu, and Tretyakova et al (29) made an unsuccessful search for ^{30}Mg from ^{237}Np.

^{46}Ar, ^{48}Ca, and ^{50}Ca are predicted to be emitted from several isotopes of Cm and Cf at branching ratios up to $\sim 10^{-14}$. Since Mg and Si were detected from ^{238}Pu at branching ratios $\sim 10^{-16}$, one might expect it to be possible, by filtering out the fission fragments that constitute an enormous background in Cm and Cf isotopes, to detect Ar and Ca decays. These nuclei have masses almost halfway between alpha particles and fission fragments, and their detection would provide an interesting test of the cluster models versus the fission-like models. The experimental difficulties seem, however, to be almost insurmountable. Ternary fission gives rise to fragments with a continuum of masses and energies that occupy the region of interest (38). Only by detecting all three fragments in coincidence could one decide whether an Ar or Ca ion originated in ternary fission or in two-body, heavy-particle radioactivity. The author's group has found that a filter thick enough to absorb ordinary binary fission fragments from ^{252}Cf transmits a discouragingly high flux of particles of mass and energy in the region of interest.

4. COMPARISON OF DATA WITH THEORETICAL MODELS

The last row in Table 1 gives the standard deviation of the measured value of log T from the calculated value: $(\sigma_{\log T})^2 = \Sigma (\log T_{\text{expt}} - \log T_{\text{theor}})^2/n$. Only the values of log T for which positive results had been obtained were included in the calculation of $\sigma_{\log T}$. Note that, for decays of ^{223}U to Ne, ^{234}U to Ne, and ^{238}Pu to Mg, the contributions of two isotopes to the calculated half-life were included in the comparison with the measured half-life. One cluster model (BW), one unified model (Poe.), and the carefully tuned SqW model fit the data to within better than a factor of three. This agreement is remarkable, in view of the difficulty of the experiments, the simplifying assumptions made in the models, and the dynamic range of the data: emission of fragments from ^{14}C to ^{34}Si spanning a factor 10^{17} in partial half-life and a factor 10^7 in branching ratio relative to alpha decay, for both odd and even parent nuclei. Tables 2 and 3 compare the two classes of models. For details, consult the original references.

4.1   Unified Models

The motivation for the ASAFM model was to account equally for alpha decay, heavy-particle radioactivity, and cold fission. The latter refers to a

Table 2 Unified models

	ASAFM (5, 7)	Shi & Swiatecki (27)	Pik-Pichak (35)	Shanmugam & Kamalaharan (37)
Mass coefficient	$m(r) = \mu = $ const.	$m(r) = \mu = $ const.	$m(r) = f(r)$	$m(r) = f(r)$
Radii	$1.2249A^{1/3}$	ellipsoidal	ellipsoidal	ellipsoidal
Deformation?	no	arbitrary directions	along fission axis	along fission axis
Assault frequency	$f(A_2, Q, e\text{-}o)$	$v_e = 10^{22} = 50v_0$	1×10^{21} s^{-1}	$\sqrt{Q/2\mu}/R_t \approx 10^{21}$
Prescission potential	quadratic	power law	cubic	cubic
Postscission potential	V_C	proximity potential + V_C	$V_C(r, \theta)$	Yukawa + exponential + V_C
Features	add E_v to Q inside barrier	attenuation of shell effects of fragments	volumes of fragments = const.	realistic potential

Table 3 Cluster models

	Square well (13)	Blendowske & Walliser (28)[a]	Buck & Merchant (36)[a]
Radii	$0.928A^{1/3}$	$1.233A^{1/3} - 0.978A^{-1/3}$	$R = 6.75$; adjust V_0
Assault frequency	$v_e = 40v_0$	$v/2R_t \approx 4 \times 10^{21}$	see Ref. 36
Preformation probability	$\equiv 1$	quantal; $S(A_2) \approx S_\alpha^{(A_2-1)/3}$	~ 1
Prescission potential	$\equiv 0$	Christensen-Winther potential + V_C	Buck-Dover-Vary (49)
Postscission potential	V_C only	Christensen-Winther potential + V_C	Buck-Dover-Vary (49)
Features	simplicity	quantal calculation of clustering	3 turning points

[a] Also unpublished additional calculations.

separate path in deformation space from that taken in normal fission. In cold fission the most compact shape is maintained, with the fragments emerging in their ground states with kinetic energy closely equal to Q and with masses strongly favoring closed shells. The relative rates of cold fission and conventional fission vary with the parent nuclide, and the two processes are distinguishable experimentally by the kinetic energy and mass distributions (50). It is impressive that the ASAFM model is able to account quantitatively for cold fission as well as for heavy-particle radioactivity and alpha decay (8). In order to correct for the unrealistically high barrier at R_t, Greiner et al (8) adopt the procedure of adding an energy parameter, E_v, to Q inside the barrier and taking away this parameter when the fragments penetrate the barrier. The preexponential frequency factor is set equal to $2E_v/h$. To fit the data they adjust the value of E_v depending on A_2, Q, and the even-odd nature of the parent nucleus.

The proximity-potential model (27) originally had no free parameters. Shi & Swiatecki later found that two refinements—an integration over all fission directions with respect to ellipsoidal ground-state deformations, and correction for the attentuation of shell effects of the two fragments while still close together—tended to cancel out. When a hindrance factor of ~ 50 favoring decays of even-even nuclides is introduced, their model fits the data to within a factor of 5 (27; also unpublished additional calculations).

4.2   Cluster Models

Blendowske & Walliser (28; also unpublished additional calculations) have used realistic nuclear radii (approximated as spheres) and a potential that fits heavy-ion scattering data (48). They have done microscopic calculations of the preformation factor for clusters up to ^{16}O in even-even nuclei, using many-body wave functions to compute the overlaps of the nucleon states in the cluster with those in the parent nucleus. Because of the difficulty of the calculations for large clusters, especially for clusters in odd-A (deformed) nuclei, they investigated semi-empirical approaches, and found that the statistically motivated expression $\log S(A_2) = \text{const}(A_2 - 1)$, with a properly chosen constant, fitted their calculated preformation factors and also gave good fits to the experimental data. For alpha-particle clusters they used $S(\alpha) = 6.3 \times 10^{-3}$ (even parent) and 3.2×10^{-3} (odd parent). The concept of preformed clusters with $\log S(A_2)$ proportional to $A_2 - 1$ should surely break down at some value of A_2. In fact, it is surprising that it works so well up to ^{34}Si, for which $S(A_2) = 6 \times 10^{-25}$.

The cluster model of Buck & Merchant (36; also unpublished additional calculations) is conceptually quite different: it assumes a preformation factor of unity and employs a local, effective cluster-core potential, which had successfully described energy levels, enhanced E2 transition strengths, and alpha-decay widths of certain α-like states of ^{16}O and ^{20}Ne (49). It fits C- and Ne-decay rates quite well but, in its simplest form, underestimates emission rates of some of the heavier clusters. By replacing the naive picture of a constant geometry for their cluster-core potential with a radius parameter $R = 1.04 \ (A_1^{2/3} + A_2^{2/3})^{1/2}$ fm, they obtain the results shown in column 7 of Table 1, which agree with data to within a factor of 5.

Barranco, Broglia & Bertsch (39) developed a model in which the inertial mass depends on the superfluid pairing gap of the nucleus. Theirs is the only model that includes the detailed microscopic dynamics during deformation prior to scission. In its present form, their model agrees with data less well than the ones shown in Table 1.

4.3   *Comments on Quality of Fits to Data*

The main utility of the square-well model is its demonstration that, by carefully tuning three parameters, all the partial half-lives can be reproduced to within a factor two. This performance is a benchmark showing how much improvement should be demanded of the more realistic models that now do less well. From the microscopic point of view, one could say that the value of r_0 in the square-well model is deliberately made unrealistically small, 0.928 fm, in order to compensate for the absence of a preformation factor.

It is worth noting that partial half-lives for alpha decay and spontaneous fission decay cannot be predicted nearly as well. For alpha decay a recent three-parameter model fits most but not all data to within a factor ten (36); a semi-empirical model with 24 parameters (!) fits a very extensive body of data for even-even, odd-even, even-odd, and odd-odd nuclides only to within a factor 3 or 4 (53). Part of the problem is the sensitivity of the alpha-decay rate to the centrifugal barrier and the absence of knowledge of spins for some parent and daughter nuclei. For spontaneous fission of even-even nuclides the average error in the partial half-lives is not better than a factor 10^2 to 10^3, even when both valleys (liquid drop and compact shape or cold fission) are taken into account (54); for odd-A nuclides the error is much larger.

5.   REFINEMENTS OF THE MODELS

The calculated decay rates are very sensitive to errors in the action integral, which appears in the exponential expression for penetrability. Using the square-well model, we can estimate from Equation 2 the effects of errors in Q and R_t on decay rates. For parents and daughters with masses listed by Wapstra & Audi (51), Q is known to within ~ 10 keV, which results in a negligible error in decay rate. However, there are a number of cases for which the mass of either the parent or the heavier daughter is not known and must be estimated from nuclear systematics. For example, for $^{238}U \rightarrow$ $^{34}Si + ^{204}Pt$, the decay rate is uncertain to within about a factor 10^2 because of an uncertainty of ~ 1.2 MeV in the mass of ^{204}Pt. Decays of ^{232}Th to ^{28}Mg and of ^{236}U to ^{32}Si also involve ^{204}Pt and are similarly uncertain. Examples of predicted heavy-particle radioactivities for which the parent mass is uncertain include $^{114}Ba \rightarrow ^{12}C + ^{102}Sn$ and $^{186}Bi \rightarrow ^{8}Be + ^{178}Au$. For these cases the uncertainty in Q of ~ 1 MeV leads to an uncertainty of a factor ~ 200 in decay rate.

Models such as the cluster model of Blendowske & Walliser, which predict absolute decay rates without arbitrary parameters, could probably be fine-tuned by making tiny changes in the nuclear potential without losing the attractive feature that the potential fits heavy-ion scattering data. For example, within the square-well model, an uncertainty of only 0.2 fm in R_t results in an error of a factor of 10 in decay rate.

For proton radioactivity and alpha decay, the centrifugal energy due to a change in angular momentum (Δl) represents a barrier that can be a substantial fraction of the Coulomb barrier. For a large Δl the hindrance factor due to the centrifugal barrier can be a factor of ten or more. However, in heavy-particle radioactivity the centrifugal energy at R_t is typically less than 1% of the Q value and an even smaller fraction of the Coulomb energy at R_t. The decay mode with the largest mutual angular momentum in Table 1 is that of ^{14}C decay of ^{223}Ra, with $\Delta l = 4$. Inclusion of the centrifugal barrier in this case would decrease the decay rate by less than a factor of 2. Examination of the spin changes for the decays in Table 1 reveals that including the centrifugal barrier will not improve the fits.

Greiner & Scheid (47) developed a simple model to include decays to excited states of the daughter nucleus. Since Q is reduced, the particle must tunnel through a thicker barrier, with the consequence that the rate for such decays is small except for low-lying levels. For decay to ^{208}Pb, which has no low-lying levels, the increase in decay rate is quite small; for decays to other daughters, they found increases by as much as a factor of four. For the seven cases they considered, incorporation of the decreases in partial half-life into Table 1 does not improve the fits. Since the standard deviation for the best models corresponds to a factor of two in decay rate, inclusion of effects as large as a factor of four might improve the fits even further if the model could be made more rigorous. For example, instead of assuming, as Greiner & Scheid did, that excitation of the daughter or cluster nucleus occurs with a certain probability at a fixed distance within the barrier, one could integrate over all distances.

With increasing size of the smaller fragment, it seems clear that inclusion of more deformation variables will increase the calculated decay rate, because a more favorable path can be taken in barrier penetration. Poenaru et al (52) recently compared values of the penetrabilities calculated with two models of intersected spheres, one with constant radius R_2 (cluster-like shapes) and one with constant volume V_2 (more compact shapes) of the light fragment. Using a Yukawa+exponential potential and a more sophisticated prescription for the mass parameter, they conclude that, for $A_2 < 31$, the decay rate is greater for the constant-radius model, whereas, for $A_2 > 31$, the decay rate is greater for the constant-volume model.

6. FUTURE PROSPECTS AND SUMMARY

Most of the models include nuclear structure only through its effect on nuclear mass and Q. In order to fit the data to better than a factor of two, it seems clear that nuclear structure must be considered beyond simply an even-odd hindrance factor. The conclusion that decays to excited levels of a daughter or cluster nucleus may occur at rates comparable to decays to the ground state (45) could be tested with a magnetic spectrometer. For ^{14}C decay of ^{223}Ra, either the ^{14}C or the ^{209}Pb might be excited; for ^{24}Ne decay of ^{232}U, only the ^{24}Ne is likely to be excited. Both cases may be amenable to study in a long exposure at the Orsay spectrometer.

A number of new decay modes can be sought, using track detectors. Experiments in progress or being planned include searches for heavy-particle radioactivities in ^{235}U, ^{237}Np (with higher sensitivity than before), ^{242}Cm (to look for ^{34}Si decay) and ^{233}U (to look for ^{28}Mg decay at a rate $\sim 2\%$ that for ^{24}Ne decay). Searches for O, Ne, Mg, and Si decays of various isotopes of Th and U are feasible. The possibility of detecting both ^{30}Mg decay and ^{32}Si decay of ^{236}U is intriguing; in the latter case the daughter would be ^{204}Pt, whose mass is uncertain by more than 1 MeV and could be determined rather well if the partial half-life for ^{32}Si decay were measured. Using on-line track detectors at ISOLDE-III, one could look for ^{14}C decay of the odd-odd nuclide ^{220}Fr, for ^{12}C decay of ^{114}Ba, and for ^{8}Be decay of ^{186}Bi.

In the five years since the work of Rose & Jones, the sensitivity of techniques has improved by a factor of 10^7; decays involving heavy particles up through Si have been detected; and both unified and cluster models can fit the data to within about a factor of two.

ACKNOWLEDGMENTS

S. W. Barwick read the manuscript and made useful suggestions. This work was supported in part by the US Department of Energy (LBL).

Literature Cited

1. Sandulescu, A., Poenaru, D. N., Greiner, W., *Sov. J. Part. Nucl.* 11: 528 (1980)
2. Poenaru, D. N., Ivascu, M., *J. Phys. (Paris)* 45: 1099 (1984)
3. Poenaru, D. N., Ivascu, M., *Rev. Roum. Phys.* 29: 623 (1984)
4. Poenaru, D. N., Ivascu, M., Sandulescu, A., Greiner, W., *J. Phys.* G10: L183 (1984)
5. Poenaru, D. N., Ivascu, M., Sandulescu, A., Greiner, W., *Phys. Rev.* C32: 572 (1985)
6. Poenaru, D. N., et al., *At. Data Nucl. Data Tables* 34: 423 (1986)
7. Poenaru, D. N., et al., *Publ. NP-54-86,* Inst. Central de Fizica, Bucharest, Romania (1986)
8. Greiner, W., Ivascu, M., Poenaru, D. N., Sandulescu, A., In *Treatise on Heavy Ion Science,* ed. D. A. Bromley. New York: Plenum (1989)
9. Aleksandrov, D. V., et al., *JETP Lett.* 40: 909 (1984)
10. Rose, H. J., Jones, G. A., *Nature* 307: 245 (1984)
11. Wang, S., et al., *Phys. Rev. C, Rapid Commun.* In press (1989)
12. Price, P. B., Barwick, S. W., In *Particle Emission from Nuclei,* ed. D. N. Poenaru, M. Ivascu, Vol. II, Chap. 8. Boca Raton, Fla: CRC Press (1988)
13. Price, P. B., Stevenson, J. D., Barwick, S. W., Ravn, H. L., *Phys. Rev. Lett.* 54: 297 (1985)
14. Barwick, S. W., et al., *Phys. Rev.* C34: 362 (1986)
15. Gales, S., et al., *Phys. Rev. Lett.* 53: 759 (1984)
16. Hourani, E., et al., *Phys. Lett.* 160B: 375 (1985)
17. Kutschera, W., et al., *Phys. Rev.* C32: 2036 (1985)
18. Fleischer, R. L., Price, P. B., Walker, R. M., *Nuclear Tracks in Solids.* Berkeley, Calif: Univ. Calif. Press (1975)
19. Hasegan, D., Tretyakova, S. P., See Ref. 12, Chap. 9
20. Wang, S., et al., *Nucl. Instrum. Methods* B35: 43 (1988)
21. Wang, S., et al., *Phys. Rev. C. Rapid Commun.* 36: 2717 (1987)
22. Moody, K. J., et al., *Phys. Rev. C. Rapid Commun.* 36: 2710 (1987)
23. Balysh, A. Ya., et al., *Sov. Phys. JETP* 64: 21 (1986)
24. Tretyakova, S. P., et al., *JINR Rapid Commun.* 7: 23 (1985)
25. Moody, K. J., Hulet, E. K., Wang, S., Price, P. B., *Phys. Rev. Brief Comments.* In press (1989)
26. Poenaru, D. N., et al., *Z. Phys.* A325: 435 (1986)
27. Shi, Y.-J., Swiatecki, W. J., *Nucl. Phys.* A438: 450 (1985); *Nucl. Phys.* A464: 205 (1987)
28. Blendowske, R., Walliser, H., *Phys. Rev. Lett.* 61: 1930 (1988)
29. Tretyakova, S. P., et al., *JINR Rapid Commun.* 13: 34 (1985)
30. Sandulescu, A., et al., *Iz. Akad. Nauk SSSR (Fiz.)* 49: 2104 (1985)
31. Barwick, S. W., Price, P. B., Stevenson, J. D., *Phys. Rev.* C31: 1984 (1985)
32. Barwick, S. W., PhD Thesis, Univ. Calif., Berkeley (1986)
33. Poenaru, D. N., Ivascu, M., See Ref. 12, Chap. 5
34. Paul, M., Ahmad, I., Kutschera, W., *Phys. Rev.* C34: 1980 (1986)
35. Pik-Pichak, G. A., *Sov. J. Nucl. Phys.* 44: 923 (1986)
36. Buck, B., Merchant, A. C., Dept. Theor. Phys., Oxford Univ., *Preprint 88/88* (1988)
37. Shanmugam, G., Kamalaharan, B., *Phys. Rev.* C38: 1377 (1988); also submitted to *Phys. Rev. C* (1989)
38. Schall, P., Heeg, P., Mutterer, M., Theobald, J. P., *Phys. Lett.* B191: 339 (1987)
39. Barranco, F., Broglia, R. A., Bertsch, G. F., *Phys. Rev. Lett.* 60: 507 (1988); also unpublished additional calculations
40. Petrascu, M., Buta, A., Simion, V., Central Inst. Phys., Bucharest, *Preprint NP-44-1985* (1985)
41. Landowne, S., Dasso, C. H., *Phys. Rev.* C33: 387 (1986)
42. Grashin, A. F., Efimenko, A. D., *Sov. J. Nucl. Phys.* 43: 854 (1986)
43. de Carvalho, H. G., Martins, J. B., Tavares, O. A. P., *Phys. Rev.* C34: 2261 (1986)
44. Iriondo, M., Jerrestam, D., Liotta, R. J., *Nucl. Phys.* A454: 252 (1986)

45. Rubchenya, V. A., Eusmont, V. P., Yavshits, S. G., *Iz. Akad. Nauk SSSR (Fiz.)* (Transl.) 50: 184 (1986)
46. Malik, S. S., Gupta, R. K., Submitted to *Phys. Rev.* C (1989)
47. Greiner, M., Scheid, W., *J. Phys.* G12: L229 (1986)
48. Christensen, P. R., Winther, A., *Phys. Lett.* 65B: 19 (1976)
49. Buck, B., Dover, C. B., Vary, J. P., *Phys. Rev.* C11: 1803 (1975)

50. Hulet, E. K., et al., *Phys. Rev. Lett.* 56: 313 (1986)
51. Wapstra, A. H., Audi, G., *Nucl. Phys.* A432: 1 (1985)
52. Poenaru, D. N., et al., Inst. for Theor. Phys., Univ. Frankfurt, *Preprint 226* (1988)
53. Ivascu, M. S., Poenaru, D. N., See Ref. 12, Chap. 4
54. Moller, P., Nix, J. R., Swiatecki, W. J., *Nucl. Phys.* A409: 1 (1987)

Annu. Rev. Nucl. Part. Sci. 1989. 39: 43–71

HIGHEST ENERGY COSMIC RAYS

J. Wdowczyk

Institute of Nuclear Studies, Lodz, Poland

A. W. Wolfendale

Physics Department, University of Durham, Durham DH1 3LE, United Kingdom

KEY WORDS: air showers, origin of cosmic rays, composition of cosmic radiation, acceleration mechanisms.

CONTENTS

INTRODUCTION

Although cosmic rays were discovered as early as 1912, they still pose many unanswered questions, especially about their nature at high energies and their origin. In fact, even their name, cosmic *rays*, is a misnomer

43

0163–8998/89/1201–0043$02.00

because the entities themselves are not "rays" on the whole; they are atomic *particles* of various kinds rather than electromagnetic waves.

This chapter analyzes the characteristics of the most energetic cosmic rays. It will be no surprise to learn that the problems in this region are severe and the conclusions necessarily very imprecise. Nevertheless, the topic is of considerable interest, not least because we are dealing with particles—largely protons and heavier nuclei—having the highest individual energies known.

In defining "highest energies" we set the lower limit as 10^{17} eV. We choose this value because there is evidence that extragalactic particles probably take over from Galactic ones fairly soon above this energy, and the top three decades of energy out of a nominal 12 decades in total (10^8–10^{20} eV) probably comprise a reasonable bracket for the "highest energies."

We begin here with a general discussion of cosmic rays, their properties at low energy, and various models of their origin. The importance of energy density arguments is stressed. Continuing to the "highest energies," we examine the experimental techniques involved in their study and discuss the crucial role of fluctuations in a variety of contexts. The energy spectrum of the particles is considered, along with the information about masses and anisotropies. Problems and possibilities associated with the character of high energy interactions are discussed. The crucial quest for information about origin and propagation in Galactic and extragalactic space is addressed in some detail. The article concludes with a brief summary.

GENERAL DISCUSSION OF COSMIC RAYS

A Brief History

The balloon flights of Hess (1), which led to his famous "recourse to a new hypothesis either invoking the assumption of the presence at great altitudes of previously unknown matter, or the assumption of an extra-terrestrial source of penetrating radiation," were followed by hectic activity among some of the greatest physicists of the early decades of this century. The initial view that some form of ultra-gamma radiation was responsible gave way to the contention, now universally accepted, that atomic particles were responsible. Clay's observation (2) of a "latitude effect" played a key role in indicating that charged particles were involved. The curved path of a charged particle in a magnetic field indicated that particles make up the cosmic rays (CR); but it is the same magnetic deflection—this time caused by the tangled magnetic fields in the interstellar medium (ISM) in the Galaxy—that makes it so difficult to use CR arrival directions to throw light on the origin of the particles.

An early observation of great relevance to this review was the discovery by Auger & Maze (3) of extensive air showers (EAS), those great cascades of particles spreading over 100,000 m² or more at ground level and generated by very energetic particles incident on the top of the atmosphere. It is from the study of EAS alone that we are able to say anything at all about the highest energy cosmic rays.

Cosmic Rays Below 10^{14} eV

MASS COMPOSITION   Starting from the minimum energy for a cosmic ray of 10^8 eV per nucleon (a kinetic energy equal to one tenth of the rest mass), measurements have been made directly to higher and higher energies as time has progressed. Insofar as some important facts can be learned from an analysis of the mass composition at these low energies, we present the elemental abundances in Figure 1. Also presented are the universal abun-

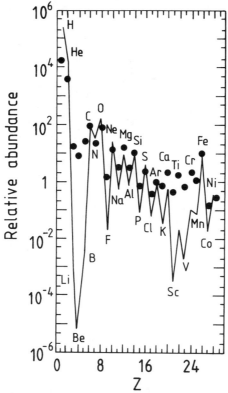

Figure 1   Comparison of the cosmic-ray abundances (*full circles*) and universal abundances (joined by *full lines*). The data are normalized at carbon. The CR results refer to the range 170–280 MeV per nucleon and are from the summary by Meyer (4).

dances. A comparison between the two shows some remarkable differences, most notably the large excess of the three elements just below carbon (Li, Be, and B) and the five elements just below iron (Sc, T, V, Cr, and Mn).

The generally accepted explanation is that these "excess" elements arise from the fragmentation of the heavier cosmic rays by collisions with the nuclei of the ISM. Knowledge of the cross sections for fragmentation allows one to estimate the average mass per unit area of the ISM traversed by these low energy cosmic rays: ~ 6 g cm^{-2} at energies below 1 GeV per nucleon.

An interesting feature is that the ratio of the flux of secondary (i.e. fragmented) nuclei to those of the primary nuclei falls with energy, or more precisely with rigidity, R $(R = pc/Ze)$. Specifically, the ratio is approximately proportional to $R^{-0.5}$ for $R > 6$ GV. One explanation of this fact is that the lifetime of the cosmic rays against escape from the Galaxy varies as $R^{-0.5}$. However, it is also possible that this dependence is due to the manner in which cosmic rays are accelerated in the Galaxy; the continuous acceleration model of Cowsik (5), Silberberg et al (6), Giler et al (7) and others can give this R dependence with a reduced escape lifetime–rigidity dependence or even none at all. There is as yet no general agreement on this point. For reasons explained below, we subscribe to the view that the lifetime is, in fact, constant to 10^{15} eV, the continuous acceleration process being responsible for the changing secondary-to-primary ratio.

After allowance is made for the fragmentation losses, there are still some residual differences and these have great relevance to the manner in which the particles are generated and accelerated. Presumably we are not dealing simply with typical ISM nuclei accelerated by, for example, supernova shocks but rather a preaccelerated sample. It seems likely that strong stellar winds comprise such partially accelerated particles.

CR LIFETIME   The standard method used to determine the mean lifetime of cosmic rays in the Galaxy is to measure the fraction of unstable ^{10}Be $(T = 2.7 \times 10^6$ y) in the primary beam. Under the assumption of a leaky-box situation in equilibrium, a mean lifetime of $\sim 2 \times 10^7$ y is indicated at a rigidity of 1 GV (8). However, this value must be regarded as very approximate at best, in view of the unlikelihood of the leaky-box model being valid (for example, continuous acceleration plays a distorting role, as do variations of diffusion coefficient in space and time). It is not inconceivable that the mean lifetime against escape from the Galaxy is considerably greater than 10^7 y at 1 GV.

Energy Spectra

A detailed analysis of the energy spectra of the various low energy components is beyond the scope of this article, but insofar as there is relevance to the highest energies a brief treatment is justified. Figure 2 shows the situation below 10^{17} eV. Note that there are some significant differences in slope of the various components. Furthermore, in the region of 10^{15} eV (Figure 3) there is considerable uncertainty in the magnitude of the total intensity because this is the lower end of the extensive air shower region where showers have rather few particles and energy estimates are difficult to make.

The whole energy range is indicated in Figure 3 and the relative intensity of the γ, e, and nuclear components can be appreciated.

CR Origin at Low Energies

Although it is likely that supernova remnant (SNR) shocks are responsible for most of the energy achieved by low energy CR particles (say $E < 10^{14}$ eV) [see, for example, the γ-ray analysis of Bhat et al (12)], the case is by no means proven and it is necessary to proceed with caution. What can be said with more confidence is that the bulk of the CR are almost certainly accelerated in our own Galaxy rather than having been generated in

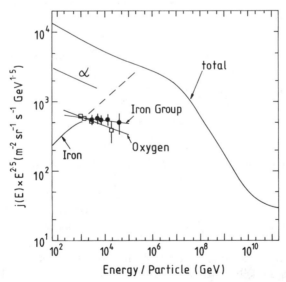

Figure 2 Energy spectrum of some of the more common cosmic-ray nuclei and the total all-particle spectrum. The latter is from the summary by Linsley (9). The points for oxygen and the iron group ($Z = 25, 26, 27$) are from the Spacelab 2 observations of Grunsfeld et al (10). The dashed line is an extrapolation of the previously measured low energy spectrum of the iron group. The "turnover"—magnified because of the manner in which the data are plotted—is caused in part by a diminution of fragmentation losses.

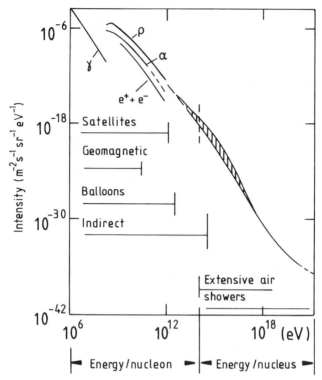

Figure 3   Energy spectrum of protons, α particles, electrons, and (diffuse) γ rays showing the techniques used in the various energy regions [after Wolfendale (11)].

"universal" processes and thus being of extragalactic origin. (This is not to say that the very highest energy particles are produced in the same way—they probably are not.)

Figure 4 summarizes possible Galactic sources. Note that there is no shortage of mechanisms. Where the shortage does occur is in processes to generate particles much above 10^{14} eV. SNRs appear to have a natural upper limit of 10^{13}–10^{14} eV (e.g. 13), and although one can probably extend the value somewhat by invoking multiple supernova and magnetic field effects, it is unlikely that particle energies much higher than 10^{15} eV can be achieved. This topic is discussed again below.

CR Energy Density

The total energy density of CR, as distinct from the energies of individual particles, is an important topic both from the standpoint of the origin of

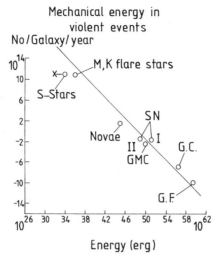

Figure 4 Mechanical energy in violent events in the Galaxy from which cosmic rays may achieve energy [after Wolfendale (11)]. The line represents the CR energy needed.

S-Stars: stars like the Sun (the cross indicates the actual energy going into CR in a solar flare). *SN I, II*: supernovae of the two types. *GMC*: giant molecular clouds (loss of gravitational potential energy on formation). *G.C.*: galactic center explosions such as invoked by Giler (40). *G.F.*: galaxy formation (loss of gravitational potential energy on formation).

It is apparent that, apart from S-Stars, several mechanisms are contenders for sources of the cosmic-ray energy budget, namely those mechanisms with symbols near to, and preferably to the right of, the line.

CR and of the interaction of the particles with the plasma in the ISM and in intergalactic space. Figure 5 indicates the situation. Also indicated in the figure are energy densities for a variety of phenomena as itemized in the caption. The energy density above 10^{17} eV is $\sim 10^{-6}$ eV cm^{-3}, a value close to that in the extragalactic γ-ray flux above 10^8 eV.

CR Anisotropies

A strong clue to the origin of cosmic radiation would be the observation of excesses of CR intensity from specific regions of the sky. Such anisotropies have been sought for many decades, and the experimental situation is gradually becoming clearer. Figure 6 summarizes the situation and shows a near constancy in amplitude of the first harmonic and a phase up to about 10^{14} eV and a little beyond, followed by a rising amplitude with increasing energy. Such an increase is consistent with a Galactic origin for the particles and a relatively rapidly falling mean lifetime within the Galaxy for energies above 10^{15} eV.

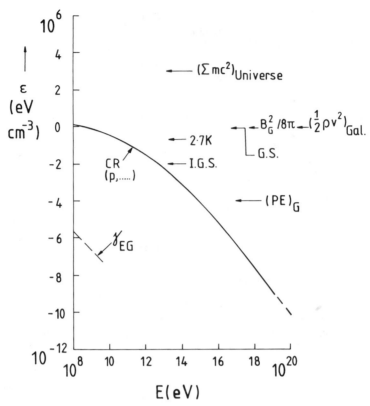

Figure 5   Cosmic-ray energy density as a function of the threshold CR particle energy. Curves are shown for both particles and γ rays. The arrows relate to energy densities for a variety of other phenomena: $(\Sigma mc^2)_{universe}$ = rest energy of total mass in the universe; $B_G^2/8\pi$ is the Galactic magnetic field; $(\frac{1}{2}\rho v^2)_{Gal}$ represents Galactic gas clouds; G.S. is Galactic starlight; I.G.S. is intergalactic starlight; and $(PE)_G$ is the gravitational potential energy of galaxies.

The phase of the first harmonic, indicating the preferred direction of flow, changes in a more complicated manner than the amplitude. As described below, we have put forward an explanation in terms of a specific topography for the Galactic Magnetic Field (14, 15). At this stage we draw attention to the apparent change of phase at 10^{18} eV.

A Mixed Origin Model for Intermediate Energies

The foregoing has been concerned largely with particles below about 10^{14} eV; most of what follows concerns energies above 10^{17} eV—the highest energies—so that a brief mention is necessary of the intermediate energy

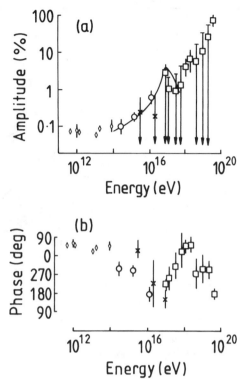

Figure 6 Amplitude and phase of the first harmonic anisotropy of cosmic rays. It will be noted that the amplitude and the phase are nearly constant below about 10^{14} eV.

region, particularly insofar as there is relevance to the highest energies. Virtually every aspect is controversial, but a model with which we have sympathy is indicated in Figure 7. There are essentially two components, the near-universal abundances at the lower energies with a cutoff associated with rigidity (source property) at $R_0 \approx 10^{14}$ eV/nucleus, and a "new" proton component that predominates at energies above 2×10^{14} eV/nucleus and that has a rather sharp Galactic modulation starting at $\sim 10^{15}$ eV. This latter component is important and has a production spectrum with differential exponent $\gamma = 2$, where we write the spectrum as $j(E)\,dE = AE^{-\gamma}\,dE$.

Our currently preferred model has the lower energy, mixed composition derived from SNR: the lifetime against escape is essentially energy independent and the shape of the energy spectrum is determined by a production spectrum with γ somewhat greater than 2 and by the effect of energy-dependent continuous acceleration. The flatter spectrum (the

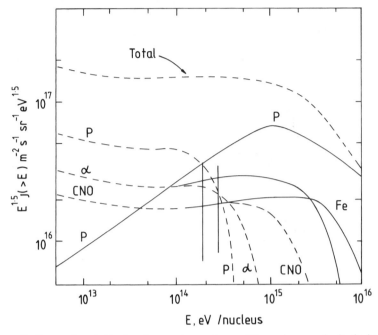

Figure 7   Our preferred situation for cosmic-ray fluxes of the various species in the intermediate energy region, 10^{13}–10^{16} eV. After Gawin et al (16).

"new" proton spectrum) has quite a different source, pulsars are one
possibility and Galactic shocks (e.g. 17) are another.

EXPERIMENTAL TECHNIQUES

Extensive Air Shower Arrays

It was mentioned in the introduction that EAS were discovered in the late
1920s. Since that time there have been a succession of EAS arrays at
various points on the Earth's surface and at various altitudes. Some of
these arrays were operated and then closed down, others have been
enlarged from time to time, and yet others were constructed very recently.
Indeed, workers are still engaged in designing and constructing such arrays.
    A detailed list of arrays operating in 1984 (and important ones that had
ceased operation by that date) was given by Wolfendale (18). Here we give
a less detailed list of those related to the highest energy particles of current
concern. Table 1 gives these details.

Table 1 Extensive air shower arrays

	Volcano Ranch[a] (26)	Haverah Park[a] (24)	SUGAR[a] (25)	Yakutsk (23)	Fly's Eye (28)	Akeno (27)
Longitude	35°N	54°N	31°S	62°N	41°N	36°N
Latitude	107°W	2°W	150°W	130°E	112°W	139°E
Altitude (m.a.s.l.)	1770	220	250	100	850	900
Type of detector	plastic scintilators	water Čerenkov	muon detectors	plastic scintilators	fluorescent atmospheric light	plastic scintilators
Typical detector separation (m)	400–800	500–1000	1600	500	—	1000
Area (km²)	8	12	55	18	~60	20
Exposure (km² y)						
>10^{19} eV	60	130	600	70	~90	37
>10^{20} eV	100	320	1000	200	~145	37
No. of EAS						
>10^{19} eV	44	108	423	147	63	19
>10^{20} eV	1	5	8	0	0	0

[a] Array no longer operational.

Remarks about the Method of Analysis

A number of factors conspire to cause difficulty in converting EAS data to primary energy. Uncertainty in the nature of the high energy interaction mechanism and the uncertain mass composition are clearly important factors. One by-product of this uncertainty is that differences may appear between the claimed energy spectra from different arrays (e.g. in opposite geographical hemispheres) that are due not to astrophysical effects but to the arrays responding differently to the constituent detected particles. For example, the SUGAR array responded to muons only, whereas the Volcano Ranch array's response was largely to the electron component.

Another factor concerns the different propagation assumptions made by the different workers; most important is the assumed lateral distribution of particle density versus distance from the axis of the shower. A further factor causing discrepancy between the claimed energy spectrum and the true spectrum is the effect of fluctuations in various elements of the shower propagation and the nature of the response of the particle detectors. We have examined some of these fluctuation effects over a number of years but there are still problems.

It should be mentioned that the showers at the highest energies are detected using arrays with separation of detectors of order 1000 m. The nearest detector is then situated a few hundred meters from the core. At these distances the particle densities are very low and the energy carried by them is only a very small fraction of the primary energy, so that in converting from the measured parameters to the primary particle energy, we have to rely on model calculations and extrapolations from lower energies (where the separation of the detectors is much smaller).

The point of the above is that some of the differences that arise between one experiment and another and between observation and expectation in claimed primary energy spectra are not genuine and are not necessarily due to astrophysical phenomena.

ANISOTROPIES AT THE HIGHEST ENERGIES

Reality of the Anisotropy Claims

As remarked above, studies of anisotropies of arrival directions are important in helping to elucidate the origin problem; certainly the basic argument of Galactic versus extragalactic origin should be resolvable from anisotropy analyses.

Inspection of Figure 6 shows that at energies below about 10^{15} eV the amplitudes are rather precise, but at higher energies there are many upper limits and one worries about the reality of the frequently claimed rise in amplitude with energy. However, there is more information available than just the first harmonic, and a significant improvement in precision can be achieved by combining data from different experiments and taking the first and second harmonics simultaneously. In an analysis of all the data then available, we (15) derived the results shown in Figure 8. There is good evidence for the claimed anisotropies being real in one hemisphere or another (and often in both) at all energies, although there is an interesting region near 10^{18} eV where the significance almost disappears.

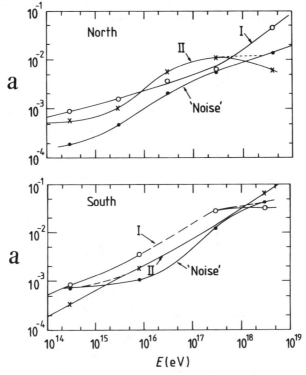

Figure 8 Amplitudes of the first (*I*) and second (*II*) harmonics of anisotropy measurements averaged over successive decades of energy. The sources of data are described in the text. The measurements in the Northern Hemisphere (*upper part*) were from extensive air shower arrays located at latitudes from 43°N to 62°N; the average latitude is about 53°N. Those in the Southern Hemisphere (*lower part*) cover the range 16°S to 37°S and the average here is about 30°S. The curve denoted *Noise* relates to expectation from a true distribution with zero anisotropy, measured isotropies arising simply by chance. The significance level is one standard deviation, i.e. the chance of an apparent anisotropy appearing from isotropy is, on this curve or above, one in six [after Wdowczyk & Wolfendale (15)].

Galactic Plane Enhancement

A detailed inspection of the data has led us to interpret the results below 10^{19} eV in terms of an enhancement of intensity from the general direction of the Galactic plane. The trend of phase of the first harmonic with energy is in the direction expected from changes in the average local magnetic field direction (averaged over increasing distances corresponding to the increasing Larmor radius for the particles). Concentrating on the energy region above 10^{18} eV, Szabelski et al (19) examined the magnitude of the enhancement as a function of energy. They describe the intensity as a function of Galactic latitude, b, by a simple function: $I(b) = I_0[(1-f_E)+f_E\exp(-b^2)]$ and plot f_E against E; the result is shown in Figure 9.

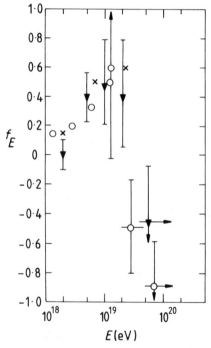

Figure 9 Dependence of the Galactic latitude enhancement factor, f_E [introduced by Wdow-czyk & Wolfendale (14) and defined in the text] on energy. Points for Haverah Park (*triangles*) and Sydney (*open circles*) have been calculated by us from the data given by the respective authors (references given in 14). The crosses refer to the Yakutsk data (see 14). The increase in f_E with energy seen as far as 2×10^{19} eV supports a Galactic origin for these particles. The dramatic fall at higher energies indicates a Galactic plane avoidance and suggests an extragalactic origin in this region (19).

The conclusion to be drawn from this figure is clear; there is evidence for an initial growth of f_E with E, as would be expected if the particles are of Galactic origin because the sources are largely along the Galactic plane, but an abrupt decrease as these sources switch off above 10^{19} eV. Clearly, the results are consistent with the highest energy particles being extragalactic in origin. Taken literally, the plot would indicate that the extragalactic particles are not distributed isotropically but rather come preferentially from high latitudes (a preference for directions from the central region of the local supercluster of galaxies is one possibility, and this is discussed below).

Although little new information has accumulated since 1986, mention should be made of recent results from the Fly's Eye Experiment (20). These authors present evidence for an excess of EAS from the general direction of Cygnus X-3 at energies above 5×10^{17} eV. The probability of this being a fluctuation from an isotropic distribution is quoted as 6.5×10^{-4}. Insofar as Cygnus X-3 is close to the Galactic plane, this observation may be an example of the phenomenon we are suggesting, although it is disturbing that the Haverah Park group has not confirmed the Fly's Eye observation despite having considerably more data (A. A. Watson, private communication).

MASS COMPOSITION AT THE HIGHEST ENERGIES

Uncertainty in the mass composition increases rapidly with energy above 10^{14} eV, the limit of direct measurements. It is true that there are quite precise measurements of EAS characteristics at energies as high as 10^{19} eV or so, but the necessary information about the characteristics of the interactions of cosmic-ray nuclei with air nuclei is lacking. In fact, there has been considerable argument about the mass composition at energies only a decade or so above the upper limit of direct measurements. The extreme positions have been that at 10^{15} eV iron nuclei dominate or, alternatively, protons are the main component. We believe in the latter position and have pointed out in a number of papers (e.g. 21) that one consequence of a protonic component is a serious breakdown of the so-called Feynman scaling. Recent accelerator work, primarily the observation of the so-called minijets in the SPS collider at CERN, have confirmed scaling breakdown and given us more confidence in the idea of protonic dominance in the 10^{15}-eV region. Such considerations are, in part, behind the spectral components indicated in Figure 7.

Turning to energies above 10^{15} eV and extending to the highest energies of main concern here, our remarks can be based on two lines of evidence: extrapolation from lower energies and fluctuation studies. These two are considered in turn.

If, as seems likely, the rise in cross section for minijets continues unabated, and if there are no surprises in the general behavior of proton-air-nucleus collisions, then the evidence for proton dominance will continue to be strong. New types of interaction may occur at energies in the range 10^{15}–10^{20} eV, but the fraction of the available interaction energy going into these processes is presumably not large. Even if new interactions dominate, then the idea of proton dominance is not nullified; rather, the mass composition then becomes indeterminate.

Concerning fluctuation studies, there has been considerable work over several decades. The technique is to determine the total number of muons (N_μ) and electrons (N_e) in individual EAS and to examine the frequency distribution of N_e for showers of fixed N_μ and of N_μ for fixed N_e. An important summary of data from the major EAS groups was given recently by J. N. Stamenov (private communication, 1988) and is presented in Figure 10. One can see that protons are dominant at all energies although it is also true that the nonprotonic component is somewhat higher here than in our Figure 7. However, we believe that there are technical reasons leading to an underestimate of the proton fraction from fluctuation studies.

ENERGY SPECTRUM AT THE HIGHEST ENERGIES

The Form of the Spectrum

A collection of differential spectra from the various arrays is given in Figure 11 [see also the review by Khristiansen (23)].

It will be noted that there is structure in the sense that the data from Haverah Park (HP), Sydney (S), Yakutsk, and Akeno all show a "dip" [when the intensity $j(E)$ is multiplied by E^3] at $E \approx 10^{19}$ eV. None of the other results is inconsistent with this observation. Above about 5×10^{19} eV there may well be a difference between the HP and S results, with the former rising and the latter falling. There are differences between the two arrays that may be of relevance; firstly, S responded to muons only, whereas HP detected a mixture of muons and electrons. Secondly, HP is in the Northern Hemisphere whereas S is in the Southern—the region of the Galaxy covered is thus different. These differences are considered in the next section.

If we ignore the difference in spectra above 5×10^{19} eV (which may not be significant in any case), the global average spectrum of Figure 12 results (the significance of the other spectra in Figure 12 will become clear later).

Interpretation of the Spectrum

THE RESPECTIVE GALACTIC AND EXTRAGALACTIC COMPONENTS   As mentioned above, a number of observed features lead us to suggest that the

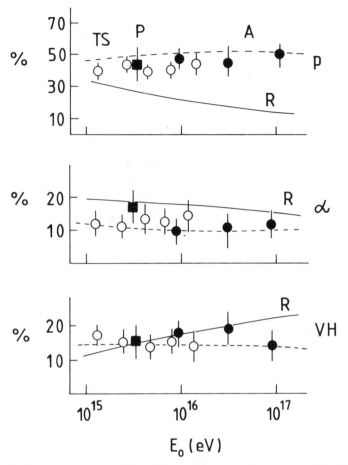

Figure 10   Summary of composition estimates for protons (P), alpha particles (α), and heavy nuclei (VH: the iron group) from fluctuation studies, after J. N. Stamenov (private communication, 1988). *TS*: Tien Shan; *A*: Akeno; *P*: Pamir. *R* indicates expectation for rigidity-dependent modulation, and the dotted line is that for a rigidity-independent composition.

bulk of the particles are Galactic below 10^{19} eV or so and extragalactic above. With this hypothesis, there are two main problems:

What are the Galactic sources that enable acceleration as high as 10^{19} eV? (We have shown that SNR cut off at 10^{15} eV.) What are the extra-galactic sources at the higher energies and how do these particles propagate through the Universe?

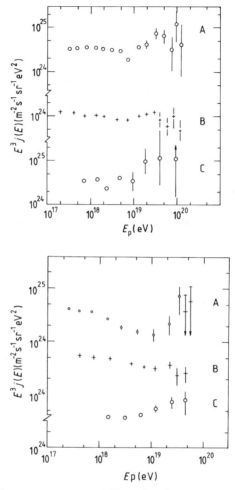

Figure 11   (*Top*) Primary energy spectra. *A*: Haverah Park (24). *B*: Sydney (25). *C*: Volcano Ranch [after Cunningham et al (26)]. The summary is from Wolfendale (18); this author carried out some grouping of points.

   (*Bottom*) Primary energy spectra. *A*: Akeno (27). *B*: Yakutsk (23). *C*: Fly's Eye (28). We have averaged successive pairs of points.

Starting with the Galactic sources for the more energetic particles (i.e. the higher energy proton component (P) in Figure 7) pulsars are a possibility. Although the simple models give an upper limit of $\sim 2 \times 10^{16}$ eV (29), recent work by Michel et al (30) and Gaisser (31) suggests higher maximum energies. These workers consider a model in which a pulsar

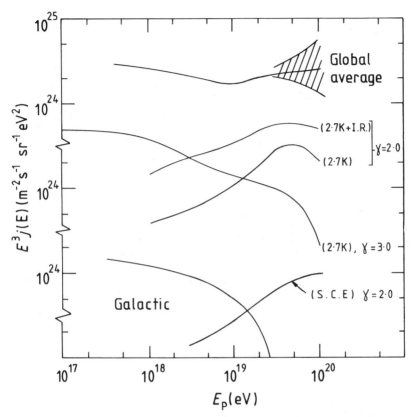

Figure 12 The "global average" primary spectrum derived by averaging the spectra of Figure 11, top and bottom (with some allowance for differences of calibration). The lines below in the central part represent predicted shapes for production spectra with the differential exponents indicated and for the two radiation fields (see Figure 13 for the details). The bottom spectra represent (S.C.E.) (and $\gamma = 2.0$), i.e. the situation with supercluster enhancement (see Figure 13) for the extragalactic component and a Galactic component falling rapidly above 10^{19} eV. The heights of Galactic and extragalactic spectra are arbitrary.

drives a wind that interacts with ejecta from the initial explosion. The result is a value for the maximum particle energy of $E_{max} = 10^5 B_{12} p_{10}^{-2}$ TeV, where B_{12} is the magnetic field of the pulsar at its surface (in 10^{12} gauss) and p_{10} is the period in units of 10 ms. The pulsars known in our own Galaxy commonly have $B_{12} = p_{10} = 1$ yielding $E_{max} = 10^{17}$ eV, a value somewhat higher than E_{max} commonly adopted for direct particle acceleration by pulsars. The process is therefore encouraging. Insofar as the physics of pulsars is not fully understood and our pulsar sample is by no means complete, it is conceivable that there are pulsars with larger $B_{12}/$ p_{10}^2 ratios, in which case values of E_{max} as high as 10^{18}–10^{19} eV could result.

Another possibility is to invoke the Galactic-halo termination-shock model of Jokipii & Morfill (17) in an amended form. In its original form the whole of the cosmic-ray spectrum is attributed to this mechanism, but we prefer to consider it as a mechanism for the higher energy particles only (namely $> 10^{15}$ eV). Problems with providing the total energy content of the cosmic rays are clearly reduced with the amended model, although there may be difficulties with the anisotropies.

Turning to the extragalactic particles, the discussion falls naturally into two parts concerned successively with the characteristics of propagation of the particles in extragalactic space and with the mode of actual acceleration of the particles.

Inevitably the manifestation of the Greisen/Zatsepin cutoff caused by interactions between CR protons and the microwave background radiation is a sought-after characteristic and this is considered first.

MICROWAVE CUTOFF   Many workers have drawn attention to the modulation expected for CR protons passing through the 2.7-K microwave background. In fact (and as might have been expected) the situation is not completely clear for a number of reasons. These can be listed as follows.

1. The form of the energy spectrum before modulation is not known.
2. There may be a rather high infrared flux at a temperature above 2.7 K, which acts with the 2.7-K flux (and modifies the cutoff severely).
3. The spatial distribution of extragalactic sources is not known and is thus another variable; indeed there is probably an excess of CR sources locally (the "supercluster" enhancement, SCE). Furthermore, extragalactic sources may have outputs that are functions of time, e.g. the presence of bursts and of cosmological effects.

Incorporating the first two points, Figure 13 shows the straightforward modulation factor F (defined as the ratio of the intensity after propagation through the 2.7-K radiation field to that in the absence of the field), for a uniform distribution of CR sources in the universe and the 2.7-K radiation field alone (denoted 2.7 K) and for a particular enhancement of the infrared field (denoted 2.7 K + I.R.). Figure 12 (middle part) shows the effect of these modulation factors on universal production spectra having differential exponents $\gamma = 2.0$ for both cases, and for a value of $\gamma = 3.0$ with 2.7 K.

Figure 12 also shows that production spectra with $\gamma = 2.0$ are favored in many acceleration models. Insofar as SCE is the most likely modulation factor, a possible situation is that shown in the bottom part of the figure:

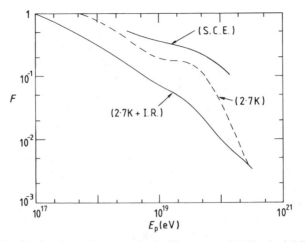

Figure 13   Modulation factor for extragalactic CR protons. (*2.7 K*): the 2.7-K radiation field alone and a universal distribution of sources (32). (*2.7 K+I.R.*): with the 2.7-K field enhanced by a strong infrared field, as postulated by Puget & Stecker (33) [Wdowczyk & Wolfendale (34)]. (*S.C.E.*): with the 2.7-K field alone but emphasizing the contribution from sources in the local supercluster (32).

a Galactic component falling rapidly above 10^{19} eV and an enhanced contribution from sources in the local supercluster.

In an effort to resolve the third problem listed above, a number of workers have examined the effect of the 2.7-K radiation field on extragalactic protons for a variety of assumptions about the spatial and temporal distribution of source yields. Hillas (35) took a model in which the steepening in the proton spectrum at $\sim 10^{15}$ eV occurs because the extragalactic source peaked in emission at a red shift of $z = 5$ and the onset of losses was displaced to a lower proton energy than at present. Similar studies of cosmological aspects were made by Rana & Wolfendale (36); not surprisingly the predicted contemporary spectra show modulation factors falling off with increasing energy at least as strongly as the 2.7-K curves in Figures 12 and 13. Thus, invoking cosmological effects does not help to smooth out the predicted extragalactic spectrum above several 10^{19} eV and in turn does not help to explain the observations (summarized at the top of Figure 12) that in our view give no support whatever for the contention that the microwave cutoff has been observed. (The only suggestion for such a cutoff is from the Yakutsk data—Figure 11, bottom—and this is of poor statistical accuracy. To be consistent with the Haverah Park spectrum, at most three showers would have been expected whereas none was observed—a not unlikely situation.)

Recent calculations by Hill & Schramm (37) and Berezinsky & Grigoreva (38) demonstrate the dependence of spectral shape on z distribution of the sources. Figure 14 shows the modulation factor $\eta(E, z_{max})$ (using the authors' notation—equivalent to F as used in this chapter) when sources are distributed uniformly in a co-moving volume out to z_{max}, i.e. for no evolutionary increase. The modulation factor referred to as 2.7 K in Figure 13 is approximately the limiting case of Figure 14 (large z_{max}).

Of the many other possible situations where extragalactic sources are distributed nonuniformly in space, the model in which there is a super-cluster enhancement is probably the most likely. Insofar as our Galaxy is a member of an association—a supercluster, centered on the Virgo galaxy cluster—the average extragalactic cosmic-ray intensity may be assumed to be higher in the supercluster than on average in the Universe as a whole. In turn, the fraction of extragalactic cosmic rays detected on Earth from the supercluster regions will be higher than expected on the basis of a constant density of extragalactic sources everywhere. If the losses by interaction with the 2.7-K radiation are small for these particles, the modulation factor will not be as serious as for 2.7 K. Figure 13 shows

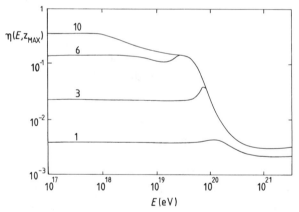

Figure 14 Modulation factors for extragalactic protons interacting with the 2.7-K radiation field as given by Berezinsky & Grigoreva (38) for various cosmological production models. Production is assumed to be uniform in the co-moving frame out to the value $z = z_{max}$ (transit time $= t_{max}$); the key to the labels on the curves is

	z_{max}		t_{max}	
1	$= 3.8 \times 10^{-3}$;		$= 5 \times 10^7$ y	
3	$= 8.5 \times 10^{-2}$		$= 3 \times 10^8$ y	
6	$= 0.19$		$= 2 \times 10^9$ y	
10	$= 1.17$		$= 6 \times 10^9$ y.	

The production spectrum has a differential exponent $\gamma = 2.1$. As expected, case 10 is not far from 2.7 K in Figure 13.

the modulation factor for the earlier supercluster enhancement model of Strong et al (32), which demonstrated that the effect of the 2.7-K cutoff is much reduced. Clearly, from inspection of Figure 12, which shows the global average spectrum, there is no objection to the supercluster enhancement model for the extragalactic component.

In fact, the supercluster enhancement model referred to above is an oversimplification of the actual situation; the distribution of likely sources within the supercluster should be taken into account. We have done this (32a) for a model in which Virgo sources are strong and there is energy-dependent diffusion between Virgo and the Galaxy. The results are shown in Figure 15; again it is seen that there can be a reasonable fit to observation.

CONVENTIONAL ACCELERATION MECHANISMS

Turning to the mechanisms for acceleration as distinct from propagation characteristics, we start with what might be termed conventional mech-

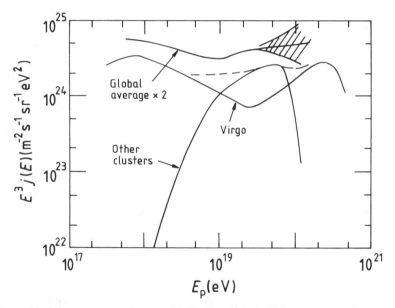

Figure 15 Primary spectrum for a model (32a) in which the highest energy particles come from Virgo cluster galaxies. The specific shape arises from the adoption of an energy-dependent diffusion coefficient favoring the particles of highest energy. Other clusters contribute a nonnegligible fraction of the flux near 10^{19} eV. In this model, Galactic particles would cease above about 10^{17} eV. The global average is the average observed spectrum from Figure 12.

anisms, processes that are seen to occur in our own Galaxy but are perhaps modified in other environments. [Reference can be made here to the useful summary of possible mechanisms by Hillas (39).] As mentioned earlier, supernova remnants are useful to perhaps 10^{15} eV but no further. It is difficult to think of SNRs of much increased energy, which might occur preferentially in other galaxies, or of plausible super-SNRs. It is true that giant SNRs have been postulated in our own Galaxy (e.g. Loop I might be such a phenomenon) but the increased energy is much more likely to manifest itself in terms of an increased energy per SNR rather than energy per individual particle. More important still, perhaps, is the problem of expected flux. In order to achieve a measurable flux of CR from other galaxies their yield must be much higher than our own Galaxy and this does not seem at all likely for SNR-type events.

It is much more likely that we are dealing with extragalactic phenomena not occurring in our Galaxy at the present time. Under conventional mechanisms one should include galactic center activity of the type considered for our Galaxy by Giler (40). In this model, although the Galactic center is inactive now, outbursts occur from time to time and the extragalactic flux is supposed to arise from such outbursts from other galaxies. It is not unreasonable to assume that very energetic particles are produced in the very active phases of the outbursts. M87 is an example of such an outburst and no doubt others have occurred in the "correct" time window in the Virgo cluster; the apparent excess of particles from high Galactic latitudes (see above) is often quoted as support for this hypothesis.

For very large objects and low fields, coherent magnetic fields of several microgauss acting over tens of kpc correspond to 10^{20} eV. Now, as already remarked, the contemporary Galactic magnetic fields are most unlikely to be smooth enough, over such dimensions, to trap particles or the shocks to be sufficiently strong to allow acceleration to this high energy. However, during galaxy formation, the situation is different. It has already been remarked that the total energy for particle production is 10^{-4} eV cm^{-3} if all the collapse energy goes into cosmic rays. This is more than sufficient to give the 10^{-6} eV cm^{-3} needed if the differential exponent is $\gamma = 2$ (to 10^{20} eV) for extragalactic particles (Figure 12). The question of the maximum energy is covered, in principle, by the parameters indicated. Presumably betatron acceleration is a possibility for the actual mechanism, the magnetic field arising at an early stage in the galaxy's life. In this model our own Galaxy would have formed very energetic cosmic rays early on but none would be left by now.

Another interesting possibility concerns galaxy-galaxy collisions. Recent studies of far-infrared (FIR) radiation (largely by way of results from the IRAS satellite) have drawn attention to such collisions. The FIR emission from these galaxy pairs is considerable and there is also enhanced extended radio emission. In this latter context, Hummel & van der Hulst (41) have suggested that particle acceleration in shocks between the colliding galaxies may be responsible for the high electron fluxes. It appears that the frequency of such collisions may be sufficient and the energetics appropriate for significant energy densities of protons to be achieved. Most importantly, we can hypothesize that the recombination of field lines in colliding spiral galaxies with oppositely directed spiral fields might be of sufficiently large scale to accelerate protons to 10^{20} eV or so.

UNCONVENTIONAL MECHANISMS

Perhaps the most exotic mechanism suggested is that of superconducting cosmic strings. Witten (42) and others have shown that such strings, predicted in some elementary particle theories, can have significant cosmological effects. For example, Ostriker et al (43) associated the voids observed in the large-scale distribution of galaxies with bubbles caused by low frequency radiation from loops of superconducting strings. Vilenkin (44; and private communication) has considered the possibility of γ rays from such superconducting strings and pointed out that a detector sensitive to 10^{14} eV γ rays might be able to see the pattern of the loops in the sky.

Such γ-ray loops have not (yet) been seen, but the limits set on the loop frequency and characteristics are not very stringent. It is thus useful to consider their properties further, with particular reference to the acceleration of very energetic particles. Vilenkin has pointed out that oscillating loops of string tend to develop cusps where the speed of light is momentarily reached by the string. The model then leads to the production of heavy particles whose decay products will appear as very energetic cosmic rays. The implied flux of particles of energy m is given by the expression

$$F = 10^{-19} B_{-9}^2 K m_{12}^{-1} \, \mathrm{cm}^{-2} \, \mathrm{s}^{-1},$$

where B_{-9} is the adopted large-scale primordial magnetic field in 10^{-9} gauss, K is of order unity, and m_{12} is the energy in units of 10^{12} GeV.

In the simple model, m is the mass of a hypothesized heavy particle that presumably decays down to protons carrying the bulk of the initial energy. The expression for F can then be regarded as, roughly, the expected flux of protons above this energy. Inspection of Figure 3 shows that such a flux is of the order of the observed CR flux above 10^{20} eV (it is presumably designed so to be).

If, as seems likely, a roughly equal amount of energy finds its way into the e^+e^- component, then the resulting cascade in the universe would be of the type described in detail by Strong et al (32). The result would be an eventual γ-ray spectrum of shape not too dissimilar from the actual iso-tropic γ-ray flux observed. Use of the value given above yields an expected γ-ray energy density of 10^{-6} eV cm^{-3}. Such a value is by no means impossible: the observed isotropic γ-ray flux has an energy density above 1 keV of 10^{-4} eV cm^{-3} and above 100 MeV $\sim 10^{-6}$ eV cm^{-3}. Indeed, the equality of the observed and predicted γ-ray energy densities gives a mea-sure of support for the superconducting string model, although it must be remarked that the model itself seems a priori unlikely.

An interesting feature of this class of models, which has relevance to the question of particle propagation in the universe (as distinct from particle acceleration), is the assumption of a primordial magnetic field. In our paper (45) describing the propagation of electrons of initial energy 10^{19} eV in the universe, we pointed out that, for fields greater than about 10^{-10} gauss, synchrotron radiation would be important. In fact, such radiation transfers the electron energy down to γ rays of 10^{12} eV and below very quickly.

Another important consequence of a field as large as 10^{-9} gauss would be the slowness of proton diffusion if, as is almost certain, the field were tangled. At 10^{-9} gauss the Larmor radius of a proton of energy 10^{20} eV is 100 Mpc. The collection volume of particles would be reduced over the dimension of the universe, although the trapping effect would not be expected to raise the necessary energy input. Such a restriction of the "sphere of collection" does not help the cutoff problem if the diffusion coefficient is independent of energy; however, an energy dependence can be adopted (46) that favors the higher energies and results in a significant flattening of the modulation factor (F of Figure 13), as mentioned earlier, and leads to a good fit with the observed spectrum (Figure 15).

SYNTHESIS AND CONCLUSIONS

Starting with the actual measurements, there appears to be no great differ-ences among the energy spectra from different laboratories. There may be small differences above about 3×10^{19} eV; if true, this can readily be explained in terms of an excess of extragalactic particles from high latitudes in the Northern Hemisphere, but the evidence is not yet sufficiently strong to enable such a claim to be realistically made.

Such evidence as there is on the nature of the particles favors a com-position dominated by protons.

Turning to the origin of the particles, there seems little doubt that the bulk of the particles are of Galactic origin, and here we favor the two-component model: SNR acceleration below 10^{14} eV and another source (pulsars and/or galactic shocks) at higher energies. Above 10^{19} eV we support the contention that the particles are mainly extragalactic, although it is true that the case is not overwhelming. The origin of these particles would then be galactic phenomena that are rare and of a large scale and that involve a process not occurring in our Galaxy at the present time. A number of possibilities have been enumerated. Figure 16 gives a schematic representation of the production and ambient spectra and the Galactic lifetime of cosmic radiation as a function of energy. The extent to which the microwave cutoff does not destroy the flattening of the total spectrum above 10^{19} eV—because of the very flat production spectrum—can be seen in the figure. Under these circumstances the onset of the rapid fall of intensity is delayed until somewhat above 10^{20} eV.

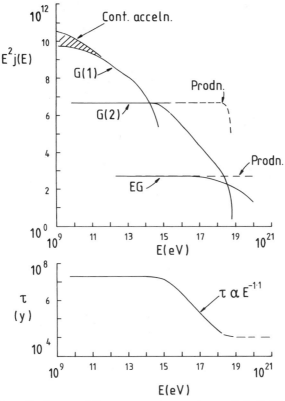

Figure 16 Schematic diagram of the energy spectrum and mean Galactic lifetime for our preferred model for the sources of the cosmic radiation. $G(1)$ and $G(2)$ are the two Galactic sources (SNR and pulsars/Galactic shocks?, respectively) and EG is the extragalactic component. The fall in mean lifetime with increasing energy above 10^{15} eV is due to increasing inefficiency of Galactic containment.

The array at Akeno and the Fly's Eye array are both being extended and a new giant array will be constructed in the USSR. Data from these new experiments will surely add to our understanding of the ideas put forward in Figure 16.

ACKNOWLEDGMENTS

The authors wish to thank Professor A. A. Watson and colleagues for making available their final high energy Haverah Park spectrum prior to publication.

Literature Cited

1. Hess, V. F., *Phys. Z.* 13: 1084 (1912)
2. Clay, J., *Proc. K. Ned. Akad. Wet.* 30: 1115 (1927)
3. Auger, P., Maze, R., *Comptes Rendus* 208: 1641 (1939)
4. Meyer, P., In *Origin of Cosmic Rays*, IUPAP/IAU Symp No. 94, ed. G. Setti, G. Spada, A. W. Wolfendale, p. 7. Dordrecht: Reidel (1981)
5. Cowsik, R., *Astrophys. J.* 241: 1195 (1980)
6. Silberberg, R., Tsao, C. H., Letaw, J. R., Shapiro, M. M., *Phys. Rev. Lett.* 51: 1217 (1983)
7. Giler, M., Szabelska, B., Wdowczyk, J., Wolfendale, A. W., In *Proc. 19th Int. Cosmic Ray Conf.*, La Jolla, ed. F. C. Jones. Washington: NASA (1985), 3: 234
8. Garcia-Munoz, M., Simpson, J. A., Wefel, J. P., In *Proc. 17th Int. Cosmic Ray Conf.*, Paris, ed. F. C. Jones. Washington: NASA (1981), 2: 72
9. Linsley, J., In *Proc. 18th Int. Cosmic Ray Conf.*, Bangalore, ed. F. C. Jones. Washington: NASA (1983), 12: 135
10. Grunsfeld, J. M., L'Heureux, J., Meyer, P., Muller, D., Swordy, S. P., *Astrophys. J.* 327: L31 (1988)
11. Wolfendale, A. W., *Q. J. R. Astron. Soc.* 24: 122 (1983)
12. Bhat, C. L., Issa, M. R., Mayer, C. J., Wolfendale, A. W., *Nature* 314: 515 (1985)
13. Lagage, P. O., Cesarsky, C., *Astron. Astrophys.* 118: 223 (1983)
14. Wdowczyk, J., Wolfendale, A. W., *J. Phys.* G10: 1453 (1984)
15. Wdowczyk, J., Wolfendale, A. W., *J. Phys.* G10: 1599 (1984)
16. Gawin, J., Kempa, J., Wdowczyk, J., *Acta Univ. Lodziensis, Folia Physica* 7: 59 (1984)
17. Jokipii, J. R., Morfill, G. E., See Ref. 7, 3: 132
18. Wolfendale, A. W., *Rep. Prog. Phys.* 47: 655 (1984)
19. Szabelski, J., Wdowczyk, J., Wolfendale, A. W., *J. Phys.* G12: 1433 (1986)
20. Cassiday, G. L., et al., To be published (1989)
21. Wdowczyk, J., Wolfendale, A. W., *Nature* 306: 347 (1983)
22. Deleted in proof
23. Khristiansen, G. B., In *Proc. 20th Int. Cosmic Ray Conf.*, Moscow, ed. V. A. Kozyarivsky, et al. Moscow: NAUKA (1987), 8: 54
24. Lawrence, M. A., Reid, R. J. O., Watson, A. A. Submitted to *J. Phys. G* (1988)
25. Horton, L., McCusker, C. B. A., Peak, L. S., Ulrichs, J., Winn, M. M., See Ref. 9, 2: 128
26. Cunningham, G., Pollock, A. M. T., Reid, R. J. O., Watson, A. A., In *Proc. 15th Int. Cosmic Ray Conf.*, Plovdiv, ed. B. L. Betev. Sofia: Bulgarian Acad. Sci. (1977), 2: 303
27. Teshima, M., et al., See Ref. 23, 1: 404
28. Baltrusaitis, R. M., et al., See Ref. 23, 1: 409
29. Karakula, S., Osborne, J. L., Wdowczyk, J., *J. Phys.* A7: 437 (1974)
30. Michel, F. C., Kennel, C. F., Fowler, W. A., *Science* 238: 938 (1987)
31. Gaisser, T. K., In *Proc. Volcano Workshop*, May 1988
32. Strong, A. W., Wdowczyk, J., Wolfendale, A. W., *J. Phys.* A7: 1767 (1974)
32a. Giler, M., Wdowczyk, J., Wolfendale, A. W., *J. Phys.* G6: 1561 (1980)
33. Puget, J. L., Stecker, F. W., In *Proc. 14th Int. Cosmic Ray Conf.*, Munich, ed. H. Bause, et al. Garching, Munich: Max Planck Inst. Extraterrestrische Phys. (1975), 2: 734
34. Wdowczyk, J., Wolfendale, A. W., *Nature* 258: 217 (1975)

35. Hillas, A. M., *Can. J. Phys.* 46: S623 (1968)
36. Rana, N. C., Wolfendale, A. W., *Vistas Astron.* 27: 199 (1984)
37. Hill, G. T., Schramm, D. N., *Phys. Rev.* D31: 564 (1985)
38. Berezinsky, V. S., Grigoreva, S. I., *Astron. Astrophys.* 199: 1 (1988)
39. Hillas, A. M., *Annu. Rev. Astron. Astrophys.* 22: 425 (1984)
40. Giler, M., *J. Phys.* G9: 1139 (1983)

41. Hummel, E., van der Hulst, J. M., *Astron. Astrophys.* 155: 151 (1986)
42. Witten, E., *Nucl. Phys.* B249: 557 (1985)
43. Ostriker, J. P., Thompson, C., Witten, E., *Phys. Lett.* B180: 231 (1986)
44. Vilenkin, A., *Nature* 332: 610 (1988)
45. Wdowczyk, J., Wolfendale, A. W., *Nature, Phys. Sci.* 236: 29 (1972)
46. Wdowczyk, J., Wolfendale, A. W., *Nature* 281: 356 (1979)

Annu. Rev. Nucl. Part. Sci. 1989. 39: 73–111

SEARCHES FOR FRACTIONAL ELECTRIC CHARGE IN TERRESTRIAL MATERIALS

P. F. Smith

Rutherford Appleton Laboratory, Chilton, Oxfordshire, OX11 0QX, England

KEY WORDS: free quarks, stable particles, levitation technique, charge measurement, mass spectrometry.

CONTENTS

1. INTRODUCTION: MOTIVATION FOR FRACTIONAL CHARGE SEARCHES

Experimental investigations of the possible existence of elementary particles with electric charge smaller than that of the electron were first

73

0163–8998/89/1201–0073$02.00

motivated by the quark model, proposed in 1963 by Gell-Mann (1) and independently by Zweig (2) to explain the observed families of strongly interacting particles (baryons and mesons). An interesting narrative account of the origins and early history of this model is given by Crease & Mann (3). Gell-Mann had pointed out that the simplest way of obtaining the observed particle multiplets would be to construct them from constituent "quarks" with electric charge $\frac{2}{3}e$ and $-\frac{1}{3}e$, but he believed that such constituents would be only "mathematical entities" and would not manifest themselves as real particles. His paper nevertheless recommended the search for fractionally charged particles in matter and in accelerator experiments in order to "reassure ourselves of the non-existence of real quarks."

It was quickly found that the model of baryons as bound states of three quarks (and mesons as bound states of quarks and antiquarks) was remarkably successful in explaining the quantum numbers of the known particle multiplets, the regularities in the mass spectrum, the measured magnetic moments of the baryons, and many other observations (e.g. 4), in terms of relatively few input parameters (the quark masses and their binding potential). Nevertheless it was recognized that a model involving apparently fractionally charged constituents could have several possible interpretations.

(a) The quarks might be purely "mathematical entities," as originally suggested by Gell-Mann, arising from the group structure of the particle multiplets but not corresponding to, or manifesting themselves as, real particles.

(b) The fractional charge might be the result of charge exchange or charge sharing between integrally charged constituents (just as, for example, an H_3^{2+} ion might be thought of as three constituent atoms of average charge $\frac{2}{3}e$, while in reality consisting of three positively charged nuclei sharing two electrons). The "three-triplet" model (5), while not specifically postulating charge sharing, was physically equivalent to this viewpoint, since each constituent was an appropriate superposition of charge-1 and charge-zero states.

(c) The quarks might be new fractionally charged particles capable of existing in the free state but have a large rest mass, nearly cancelled in the bound state by a large negative binding energy. In a sufficiently energetic collision the hadrons might then dissociate into these heavy constituents. Alternatively, even if the quarks themselves are permanently confined, the unit of electric charge may nevertheless be $\frac{1}{3}e$ and other particles may exist in the free state with fractional electric charge.

(d) The quarks might exist as particles only in integrally charged combinations, bound together by an indefinitely rising potential and thus permanently "confined" and never observable as separate particles.

(e) The constituents may be quasi-particles (analogous to the elementary excitations in crystals and superconductors) that cannot exist outside the hadronic environment—just as, for example, "rotons" exist within super-fluid helium but have no independent existence. This analogy has been strengthened in recent years by the recognition that soliton discontinuities in molecular chains can under some circumstances possess fractional electric charge (for a review, see 6) dependent on the periodicity or symmetry of the molecule, and in some cases with the values $\frac{1}{3}e$ or $\frac{2}{3}e$ (7).

If (c) is correct and quarks can exist as free particles, at least one fractional charge state would be completely stable (from charge conservation). Thus, although free quarks are not seen in particle collisions at accelerator energies, the possibility exists that they might be produced by high energy cosmic-ray interactions in the atmosphere and accumulate in terrestrial matter to a measurable concentration during the lifetime of the Earth. Alternatively a primordial concentration of isolated quarks might exist, if a small proportion had failed to combine into baryons during the initial stages of the universe.

Although the majority theoretical viewpoint did not favor the possibility of free quarks, it was nevertheless important to investigate this hypothesis experimentally, and in the late 1960s many groups began searches for fractionally charged particles, both at accelerators and in terrestrial materials. Moreover, in subsequent years the reality of quarks within hadrons was clearly demonstrated by three types of experiment:

1. Measurements of the inelastic scattering of electrons, muons, and neutrinos from nucleons demonstrated the existence of point-like constituents and were also consistent with the fractional electric charges predicted by the quark model (8, 9).

2. New particles were discovered that showed the existence of several generations of quarks. The properties and excitation spectrum of bound quark-antiquark states (in particular the psi and the upsilon) were found to be well described by a positronium-like model (10); magnetic moments and electromagnetic mass differences of baryons showed the constituents to behave as individual fractionally charged Dirac particles (11), while the spin dependence of the baryon masses was found to be well described by a semirelativistic Fermi-Breit Hamiltonian summed over the interactions between each pair of constituents (12).

3. In the production of hadrons in electron-positron collisions, the magnitude of the steps in the cross section with increasing energy (13) was clearly consistent with the pair creation of new generations of quarks with

charge $\frac{2}{3}e$ or $\frac{1}{3}e$ (although these never appear as free particles in such experiments).

As a result of these experiments, fractionally charged quarks can clearly be regarded as being observable entities, having a physical reality and a particle-like behavior within hadrons (14). This strengthens the case for the confined-quark hypothesis (*d*), but the free-quark hypothesis (*c*) still remains as a possibility and continues to motivate free-quark searches. The experimental evidence may also be regarded as entirely consistent with the quasi-particle analogy (*e*), although at present most theorists believe that this is merely a coincidental similarity of behavior and not directly relevant or applicable to the quark model. An alternative viewpoint, however, is that the principle by which localized regions of fractional charge can arise in a medium composed of positive and negative integral charges is fundamentally instructive (6) and could provide an important clue to the nature and origin of the quark charges. Some further comments on this are given at the conclusion of this paper.

This review summarizes the experiments that have searched for fractional electric charge in natural materials, principally terrestrial (but also including moon rock and meteorites). We begin in Section 2 by reviewing the results of the many searches made up to about 1981. These are of two main types: (*a*) searches for particles of nonintegral charge in ion beams (which cover a limited mass range), and (*b*) measurements of the residual electric charge on neutralized samples isolated by levitation (or free-fall). These experiments revealed two dominant problems requiring further investigation:

1. One experiment had reported positive results over a period of several years, which were claimed as unambiguous evidence for the existence of electric charges $\pm \frac{1}{3}e$. This experiment is discussed in Section 3, together with subsequent independent experiments designed to check the claim but yielding only null results.

2. Although many fractional charge searches had been carried out, none had reached the low concentration levels that might be expected from cosmic-ray production processes (bearing in mind the production limits already set by direct observation). Thus the null results obtained were relevant to the possibility of a geochemically enriched abundance in specific materials, but not to the more general possibility of a low abundance in common materials at the Earth's surface. Because of this, several ideas have been proposed for testing larger quantities of material. Progress in developing such tests is described in Section 4, together with prospects for improved concentration limits.

Some other topics, including searches in materials of nonterrestrial origin and tests on material exposed to accelerator beams, are discussed in Section 5. The review concludes with some comments on the scientific justification for further searches.

Fractional charge searches in matter and in accelerator experiments prior to 1982 have been reviewed by Kim (15), Landsberg (16), Jones (17), and Lyons (18). A detailed account of the magnetic levitation technique by Marinelli & Morpurgo (19) in 1982 includes a review of previous levitation searches, and a 1985 review of accelerator quark searches by Lyons (20) also includes a section on matter searches.

2.   PRINCIPLES OF FRACTIONAL CHARGE SEARCHES AND CONCENTRATION LIMITS ACHIEVED

2.1   *Summary of Concentration Limits Achieved in Free-Quark Searches*

A free quark, if positively charged, would be expected to acquire an electron to form a hydrogen-like state with charge $-\frac{2}{3}e$ or $-\frac{1}{3}e$. If negatively charged it could form a bound state within the inner shell of a normal atom or, if strongly interacting, may be bound within the nucleus itself. The chemistry of the resulting fractionally charged atoms has been considered in some detail by Lackner & Zweig (21), who specified some host materials likely to provide favorable sites for the accumulation of such anomalous atoms.

However, experimental searches have been designed to be essentially independent of any assumptions regarding the nature of the quark binding, the chemistry of the resulting atom, or the most likely materials, and a policy of investigating a wide range of the most experimentally accessible materials has been adopted. Nevertheless, the need to use materials compatible with the available experimental techniques, while at first sight covering a representative range of elements and compounds, has in fact resulted in the omission of some of the most likely materials recommended by Lackner & Zweig (21). Some comments on the prospects for testing the neglected materials are given in Sections 4 and 5.

With one exception (discussed in the next section), all experiments have given null results. Figure 1 summarizes the concentration limits achieved in the most important searches (19, 22–44) carried out during the 24 years since the quark hypothesis was proposed. The concentration limit shown is simply the reciprocal of the total amount of material tested (allowing for estimated efficiencies of enrichment and detection). It can be seen that these experiments fall into two main categories: ion beam experiments that, for the reasons discussed below, cover a restricted range of ion mass;

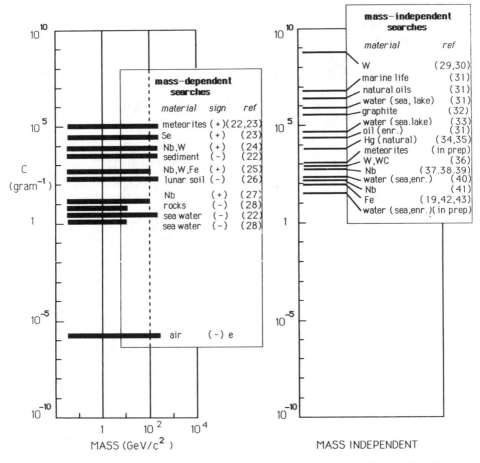

Figure 1 Overall summary of approximate concentration limits achieved for fractionally charged ions in various materials. The left-hand chart refers to searches using ion beams, which were sensitive to a specified charge sign and to masses up to the approximate limit shown. The right-hand chart refers to experiments based on direct measurement of electric charge, giving limits independent of charge sign or particle mass.

and levitation or free-fall experiments that directly measure charge and are thus mass independent.

2.2 *Ion Beam Experiments*

The basic principle of an ion beam experiment is shown in Figure 2a. The sample to be tested is vaporized and ionized, and the individual ions of

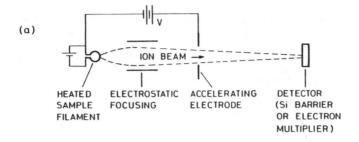

(a)

HEATED   ELECTROSTATIC   ACCELERATING   DETECTOR
SAMPLE   FOCUSING   ELECTRODE   (Si BARRIER
FILAMENT   OR ELECTRON
                                                MULTIPLIER)

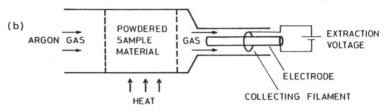

Figure 2   (*a*) Principle of ion beam searches for fractional charge. The system discriminates between particles of different energy qV, and hence between different values of q. (*b*) Typical technique used for preextraction of anomalous ions from large volume sample, and concentration onto collecting filament.

charge q are then accelerated by a potential V to a kinetic energy $E_k = qV$. The particle is brought to rest in a detector capable of measuring E_k and hence q. With a sufficiently low background, it would be possible to observe events with abnormal values of E_k corresponding to fractional values of q/e. To obtain higher sensitivity, recent experiments have added refinements to this basic principle. The experiments of Milner et al (23, 24) use postacceleration stripping of electrons followed by electrostatic selection of specific E_k/q values (Figure 3) to reduce background levels. Because of limitations on ion beam current, these techniques can be used directly only for small quantities of material (up to, say, 10^{-4} g), but can be extended to larger quantities by a preliminary enrichment process. For example, a quantity of air or water vapor can be passed through an electric field to extract and concentrate any charged ions on to a small filament, which is then used as the source for an ion beam experiment. In a further extension of this procedure, various minerals have also been tested—by first heating these in a stream of inert gas, which in turn is passed through an electric extraction field (Figure 2*b*)—but the efficiency with which fractionally

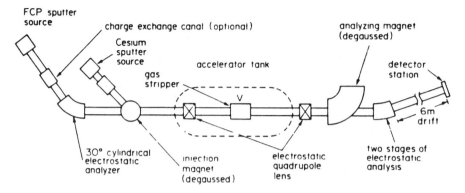

Figure 3 Low background fractional charge spectrometer developed by Milner et al (23, 24).

charged atoms would have been extracted from solid materials in this way has not been estimated.

Materials tested in these ion beam experiments include air, water, rocks, and various metals. Allowing for the estimated detection efficiency, no evidence of fractional charge has been found in quantities of material $\sim 10^{-4}$ g (tested directly), or up to ~ 1 g (using extraction and enrichment techniques). The exception is air, for which $\sim 10^6$ g have been processed. However, it should be noted that it is theoretically unlikely that a significant level of fractionally charged ions could be present in the atmosphere since they would be continuously removed by electrostatic attachment to water droplets and would subsequently re-evaporate from terrestrial water only at a rate $\sim \exp(-E_B/kT) < 10^{-9}$ relative to that of normal water atoms because of the additional surface electrostatic binding energy $E_B \approx 0.5$ eV for a charge $\frac{1}{3}e$ ion (26). Thus the limit $\sim 10^{-6}$ g^{-1} for air in fact has less significance than the limit ~ 1 g^{-1} for sea water.

A significant disadvantage of all ion beam experiments is that they cover a restricted mass range—typically up to 100 proton masses. For larger masses the ion velocity (for a fixed accelerating potential) would fall below the critical value ($\sim 10^7$ cm s^{-1}) for detection by ionization. This has been analyzed in detail by Lewin & Smith (45) for several illustrative experiments, and the more recent papers now specify limits as a function of mass. An approximate indication of the mass range covered by the various ion beam experiments is shown in Figure 1. In practice the experimental sensitivity varies with the type of ion assumed and its charge; for further details of the concentration limits and their mass dependence, reference should be made to the original papers and to the calculations of Lewin & Smith (45).

In the early quark searches, the mass range covered seemed quite adequate, since the discovery of families of particles heavier than the proton was still relatively new, and even a hypothetical free-quark mass of ~ 10 GeV appeared to be large. Twenty years later, there has been a dramatic revision in our ideas regarding possible mass scales, with some particles already known up to ~ 100 GeV in mass, and conjectured unification scales $\sim 10^{15}$ GeV or more. Thus, to be confident of a concentration limit for the absence of free quarks or other fractionally charged ions in a given material, it is important to carry out searches that are independent of particle mass—that is, to use techniques based on the direct measurement of electric charge.

2.3 Experiments Based on Levitation

Methods of measuring an electric charge on a sample, to a precision better than a single electron, are in general based on the following two principles: (a) the sample is isolated from surrounding materials to ensure a constant electric charge (since any material contact with surrounding materials will result in charge fluctuations $\Delta q^2 \approx akT$, where a is the sample radius); and (b) the force qE produced by an applied electric field E is measured by observing changes in the position or motion of the sample, which allows the value of q to be determined.

The classic experiment of this type is the well-known Millikan technique, in which the small liquid drops or solid grains are isolated by levitation in an electric field and their speed of fall is used to measure their charge to an accuracy $\ll e$ (Figure 4a). Some fractional charge searches have in fact

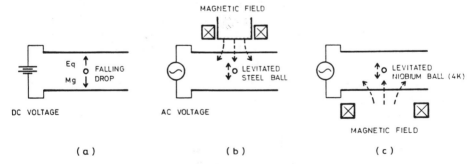

Figure 4 Measurement of electric charge on isolated samples: (a) Falling drop technique ("Millikan experiment"). (b) Ferromagnetic levitation and electric deflection of steel ball. (c) Diamagnetic levitation and electric deflection of superconducting ball.

been carried out in this way, but the method is inherently limited by the fact that electric field levitation restricts the mass of the individual particles of material to $\sim 10^{-9}$ g, and hence the total mass tested to typically $\sim 10^{-7}$ g. This technique has in fact recently been considerably extended by automation, allowing testing rates up to ~ 1 s^{-1} (total mass up to $\sim 10^{-4}$ g day^{-1}) to be achieved (33, 34), but it would be difficult to improve further on this.

Because of the inherent limitations of electrostatic levitation, most of the mass-independent experiments have been based on magnetic levitation, by which much larger samples can be tested. This technique was used for some of the earliest quark searches, and development work continues to the present day. Two methods of magnetic levitation can be used, ferromagnetic or diamagnetic (superconducting). In a ferromagnetic levitation system, the sample is typically a small steel ball (or a small non-magnetic sample coated with steel) about 0.25 mm in diameter, levitated in a vacuum chamber at room temperature by means of a shaped magnetic field (Figure 4b). In the diamagnetic scheme, the sample is a small superconducting sphere levitated in vacuum, at a temperature ~ 4 K, by means of superconducting coils (Figure 4c). In the latter case, the magnetic field provides stable levitation in both vertical and horizontal directions. In the former case, one coordinate direction is always unstable, and additional coils and feedback techniques are used to achieve full position stability. In either case a pair of conducting plates, separated by 1–3 cm, are used to apply an alternating electric field; this causes the sample to execute damped oscillations with an amplitude proportional to its electric charge.

To check for the presence of fractional charge, one begins with a negative charge on the sample, and progressively removes electrons by illumination with ultraviolet light (or with a movable radioactive source) until only a few surplus electrons remain. These electrons are then removed singly to change the charge from, say, $-6e$ to $+6e$, and the oscillation amplitude (measured optically or magnetically) is plotted as a function of time (Figure 5a). If there is no fractional charge, the amplitude should pass through zero. The presence of fractional charge $\frac{2}{3}e$ or $\frac{1}{3}e$ would give an offset $\pm\frac{1}{3}e$ as illustrated in Figure 5b (since the absolute charge is always uncertain by an integral multiple of e, the charge offset can only be defined between $-0.5e$ and $+0.5e$). Typically 6–10 hours of data are required to reduce the statistical errors to $<0.05e$, sufficient to exclude the possibility of an offset of $0.33e$ arising as a chance fluctuation in the data.

However, problems have arisen from several sources of systematic error that are capable of producing a zero shift of sufficient magnitude to simulate the presence of fractional residual charges $\sim\frac{1}{3}e$. These effects were investigated and analyzed in detail by Marinelli & Morpurgo (19) during

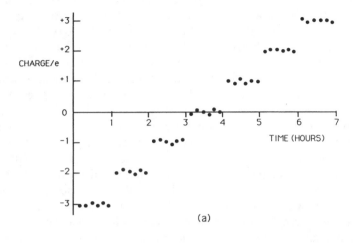

(a)

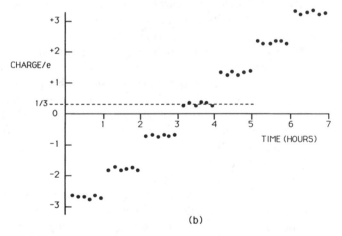

(b)

Figure 5 Typical appearance of data from levitation experiment, showing effect of removing single electrons by exposure to weak ultraviolet light pulses at intervals of 1 hour. Each point is calculated from the average oscillation amplitude of the levitated sample over a period of 10 minutes. (*a*) Ideal appearance of data for zero residual charge and no systematic zero error, giving an oscillation amplitude passing through zero. (*b*) Data offset by fractional residual charge (or by zero shift produced by "patch effect").

their development of the ferromagnetic levitation technique over a period ~ 10 years. The errors result from additional forces arising from the interaction of intrinsic and induced electric and magnetic dipole moments on the sample with the magnetic levitating field and with any non-uniformities in the electric field. Some of these sources of error can be eliminated by spinning the sample about the electric field axis (using additional coils to create an induction motor effect on the levitated sample), or they can be eliminated by additional measurements to determine the correction terms. However, one effect cannot be eliminated in this way and must simply be made sufficiently small or demonstrated to remain constant over a period of several measurements. This is the "patch effect," which is a force resulting from small electric field gradients from impurity patches on the electric field plates. These field gradients interact with the induced electric dipole moment on the sample to produce a force on the sample even when it is electrically neutral—and hence imitates a residual fractional charge. The effect can only temporarily be removed by cleaning the plate surfaces since further impurity layers form on the plates and change with time, despite care in the design and operation of the vacuum system.

The remedy is to design the apparatus and experimental procedure to ensure firstly that the magnitude of this effect is reasonably small (certainly $<0.3e$, and preferably $<0.1e$) and secondly that it varies only slowly with time. Then for two successive measurements it can simply be regarded as a constant zero shift, and subtraction of successive measurements on different samples would give any difference of residual charge between the two samples. The magnitude of the effect is inversely proportional to the square of the plate separation and can be typically as large as 0.5–$0.8e$ for a 1-cm plate separation, reducing to $<0.1e$ for a 2.5-cm separation. It is also proportional to the cube of the sample diameter, and this limits sample size typically to 0.3 mm ($\sim 10^{-4}$ g).

Using this technique, tests can be made at a rate ~ 1 per day, so that in principle several hundred samples per year can be tested with one apparatus. This still limits the total quantity of material that can be directly examined in this way to <0.1 g. Various ideas for improving on this by many orders of magnitude are discussed in Section 4.

With one exception, all of the experimental searches summarized in Figure 1 gave null results, the ion beam searches giving no evidence of nonintegrally charged ions and the levitation experiments being consistent with zero residual charge on all samples. The exception was the super-conducting levitation experiment carried out by Fairbank et al at Stanford, which reported residual charge values on niobium samples consistent with the values zero and $\pm \frac{1}{3}e$ over a period of several years. However, subsequent independent experiments on niobium and related materials at a higher level of sensitivity have not confirmed this result. The evolution and current status of this positive claim are discussed in the next section.

3. THE POSITIVE STANFORD RESULT AND ITS INVESTIGATION

3.1 Discussion of the Stanford Experiment

The results of the Stanford levitation experiment were summarized in three papers between 1977 and 1981 (37–39), with full experimental detail given in the theses of Hebard (46), LaRue (47), and Phillips (48). Because of the attention that had been paid to the problem of removing sources of systematic error, these observations were claimed as "unambiguous" evidence for fractional electric charge.

This remarkable claim was based on 40 measurements on 13 niobium samples (0.28-mm spheres) for which it was believed that all systematic errors, including the patch effect mentioned in Section 2, had been taken into account and reduced to levels well below $0.1e$, and hence could not account for observations of values in the region $\pm \frac{1}{3}e$. The charge measurements (actually charge differences between two samples) and quoted errors are summarized in the now well-known diagram from LaRue et al (39) reproduced in Figure 6a. In contrast to this, the results of a Genoa experiment (19) for a series of tests on 0.2-mm steel balls were all consistent with zero residual charge, as shown in Figure 7. Taken at face value the two sets of data may appear to constitute quite impressive evidence for fractional charge values. The results of Figure 7 show that the levitation technique is capable of giving reliable zero values, without spurious results. The results of Figure 6a (on a different material) also show a number of values accurately grouped around zero, but this time accompanied by a number of results grouped in the vicinity of $\pm \frac{1}{3}e$. Any conceivable source of error should give a continuous spread of values and thus cannot account for the grouping of the results.

However, there were several difficulties associated with the Stanford data. The charge on a given sample did not remain constant on repetition. Three samples showing fractional charge values also gave a result of zero charge in another test (Figure 6b). One of these gave values of zero, $-\frac{1}{3}e$, and $+\frac{1}{3}e$ when retested several times, which implies that fractional charges were being easily gained or lost from the sample surface when placed in contact with other matter between tests. This would in turn suggest that fractional charge was not specifically associated with niobium but should exist generally in other terrestrial materials, whereas there was no evidence at all for fractional charge at this abundance in tests on steel.

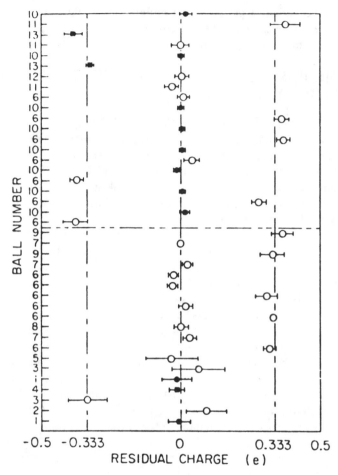

Figure 6a   Residual charge results for niobium samples, as reported by LaRue et al (39) at Stanford. Results are in chronological order from bottom to top.

Another difficulty with the Stanford data was the comparatively large patch effect. Their plate separation was only 1 cm, compared with 2 cm in the Genoa apparatus, and this, as discussed in Section 2, resulted in uncertainties in the absolute charge measurements, which were $>0.5e$. Thus the results in Figures 6a,b all represent the charge difference between two samples. However, random fluctuations of $\sim \frac{1}{3}e$ in successive tests might still occur. To investigate this, measurements were made to check the constancy of the patch field gradient, with the requirement that the patch field remain constant over a period of several days during which a

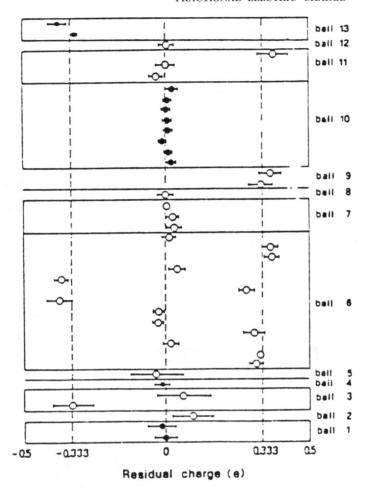

Figure 6b Stanford results regrouped by Marinelli & Morpurgo (19), showing changes in measured charge on individual samples.

test could be made on a given sample *S* preceded and followed by tests on a reference sample *R*. If the absolute result for *R* was the same before and after the test on *S*, it could be assumed that the patch field had remained constant for this sequence of measurements, and any difference between the results for *S* and *R* should then represent a genuine difference in the absolute charges (although with large patch fields the possibility of fluctuations of approximately $+0.3e$ and $-0.3e$ on successive tests cannot be excluded). If, on the other hand, the result for *R* had changed, then the patch field was not constant and the whole sequence of data must be discarded.

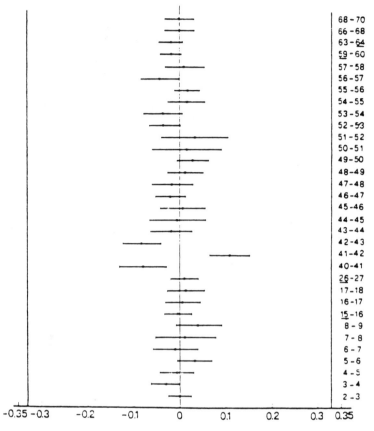

Figure 7 Residual charge *e* (*horizontal axis*) results for steel samples (ball numbers on *vertical axis*), reported by Marinelli & Morpurgo (19). Values shown are charge differences between successive samples, in chronological order.

The need to select stable data sequences in this way reduced the output of accepted results to only 40 over a period of about 5 years, making it impossible to carry out sufficient tests to investigate the origin and consistency of the apparent fractional charges. In addition, the need for selection of results created concern, since it was not clear how acceptance decisions were made in borderline cases (e.g. when the patch field had varied only slightly during the test sequence). To settle this point, a series of "blind tests" was proposed (by L. W. Alvarez), in which a random fractional charge value q_b (unknown to the experimenters) would be added to the data for each sample by an independent person. This would guarantee that the data acceptance decisions could not be biased by knowledge of the results. The residual charge value for each accepted test would afterward be obtainable by disclosing and subtracting the value of q_b.

Unfortunately this plan could not be carried through, since a number of still unresolved technical problems arose in 1982 and resulted in unsatisfactory test data sequences. No reliable residual charge measurements have been possible since then (49, 50). Furthermore, there has been concern that slight misalignments of the electric and magnetic fields, together with uncertainties regarding ball spin, could have caused some systematic errors in the original measurements. Attempts have been made to investigate and rectify these (although they could not in fact account for the grouping of the results in Figures 6a,b). As a result of these problems, the Stanford group have now modified their earlier claim of "observation of fractional charge" to "evidence for fractional charge" (50).

3.2   The RAL/IC Levitation Experiment

Since the Stanford experiment was the only one based on superconducting levitation, it was not immediately possible to carry out independent tests on the same material in any other apparatus. However, it was important to resolve the uncertainties surrounding the results of Figures 6a,b, and in 1982 work began in the UK at the Rutherford Appleton Laboratory (RAL) and at Imperial College (IC) London to develop an independent experiment capable of repeating the Stanford observations with lower systematic errors. The technique adopted was to plate 0.28-mm niobium spheres with about 0.01 mm of iron, which gave them magnetic properties similar to solid iron spheres and allowed room temperature ferromagnetic levitation to be used. Since the plating process needed ~ 100 samples per batch and there were too few of the original Stanford samples for this purpose, a new batch of 0.29-mm balls was manufactured from similar basic material (99.8% purity commercial niobium wire), enough for testing by both the RAL/IC and the Stanford groups.

The design of the apparatus took full account of the experience of the Genoa and Stanford groups and aimed at a testing rate of 1 per day with all systematic errors reduced to $< 0.1e$. A geometry was chosen with magnetic, electric field, and ball spin axes parallel (Figure 8). The plate separation was increased to 2.5 cm to reduce the patch effect of < 0.15 of the Stanford level. The result of these improvements was that the offset due to the patch field was sufficiently low and slowly varying to eliminate fully the possibility of producing any spurious $\frac{1}{3}e$ residual charges. It could in fact be estimated to a precision $\pm 0.02e$ by means of the results of the sample tests interspersed with control measurements on steel samples

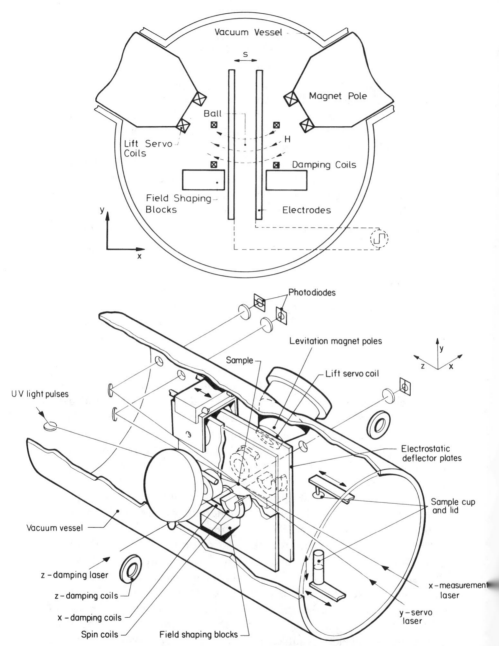

Figure 8 RAL/IC levitation system. (*Top*) Cross section of basic levitation geometry, showing parallel transverse electric and magnetic fields. (*Bottom*) Three-dimensional view of apparatus, showing laser/photodiode sample position measurement system. The electric deflector plates are 15 cm square and there is a separation of 2.5 cm.

(Figure 9, *top*).[1] Other sources of random and systematic error amounted typically to $\pm 0.02e$–$0.03e$, giving an overall accuracy better than $0.05e$. The use of iron-plated samples caused no problems, despite the resulting irregularities, since these were averaged out by the rapid ball spin (Figure 10a,b). For the same reason, cylindrical, cubic, and even completely irregular samples also gave satisfactory test data and consistent results—a valuable bonus that considerably extended the range of materials and sample types that could be tested.

Since the sample treatment and environment were not identical to those in the Stanford experiment, three hypotheses were investigated in an attempt to replicate the Stanford observations.

1. The fractional charge could have been directly associated with the niobium itself. A total of 82 tests were made on 64 different niobium spheres, all consistent with zero residual charge (41). The successive differences of the measured charges are distributed about zero with a width consistent with the typical measurement accuracy $0.02e$–$0.05e$ (Figure 9, *bottom*) and with no examples of differences $\sim \frac{1}{3}e$. This result is clearly in statistical disagreement with the Stanford result of 14 values of $\sim \frac{1}{3}e$ in 40 tests on 13 different samples.

2. The fractional charge could have been transferred to the samples from the tungsten plate used to support the Stanford samples during prior heat treatment (necessary to reduce damping effects in the superconductor). Three types of tungsten sample were tested: iron-plated commercial tungsten carbide balls, tungsten-coated steel balls, and iron-plated cylinders of tungsten wire (Figure 10c). The total amount of tungsten tested was far more than could have diffused into, or condensed onto, the Stanford samples, and all samples gave results consistent with zero charge (36).

3. The fractional charge could have been adsorbed from the helium exchange gas, or acquired by some other mechanism from the low temperature environment. To simulate this iron-plated niobium balls were suspended in a moving stream of liquid helium, with a section of the coating removed to expose the niobium surface (Figure 10d). The samples were exposed for up to 24 hours both with and without an applied electric

---

[1] Increasing the plate separation would also seem an obvious and essential improvement to the Stanford apparatus, since the resulting reduction in patch field eliminates all controversial data selection problems and avoids the need for "blind testing." Examination of our own data shows that we could not have carried out a satisfactory experiment with a plate separation as small as 1 cm, since a simulated scaling up of our systematic errors by a factor of $(2.5)^2$ shows that there would have been many examples of spurious differences $\sim(1/3)e$ between successive results, which could not have been unambiguously eliminated by any selection criterion.

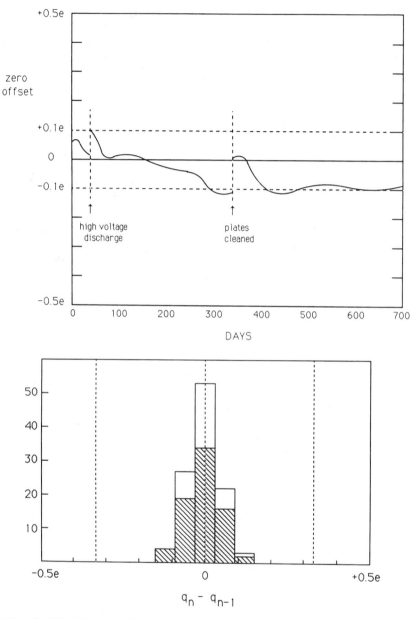

Figure 9 (*Top*) Variation with time of zero offset due to "patch effect," estimated as a moving average obtained from over 300 results on actual and control samples during the two-year period. The zero offset varied sufficiently slowly to be estimated to $\pm 0.02e$ at any given time. (*Bottom*) Histogram of measured successive charge differences for niobium samples. Shaded area: initial series. Total area: all tests, including liquid helium test series. The numbers along the vertical axis represent the number of tests.

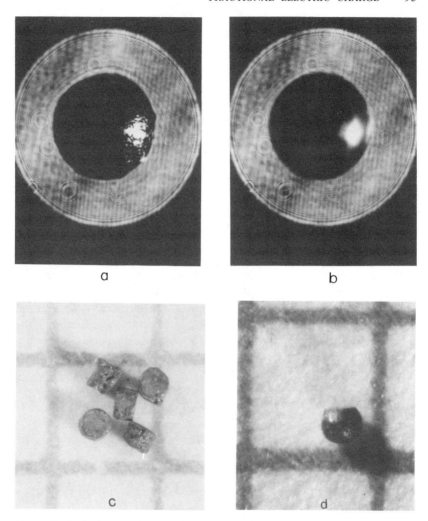

Figure 10   Levitation samples: (*a*) Levitated niobium ball (0.3 mm), showing irregular iron coating. (*b*) Levitated ball with spin, showing averaging of irregularities. (*c*) Iron-coated cylinders (0.3 mm) of tungsten wire. (*d*) Iron-coated niobium ball with section removed to expose niobium surface. The background grid is in 1-mm squares.

potential to assist the extraction of any charged ions from the helium. In some cases a coating of gold was immediately vapor-deposited on the samples (while still suspended) to retain any ions attracted to the surface during these processes. These variations covered most of the possibilities by which fractional charge might be transferred onto the samples from low temperature helium. Again, all samples tested (22 niobium balls plus 14 steel balls) gave results consistent with zero residual charge (44).

It should be noted that some elementary data selection criterion is still needed in all experiments of this type since, for example, occasional large spontaneous charge changes can occur that may curtail a test with insufficient data to estimate the charge offset. For such cases, the RAL/IC tests included the rigid selection criterion that there be at least two hours of data at charge levels between $-3e$ and $+3e$, so that no spurious offsets would arise purely as a statistical artifact from a small amount of data. In practice, very few tests failed in this way, and these were immediately repeated. Also, as already noted, no tests were rejected through changes in the patch field. As a result of this a test rate of ~ 15 per month was maintained over a period of two years, allowing a sufficiently large number of tests to be made to investigate the above hypotheses, and also allowing the technique to be extended to other types of fractional charge search, as described below in Sections 4 and 5.

It thus appears that the most likely explanation of the results of Figure 6 is that they arose as an artifact associated with systematic errors $> \frac{1}{3}e$ in the Stanford apparatus and the consequent difficulties of data selection. The less likely possibility still remains that Figure 5 is correct and the different origin and manufacturing history of the original samples could somehow be responsible. However, the Stanford group also have samples from the new batch and, if they are able to restart their test program in the future (and convincingly eliminate the systematic errors), they will be in a position to make comparative tests on the original and new samples.

3.3 Ion Beam Experiments with Stanford-Related Materials

Although the RAL/IC program represents the only attempt to investigate the Stanford claim using a mass-independent levitation technique, several ion beam experiments were also carried out specifically to look for fractional charge in the Stanford-related materials, niobium and tungsten. Schiffer et al (25) searched for fractionally charged ions (charge $+\frac{1}{3}e$ only) from heated filaments of Nb and W; their tests were sensitive to masses up to ~ 100 amu (45). Kutschera et al (27) looked for fractionally charged ions (charge $+\frac{1}{3}e$ or $+\frac{2}{3}e$) from heated filaments of Nb and W that had first been cooled to liquid helium temperatures; their tests were sensitive to masses up to ~ 1000 amu. However, both of these experiments could have missed fractional charge that was attached to nuclei with $Z > 20$ (45) or that was not released from the filament by heating. The latter problem was overcome in the more recent ion beam experiment (Figure 3) by Milner et al (23, 24), in which a sputter ion source was used to vaporize and ionize all of the sample material. The sensitivity achieved was better than the concentration level $\sim 10^{-18}$ per atom that might be inferred from the Stanford results, for masses in the range ~ 1–200 amu.

4. ATTEMPTS TO REACH LOWER CONCENTRATION LEVELS

4.1 *Concentration Levels Expected from Cosmic-Ray Interactions*

Since the primary cosmic-ray spectrum extends to very high energies, any particle can be produced (for example via electromagnetic or strong interaction processes) at some low but finite rate by cosmic-ray interactions in the atmosphere. Any stable or quasi-stable particles produced would accumulate to give a significant concentration in terrestrial material during the lifetime of the Earth. A specific example is that of cosmic-ray-produced tritium (lifetime ~ 12 years), the majority of which accumulates in the sea to give a concentration $\sim 10^{-18}$ (which would in fact be $\sim 10^{-9}$ if tritium were completely stable). Similarly, for free fractionally charged particles, at least one would be stable and would progressively accumulate in terrestrial material.

Direct observations of cosmic-ray air showers have occasionally suggested the existence of low ionizing particles, consistent with charges $<e$. These claims have not been generally accepted, but the estimated rate of such abnormal events $\sim 10^{-11}$ cm^{-2} s^{-1} (51) can at least be taken as an upper limit for the flux of fractional charge incident on the surface of the Earth. Two simple estimates for the typical resulting concentration can then be made: (*a*) Assume that the whole of the flux accumulates in the sea during the lifetime of the Earth. This gives a concentration ~ 1 g^{-1} or lower. (*b*) Alternatively, assume that the whole of the flux becomes uniformly distributed in the top 300–1000 km of surface material. This gives a concentration ~ 0.01 g^{-1} or lower.

Of course the further possibility exists that, like many elemental materials, much higher concentrations of fractionally charged ions might subsequently be produced by geochemical enrichment in association with favorable host materials (21, 52). From Figure 11 we can see that this is in fact the only hypothesis that has been covered by quark searches to date. It is sometimes stated, as evidence against the existence of free fractional charge, that "experiments have looked for free quarks and have not found them." We can see that this is an entirely misleading statement. It is remarkable fact that, after 25 years, the possibility of a cosmic-ray-produced concentration of <1 g^{-1} in common surface materials still remains unexplored.

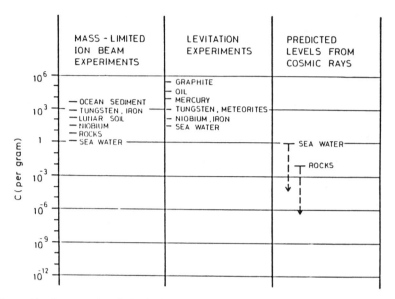

Figure 11 Concentration limits in fractional charge searches compared with estimated upper limits from cosmic-ray production.

It is also evident that one would expect the cosmic-ray-produced concentration to be strongly mass dependent (if quarks are produced by interaction of the primary particles and not a constituent of the primary flux itself). Since the primary energy spectrum falls as $\sim E^{-2.5}$ and the center-of-mass energy is proportional to $E^{0.5}$, production rates for particles of mass M_q are expected to fall as $\sim M_q^{-5}$ (53). This further reduces the likely concentration levels and reinforces the point that searches to date have not been sufficiently sensitive to have any clear significance regarding the existence of three fractional charge.

We have already seen that this problem arises from the difficulty of testing the necessary quantities of material, $\sim 10-10^3$ g or more. This requires higher testing rates and/or larger sample masses. By automation of the Millikan technique, the California State University at San Francisco (SFSU) group (33–35) has achieved testing rates for liquid drops as high as 1 s^{-1}, but since the individual drops have a mass $\sim 10^{-9}$ g, the net processing rate is only $\sim 10^{-4}$ g per day, similar to that achievable with ferromagnetic levitation at 1 sample per day. Some streamlining of the latter technique would be possible, since the statistical error for a measurement time t is $\sim 0.4e/(t \text{ minutes})^{1/2}$ and ~ 1 hour of data would normally be sufficient to check each sample. With sufficient manpower, therefore, and preferably two testing stations, a test rate of > 10 samples per day (10^{-3} g per day) could be achieved. It would also be technically possible to levitate much larger samples, but the immediate problem encountered is that the patch field error scales as sample volume/(plate separation)^2 so that a factor of 100 increase in individual sample volume would require apparatus dimensions to scale by a factor ~ 10 (with cost increased by $\sim 10^2$). Alternatively, it would also be possible to construct a large number of similar levitation systems to permit parallel testing: fractional charge searches are currently the lowest in cost of all "nonaccelerator" experiments (for a review, see 54), having typically 1% of the funding level of a proton decay experiment. Thus given the same level of resources as the latter, a parallel levitation testing system could be constructed capable of reaching ~ 1 g per month.

Although there is currently insufficient scientific justification for such a major scale-up of existing techniques, three attempts have been made to solve the problem of testing larger quantities of material at a more reasonable cost, using quite different approaches, and we now review progress on these.

4.2   The Rotor Electrometer

The rotor electrometer principle was devised and studied by Innes et al at the Stanford Linear Accelerator Center (SLAC) (55) and described in detail in the thesis by Price (56). The principle of a similar instrument is described by Williams & Gillies (57). The basic scheme is shown in Figure 12. A conducting sample container is suspended from a quartz thread, close to a rotating array of Faraday cups, which produce an ac signal proportional to the charge on the container. By measuring the signal with and without a sample in the container, the charge on the sample alone can be obtained. At first sight this may appear to violate the principle mentioned above, that any material suspension will allow charge fluctuations $> e$ to occur. However, such charge fluctuations have a frequency spectrum characterized by $R_t C_s$ (58), where C_s is the capacitance of the sample ($\propto$ radius) and R_t is the resistance of the supporting thread, so that the timescale of the charge fluctuations can in principle be made larger than the measurement time if the resistance of the suspension is sufficiently large.

Two basic objectives need to be achieved to make this technique suitable for fractional charge searches. The first is that noise levels must be made sufficiently small, and the electronics sufficiently sensitive, to measure charge changes to an accuracy of $\sim 0.1e$. The second is that the measurements with and without the sample must be made faster than significant

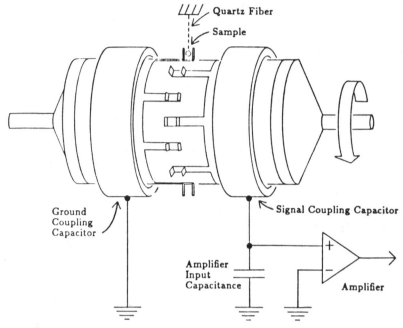

Figure 12   Rotor electrometer for fractional charge measurement on a suspended sample [from Price (56)].

changes in the background charge distribution on the container and support thread. The first came very close to being achieved: single-electron changes were observed with a resolution of $\sim 0.3e$. A statistical precision $<0.1e$ in a 200-s measurement time was in principle possible, although systematic errors of $\sim 0.25e$ remained. The second objective proved unattainable, since it did not appear possible to insert or change samples sufficiently rapidly (~ 30 s) to compete with the systematic charge fluctuations and without disturbing the background charges.

For this reason, the program has now been discontinued (W. Innes, personal communication). Nevertheless it would be of interest to consider whether mechanical suspension of the container is actually necessary, and whether it would be possible to apply the same technique to measurement of the charge on a large, magnetically levitated sample (for which the problem of charge fluctuations would be absent). Certainly, with the combined experience now available, such a system could be designed and constructed, but this would require an increased level of effort and cost difficult to justify at the present time considering the strong competing demands for particle physics resources.

4.3   *Droplet Stream Experiments*

A rapid charge measurement technique based on the electrical deflection of a fast stream of liquid droplets was proposed in 1979 by Hirsch, Hagstrom & Hendricks (59). This basic idea is currently familiar as the principle of the ink-jet printer. A piezoelectric mechanical oscillator is used to create a stream of drops emerging from an orifice in a liquid reservoir at a rate of $\sim 10^4 \, \text{s}^{-1}$, velocity ~ 1–$5 \, \text{m s}^{-1}$, and a constant diameter typically ~ 30–$100 \, \mu\text{m}$. This is equivalent to about one liter per day. For a fractional charge experiment, the drops are electrically biased to zero average charge and have a charge spread of about $\pm 20e$. They pass vertically downward (in vacuum) through a transverse electric field, which produces a deflection proportional to their charge qe (since the droplet mass is constant to better than 0.1%), and become separated into distinct integer charge positions that can be registered with a photographic or charge-coupled device (CCD) camera (Figure 13).

Although the technique is restricted to suitable liquids, it can in principle be extended to solid materials by putting the latter into suspension or solution in a suitable carrier liquid. It would thus offer the possibility of testing substantial quantities of almost any material.

However, to achieve a sensitivity of, say, $10^{-2} \, \text{g}^{-1}$, it is necessary to avoid spurious deflections to the level of 1 in 10^8 drops. Systematic effects giving nonintegral deflections can result from electrostatic forces between successive drops, contamination by macroscopic particles, anisotropic evaporation in flight, and charge changing in flight as a result of field emission from the electrodes. Development work on two experiments of this type has been reported. The apparatus of Van Polen, Hagstrom & Hirsch (60) at Argonne used a flight path of 22 m to obtain good separation of integer charges (> 1 mm), but were nevertheless limited by systematic effects to a precision of 1 in 4000 drops (with a droplet mass of 16 ng), which is approximately the same amount of material ($\sim 10^{-4}$ g) as a single sample in a levitation test.

There needs, therefore, to be an improvement factor of $\sim 10^4$–10^5 in the elimination of spurious drop deflections to reach the objective of investigating quantities of material ~ 1–10 g or more. Because a gain of this magnitude appeared unlikely to be achievable, further development of the above experiment was discontinued (J. Van Polen, personal communication). However, further studies of the systematic errors in this technique have been made by Hendricks et al (61), who claim that spurious events have now been reduced by three orders of magnitude to reach a sensitivity of one fractional charge in ~ 0.1 g liquid, although no results have yet been reported. At the time of writing (early 1989) this work was still in progress (C. Hendricks, personal communication).

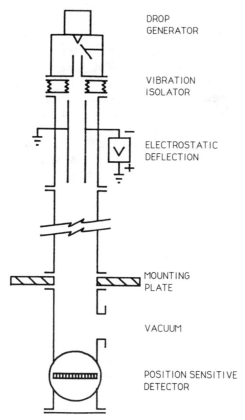

Figure 13 Apparatus for measurement of deflection of falling charged droplets [from Van Polen et al (60)].

4.4 *Enriched Sea Water Tests Using Levitation*

A specific advantage of fractional charge searches involving liquids is that a large increase in concentration can first be obtained using evaporation. This idea is similar to fractional distillation, except that in the case of permanently charged ions the difference in vapor pressure would be so large (see discussion in Section 2.2) that the loss of fractionally charged ions from an evaporating liquid would be negligible. As a result, some caution is necessary in selecting materials for fractional charge searches, and it is clear that null results are not significant in substances that have themselves been produced as distillates (the majority of commercially supplied organic liquids, purified mercury, etc). Another example of this is fresh water, which is the result of the natural distillation of sea water.

The principle of concentration by evaporation is being used by the RAL group in an attempt to test a quantity of sea water ~ 3000 cm^3. This principle was previously utilized on a smaller scale by Mitsuhashi et al (40), who carried out a Millikan experiment on small salt grains produced by the evaporation of falling drops of sea water. Because sea water contains $\sim 3\%$ dissolved solids, mainly NaCl, straightforward evaporation yields only a factor ~ 30 gain in fractional charge concentration (for a given mass of salt tested). To overcome this restriction, Mitsuhashi et al utilized an ion exchange column to separate out the constituents of sea water by charge. The technique is as follows: the sea water is first allowed to pass through a column containing an ion exchange resin that removes all the positively charged ions. The initial water (now dilute HCl) emerging from the column thus still retains any negative fractional charges, but the original Na and Mg concentration has been reduced by several orders of magnitude. To retrieve the positively charged ions, dilute HCl is then added to the column, which progressively releases the positive ions in order of charge. Analysis of the collected fluid shows localized peaks corresponding to the release of the Na and Mg from the original sea water (Figure 14, *top*). One therefore expects that any fractional charges $\frac{1}{3}e$ or $\frac{2}{3}e$ would emerge earlier than the Na peak, and any $1\frac{1}{3}e$ or $1\frac{2}{3}e$ ions would appear between the Na$^+$ and Mg^{2+} peaks.

Thus by discarding those portions of water containing the majority of the Na and Mg, there is a high probability that most of any fractional charges would remain in the resulting purified water, which can then be evaporated to obtain a small concentrated sample. In this way, Mitsuhashi et al were able to achieve a concentration factor of ~ 100 using two stages of ion exchange. Their overall sensitivity was nevertheless limited by the fact that, by using electrostatic levitation, the mass of the individual salt grains tested could be no greater than $\sim 10^{-9}$ g. The total amount of salt tested was ~ 1.5 μg, corresponding to ~ 4.5 mg sea water.

To extend this to larger quantities, the RAL group (G. Homer et al, in preparation) evaporated the salt solution onto the surface of a steel ball, producing a thick salt coating that, despite having an irregular and non-conducting surface (Figure 14, *bottom*), gives satisfactory results in a levitation test. About 30 μg of salt can be tested on a single ball—about 20 times the total quantity tested by Mitsuhashi et al. With care, a two-stage concentration factor of $\sim 10^5$ may be achievable, and an initial volume of several liters of sea water can in principle be searched within a testing period of a few months. The actual enrichment process takes considerably longer than this because of the amount of ion exchange processing and subsequent evaporation stages involved. Moreover, problems arise through contamination with organic material from the ion

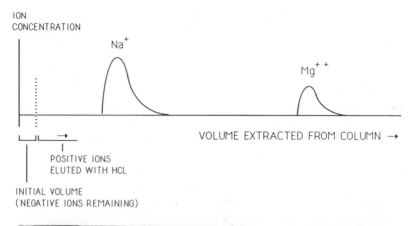

Figure 14 (*Top*) Typical ion concentration peaks for sea water eluted with HCl from ion exchange column. (*Bottom*) Steel ball (0.25 mm) with and without evaporated salt coating.

exchange column, giving a final residue containing a high proportion of suspended solids and difficult to evaporate onto the surface of a 0.25-mm ball. At the beginning of 1989 this work was still in progress, with no certainty regarding the final sensitivity that would be achieved. It will, however, reach levels below 0.1 g^{-1} (at least two orders better than the achieved droplet sensitivity) and thus probe possible cosmic-ray-produced concentrations in sea water for the first time.

With the experience now gained, it would be possible to extend the ion exchange enrichment process to larger volumes. This would be most straightforward in a negative ion search, which simply requires the water to percolate through the column to remove the dissolved solids and does not involve separation from the Na^+ and Mg^{2+} peaks as in the case of positive ions. If the problem of suspended material from the column could be solved, at least another factor of 10^3 could be obtained in this way. It should be noted, however, that a discussion by Nir (52) of the likely geological and geochemical fate of free quarks suggested that fractional charges may in fact be filtered out of the sea by trivalent ions of Fe and Al, in which case they would form a low concentration in appropriate rocks. Although, as noted in the next section, rock material can be coated with iron and tested by levitation, the concentration is likely to be too low for direct detection in this way—and the enrichment of kilogram quantities of rock (presumably still conceivable by techniques such as laser evaporation or charge extraction from chemically processed material) would appear to require a prohibitive level of resources and funding.

5.   OTHER SEARCHES AND FUTURE PROSPECTS

5.1   Searches in "Nonterrestrial" Materials

There are three available types of material of nonterrestrial origin: meteorite samples, lunar rock, and interstellar material. Some fractional charge searches have been carried out on the first two of these, although, as indicated in Figure 1, it has not been possible to achieve the concentration levels that might result from cosmic-ray production. Meteoritic material is, nevertheless, interesting in having presumably been subject to less geochemical processing than terrestrial material, and thus it could have a different and perhaps larger primordial abundance of any atoms containing free quarks.

Samples of lunar rock were tested by Stevens et al (26) using a technique similar to that of Figure 2, i.e. the extraction of ions from a heated sample by flushing with inert gas, followed by electrical concentration onto a filament that was then used as the source in an ion beam experiment. The system was sensitive to abnormal negative ions up to ~ 10–10^2 amu (45). Although the electrical concentration process can be regarded as reliable, the initial extraction is subject to some doubt, since there appears no guarantee that the fractionally charged particles Q would be released by heating unless specifically in the form of hydrogen-like atoms $Q^{+2/3}e^{-1}$ or $Q^{+1/3}e^{-1}$. They interpret their sensitivity $\sim 10^2 \, g^{-1}$ as a limit $\sim 10^{-11} \, cm^{-2}$ s^{-1} for the cosmic-ray flux of fractional charges incident on the lunar surface, which is of the same order as the limit on the terrestrial flux discussed in Section 4.

The same group had earlier tested various meteorite samples (22), but the limits set were at the much higher level of $\sim 10^7 \, g^{-1}$. A more sensitive meteorite search was made recently by Milner et al (23), using the ion beam technique of Figure 3 and employing a sputter ion source to vaporize the sample material (thus eliminating the above-mentioned uncertainty regarding extraction by heating in an inert gas stream). The detector was sensitive to masses up to ~ 250 amu and concentration limits were in the region $10^5 \, g^{-1}$ for six meteorites already known to have interesting isotopic anomalies. For mass-independent limits it is necessary to test samples in a levitation experiment, and three types of meteorite have recently been tested in this way by the RAL/IC group (W. Jones et al, in preparation), taking advantage of the previously mentioned observation that spinning the levitated sample about the measurement axis allows nonspherical samples to be tested without difficulty. Since iron meteorites are already ferromagnetic, they could be tested simply by cutting into 0.2-mm cubes. Levitation of nonmetallic materials presented a greater problem, but a technique was nevertheless devised for plating the (nonconducting) samples with iron, and these were then tested in the normal way. Since stony meteorites are not homogeneous, care also has to be taken that the samples are representative. Figure 15 shows that 0.2-mm cubes are adequate from this viewpoint in the case of the Murchison meteorite. These tests have given (mass-independent) limits $\sim 10^3 \, g^{-1}$ for the three types of meteorite tested, and, because of the limited testing rate ($< 10^{-4}$ g per day) it would be difficult to improve further on this without the development of one of the techniques discussed in Section 4.

5.2   Searches in Materials Exposed to Accelerator Beams

Searches for free fractionally charged particles produced by high energy particle beams from accelerators (and also by cosmic rays), using conventional particle detection techniques, were reviewed in detail by Lyons (20). An alternative approach of some advantage for stable particles is to test material that has been exposed to accelerator beams or to the collision products by using one or more of the techniques described in the previous sections. In this way, one has the chance of accumulating events over a long period of time to obtain higher sensitivity to rare production processes.

Several searches of this type have been carried out. The first of these was made at the Bevalac, where beams of heavy ions are directed at heavy target nuclei in an attempt to produce and study the "quark plasma." It was suggested that any free quarks produced might be stopped in suitable

Figure 15 (*Top*) Section of Murchison meteorite. (*Bottom*) Diced samples (0.2 mm) of Murchison meteorite.

materials that could subsequently be tested for fractional charge (62). The collecting materials included steel balls (for levitation experiments), copper (for ion beam experiments), mercury (for Millikan droplet tests), and a number of tanks of carbon tetrachloride, each equipped with a central metal wire (e.g. gold) maintained at a positive or negative potential with respect to the container (Figure 16a). The purpose of the latter was to obtain a large concentration factor by collecting any stable charged particles on the central wire. The surface layer of the latter was then dissolved in a small bead of mercury, which was tested in the SFSU automated Millikan apparatus (63). As a backup procedure, the remainder of the gold wire was then dissolved in mercury from the target to form a second sample. The two samples would thus have contained any fractionally charged particles stopping either directly in the mercury target or subsequently in the collecting tanks. The null results corresponded to a limit $\sim 10^{-6}$ quarks per primary interaction for particles collected in the first sample, and ~ 10 quarks per primary interaction for particles collected in the second sample.

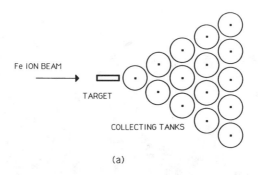

(a)

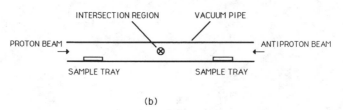

(b)

Figure 16 (a) Arrangement for collection of fractionally charged ions from heavy ion collisions [from Lindgren et al (63)]. (b) Arrangement for exposure of steel balls inside vacuum chamber of SPS collider [from Lyons et al (43)].

The same collaboration also reported a similar set of experiments carried out at the Fermilab Tevatron (64). A target consisting of six liters of mercury was exposed to an 800-GeV proton beam, and samples were tested in the automated Millikan apparatus both before and after distillation to obtain a concentration factor $> 10^5$. A second technique used a lead target, backed by tanks of liquid nitrogen containing gold-plated glass fibers with applied positive and negative potentials. As before, the gold was dissolved in mercury for testing by the Millikan technique. The null results set a limit $\sim 10^{-9}$ quarks per primary interaction.

A third experiment, carried out in the CERN SPS collider (at 630-GeV center-of-mass energy), was based on the hypothesis that free quarks may be produced in collisions but are sufficiently strongly interacting to be stopped by the accelerator vacuum chamber, and would not then be observed by external detectors. To test this possibility, several hundred steel balls were attached to copper trays and exposed near one of the intersection regions of the CERN SPS collider (Figure 16b) for a period of four months, after which the balls were tested for fractional charge in the RAL levitation apparatus (43). In the absence of confinement effects, each ball would have received a total flux of ~ 100 quarks, so that the null results obtained with 60 balls demonstrate that confinement reduces the production of free quarks by a factor $> 10^3$.

5.3   Future Prospects

In this review we have drawn attention to the fact that a significant natural (for example, cosmic-ray-produced) abundance of free fractionally charged particles could exist in terrestrial materials without having been seen by any of the free-quark searches carried out during the past 25 years. This is because (a) the achieved sensitivity has not reached even the highest concentration levels expected from cosmic-ray production (uniformly distributed in sea water or the Earth's crust), and (b) most of the minerals proposed [from the predicted chemistry of fractionally charged atoms (21)] as the most likely hosts for an enhanced concentration have not in fact been tested, owing to the lack of compatibility between these materials and the experimental techniques used. For example, until recently theoretically recommended minerals such as fluorapatite, pegmatites, and carbonaceous chondrites (21) would have been difficult to use either as the source in an ion beam experiment or as bulk samples in a levitation experiment. Extraction and concentration of any fractionally charged atoms in these materials would also present major technical problems.

A second important factor in the experimental strategy for quark searches is the mass limitations of ion beam experiments. Although the effective masses of quarks in the bound state are quite low (constituent masses lie between 0.3 and 5 GeV for the known quarks), the evidence favors a scalar form for the confining potential $V(r)$, for which the effective mass of a Dirac particle with mass M in the free state is reduced to the value $\langle M - V \rangle$ in the bound state and it is this reduced value that governs all observable properties such as magnetic moment, spin dependence of hadron masses, and decay processes (e.g. 65). It would be incorrect, therefore, to assume that the bound-state masses offer any indication of the likely free-quark masses, which could be many orders of magnitude larger. Thus the value of ion beam searches is limited by the fact that they necessarily involve a final detection stage that has a mass cutoff (typically below 10^3 GeV) and thus leaves open the possibility that the source material may still contain free quarks of higher mass. In practice we have seen that almost all the materials tested by ion beam techniques have now also been tested, at a similar or lower concentration level, by levitation methods. It is essential for any future searches in natural materials to be based on direct charge measurement, to ensure that very high masses are not excluded. Searches for particles produced at accelerators are, of course, an exception to this since the masses produced are already limited by the collision energy.

It is apparent from the preceding discussion that, with the accumulated understanding of the technical problems of fractional charge measurement, some larger-scale experiments could now be designed to search various natural terrestrial materials (in particular sea water, rocks, and minerals) at lower concentration levels than has hitherto been attempted. Nevertheless, such improved experiments now seem very unlikely to be carried out in view of the current declining level of scientific interest in this type of experiment.[2]

There are several basic reasons for this loss of motivation. The first is the common belief that the numerous null results of previous experiments have reduced the case for continuing terrestrial searches. This, as we have seen, is a mistaken viewpoint since no useful limits have yet been placed on natural cosmic-ray-produced levels. The second reason is the continuing, widespread theoretical view that the quarks are permanently confined by an indefinitely rising potential [due to the QCD interquark force forming a linear tube of flux as the separation increases (e.g. 67)] or by a confining pressure within a spherical "bag" (e.g. 68). This has not yet been proved to be a consequence of QCD or any other principle and is therefore still in the category of theoretical prejudice. However, strong support for such a viewpoint arises from the fact that analogous situations are now

[2] This is well illustrated by the 1987 Moriond Workshop on New and Exotic Phenomena. During this 6-day meeting over 100 papers were presented on all aspects of nonaccelerator particle physics experiments, but only one of these was on fractional charge searches (66).

recognized to exist in solids, where discontinuities in a periodic structure can produce mobile quasi-particles with abnormal quantum numbers (including fractional charge) that are automatically "confined" to the molecular structure and must, of course, be formed only in combinations of integral total charge (6, 7). Moreover, examination of the basic physics by which the displaced electron cloud in a molecule creates a local region of fractional charge shows that this apparently unique characteristic of quarks can occur quite trivially in any periodic array of positive and negative charges (provided the momentum transfer is insufficient to excite the individual members of the array, so that the interaction amplitudes add coherently).

It is thus clear that vacuum excitation or polarization could in principle similarly permit an integral total charge to separate spatially into two or more fractionally charged regions within a given volume. There is as yet no theory suggesting that a family of quarks could be represented in this way. Nevertheless, with the knowledge that there exists at least one simple mechanism by which confined fractional charges could arise (and indeed do arise for known examples in solid-state physics), this originally unique feature of the quarks is now seen to be less distinctive than it first appeared, and this must reduce the incentive for continuing searches for free fractional charge. Thus, although some work will continue to complete terrestrial searches still in progress, very little new activity in this field is expected in the foreseeable future. Nevertheless, free quarks will continue to be sought in accelerator and cosmic-ray experiments, and the question will remain unresolved until a theoretical explanation of the nature of the quarks and lepton families demonstrates convincingly that fractional charge can only exist in the confined state.

ACKNOWLEDGMENTS

I acknowledge numerous discussions over a period of many years with my experimental collaborators, in particular G. J. Homer, J. D. Lewin, W. G. Jones, L. Lyons, and H. E. Walford. Discussions and correspondence are also acknowledged with G. Morpurgo, W. M. Fairbank, J. D. Phillips, C. D. Hendricks, J. C. Price, W. Innes, R. Bland, and J. Van Polen.

Literature Cited

1. Gell-Mann, M., *Phys. Lett.* 8: 214–15 (1964)
2. Zweig, G., *CERN 8182/TH.401*; *CERN 8419/TH.412* (1964)
3. Crease, R. P., Mann, C. C., *The Second Creation.* New York: MacMillan (1984), pp. 280–308
4. Levin, E. M., Frankfurt, L. L., *Sov. Phys. Usp.* 11: 106–29 (1968); Zel'dovitch, Y. B., *Sov. Phys. Usp.* 8: 489–95 (1965); Kokkedee, J. J. J., *The Quark Model.* New York: Benjamin. 239 pp. (1969)
5. Bacry, H., Nuyts, J., Van Hove, L.,

Phys. Lett. 9: 279–80 (1964); Han, M. Y., Nambu, Y., *Phys. Rev.* B139: 1006–10 (1965)

6. Schreiffer, J. R., In *The Lesson of Quantum Theory*, ed. J. de Boer. Amsterdam: Elsevier (1986), pp. 59–78; Krive, I. V., Rozhavskii, A. S., *Sov. Phys. Usp.* 30: 370–92 (1987)

7. Su, W. P., Schrieffer, J. R., Heeger, A. J., *Phys. Rev.* B22: 2099–2111 (1980); Su, W. P., Schrieffer, J. R., *Phys. Rev. Lett.* 46: 738–41 (1981); Bak, P., *Phys. Rev. Lett.* 48: 692–94 (1982)

8. Perkins, D. H., *Rep. Prog. Phys.* 40: 409–81 (1977); Sloan, T., Smadja, G., Voss, R., *Phys. Rep.* 162: 45–167 (1988)

9. Hendry, A. W., Lichtenberg, D. B., *Rep. Prog. Phys.* 41: 1708–80 (1978)

10. Bykov, A. A., Dremin, I. M., Leonidov, A. V., *Sov. Phys. Usp.* 27: 321–37 (1984); Kwong, W., Rosner, J. L., Quigg, C., *Annu. Rev. Nucl. Part. Sci.* 37: 325–50 (1987)

11. Thirring, W., *Acta Phys. Austr. Suppl.* II: 205–11 (1965); Gal, A., Scheck, F., *Nucl. Phys.* B2: 110–20 (1967); Rubenstein, H. R., *Phys. Rev. Lett.* 17: 41–45 (1966)

12. De Rujula, A., Georgi, H., Glashow, S. L., *Phys. Rev.* D12: 147–62 (1975)

13. Marshall, R., *Z. Phys. C.* In press (1989); *Preprint RAL-88-049* (1988)

14. Bjorken, J. D., see Ref. 6, pp. 155–65

15. Kim, Y. S., *Contemp. Phys.* 14: 289–318 (1973)

16. Landsberg, L. G., *Sov. Phys. Usp.* 16: 251–74 (1973)

17. Jones, L. W., *Rev. Mod. Phys.* 49: 717–52 (1977)

18. Lyons, L., *Prog. Part. Nucl. Phys.* 7: 157–67 (1981)

19. Marinelli, M., Morpurgo, G., *Phys. Rep.* 85: 162–258 (1982)

20. Lyons, L., *Phys. Rep.* 129: 225–84 (1985)

21. Lackner, K. S., Zweig, G., *Lett. Nuovo Cimento* 33: 65–73 (1982); *Proc. AIP* 93: 1–14 (1982); *Phys. Rev.* D28: 1671–91 (1983)

22. Chupka, W. A., Schiffer, J. P., Stevens, C. M., *Phys. Rev. Lett.* 17: 60–65 (1966)

23. Milner, R. G., Cooper, B. H., Chang, K. H., Wilson, K., Labrenz, J., McKeown, R. D., *Phys. Rev.* D36: 37–43 (1987)

24. Milner, R. G., Cooper, B. H., Chang, K. H., Wilson, K., Labrenz, J., McKeown, R. D., *Phys. Rev. Lett.* 54: 1472–74 (1985)

25. Schiffer, J. P., Renner, T. R., Gemmell, D. S., Mooring, P. P., *Phys. Rev.* D17: 2241–44 (1978)

26. Stevens, C. N., Schiffer, J. P., Chupka, W. A., *Phys. Rev.* D14: 716–25 (1976)

27. Kutschera, W., et al., *Phys. Rev.* D29: 791–803 (1984)

28. Cook, D. D., DePasquall, G., Frauenfelder, H., Peacock, R. N., Steinrisser, F., Wattenberg, A., *Phys. Rev.* D188: 2092–97 (1969)

29. Putt, G. D., Yock, P. C. M., *Phys. Rev.* D17: 1466–67 (1978)

30. Bland, R., Bocobo, D., Eubank, M., Royer, J., *Phys. Rev. Lett.* 39: 369–70 (1977)

31. Rank, D. M., *Phys. Rev.* 176: 1635–43 (1968)

32. Morpurgo, G., Gallinaro, G., Palmieri, G., *Nucl. Instrum. Methods* 79: 95–124 (1970)

33. Joyce, D. C., Abrams, P. C., Bland, R. W., Johnson, R. T., Lindgren, M. A., et al., *Phys. Rev. Lett.* 51: 731–34 (1983)

34. Hodges, D. C., Abrams, P., Baden, A. R., Bland, R. W., Joyce, D. C., et al., *Phys. Rev. Lett.* 47: 1651–53 (1981)

35. Savage, M. L., et al., *Phys. Lett.* B167: 481 (1986)

36. Smith, P. F., Homer, G. J., Lewin, J. D., Walford, H. E., Jones, W. G., *Phys. Lett.* B197: 447–51 (1987)

37. LaRue, G. S., Fairbank, W. M., Hebard, A. F., *Phys. Rev. Lett.* 38: 1011–14 (1977)

38. LaRue, G. S., Fairbank, W. M., Phillips, J. D., *Phys. Rev. Lett.* 42: 142–45 (1979)

39. LaRue, G. S., Phillips, J. D., Fairbank, W. M., *Phys. Rev. Lett.* 46: 967–70 (1981)

40. Mitsuhashi, Y., Goto, E., Kuroda, R., *J. Phys. Soc. Jpn.* 40: 613–20 (1976)

41. Smith, P. F., Homer, G. J., Lewin, J. D., Walford, H. E., Jones, W. G., *Phys. Lett.* B153: 188–94 (1985); B171: 129–34 (1986)

42. Liebowitz, D., Binder, M., Ziock, K. O. H., *Phys. Rev. Lett.* 50: 1640–43 (1983)

43. Lyons, L., Smith, P. F., Homer, G. J., Lewin, J. D., Walford, H. E., Jones, W. G., *Z. Phys.* C36: 363–67 (1987)

44. Smith, P. F., Homer, G. J., Lewin, J. D., Walford, H. E., Jones, W. G., *Phys. Lett.* B181: 407–13 (1986)

45. Lewin, J. D., Smith, P. F., *Phys. Rev.* D32: 1177–85 (1985)

46. Hebard, A. F., *Search for fractional charge using low temperature techniques*, Thesis, Stanford Univ., Calif. (1970)

47. LaRue, G. S., *Measurement of residual charge of superconducting niobium spheres*, Thesis, Stanford Univ., Calif. (1978)

48. Phillips, J. D., *Residual charge of niobium spheres*, Thesis, Stanford Univ., Calif. (1983)

49. Fairbank, W. M., Phillips, J. D., in *Inner*

Space/Outer Space, ed. E. W. Kolb (1986), pp. 563–67

50. Phillips, J. D., Fairbank, W. M., Navarro, J., *Nucl. Instrum. Methods* A264: 125–30 (1988)
51. McCusker, C. B. A., *Aust. J. Phys.* 36: 717–23 (1983)
52. Nir, A., *Phys. Rev. Lett.* 19: 336–39 (1967)
53. Isgur, N., Wolfram, S., *Phys. Rev.* D19: 234–38 (1979)
54. Rich, J., Lloyd-Owen, D., Spiro, M., *Phys. Rep.* 151: 240–364 (1987)
55. Innes, W., Kelin, S., Perl, M., Price, J. C., *Stanford Linear Accelerator Center preprint SLAC-PUB-2938*, 11 pp. (1982)
56. Price, J. C., *Stanford Linear Accelerator Rep. SLAC-288*, 199 pp. (1985)
57. Williams, E. R., Gillies, G. T., *Lett. Nuovo Cimento* 37: 520–24 (1983)
58. Bleaney, B. I., Bleaney, B., *Electricity and Magnetism*, Oxford Univ. Press. 2nd ed. (1965), pp. 451–54
59. Hirsch, G., Hagstrom, R., Hendricks, C., *Lawrence Berkeley Lab. Rep. LBL-9350*, 21 pp. (1979)
60. Van Polen, J., Hagstrom, R. T., Hirsch, G., *Phys. Rev.* D36: 1983–89 (1987); for additional detail see *Argonne Natl. Lab. Rep. ANL-HEP-PR-86-90*, 30 pp. (1986)
61. Hendricks, C. D., Kim, K., *IEEE Trans. Ind. Appl.* IA-21: 705–8 (1985); Hendricks, C. D., Hornady, R. S., *Proc. Int. Conf. Modern Electrostatics.* Beijing, China: IAP (1988), pp. 335–38
62. Shaw, G. L., Slansky, R., *Phys. Rev. Lett.* 50: 1967–70 (1983)
63. Lindgren, M. A., et al., *Phys. Rev. Lett.* 51: 1621–24 (1983)
64. Matis, H. S., et al., *Proc. 23rd Conf. High Energy Phys.* Berkeley (1986); Preprint *LBL-21670*, 9 pp. (1986)
65. Lipkin, H. J., Tavkhelidze, A., *Phys. Lett.* 17: 331–32 (1965); Smith, P. F., Lewin, J. D., *Nuovo Cimento* A64: 421–52 (1981)
66. Smith, P. F., in *Searches for New and Exotic Phenomena*, Proc. 7th Moriond Workshop, Les Arcs: Ed. Frontieres (1987), pp. 527–31
67. Greenberg, O. W., *Annu. Rev. Nucl. Part. Sci.* 28: 327–86 (1978); Fishbane, P. M., Kaus, P., Meshkov, S., *Phys. Rev.* D33: 852–60 (1986)
68. Close, F. E., *An Introduction to Quarks and Partons*, New York: Academic (1979), pp. 410–28

Annu. Rev. Nucl. Part. Sci. 1989. 39: 113–50

NUCLEAR SYSTEMS WITH STRANGENESS

Robert E. Chrien and Carl B. Dover

Department of Physics, Brookhaven National Laboratory, Upton, New York 11973

KEY WORDS: hypernuclei, hyperons, strangeness exchange, associated production, dibaryons, weak decay, strange particles.

CONTENTS

1. INTRODUCTION

"Even the deepest thinkers, the purest philosophers . . . were emotionally divided in their attitude toward the possibility that there existed 'a distortive glass of our distorted glebe' . . . a rainbow mist of angelic spirits, inhabitants of sweet Terra."—Vladimir Nabokov, in *Ada, or Ardor: A Family Chronicle.*

In Vladimir Nabokov's famous novel *Ada*, there is drawn a picture of a shadow world, parallel to our own real one but distorted from it in many

113

0163–8998/89/1201–0113$02.00

small and amusing details. The inhabitants of this world of Anti-Terra dream of the real world, which, to them, is a sweet and wondrous land. Nabokov uses this contrasting fantasy land of Anti-Terra to draw some pointed observations about the quality of human relationships.

In somewhat the same way we can imagine a nuclear domain in which one of the nucleons is replaced by a hyperon with distinctive properties. A subtle distortion becomes apparent; energy levels and their ordering change, new symmetries appear and old ones vanish. These distortions are not imaginary, however; they can be produced by experimentally transforming one of the nucleons, as for example in a strangeness-exchanging (K^-, π^-) reaction. From this transformation one can deduce features of elementary hadronic interactions that are forever obscured in ordinary nuclei. These properties may be derived from a study of hypernuclei and their properties.

The goal of this article is to review the experimental and theoretical status of strange-particle nuclear physics, with emphasis on the production mechanisms, spectroscopy, and decay modes of hypernuclear states. A hypernucleus consists of one or more hyperons bound to a nuclear core. In the SU(3) classification, the Λ, Σ, and Ξ hyperons occupy the same octet representation of spin-parity $1/2^+$ baryons as the neutron and proton, the familiar constituents of ordinary nuclei. The Λ and Σ^0 possess strangeness $S = -1$ and isotopic spin $I = 0$, 1, respectively, while the Ξ has $S = -2$, $I = 1/2$. In the underlying quark picture, the Λ and Σ have the flavor structure $s(ud)_{I=0,1}$, compared to the combination uud for the proton. The strange quark, s, carries the strangeness quantum number $S = -1$, which makes it distinguishable from the $S = 0$ u and d quarks. The lowest-lying hyperon is the Λ, with a mass of 1115.6 MeV/c^2, some 177 MeV/c^2 heavier than the proton. The study of the behavior of a hyperon embedded in the nuclear medium, through the theoretical analysis of level spectra, sheds light on the nature of the hyperon-nucleon (YN) effective interaction, i.e. on the role of the strange quark in strong interactions.

A single Λ behaves essentially as a distinguishable particle in the nucleus: there are no discontinuities in the binding energy B_Λ as a function of A, a signal of shell effects in ordinary nuclei. This property of the Λ implies additional dynamical symmetries for hypernuclear states that are not allowed for ordinary nuclei because of the Pauli principle. The Λ provides a superb example of single-particle structure in a many-body system. Deeply bound nucleon-hole states are very broad, a reflection of the large spreading width $\Gamma_\downarrow$ caused by admixtures with more complicated configurations. For the Λ, even for s-states in heavy systems, $\Gamma_\downarrow$ is rather small, which indicates the rather weak ΛN residual interaction. The existence of a well-defined set of single-particle states of different orbital angular momentum ℓ in a given hypernucleus enables one to extract information on the well depth, geometrical shape, and effective mass, all of which characterize the Λ-nucleus potential.

The decay modes of hypernuclear states provide additional information of a fundamental nature. The ground state of a Λ hypernucleus decays by weak interactions, either via the $\Lambda \to N\pi$ process as in free space, or by the nucleon-induced nonmesonic reaction $\Lambda N \to NN$. The latter, which is the $\Delta S = 1$ strangeness-changing analogue of the $\Delta S = 0$ parity-violating $NN \to NN$ reaction, is uniquely accessible from the hypernucleus. The γ decays of excited states, which can be measured with good energy resolution, yield constraints on the spin-dependent components of the ΔN interaction.

This review does not attempt to cover the earlier history of hypernuclear research carried on by emulsion and bubble chamber techniques. An excellent summary of the early work has been provided by Davis & Sacton (1). Povh's reviews (2) cover the more recent work up to 1981. Other review articles that merit attention are those of Davis & Pniewski (3) and Gal (4). The reader may also consult several conference proceedings (5–10). The present discussion is meant to cover the more recent research in the field and the outlook for future developments.

2. HYPERNUCLEAR PRODUCTION AND RESEARCH

Modern experiments on hypernuclei take advantage of the secondary meson beams available at proton synchrotrons operating above 10 GeV/c. At the present time, hypernuclear experiments are done with the separated kaon beams at the Brookhaven National Laboratory (BNL) Alternating Gradient Synchrotron (AGS) and the Japanese National Laboratory for High Energy Physics (KEK).

2.1 Production Mechanisms for Hypernuclei

The principal production mechanisms for hypernuclei are strangeness exchange and associated production, corresponding to the elementary processes

(a) $\mathrm{K}^- + \mathrm{n} \to \Lambda + \pi^-$

$\mathrm{K}^- + \mathrm{p} \to \Lambda + \pi^0$

$\mathrm{K}^- + \mathrm{n} \to \Sigma^{0,-} + \pi^{-,0}$

$\mathrm{K}^- + \mathrm{p} \to \Sigma^{\pm} + \pi^{\mp}, \Sigma^0 + \pi^0$

(b) $\pi^+ + n \rightarrow \Lambda + K^+$

$\pi^+ + n \rightarrow \Sigma^{0,+} + K^{+,0}$

$\pi^+ + p \rightarrow \Sigma^+ + K^+.$

Of these reactions, only those involving charged mesons in the exit channel have been studied extensively. In the case of nuclear species, the n,p above are replaced with the corresponding target nuclide. A special case of (a) above is the set of experiments in which the kaon is captured at rest ($p_K = 0$) from a K^- atomic orbit. Rates of Λ- and Σ-hypernuclear formation have been calculated by Gal & Klieb (11) and Morimatsu & Yazaki (12) and are typically a few percent for Σ and a few tenths of a percent for Λ production.

The characteristic differences between the $(\bar{K},\pi)$ and (π,K) production modes have been discussed by Dover, Ludeking & Walker (13) and by Bandō & Motoba (14), among others. The $(\bar{K},\pi)$ reaction at small angles is a process with small momentum transfer q. It preferentially populates "substitutional" hypernuclear states, i.e. those states in which a nucleon in a shell model orbit with orbital angular momentum l and total spin j is replaced by a Λ particle in the same orbit (l,j). This transition is characterized by an orbital angular momentum transfer of $\Delta L = 0$. For the $(\bar{K},\pi)$ reaction, there exists a "magic momentum," i.e. a value of laboratory kaon momentum p_K for which q vanishes at $\theta = 0°$. This corresponds to $p_K = 530$ MeV/c for Λ and 290 MeV/c for $\Sigma^°$ production. Note that q at $\theta = 0°$ remains less than 100 MeV/c for $p_K \leq 800$ MeV/c. At or near the magic momentum, the substitutional transitions with $\Delta L = 0$ are particularly enhanced with respect to those for which $\Delta L \neq 0$. The Λ hypernuclear states populated in $\Delta L = 0$ transitions tend to occur near, or above, zero Λ binding. The (K^-,π^-) process involves strong absorption of both the incoming and outgoing waves and consequently is localized in the nuclear periphery.

For the associated production (π^+,K^+) reaction, there is no magic momentum and $q \geq 350$ MeV/c at the elementary cross-section maximum near $p_K \approx 1050$ MeV/c. The formation of high spin states is favored by the form factor of the (π^+,K^+) reaction (13). Furthermore, because of the long mean free path in nuclear matter for the K^+, the distortion for the outgoing wave is reduced; the reaction is less peripheral in nature. Figure 1 nicely illustrates this comparison between (K^-,π^-) and (π^+,K^+) strength distributions. One sees the complementarity of these two reactions in exciting the low and high spin parts, respectively, of the hypernuclear spectrum.

The stopped kaon reaction is characterized by sizable q values (254 MeV/c for Λ, ~ 180 MeV/c for Σ's for elementary capture processes),

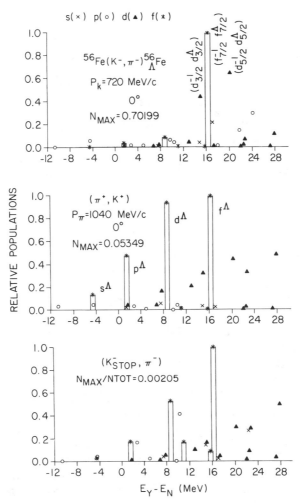

Figure 1   Relative strengths for the population of states in $^{56}_{\Lambda}$Fe by (K^-,π^-), (π^+,K^+), and stopped K^- reactions. The neutron-hole states, s, p, d, and f are designated by symbols as shown. The selectivity of the in-flight (K^-,π^-) reaction for low spin substitutional states and the tendency for the (π^+,K^+) reaction to populate a series of high spin states $(f_{7/2}^{-1} \otimes j_\Lambda)$ are evident. The figure has been adapted from Bandō & Motoba (14); the $f_{7/2}^{-1}$ series is highlighted.

intermediate between the in-flight reactions (*a*) and (*b*). The (K^-,π^-) strength at rest is broadly distributed, with no particular selectivity, as Figure 1 indicates. Cross sections $d\sigma/d\Omega$ for hypernuclear excitation are conventionally given in terms of the effective neutron number N_{eff} via the relation

$$\frac{d\sigma}{d\Omega} = \left(\frac{d\sigma}{d\Omega}\right)_{2\text{-body}} N_{\text{eff}},$$

where $(d\sigma/d\Omega)_{2\text{-body}}$ is the $K^-n \to \pi^-\Lambda$ or $\pi^+n \to K^+\Lambda$ differential cross section at zero degrees. In Figure 1, N_{MAX} represents the strength of the largest transition.

2.2 Facilities for Hypernuclear Research

Kaon-induced reaction studies of hypernuclei require secondary particle beams enriched in K mesons, derived from a primary proton beam impacting a heavy metal target.

All available separated beams have similar optical characteristics (15). A dipole and quadrupole doublet are provided to disperse the beam in momentum and focus it at the separator section that follows. The separator section consists of crossed electric and magnetic fields, forming a Wien Filter, so that, while the magnetic displacement depends only on momentum, the displacement due to the electric field depends on the length of time spent by the particle within the field volume, and therefore on the velocity. Following the separator, a second quadrupole doublet brings the beam to a dispersed double focus at the "mass slit." The mass slit is so named because it selects particles of different velocity, hence differing mass, at the same momentum.

Separated kaon lines have been used for kaon and hypernuclear studies at the CERN PS, the BNL AGS, and KEK in Japan. At the AGS, the in-flight kaon hypernuclear measurements use the Moby Dick Spectrometer, a QQDQQ (Q = quadrupole, D = dipole) configuration following the Low-Energy Separated Beam I (LESB-I) beam line. For such experiments, the QQDQ configuration after the mass slit of the separator is used to analyze the incoming particle, while Moby Dick analyzes the outgoing particle. The Moby Dick spectrometer is mounted on a rotatable platform, so scattering angles of $-5° < \theta < 45°$ can be accommodated. Moby Dick is used for both $(\bar{K},\pi)$ and (π,K) reactions; for the latter, the roles of the entrance and exit spectrometer are interchanged.

At KEK, the "SKY" spectrometer (16, 17) is a compact, low momentum spectrometer used for stopped kaon hypernuclear spectroscopy at the K3 (600-MeV/c K$^-$) beam line. The device, formerly used in a heavy neutrino search from $K_{\mu2}$ decay, has been extensively used recently for Σ^- hypernuclear studies and for studies of pion emission and γ rays from hyperfragments. The "PIK" spectrometer has been designed for associated production studies of Λ hypernuclei through the (π^+,K^+) reaction (16). The spectrometer is located in the K2 (1–2-GeV/c K$^-$) beam line. At this time, the LESB-I at the AGS and the K2 and K3 lines at KEK are the only ones used for hypernuclear studies. The parameters for these spectrometers and beam lines are shown in Table 1.

All such beams suffer from the compromises necessary to keep the lines short for kaon survival (e.g. only 8% of the production target kaons survive to the reaction target of LESB-I at 800 MeV/c, a distance of 15 meters). In order to keep the line short, the portion of LESB-I following the mass slit is used to analyze the incoming particles. For this reason, position-sensitive detectors, wire chambers, and hodoscopes are placed at the mass slit. Figure 2 shows the layout of magnetic elements and particle detectors at the Moby Dick spectrometer at the LESB-I.

Kaon beam purities from typical separators range from 30% for K^+ particles to about 10% for K^- particles, for the configurations shown. The balance of the pion contamination is identified and removed by time-of-flight and Čerenkov counters. For LESB-I, the time-of-flight difference over the six meters from the mass slit to the reaction targets between pions and kaons is 3.2 ns (at 800 MeV/c), which allows effective discrimination with timing scintillators with $\sigma \approx 0.2$ ns. With lucite Čerenkov counters the K's and π's can readily be discriminated in this momentum range. A major experimental difficulty in the magnetic spectroscopy of kaons is presented by the kaon decay probability. The mean length for kaon decay, $l = \gamma\beta c\tau$ is 7.513pc meters (for pc in GeV). The effects of kaon decay background can be minimized by either or both of two methods:

Table 1   Characteristics of hypernuclear facilities (16, 17)

	Moby Dick/ LESB-I AGS	SKY/K3 KEK	PIK/K2 KEK
P_{max} (GeV/c)			
Beam	1.05	1.0	2.2
Spectrometer	1.05	0.3	1.0
$\Delta p/p$ beam (%)	± 2	± 2.5	± 3
Beam separation	10 (π^-/K^-) at 800 MeV/c		
Intensities per 10^{12} protons	4.5×10^4 (K^- on spectrometer target) at 800 MeV/c	1.5×10^3 (stopped K^- at 650 MeV/c)	2×10^7 (π^+ at 2 GeV/c)
Spectrometer momentum bite	13%	100–300 MeV/c	650–1000 MeV/c
Spectrometer solid angle	18 msr	100 msr	$\pm 4°$ θ acceptance $\pm 2.5°$ ϕ acceptance

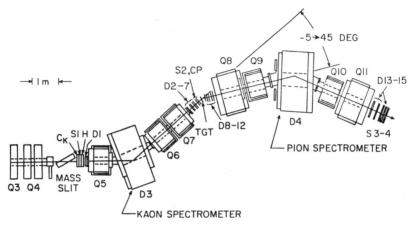

Figure 2   The Moby Dick spectrometer at the Brookhaven AGS beam line LESB-I, configured for ($\bar{K},\pi$) reactions. S = scintillator, D = drift chamber or dipole, Q = quadrupole, C = Čerenkov counter, H = hodoscope. The target (TGT) can be seen in the center.

1. the use of a veto Čerenkov counter near the production target, which eliminates pions from the $K^- \to \pi^-\pi^0$ decay branch, and
2. the use of phase space cuts in the data that take advantage of the differing distributions in phase space between the $K^- \to \pi^-\pi^0$ decay, the $K^- \to \mu^-\nu$ decay, and the pions arising from the (K,π) reaction to be studied. The $K \to \mu$ decays tend to be concentrated at large angles to the incident beam and are not so bothersome as the $K \to \pi$ decays, which peak much closer to the beam forward direction.

For (π^+,K^+) reaction studies, the restrictions imposed by the use of beam separators and decay backgrounds are not so severe. The principal problem in such studies is the relatively smaller cross sections for (π^+,K^+) as compared to the (K^-,π^-) production of substitutional states. To compensate for the lower cross section, incident beam intensities of up to 10^7 pions per second have been used in the Moby Dick configuration. The problem then reduces to the difficulties of handling particle intensities in that range. The rather significant contamination of pion beams with a positron component presents further difficulties with such beams, as it becomes difficult to separate positrons and pions at momenta of 1 GeV/c and above.

2.3   *Recent Experiments on Hypernuclei*

2.3.1 $^{12}_\Lambda$C AND $^{13}_\Lambda$C STUDIES AT MOBY DICK   For ($\bar{K},\pi$)-induced substitutional transitions, one has zero orbital angular momentum transfer, i.e. $\Delta L = 0$, dominating for $\theta = 0°$. At nonzero angles, higher transfers become important. A simple example of the usefulness of Moby Dick in studying the angular distributions, and therefore separating differing reaction components, is afforded by $^{12}_{\Lambda}$C (18, 19). Here, the (K^-, π^-) spectrum displays two major peaks, one corresponding to the ground state, and the second to the p-shell Λ, at 11 MeV of excitation energy. At $\theta = 0°$ the ground state, which has the neutron hole–Λ particle configuration $(p_{3/2}^{-1} \otimes s_{1/2})$ and therefore requires $\Delta L = 1$, is weakly populated. At larger angles, the differential cross section rises as expected. The "11-MeV" peak is more complicated, and different components come into play at various angles. Near $0°$, where $\Delta L = 0$, the substitutional configuration $(p_{3/2}^{-1} \otimes p_{3/2})0^+$ is populated, but at nonzero angles, the $(p_{3/2}^{-1} \otimes p_{1/2,3/2})2^+$ states are excited with $\Delta L = 2$. The angular distribution shows clearly the effect of the 2^+ contributions, which are relatively important above 11 degrees.

A more complex analysis is involved in a companion experiment undertaken on $^{13}_{\Lambda}$C, where a combination of s-shell Λ and p-shell Λ coupled to the spectrum of excited states of the ^{12}C core is observed. This is discussed in detail in Section 3.2.1.

2.3.2   THE DETECTION OF HYPERNUCLEAR GAMMA RAYS   The utility of identifying radiative transitions connecting hypernuclear energy levels lies in the potentially superior resolution available from γ-ray spectrometers as compared to magnetic spectrometers. This resolving power is particularly important in view of the small spin dependences displayed by the Λ-nuclear interaction; splittings of Λ spin multiplets are usually too small to be directly observable.

Gamma rays have been detected both from stopped kaon capture and from the $(\bar{K}, \pi)$ in-flight reaction. The latter combines magnetic definition of the region of hypernuclear excitation (with relatively low resolution) with a high resolution determination of the energy differences of the initial and final states of the radiative transition. The $(\bar{K}, \pi\gamma)$ in-beam γ spectroscopy must be carried out under extremely adverse conditions: the relatively low beam intensities available, and the low beam quality of those beams, with their 90% contamination of pions.

From stopped kaon experiments, γ rays are observed in coincidence with the reaction pions. From lithium targets a large number of $A = 4$ hyperfragments can be produced. The $^4_{\Lambda}$He and $^4_{\Lambda}$H hyperfragments can be distinguished by their different decay modes:

$$^4_{\Lambda}\text{H}^* \rightarrow {}^4_{\Lambda}\text{H} + \gamma; \quad {}^4_{\Lambda}\text{H} \rightarrow {}^4\text{He} + \pi^- \quad (E_{\pi^-} = 53 \text{ MeV})$$

$$^4_{\Lambda}\text{He}^* \rightarrow {}^4_{\Lambda}\text{He} + \gamma; \quad {}^4_{\Lambda}\text{He} \rightarrow {}^4\text{He} + \pi^0 \quad (E_{\pi^0} = 57 \text{ MeV}).$$

The first three entries of Table 2 list the energy measurements (20–22) with stopped kaons on lithium targets. These measurements established the $1^+ \to 0^+$ transitions of $^4_\Lambda$H, 1.07 MeV and of $^4_\Lambda$He, 1.15 MeV. These M1 spin-flip γ rays provide important information relating to the spin dependence of ΛN forces. The average 1.1-MeV spacing between the 1^+ excited state and the 0^+ ground state of the $A = 4$ system is naively attributed to the spin-spin ΛN interaction, but an important role can be played by $\Lambda N \to \Sigma N \to \Lambda N$ couplings (24) or by the spin-dependent part of the $\Lambda N N$ force (25). The mirror hypernuclei $^4_\Lambda$H and $^4_\Lambda$He are the main source of information on the ΛN charge-symmetry-breaking interaction. We refer the reader to the papers of Gibson & Lehman (26) and Bodmer & Usmani (27).

The last three entries of Table 2 show the γ rays observed in a $(\bar{K}, \pi\gamma)$ coincidence experiment at the AGS (23) carried out with an array of 7.6×12.7-cm NaI detectors. Figure 3 shows the experimental spectra associated with different regions of excitation energy in $^9_\Lambda$Be.

In a recent high resolution germanium detector experiment (28, 29), an attempt was made to detect the M1 spin-flip transition between the members of the ground-state doublets in $^{10}_\Lambda$B and $^{16}_\Lambda$O. The fact that these transitions were not seen establishes an upper limit of 100 keV for the $^{10}_\Lambda$B and $^{16}_\Lambda$O doublet energy separations (29).

The analysis of hypernuclear spin splittings has been carried out by Gal et al (30, 31) and by Millener et al (32). They express the ΛN interaction in terms of the five radial integrals V, Δ, S_Λ, S_N, and T, assumed to be constant across the p-shell, and associated with the central, spin-spin, Λ-spin-orbit, induced-nucleon spin-orbit, and tensor terms, respectively, in the potential

$$V_{\Lambda N}(r) = V_0(r) + V_\sigma(r)\mathbf{s}_N \cdot \mathbf{s}_\Lambda + V_\Lambda(r)\mathbf{l}_{N\Lambda} \cdot \mathbf{s}_\Lambda$$

$$+ V_N(r)\mathbf{l}_{N\Lambda} \cdot \mathbf{s}_N + V_T(r)S_{12}, \quad 1.$$

Table 2   Energies of observed hypernuclear gamma rays

Target	E_γ (MeV)	Identification	Reference
^{6}Li, ^{7}Li	1.09	$1^+ \to 0^+$; $^4_\Lambda$H; $^4_\Lambda$He	20
^{7}Li	1.09 ± 0.03	$1^+ \to 0^+$; $^4_\Lambda$H	21
^{6}Li, ^{7}Li	1.04 ± 0.04	$1^+ \to 0^+$; $^4_\Lambda$H	22
	1.15 ± 0.04	$1^+ \to 0^+$; $^4_\Lambda$He	22
^{7}Li	1.108 ± 0.01	$1^+ \to 0^+$; $^4_\Lambda$H; $^4_\Lambda$He	23
^{7}Li	2.034 ± 0.023	$5/2^+ \to 1/2^+$; $^7_\Lambda$Li	23
^{9}Be	3.079 ± 0.04	$5/2^+, 3/2^+ \to 1/2^+$; $^9_\Lambda$Be	23

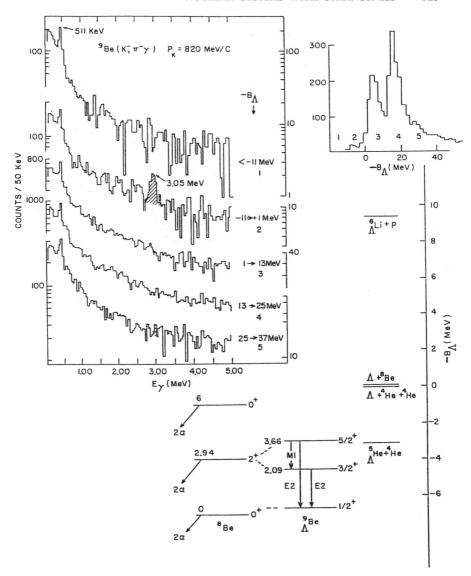

Figure 3   A set of γ-ray spectra from the ⁹Be(K⁻,π⁻γ) reaction (23) is shown at the upper left. Each spectrum is labeled by a region of binding energy (1–5) in ᵪBe, indicated at the upper right. The hypernuclear γ ray seen in region 2 is shaded; its energy of 3.05 MeV is close to that of the 2⁺→0⁺ core transition in ⁸Be, as shown in the level diagram at the lower right, taken from Dalitz & Gal (31). This γ line corresponds to the deexcitation of the first excited (3/2⁺, 5/2⁺) doublet to the ᵪBe ground state. Failure to resolve the doublet places a strong constraint on the spin-orbit matrix element S_Λ.

where $S_{12} = 3(\boldsymbol{\sigma}_N \cdot \hat{r})(\boldsymbol{\sigma}_\Lambda \cdot \hat{r}) - \boldsymbol{\sigma}_N \cdot \boldsymbol{\sigma}_\Lambda$. There have been several attempts to fit the radial integrals to the data (30–32). Early work (30, 31), based mostly on ground-state binding energies, predicted sizable doublet splittings in a number of cases, for instance in $^9_\Lambda$Be (Figure 3). These are not observed. The more recent estimates of Millener et al (32), which are based on the realistic one-boson-exchange model of Nagels et al (33), foresee small doublet splittings of the order of 100 keV in the p-shell, except for $^7_\Lambda$Li where 600 keV is predicted. The data indicate an even weaker spin dependence: the observed transition energies in $^7_\Lambda$Li, $^8_\Lambda$Li, and $^9_\Lambda$Be are 2.034, 0.441, and 3.079 MeV, respectively, very close to the energy separations 2.18, 0.48, and 2.94 MeV of the core states in ^{6}Li, ^{7}Li, and ^{8}Be.

The "standard interaction" proposed by Millener et al (32) has $\Delta = 0.50$, $S_\Lambda = -0.04$, $S_N = -0.08$, and $T = 0.04$ (all in MeV). The value of S_Λ is consistent with the single-particle spin-orbit splitting $\varepsilon_p(\Lambda) = 0.36 \pm 0.3$ MeV obtained from the analysis of the ^{13}C$(K^-,\pi^-)^{13}_\Lambda$C experiment (34, 35), i.e. $\varepsilon_p(\Lambda) = -6S_\Lambda$, yielding $S_\Lambda = -0.06 \pm 0.05$ MeV. The observation of a single peak at 3.079 MeV in $^9_\Lambda$Be (Figure 3) yields the constraint $|S_\Lambda| \leq 0.04$, also consistent with the work of Millener et al (32). However, the $^{10}_\Lambda$B ground-state doublet separation is predicted to be 190 keV, which is clearly ruled out by the high resolution germanium experiment. In addition, the $1^-_2 \rightarrow 1^-_1$ transition in $^8_\Lambda$Li, which would most naturally correspond to the 441-keV line seen by Chrien et al (29), is predicted at 855 keV. However, the branching ratios for the γ decay of the 1^-_2 level in $^8_\Lambda$Li are a sensitive function of the mixing angle of the 1^-_1 and 1^-_2 levels, so this identification could be incorrect.

In summary, it appears that theoretical expectations for the ΛN spin dependence (32), based on a one-boson-exchange model (33), are not borne out by experiment. Several key measurements are still required: the $3/2^+ \rightarrow 1/2^+$ doublet transition in $^7_\Lambda$Li provides the best measure of Δ, while the $1^- \rightarrow 0^-$ transition in $^{16}_\Lambda$O is sensitive to the tensor strength T. The possibility remains that the uniform shell-model approach, with fixed parameters across the p-shell, is inappropriate. This could reflect the presence of significant spin-dependent three-body ΛNN forces, or state-dependent effects of ΣN coupling.

2.3.3   ASSOCIATED PRODUCTION WITH (π^+,K^+) The associated production of Λ hypernuclei by the (π^+,K^+) reaction was first studied theoretically by Dover et al (13) and demonstrated in experiments carried on at the Moby Dick spectrometer at the AGS (36–38).

The experimental arrangement for (π^+,K^+) differs in several respects from the corresponding (K^-,π^-) reactions. The elementary (π,K) cross section has a maximum near $p_\pi = 1050$ MeV/c, which is at the upper limit for the LESB-I line, and well above the 700–800 MeV/c usually chosen for the strangeness-exchanging (K^-,π^-) reaction. The roles of the kaon and pion spectrometers are interchanged, and the beam separator is tuned for pion transmission. Čerenkov counters are added after the target to veto pions in the rear spectrometer.

The cross sections obtained by Milner et al (36) for the 1^- ground-state and the $0^+,2^+,2^+$ excited-state multiplet in $^{12}_\Lambda$C near 11 MeV agree well with the DWBA calculated cross sections, after Fermi motion is folded in and realistic optical model parameters for the π^+ and K^+ distorted waves are used. It is interesting to observe that the lower cross sections characteristic of the (π^+,K^+) reaction on light hypernuclei, compared to those for (K^-,π^-), are more than compensated by the increased particle flux available for pions as compared to kaons.

For hypernuclei beyond the p-shell, $(\bar{K},\pi)$ reactions become less effective in populating bound states of hypernuclei because of the increasingly higher angular momentum of the valence-shell neutrons. The coupling of the Λ to the high spin neutron-hole produces an advantageous momentum matching for the (π^+,K^+) reaction, as discussed by a number of authors (13, 39–43). The high spin selectivity preferentially highlights a series of states in which the Λ-shell model orbitals (s, p, d, f, . . .) couple to the valence neutron-hole. The lack of spin dependence in the Λ-nuclear interaction produces a set of regularly spaced, narrow, single-particle excitations that dominate over other more complicated particle-hole excitations. This strikingly simple sequence of levels was illustrated by Bandō & Motoba (14) for $^{56}_\Lambda$Fe (see Figure 1). The cross sections can be calculated by a distorted-wave impulse approximation (DWIA) and are measurable, even for deeply lying orbitals of heavy hypernuclei. In the second (π^+,K^+) experiment (37, 38), eight targets were examined at $\theta = 10°$: ^{9}Be, ^{12}C, H$_2$O, Si, ^{13}C, ^{51}V, Ca, and ^{89}Y. For the latter four targets, beam particle intensities up to 10^7 pions per second were found usable with targets of reasonable size, namely 2–4 g/cm^2. The (π^+,K^+) spectra are observable over an irreducible background level of less than 100 nb/sr/MeV, which allows a clear observation of the s_Λ ground-state peak for $^{89}_\Lambda$Y, expected to be populated at a level of 0.5 μb/sr. Figure 4 shows the data obtained for $^{89}_\Lambda$Y.

These two experiments have demonstrated that the (π,K) reaction is the method of choice for producing all but the lightest hypernuclei, preferable to $(\bar{K},\pi)$ for directly accessing deeply lying hypernuclear states. The interpretation of the Λ single-particle spectra in terms of the Λ nucleus mean field is discussed in Section 3.1.

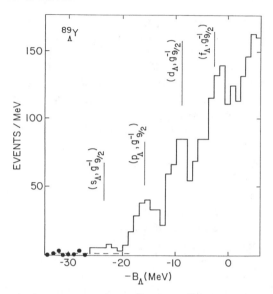

Figure 4 The excitation spectrum for the $^{89}Y(\pi^+,K^+)^{89}_\Lambda Y$ reaction at 1.05-GeV/c and $\theta_{K^+} = 10°$, from an AGS experiment (37). The predicted (13) Λ binding energies B_Λ for single-particle configurations $(g^{-1}_{9/2} \otimes \ell_\Lambda)$ are indicated. The black dots below the ground state indicate the measured background.

2.3.4 STOPPED KAON STUDIES A series of measurements with stopped kaons has been recently undertaken at KEK (44, 45). Efforts at identifying Λ- and Σ-hypernuclear states from reaction pions associated with kaon capture from atomic orbits suffer from an underlying pionic background associated with kaon reactions and kaon decay. In principle it is possible to enhance hypernuclear states by the use of decay tags and patterns of energy deposition; however, no convincing experimental enhancement of such states has yet been demonstrated (46, 47).

The production of Λ hyperfragments following K capture can, however, be uniquely identified by a measurement of the decay pion momentum. For example, Tamura et al (17) show a sizable pion peak attributable to the decay of $^4_\Lambda H$; $^4_\Lambda H \rightarrow {}^4He + \pi^-$ (133 MeV/c). They are able to deduce the formation probability per stopped K^- of this copiously produced hyperfragment from the spallation of $^7_\Lambda Li$, $^9_\Lambda Be$, $^{12}_\Lambda C$, and $^{16}_\Lambda O$.

2.4 Σ *Hypernuclei*

There is no reason to believe a priori that narrow Σ excitations can exist in nuclei because of the $\Sigma + N \rightarrow \Lambda + N$ strong conversion process. There was therefore considerable interest generated by the data of the Heidelberg-Saclay collaboration, who reported narrow peaks attributed to Σ states in

$(K^-,\pi^{\pm})$ reactions on ^{9}Be, ^{12}C, and ^{16}O at various incident momenta from 400 to 720 MeV/c (48–50). The interest was reinforced by the publication of π^0-tagged spectra for stopped kaons on ^{12}C, where narrow peaks were also claimed (44). Some of the data from the Heidelberg-Saclay collaboration at CERN are shown in Figure 5. There were also suggestions of rather broad enhancements reported from the Moby Dick spectrometer at the AGS at momenta near 720 MeV/c for (K^-,π^+) reactions on ^{6}Li and ^{16}O (51).

It is interesting to note that all of the proposed peaks mentioned above lie in the continuum, i.e. at or above the threshold for Σ binding. Chrien et al (52) have suggested an interpretation of broad structure in the continuum region in terms of quasi-free Σ production. Other authors, such as Wünsch & Žofka (53), Kohno et al (54), and Halderson & Philpott (55), have applied a continuum shell-model approach to studying the Σ structures, which can arise from threshold effects as well as genuine resonances, depending on the depth of the Σ nucleus potential. Enhancements near the thresholds of Σ + core − hole states accompanying $\Sigma \rightarrow \Lambda$ conversion are also seen by Tang et al in their recent (K^-,π^+) measurements on ^{12}C and ^{7}Li targets (56) and in the KEK π^0-tagged spectra (44). Iwasaki (47) suggested that such enhancements can be the result of the recapture of a slow Σ by a second nucleus. Such an effect would bias the shape of any tagged spectrum and complicate its theoretical interpretation.

A recent measurement suggesting the presence of a Σ-bound state resulting from the (K^-,π^-) reaction on ^{4}He has been reported by Hayano et al (57). A shoulder is observed clearly below the $\Sigma^{0,+}$ thresholds, as shown in Figure 6. It is not yet clear whether this enhancement, which was also seen in bubble chamber work (58), is attributable to a genuine bound state of $^4_\Sigma$He (59) or to final-state interactions, as suggested by Roosen et al (58).

The existence of an observable spectroscopy of single-particle Σ states depends critically on the density dependence of the Σ-nucleus potential $V_\Sigma(r)$. The well depth extracted from Σ^--atom studies (60) is deep enough to support nuclear bound states, if one assumes that $V_\Sigma(r)$ is linear in $\rho(r)$. However, the Σ^- in an atomic state is localized in the far tail region of the density $\rho(r)$, while the Σ states formed in the $(K^-,\pi^{\pm})$ reactions are localized in the surface region ($r \approx R$). In the analysis of the ^{12}C$(K^-,\pi^+)^{12}_\Sigma$Be spectrum (49) by Morimatsu & Yazaki (12) and Iwasaki (47), a best fit was obtained with a very shallow Σ well, which suggests a strong density dependence of $V_\Sigma(r)$ and renders unlikely the existence of narrow Σ resonances in the continuum.

Further experiments are clearly necessary, preferably at low q. Dover et al (61) showed that the peaks claimed in the data of Bertini et al (49, 50) and Yamazaki et al (44) elude a consistent theoretical interpretation.

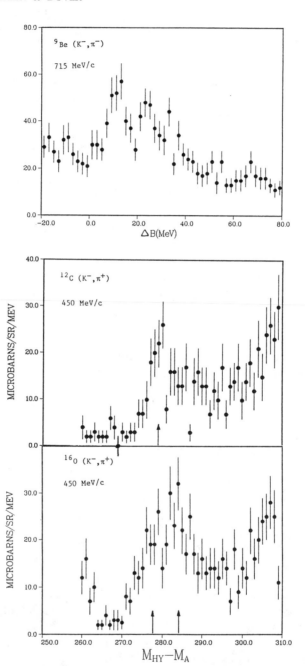

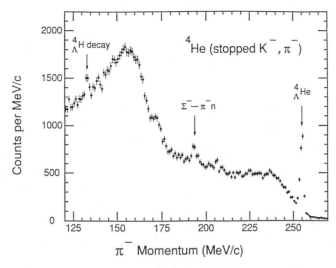

Figure 6 The π^- momentum spectrum obtained at KEK from stopped kaons on a ^{4}He target (57). The small shoulder near a pion momentum of 170 MeV/c has been interpreted as the formation of the ${}^4_\Sigma$He ground state (57, 59).

If some narrow Σ states are confirmed by further experiments, interesting theoretical questions of isospin purity and $\Sigma N \to \Lambda N$ conversion width will arise. The strong isospin dependence of the two-body ΣN potential is reflected in a sizable Σ nucleus symmetry potential, which tends to produce Σ states of good isospin (61). The width of the Σ in the nucleus can be suppressed by a number of effects (Pauli blocking, binding effects, spin dependence of $\Sigma N \to \Lambda N$ conversion, etc) to observable values of order 5–10 MeV (62). The strength of the Σ-nucleus spin-orbit splitting $\varepsilon_p(\Sigma)$ should be regarded as unknown, although values comparable to or larger than that for N nucleus have been claimed (44, 50). Early quark model estimates (63) implied $\varepsilon_p(\Sigma) \approx \varepsilon_p(N)$, whereas more realistic estimates (64, 65) give smaller values for $\varepsilon_p(\Sigma)$, similar to those predicted by one-boson-exchange models (66). These theoretical questions and others are discussed in a recent review (67).

Figure 5 $(K^-, \pi^\pm)$ spectra at $0°$ for ^{9}Be (48), ^{12}C (49), and ^{16}O (50) targets in the Σ excitation region, from the Heidelberg-Saclay collaboration at CERN. In the ^{12}C and ^{16}O spectra, the arrows indicate the position of the $p_{3/2}$ and $(p_{1/2}, p_{3/2})$ Σ substitutional states, respectively, claimed by Bertini et al (49, 50). The absolute cross sections for ${}^{12}_\Sigma$C and ${}^{16}_\Sigma$O are taken from Walcher (50). For the ${}^9_\Sigma$Be spectrum, the label $\Delta B = B_N - B_{\Sigma^0} = M_{HY} - M_A - (M_{\Sigma^0} - M_N)$; narrow Σ peaks at $\Delta B \approx 14$ and 24 MeV have been claimed by Bertini et al (48).

3. ASPECTS OF HYPERNUCLEAR STRUCTURE

This section is devoted to the theoretical interpretation of the data on hypernuclear level schemes. We start with a discussion of the Λ-nucleus mean field. The hypernuclei $^{13}_{\Lambda}C$ and $^{9}_{\Lambda}Be$ are then discussed as examples from which one is able to extract information on the ΛN residual interaction. Finally, we delineate the theoretical questions that arise regarding the spectroscopy of doubly strange hypernuclei and dibaryons.

3.1   The Λ-Nucleus Mean Field

One of the oldest problems in hypernuclear physics is that of the well depth D_{Λ} for a Λ particle in nuclear matter. A review of the older attempts to obtain D_{Λ} by extrapolating the data on the ground-state binding energies B_{Λ} of light ($A \leq 15$) systems has been given by Gal (4), who cites a value $D_{\Lambda} = 30 \pm 3$ MeV. A simple estimate of D_{Λ} in terms of a density-independent two-body potential $V_{\Lambda N}(r)$ fit to the free-space ΛN scattering data at low energy leads to the well-known overbinding problem:

$$D_{\Lambda} = \rho_0 \int V_{\Lambda N}(r) \, d^3r \approx 60 \text{ MeV}. \qquad 2.$$

Here $\rho_0 \approx 1/6$ fm^{-3} is the central density of nuclei. The problem of hyperon-nucleus single-particle potentials has been reviewed by Dover & Gal (68) and Yamamoto & Bandō (69). Recent theoretical work has been directed toward calculations of the density-dependent ΛN G-matrix (70–72), inclusion of repulsive three-body ΛNN forces to cure the overbinding problem (25, 73–76), folding model calculations with phenomenological density-dependent ΛN interactions (77, 78), and attempts to write semi-empirical mass formulae for B_{Λ} (79–82).

The G-matrix approach (70–72) uses a realistic one-boson-exchange (OBE) potential as input, typically Model D of the Nijmegen group (33), which reproduces the free-space ΛN scattering data. The Λ-nucleus potential $V_{\Lambda}(r, E)$ is obtained as a convolution of $G_{\Lambda N}(\rho, E)$ with the nuclear density $\rho(r)$, either in momentum space or in coordinate space via a local density approximation:

$$V_{\Lambda}(r, E) = \int d^3r' \rho(r') G_{\Lambda N}(r - r', \rho, E). \qquad 3.$$

The ρ dependence of $G_{\Lambda N}$ is reflected in a strong density dependence of V_{Λ} and leads to a well depth $D_{\Lambda} = V_{\Lambda}(r = 0, E = 0)$ of order 30–40 MeV, in qualitative agreement with experiment.

Another approach, developed primarily by Bodmer et al (25, 73, 74), starts with phenomenological ΛN and ΛNN potentials, and tries to explain Λp scattering, B_Λ values for s- and p-shell hypernuclei, and D_Λ in a single consistent framework. The variational Monte Carlo method is applied, including two- and three-body short-range correlations (25). This approach is quite successful, providing one introduces strongly repulsive ΛNN forces [similar to the large repulsive $\rho^2(r)$ terms in the G-matrix approach]. Both spin-dependent ΛNN forces of dispersive character and two-pion-exchange ΛNN terms of about the expected strength are required. In this picture, about one-third of the 1-MeV splitting of the 0^+-1^+ doublet in the $A = 4$ hypernuclear system is due to the spin dependence of the ΛNN potential. Further, the central part of the ΛN interaction in the relative p-state has only about half the strength of the s-state interaction [i.e. a dependence $(1-\varepsilon+\varepsilon P_x)$, with $\varepsilon \approx 1/4$].

The phenomenological ΛNN interactions required to overcome the "overbinding" problem (e.g. $^5_\Lambda$He) subsume a number of complicated dynamical effects. For instance, such terms can result from eliminating Σ degrees of freedom in a coupled-channel ($\Lambda N \rightarrow \Sigma N \rightarrow \Lambda N$) approach or from density-dependent modifications of the two-body interaction (à la the G-matrix). At a more microscopic level, quark antisymmetrization

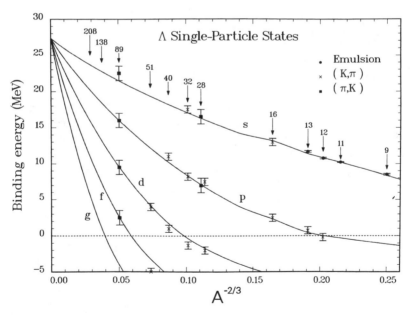

Figure 7 Data on binding energies of s, p, d, and f single-particle states of the Λ as a function of $A^{-2/3}$, from Millener et al (85). The curves correspond to a nonlocal Λ-nucleus potential with ρ^2 density dependence, as explained in the text.

generates a repulsive "Pauli pressure" equivalent to an effective ΛNN force (83, 84).

Recently, through the (π^+, K^+) studies at Brookhaven National Laboratory (37, 38), discussed in Section 2.3.3, it has become possible to track the evolution of Λ binding energies (ground and excited states) as a function of A, up to $A \approx 90$. The results for s_Λ, p_Λ, d_Λ, and f_Λ single-particle binding energies, as obtained from (π^+, K^+), (K^-, π^-), and emulsion measurements, are plotted in Figure 7, taken from Millener et al (85). The A dependence of the Λ level spacings enables us to constrain the geometry of the Λ-nucleus potential and also the well depth D_Λ.

A useful description of the data can be obtained (85, 86) in the Skyrme-Hartree-Fock approach (87), first used for hypernuclei by Rayet (88). Here, the nonlocality of the Λ-nucleus potential is parameterized in terms of an effective mass $m_\Lambda^*(r)$. The equivalent energy-dependent local potential $V_\Lambda(r, E)$ is of the form (85, 89)

$$V_\Lambda(r, E) = \frac{m_\Lambda^*(r)}{m_\Lambda} U(r) + \left[1 - \frac{m_\Lambda^*(r)}{m_\Lambda}\right] E$$

$$U(r) = t_0 \rho(r) + \frac{3}{8} t_3 \rho^2(r) + \frac{1}{4}(t_1 + t_2) T(r)$$

$$T(r) = \frac{3}{5}\left(\frac{3\pi^2}{2}\right)^{2/3} \rho^{5/3}(r)$$

$$\hbar^2/2m_\Lambda^*(r) = \frac{\hbar^2}{2m_\Lambda} + \frac{1}{4}(t_1 + t_2)\rho(r). \qquad 4.$$

The $t_3\rho^2(r)$ term in Equations 4 can be used to adjust the potential radius, while the $(1 - m_\Lambda^*/m_\Lambda)E$ term serves to spread out the single-particle levels; this enables one simultaneously to fit the spectra of light ($^{16}_\Lambda O$) and heavy ($^{89}_\Lambda Y$) systems. The choice

$$t_0 = -402.6 \text{ MeV fm}^3$$

$$t_1 + t_2 = 103.4 \text{ MeV fm}^5$$

$$t_3 = 3394.6 \text{ MeV fm}^6 \qquad 5.$$

leads to the fit displayed in Figure 7. These values correspond (85) to

$$m_\Lambda^*(r = 0)/m_\Lambda \approx 0.8$$

$$D_\Lambda \approx 27.5 \text{ MeV}. \qquad 6.$$

The calculations of Yamamoto et al (86), which incorporate self-consistency and rearrangement energies [omitted by Millener et al (85)], are consistent with the need for a strong repulsive $t_3\rho^2(r)$ term and a modest degree of nonlocality $[m_\Lambda^*(0)/m_\Lambda \approx 0.8]$.

As in the variational calculations of Bodmer et al (25, 73, 74) and the G-matrix approach (69–72), the "overbinding" problem is cured by a strong nonlinear density dependence of the Λ-nucleus potential. The well depth estimated from the linear term in Equations 4, i.e. $D_\Lambda \approx \rho_0 t_0 \approx 67$ MeV, is in fact consistent with the value from Equation 2 based on free-space ΛN potentials. The balance between $\rho(r)$ and $\rho^2(r)$ terms in the Skyrme-Hartree-Fock approach is thus very similar to that between the ΛN and ΛNN contributions in the work of Bodmer et al (25, 73, 74).

A strong $t_3\rho^2(r)$ term is also required to fit nucleon binding energies (87, 89). However, within the Skyrme-Hartree-Fock model, one cannot simultaneously describe nucleon levels near the Fermi surface [which requires $m_N^*(0)/m_N \approx 1$] and deeply bound levels [for which $m_N^*(0)/m_N \approx 1/2$]. For the Λ, on the other hand, a description of all levels is possible with a single value of $m_\Lambda^*(0)/m_\Lambda$.

The Λ behaves as a distinguishable particle in the nucleus. Unlike deeply bound nucleon-hole states, which are very broad, deeply bound Λ single-particle states remain well defined. The possibility that strange quarks in the nucleus are partially deconfined (90, 91) is an intriguing one, but the signature of this effect in the Λ binding energies is likely to be subtle and easily masked by the complicated (but conventional) dynamics of density-dependent interactions.

3.2 Hypernuclear Structure: Selected Examples

There have been numerous attempts to calculate the energy spectra of hypernuclei, both in the context of the hypernuclear shell model (35, 92, 93) and the cluster model (94–103). We do not attempt to review all of these efforts in detail. Instead, we focus on two representative examples, namely $^{13}_\Lambda$C and $^9_\Lambda$Be, where the measured excitation spectra in (K^-,π^-) and (π^+,K^+) production reactions have been used to extract properties of the ΛN effective interaction, and we elucidate the relation of the shell model and cluster descriptions of excited states.

3.2.1 SPECTROSCOPY OF $^{13}_\Lambda$C   The experimental ^{13}C$(K^-,\pi^-)^{13}_\Lambda$C excitation spectrum of May et al (34) at pion laboratory angles $\theta_{\text{lab}} = 4°$ and $15°$ is shown at the top of Figure 8. The bottom part of this figure displays the theoretical spectrum obtained by Auerbach et al (35) in the shell model, including ^{12}C core excitations and the configuration mixing induced by the ΛN residual interaction. The shell-model description is seen quantitatively to reproduce the experimental results, both in absolute cross section and in number of peaks. The changes with angle of relative cross sections

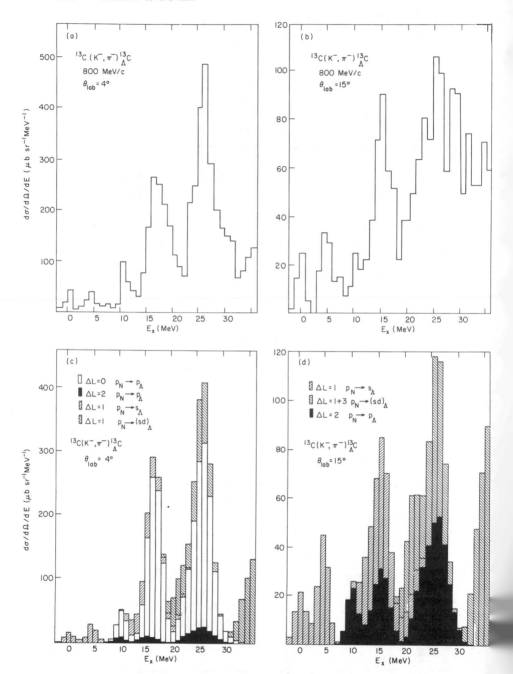

and peak positions seen in Figure 8 reflect the momentum transfer (q) dependence of the form factors and the effects of the ΛN residual interaction. We now discuss how this information is extracted.

The general features of the $^{13}_\Lambda C$ spectrum can be understood in terms of the distribution of neutron pickup strength. By coupling a Λ in an s-orbit to the ^{12}C core states, we anticipate $p_N \to p_\Lambda$ ($\Delta L = 1$) strength in $^{13}C(K^-,\pi^-)^{13}_\Lambda C$ at $E_x = 0$, 5, and 12–16 MeV. Coupling p_Λ to this same set of states leads to $p_N \to p_\Lambda$ ($\Delta L = 0, 2$) strength 10–11 MeV higher in E_x. Above $E_x \approx 20$ MeV, we expect $p_N \to (sd)_\Lambda$ and $s_N \to s_\Lambda$ transitions to enter. Thus in a weak coupling picture (no ΛN interaction), the origin of the five peaks below 30 MeV in the $^{13}_\Lambda C^*$ spectrum is clear. At forward angles, the population of $1/2^-$ via $\Delta L = 0$ dominates; $\Delta L = 1$ transitions to $(1/2^+, 3/2^+)$ states and $\Delta L = 2$ transitions to $(3/2^-, 5/2^-)$ states are important at $\theta_{lab} = 15°$, and the $\Delta L = 0$ part is small.

If there were good agreement between experiment and the weak coupling limit within experimental resolution and errors, we would learn essentially nothing about the ΛN interaction. However, several features of the 10- and 16-MeV peaks signal a departure from weak coupling: (*a*) The cross-section ratio of the second and first $1/2^-$ states is 5.5 at $\theta_{lab} = 4°$, whereas the ratio of pickup strengths is only 1.5–2. (*b*) At 4°, the spacing of the two peaks is 6.0 ± 0.4 MeV, clearly larger than the 4.4-MeV spacing (0^+-2^+) of the core states. (*c*) The 16-MeV peak is shifted downward by 1.7 ± 0.4 MeV as θ_{lab} changes from 4° to 15°. (*d*) The 10-MeV peak is shifted downward by 0.36 ± 0.3 MeV from 4° to 15°.

The shift (*d*) is essentially a measure of the Λ-nucleus spin-orbit splitting ε_p^Λ, since we excite dominantly the $^{12}C(0^+) \otimes p_{1/2}^\Lambda$ configuration at 4° and $^{12}C(0^+) \otimes p_{3/2}^\Lambda$ at 15°. The result

$$\varepsilon_p^\Lambda = 0.36 \pm 0.3 \text{ MeV} \qquad\qquad 7.$$

shows that the Λ spin-orbit coupling is very small. The value of ε_p^Λ was first shown to be much less than the typical value $\varepsilon_p^N \approx 6$ MeV for a nucleon by an analysis of the $^{16}O(K^-,\pi^-)^{16}_\Lambda O$ results (104–106).

The deviations (*a*), (*b*), and (*c*) from weak coupling have been analyzed (35) to obtain constraints on the ΛN residual interaction $V_{\Lambda N}$, taken to have the form

---

Figure 8   Differential cross section $d\sigma/d\Omega dE$ at $\theta_{lab} = 4°$ and 15° as a function of excitation energy E_x for the $^{13}C(K^-,\pi^-)^{13}_\Lambda C$ reaction at 800 MeV/c. The experimental data of May et al (34) are shown in the upper half, and the DWIA shell-model calculations of Auerbach et al (35) are shown below. The contributions of different orbital angular momentum transfers ΔL for various single-particle transitions $p_N \to s_\Lambda, p_\Lambda, (sd)_\Lambda$ are separately indicated.

$$V_{\Lambda N}(r) = V_0(r)(1-\varepsilon+\varepsilon P_x)(1+\alpha\boldsymbol{\sigma}_N\cdot\boldsymbol{\sigma}_\Lambda) \qquad\qquad 8.$$

plus symmetric and antisymmetric spin-orbit parts. Expanding $V_0(\mathbf{r}_N-\mathbf{r}_\Lambda)$ in terms of multipoles $V_k(r_N, r_\Lambda)$, one finds that the p-shell interaction ($\ell_N = 1, \ell_\Lambda = 1$) is characterized by two Slater integrals $F^{(0)}$ and $F^{(2)}$, where

$$F^{(k)} = \int R_{\ell_N}^2(r_N)R_{\ell_\Lambda}^2(r_\Lambda)V_k(r_N, r_\Lambda)\,dr_N\,dr_\Lambda, \qquad\qquad 9.$$

where R_{ℓ_N} and R_{ℓ_Λ} are the radial wave functions for the N and Λ, respectively. The value $F^{(0)} \approx -1.16$ MeV is already constrained from other structure analyses (107). The above deviations (a–c) from weak coupling can be accounted for by a ΛN quadrupole matrix element

$$-3.4 \leq F^{(2)} \leq -3 \text{ MeV}. \qquad\qquad 10.$$

This is consistent with the value extracted from the $^9_\Lambda$Be spectrum by Dalitz & Gal (107).

An interesting aspect of the analysis (35) is the intensity ratio I_{16}/I_{10} of the 16- and 10-MeV peaks in $^{13}_\Lambda$C at $\theta_{lab} = 4°$. In weak coupling one obtains $I_{16}/I_{10} = 1.8$, compared to the experimental value (34) of $I_{16}/I_{10} \approx 5$. The effect of $F^{(2)}$ is to induce configuration mixing and increase the ratio. One also obtains a constraint on the space exchange (P_x) strength ε from I_{16}/I_{10}. There is evidence for such a component ($\varepsilon \approx 0.25$) in the free-space Λp interaction (108).

The hypernucleus $^{13}_\Lambda$C displays a marked tendency toward higher spatial symmetry in its lowest $1/2^-$ state, in spite of the rather weak ΛN interaction. Using states of [54] or [441] symmetry instead of the weak coupling basis, one notes that [54] symmetry is forbidden by the Pauli principle for an ordinary nucleus, whereas it is the dominant component of the lowest $1/2^-$ state in $^{13}_\Lambda$C. Starting with the mostly [441] ground state of ^{13}C, the [54] symmetry is not attainable in a $\Delta L = 0$ process, which leads to a dynamical selection rule. The tendency toward increased spatial symmetry is a unique feature of the distinguishable Λ hyperon, and is signaled by the observed strong deviation of the cross-section ratio from its value in the weak coupling limit.

3.2.2 SHELL MODEL VS CLUSTER DESCRIPTION OF $^9_\Lambda$Be   The ^{9}Be$(K^-,\pi^-)^9_\Lambda$Be reaction at forward angles has been investigated at CERN (48, 104, 105, 109–111). The spectrum at 720 MeV/c (104) is shown in Figure 9, along with the excitation function of the ^{9}Be$(\pi^+,K^+)^9_\Lambda$Be reaction at 1.05 GeV/c (37, 38). The structure of $^9_\Lambda$Be was discussed in the shell-model framework by Dalitz & Gal (93, 107) and later by Auerbach et al (35). The cluster model for $^9_\Lambda$Be has been extensively developed (94–103).

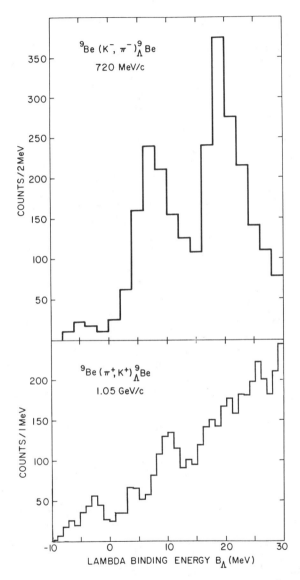

Figure 9 The measured (K^-,π^-) and (π^+,K^+) excitation spectra on a ^{9}Be target. The (K^-,π^-) forward angle data at 720 MeV/c are replotted from Bertini et al (48). The (π^+,K^+) spectrum at 1.05 GeV/c, $\theta = 10°$, has been obtained from P. Pile (38; and private communication).

The essential features of the $^9_\Lambda$Be spectrum can be seen in a coupling scheme defined by $\mathbf{J} = \mathbf{L} + \mathbf{s}_\Lambda$, $\mathbf{L} = \mathbf{J}_c + \mathbf{l}_\Lambda$, where $\mathbf{J}_c$ is the spin of the nuclear core. For an interaction independent of the Λ spin $\mathbf{s}_\Lambda$, L is a good quantum number and states with $J = L \pm 1/2$ (for $L \neq 0$) form a degenerate doublet. The structure of $^9_\Lambda$Be is similar to $^{13}_\Lambda$C, in that the LS structure of the p^4 core of ^{8}Be with [4] and [31] symmetries resembles the [44] and [431] symmetries for ^{12}C. The neutron pickup strength goes mostly to the ^{8}Be ground state, the 2$^+$ state at 2.94 MeV, a group of states between 16 and 20 MeV, and a 3$^+$ level above 19 MeV.

It is instructive to consider the ^{8}Be(0$^+$, 2$^+$) $\otimes$ p$_\Lambda$ states in $^9_\Lambda$Be in comparison with $^{13}_\Lambda$C. The 0$^+$ (ground state) and 2$^+$ (2.94 MeV) states in ^{8}Be have almost pure [4] spatial symmetry with $S = 0$. Thus the states of $^9_\Lambda$Be can have [5] or [41] symmetry, respectively. Since the ^{8}Be core is prolate while ^{12}C is oblate, the matrix element of the quadrupole operator Q for ^{8}Be is opposite in sign to that for ^{12}C; this implies an inverted order in $^9_\Lambda$Be of the $L = 1, 2, 3$ states based on the 2$^+$ core state. There is then stronger mixing in $^9_\Lambda$Be than in $^{13}_\Lambda$C of $L = 1$ states based on the 0$^+$ and 2$^+$ core states. For $F^{(2)} = -3.2$ MeV, as derived earlier, this mixing is strong enough so that the eigenstate approaches the limit of good spatial symmetry $|[5] L = 1\rangle$. An important consequence is that the ^{9}Be(K$^-$, π^-)$^9_\Lambda$Be cross section at 0°, leading to the lowest 3/2$^-$ state ($L = 1$), becomes very small relative to the second 3/2$^-$ state ($L = 1$), because the transition [41] $\to$ [5] is forbidden for $\Delta L = 0$. The remainder of the $\Delta L = 0$ strength is found in two states with $T = 0, 1$, about 12.5 MeV above the second 3/2$^-$ level. This is in excellent agreement with the (K$^-$, π^-) data shown in Figure 9, which display two strong peaks (at 7 and 19 MeV) separated by about 12 MeV.

The simplest version of the cluster model for $^9_\Lambda$Be, namely $\alpha + \alpha + \Lambda$, does not explain the third peak at 19 MeV in the (K$^-$, π^-) spectrum of Figure 9, since this involves a strong contribution of isospin-one core excited states of ^{8}Be. The cluster model was recently extended (101) to include $\alpha + \alpha^* + \Lambda$ configurations, where α^* is the intrinsic excited state of the α particle. The distribution of (K$^-$, π^-) strength for this model, shown in Figure 10, is rather similar to that obtained in the shell model (35, 107).

The (π^+, K$^+$) reaction, in contrast to (K$^-$, π^-) at 0°, favors the excitation of the higher spin states in $^9_\Lambda$Be. The predicted strength distribution (101) of the ^{9}Be(π^+, K$^+$)$^9_\Lambda$Be reaction at 1.05 GeV/c is shown in Figure 10. In contrast to the (K$^-$, π^-) spectrum, one expects a measurable cross section to the "supersymmetric" 3$^-$ state near binding energy $B_\Lambda \approx -4$ MeV; these states were first discussed by Dalitz & Gal (93). Secondly, a peak near $-B_\Lambda \approx 12$ MeV, corresponding to ^{8}Be*(2$^+$,1$^+$,3$^+$) $\otimes$ s$_\Lambda$, should be seen in (π^+, K$^+$), but not in (K$^-$, π^-). The experimental (π^+, K$^+$) spectrum of

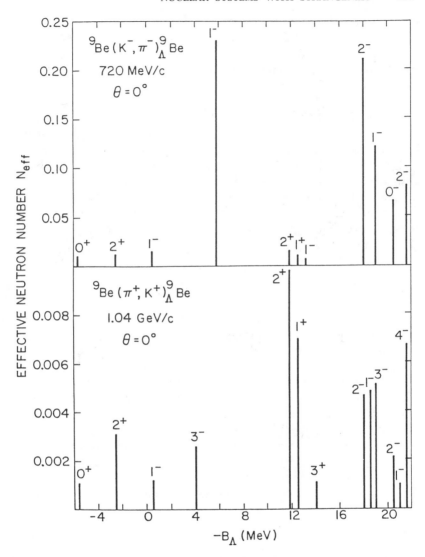

Figure 10   The effective neutron number N_{eff} for the (K^-,π^-) and (π^+,K^+) reactions on ^{9}Be, as calculated by Yamada et al (101). The forward differential cross section for each hypernuclear state with binding energy B_Λ is given by $N_{\text{eff}}(d\sigma/d\Omega)_{\text{2-body}}$ in terms of the $K^-n \rightarrow \pi^-\Lambda$ or $\pi^+n \rightarrow K^+\Lambda$ two-body $0°$ cross section $(d\sigma/d\Omega)_{\text{2-body}}$. The states in $^9_\Lambda$Be are labeled by orbital angular momentum L and parity π, in a coupling scheme where $\mathbf{J} = \mathbf{L}+\mathbf{s}_\Lambda$. Thus each line with $L \neq 0$ corresponds to a doublet for which the spin splitting is neglected. Note the strong selectivity of both the (K^-,π^-) reaction, which has low q and favors $\Delta L = 0$ transitions, and the (π^+,K^+) process, which has $q \approx 350$ MeV/c and preferentially excites the higher spin $^9_\Lambda$Be states.

Figure 9 indicates three peaks below $-B_\Lambda = 15$ MeV, consistent with dominant excitation of $L^\pi = 2^+$ ($B_\Lambda = 2$ MeV), 3^- ($B_\Lambda = -4$ MeV), and 2^+ ($B_\Lambda = -12$ MeV) states. The hypernucleus $^9_\Lambda$Be affords an excellent example of the complementary use of both (K^-, π^-) and (π^+, K^+) to obtain a more complete picture of the hypernuclear spectrum.

3.3 $S = -2$ Hypernuclei and Dibaryons

In emulsion experiments with K^- beams, there are two events corresponding to the formation of doubly strange $\Lambda\Lambda$ hypernuclei, namely $^6_{\Lambda\Lambda}$He (112) and $^{10}_{\Lambda\Lambda}$Be (113). There are also seven events that have been attributed to the formation of a Ξ^- hypernucleus; the data and the consistency of this interpretation were discussed by Dover & Gal (114). The separation energy $B_{\Lambda\Lambda}$ of two Λ's from the core nucleus is given by 10.92 ± 0.6 MeV for $^6_{\Lambda\Lambda}$He (112) and 17.71 ± 0.08 MeV for $^{10}_{\Lambda\Lambda}$Be (113). A quantity more directly related to the $\Lambda\Lambda$ interaction is

$$\Delta B_{\Lambda\Lambda} = B_{\Lambda\Lambda} - 2B_\Lambda = \begin{cases} 4.68 \pm 0.6 \text{ MeV } (^6_{\Lambda\Lambda}\text{He}) \\ 4.29 \pm 0.1 \text{ MeV } (^{10}_{\Lambda\Lambda}\text{Be}), \end{cases} \qquad 11.$$

where B_Λ is the separation energy of a single Λ from the core. Recently, there have been several attempts (74, 96, 102, 103, 115–119) to understand these data, and also the neighboring single hypernuclei $^5_\Lambda$He and $^9_\Lambda$Be, in the context of the cluster model ($n_\alpha \alpha + n_\Lambda \Lambda$, with $n_\alpha = 1,2$ and $n_\Lambda = 1,2$). For instance, Bodmer et al (74, 102, 115) are able to understand the $^5_\Lambda$He, $^9_\Lambda$Be, $^{10}_{\Lambda\Lambda}$Be binding energies and the Λ well depth D_Λ with ΛN forces consistent with free-space scattering and a repulsive ΛNN interaction. The latter implies a repulsive $\alpha\alpha\Lambda$ potential that corrects an overbinding problem in $^9_\Lambda$Be. The ΛN and $\Lambda\Lambda$ potentials are parameterized in terms of a short-range repulsive core of Woods-Saxon form and an attractive piece with a range characteristic of two-pion exchange. An important result of the cluster calculations is the linear relation between $^6_{\Lambda\Lambda}$He and $^{10}_{\Lambda\Lambda}$Be binding energies:

$$B_{\Lambda\Lambda}(^6_{\Lambda\Lambda}\text{He}) \approx -\alpha + \beta B_{\Lambda\Lambda}(^{10}_{\Lambda\Lambda}\text{Be}), \qquad 12.$$

with $(\alpha, \beta) = (4.17 \text{ MeV}, 0.76)$ from Bodmer et al (103, 115) and (4.86 MeV, 0.8) from Wang et al (119). This relationship is not satisfied by the data: if we fit $^6_{\Lambda\Lambda}$He, then $^{10}_{\Lambda\Lambda}$Be is overbound by several MeV, while if $^{10}_{\Lambda\Lambda}$Be is fitted, then $^6_{\Lambda\Lambda}$He is underbound. The significance of this discrepancy is not clear; it does, however, point up the urgent need for more data on $\Lambda\Lambda$ hypernuclei.

The $\Lambda\Lambda$ interaction extracted from the $^{10}_{\Lambda\Lambda}$Be event is strongly attractive, corresponding to a $^1S_0$ scattering length (74, 102, 115) in the range

$-5 \leq a_{\Lambda\Lambda} \leq -2$ fm. This is comparable to the $^1S_0$ ΛN and NN inter-actions, for which $a_{\Lambda N} \approx -3$ fm and $a_{NN} \approx -4$ fm, the latter with one-pion exchange subtracted out. It appears that the attraction is not quite sufficient to form a $^1S_0$ $\Lambda\Lambda$ bound state, however.

One can form $\Lambda\Lambda$ hypernuclei via the double strangeness exchange (K^-,K^0) or (K^-,K^+) reactions or by Ξ^- capture. Baltz et al (120) predict forward cross sections of order 2–10 nb/sr for the excitation of high spin 3^- and 4^+ states in the $^{16}O(K^-,K^+)^{16}_{\Lambda\Lambda}C^*$ reaction at 1.1 GeV/c. To populate low spin states, including the ground state, Ξ^- capture from an atomic orbit is more favorable; May (121) has considered the reaction $\Xi^- + {}^6Li \rightarrow n + {}^6_{\Lambda\Lambda}He$, for instance.

The $(K^-,K^{0,+})$ reactions can also be used to form Ξ hypernuclei in the one-step process $K^-N \rightarrow K\Xi$, with cross sections in the range 0.1–1 μb/sr at $0°$ (114). These could be measured at the 2-GeV/c kaon line under construction at Brookhaven. Theoretical studies of the Ξ-nucleus single-particle potential (114, 122–124) suggest a well depth of 20–25 MeV, sufficient to support a number of bound states. The key question is whether the Ξ states, which decay strongly via $\Xi N \rightarrow \Lambda\Lambda$ conversion, are sufficiently narrow to be observed. Dover & Gal (114) provide an estimate $\Gamma_\Xi \leq 5$–10 MeV for the Ξ width, while Yamamoto (123) gives a more optimistic value $\Gamma_\Xi \leq 2$ MeV. In any case, the conversion width for $\Xi N \rightarrow \Lambda\Lambda$ should be less than that for $\Sigma N \rightarrow \Lambda N$, since the former has far less phase space available and the initial ΞN system is restricted to $^1S_0$, $I = 0$ quantum numbers because of the Pauli principle for $\Lambda\Lambda$. Experiments are necessary in order to expose the potentially rich single-particle spectroscopy of Ξ's in nuclei.

The other aspect of $S = -2$ systems meriting particular attention is the possible production of six-quark dibaryons. In the MIT bag model, Jaffe has predicted (125) a stable dibaryon, the H. More generally, Mulders et al (126) considered the spectrum of such objects for $S = 0, -1, -2$ formed from clusters $(Q_c^m \otimes Q_c^n)$, with c being the SU(3) color index and $m+n = 6$. No such objects have as yet been convincingly identified.

The (Q^6) SU(3) flavor-singlet state H with $J^\pi = 0^+$, $I = 0$, $S = -2$ is thought to be the most stable of such systems, possibly even lying below the $\Lambda\Lambda$ threshold. This object has high symmetry with 2u, 2d, and 2s quarks coupled in relative s-orbits. The $S = -1$ objects are unbound with respect to the ΛN threshold. However, several of these states of the form $(Q_3^4 \otimes Q_3^2)$ couple to the ΛN system in a relative p-state, and lie below the $\Lambda N\pi$ pionic decay channel so they could be narrow (127). The triplet ^3P and singlet ^1P states with $S = -1$ are split by about 40 MeV/c^2 in the bag model calculation (126). Aerts & Dover (128) estimated the production cross section of these $S = -1$ dibaryons in (K^-,π^-) reactions on deuterium. A recent discussion of the preliminary data from Brookhaven on the $d(K^-, \pi^-)\Lambda p$ reaction has been given by Piekarz (129), while the search for $S = -1$ dibaryons in the $pp \to K^+X$ reaction is discussed by Frascaria (130). A review of theoretical considerations relevant to $S = -1, -2$ dibaryons can also be found in (131).

The H dibaryon has been searched for in pp collisions (132) and $\bar{p}$-nucleus annihilations (133), but without the requisite sensitivity. An event attributed to H formation in proton-nucleus collisions has been claimed by Shahbazian et al (134). A triggered emulsion study by Aoki et al (135) is under way at KEK.

Several experiments (136) are now under construction at the Brookhaven AGS with sensitivities comparable to theoretical estimates (137) for H-particle production. The use of a new separated kaon beam line, LESB-III, which will be ready in 1990, will provide 10^6 negative kaons per second on target at 1.6 GeV/c. Two experimental approaches are being used:

1. Twin-target method

$$K^- + p \to \Xi^- + K^+$$
$$\Xi^- + d \to H + n$$

2. Single-target method

$$K^- + {}^3He \to H + K^+ + n$$

In the twin-target approach, a two-step reaction mechanism is used to provide an H through a succession of two-body interactions. The K^- beam is momentum analyzed, and the outgoing K^+ signals the formation of a low momentum cascade particle, which is degraded in energy and stopped in a liquid deuterium target. A (Ξ^-d) atom is formed and subsequently decays to the H, with the emission of a monoenergetic neutron. The neutron energy is uniquely determined by the H binding energy. The branching ratio for (Ξ^-d) decay to the H is predicted to be a strong function of H binding, ranging from close to unity for a weakly bound H to less than 0.1 for an H binding more than 80 MeV (138).

In the single-target reaction, the process of H production occurs within the ^{3}He nuclear target. The emission of the H is accompanied by a spectator neutron. Since it is a three-body reaction, the neutron is not monoenergetic. However, because the K^+ spectrum is predicted to peak near the maximum allowed momentum, the neutron spectrum is also peaked (137). The cross section for Reaction 2 above is not very sensitive to the H mass, unlike that for Reaction 1.

4. WEAK DECAYS OF HYPERNUCLEI

Λ hypernuclei in their ground states decay via weak interactions, the principal modes being $\Lambda \to N\pi$ and $\Lambda N \to NN$. We consider these in turn.

4.1   *Pionic Decay Modes*

The Λ lifetime in free space is $\tau_\Lambda = \Gamma_\Lambda^{-1} = 2.63 \times 10^{-10}$ s, and the dominant decay modes are $\Lambda \to p\pi^-$ (64%), $n\pi^0$(36%), the approximate 2:1 ratio reflecting the $\Delta I = 1/2$ rule (139) for the isospin change ΔI. In the nucleus, binding effects and Pauli blocking serve to suppress the rate for $\Lambda \to N\pi$, particularly in heavier systems. The early work on pionic decays of hypernuclei was reviewed by Dover & Walker (140).

More recently, there have been extensive calculations of this process (141–143), employing realistic wave functions for the Λ and incorporating the effects of pion final-state interactions. The width Γ_π for pion emission is sensitve to the pion wave function in the nuclear interior, which is not well determined by the more peripheral pion elastic scattering process. Thus the $\Lambda \to N\pi$ decay in a nucleus affords a novel test of the low energy π-nucleus optical potential. The effect of distortions is to increase Γ_π, through high momentum components induced in the pion wave function. This permits the decay strength to be distributed over higher-lying shell-model states, thereby weakening the Pauli blocking effect. Enhancements in Γ_π due to pion interactions in nuclear matter have also been discussed by Oset & Salcedo (144) and those due to short-range ΛN correlations by Kurihara, Akaishi & Tanaka (145).

Although Γ_π generally decreases with A as a result of the increased Pauli blocking, there are marked variations of $\Gamma_{\pi^-}/\Gamma_{\pi^0}$ that reflect shell structure. For instance, in $^8_\Lambda$Be and $^{12}_\Lambda$C, Motoba et al (143) predict $\Gamma_{\pi^0} > \Gamma_{\pi^-}$, compared to $2\Gamma_{\pi^0} \approx \Gamma_{\pi^-}$ in free space, since low-lying $T = 0$ states of ^{8}Be and ^{12}C can be reached in π^0 decay, whereas π^- decay leads only to high-lying $T = 1$ configurations in ^{8}B or ^{12}N. This effect in reverse leads to the enhancement of π^- decays in $^8_\Lambda$Li and $^{12}_\Lambda$B.

The available data are not sufficiently accurate to test the enhancement in Γ_π due to pion distortion, nor the predictions for $\Gamma_{\pi^-}/\Gamma_{\pi^0}$. Barnes (146) quotes

$$\Gamma_{\pi^0}/\Gamma_\Lambda = \begin{cases} 0.16^{+0.34}_{-0.21} & \text{for} \quad ^5_\Lambda\text{He} \\ 0.06^{+0.08}_{-0.05} & \text{for} \quad ^{12}_\Lambda\text{C} \end{cases}$$

$$\Gamma_{\pi^-}/\Gamma_\Lambda = \begin{cases} 0.43 \pm 0.10 & \text{for} \quad ^5_\Lambda\text{He} \\ 0.05^{+0.06}_{-0.03} & \text{for} \quad ^{12}_\Lambda\text{C} \end{cases} \qquad 13.$$

compared to the predictions (142, 143) with the Michigan State University (MSU) pion potential (147) or with undistorted plane waves (PW) for the pion:

$$\Gamma_{\pi^0}/\Gamma_\Lambda = \begin{cases} 0.15(\text{MSU}), & 0.09(\text{PW}) & \text{for } ^5_\Lambda\text{He} \\ 0.17(\text{MSU}), & 0.08(\text{PW}) & \text{for } ^{12}_\Lambda\text{C} \end{cases}$$

$$\Gamma_{\pi^-}/\Gamma_\Lambda = \begin{cases} 0.32(\text{MSU}), & 0.18(\text{PW}) & \text{for } {}^5_\Lambda\text{He} \\ 0.13(\text{MSU}), & 0.06(\text{PW}) & \text{for } {}^{12}_\Lambda\text{C} \end{cases}. \qquad 14.$$

4.2 Nonmesonic Decays

Hypernuclei provide a unique laboratory for the study of the nonmesonic weak process $\Lambda N \to NN$. The light systems ${}^3_\Lambda\text{H}$, ${}^4_\Lambda\text{H}$, ${}^4_\Lambda\text{He}$, and ${}^5_\Lambda\text{He}$ play a central role in this investigation, since they enable us to extract information on the spin-isospin dependence of $\Lambda N \to NN$ conversion. Several reviews of nonmesonic decays have appeared (140, 146, 148–150), so our discussion here is brief.

The $\Lambda N \to NN$ reaction at rest yields nucleons with $q \approx 420$ MeV/c, corresponding to a distance scale of order 0.5 fm. Thus we do not expect long-range one-pion exchange (OPE) to dominate. Indeed, OPE by itself appears to fail qualitatively in describing the data: short-range effects are crucial. These have been included as heavy meson exchanges (151–153) or in terms of six-quark wave functions (154, 155). Nevertheless, existing models are still deficient in several key aspects, as we now show.

The phenomenological analysis of Block & Dalitz (156) led to the conclusion that transitions to isospin $I = 1$ final states dominate the $\Lambda N \to NN$ process, the strongest being the ${}^3S_1 \to {}^{33}P_1$ parity-changing transition. Pion exchange, on the other hand, implies a dominant ${}^3S_1 \to {}^{13}D_1$ parity-conserving transition, driven by the tensor part of the potential. The difference appears in the ratio Γ_n/Γ_p of neutron-induced ($\Lambda n \to nn$) to proton-induced ($\Lambda p \to np$) nonmesonic decay rates. For isovector ($\pi + \rho$) exchange, one expects (150) $\Gamma_n/\Gamma_p \approx 1/9$. The latest experiment (157) in ${}^5_\Lambda\text{He}$ reports a value

$$\frac{\Gamma_n}{\Gamma_p} \approx 1.30^{+0.65}_{-1.3}, \qquad 15.$$

in rough agreement with the range of values 0.8–2 given by Miller et al (158) and the limit $\Gamma_n/\Gamma_p \geq 1.4$ found by Kenyon et al (159). Although Equation 15 does not rule out the π exchange model, the earlier results (158, 159) appear to do so.

The inclusion of heavy meson exchange, most importantly the K, augments the ${}^3S_1 \to {}^{33}P_1$ rate and also Γ_n/Γ_p. Dubach (148) gives $\Gamma_n/\Gamma_p \approx 1/3$ for s-wave ($L = 0$) ΛN pairs with spin-isospin weights appropriate to an $N = Z$ core. A similar result $\Gamma_n/\Gamma_p \approx 0.37$ is obtained with a hybrid quark model description (155). Note, however, that the experimental error bar in Equation 15 is very large, which reflects the difficulty in ascertaining that the observed nn pairs actually arise from hypernuclear decay. Thus it is perhaps premature to draw a firm conclusion regarding the failure of a meson exchange approach.

The dynamics of $\Lambda N \to NN$ weak conversion presents an important open problem, in view of the unsatisfactory state of existing meson exchange or quark model calculations. Although calculated total nonmesonic rates agree with the data (160) within a factor of two (140, 149–155), the balance between parity-conserving and parity-violating transitions is not understood. The validity of the $\Delta I = 1/2$ rule also remains unclear. Better data on Γ_n/Γ_p are needed, particularly for $A = 3, 4, 5$ hypernuclei. In the future, the study of weak decays of polarized hypernuclei (161, 162) may give access to terms involving the interference of parity-conserving and -violating amplitudes (e.g. $^3S_1 \to {}^{13}D_1$ and $^3S_1 \to {}^{33}P_1$). One also anticipates that the interesting measurements of the weak decay lifetimes of heavy hypernuclei formed in antiproton-nucleus annihilations (163, 164) will continue at CERN.

5.   FUTURE DIRECTIONS

In the next decade, new facilities producing high energy beams for nuclear physics will be constructed and used for the study of nuclear systems with strangeness.

At Brookhaven National Laboratory, a new beam line is scheduled for commissioning in 1990; this line, called LESB-III, will be capable of producing 10^6 kaons in the momentum region of 1–2 GeV/c (165). In this region, the elementary (K^-, K^+) reaction cross section reaches a maximum. This double strangeness and charge exchange reaction deposits two units of strangeness in the nuclear medium and is thus capable of producing hypernuclei containing two Λ hyperons, cascade (Ξ^-) hypernuclei, and $S = -2$ dibaryons. The $\Lambda\Lambda$ hypernucleus offers the only practical milieu for the study of the interaction between Λ hyperons.

A vigorous program of stopped kaon and (π, K) reaction studies for hypernuclei is envisioned for the near future at KEK, with the installation of two new kaon beam lines (K5 and K6). A high acceptance, superconducting, toroidal spectrometer with 12 gaps will be used mainly for stopped kaon experiments for Λ hypernuclei and kaon rare decay studies. A superconducting, large-gap, dipole magnet (SKS), covering a higher momentum range, will be useful for (π^+, K^+) and (K^-, K^+) reaction studies. These devices will be able to take advantage of the installation of a new 1-GeV, high intensity, proton linear accelerator injector into the KEK proton synchrotron. An order of magnitude improvement over the present events rates is anticipated (166).

At the Continuous Electron Beam Accelerator Facility (CEBAF) at

Newport News, Virginia, it will be possible to explore the process $(e,e'K^+)$ for the production of hypernuclei (167–170). The electroproduction reaction has certain features in common with (π^+,K^+), for example, a high momentum transfer. The electroproduction operator, however, has a sizeable spin-flip component, which allows the production of both natural and unnatural parity states of the recoiling hypernuclear system. Plans are now being formulated to supplement one of the planned CEBAF spectrometers with a kaon spectrometer capable of high resolution hypernuclear studies.

A "kaon factory," with much higher energies for secondary and tertiary beams, promises qualitative advances in the domains of nuclear and particle physics and their interface. Two major facilities of this kind, at Los Alamos and at Vancouver, have been proposed. We would like to emphasize the fundamental importance of investigations of strange-particle interactions in nuclei as an integral part of the physics program at a "kaon factory."

Hypernuclear physics research is still in a primitive phase in which a few key experiments can lead to major progress in our understanding. A brief list of the open questions would certainly include the following:

1. What are the properties [level spacings, $B(E2)$ values, magnetic moments, etc] of hypernuclear excitations in heavy hypernuclei?
2. How are nuclear collective motions (moments of inertia, vibrational frequencies, etc) affected by the presence of a hyperon?
3. Do narrow Σ- or Ξ-nuclear excitations exist? What are the spin-orbit and symmetry potentials for the Σ and Ξ?
4. What are the energy splittings of ground-state doublets in Λ hypernuclei?
5. What are the spin-dependent components of the ΛN effective interaction? What is the role of three-body ΛNN forces?
6. What dynamical symmetries occur in $\Lambda\Lambda$ hypernuclei? Can we extract the $\Lambda\Lambda$ effective interaction?
7. Do the H particle and other strange dibaryons exist?
8. Can strangeness be used as a measure of quark deconfinement in nuclear matter?
9. What are the consequences of the Pauli principle at the quark level for a hyperon in the nucleus?
10. What is the microscopic origin of the strong spin-isospin selectivity apparently seen in the weak process $\Lambda N \to NN$?
11. How can we produce multiply strange nuclear systems? What are their properties?

The question of the existence and stability of multiply strange systems, perhaps in the form of "strange matter" (171, 172) or "strangelets" produced from the quark-gluon plasma (173), is a fascinating one. These and many other questions remain for future generations of machines and scientists to answer.

"... O day and night, but this is wondrous strange. ... There are more things in heaven and earth, Horatio, than are dreamt of in your philosophy."

Hamlet, I, v.

Literature Cited

1. Davis, D. H., Sacton, J., in *High Energy Physics*, ed. E. H. S. Burhop. New York: Academic (1967), 2: 365
2. Povh, B., *Annu. Rev. Nucl. Part. Sci.* 28: 1 (1978); *Prog. Part. Nucl. Phys.* 5: 245 (1981)
3. Davis, D. H., Pniewski, J., *Contemp. Phys.* 27: 91 (1986)
4. Gal, A., *Adv. Nucl. Phys.* 8: 1 (1975)
5. Dalitz, R. H., in *Proc. Int. Conf. on Nucl. Phys.*, Berkeley, ed. R. M. Diamond, J. O. Rasmussen. Amsterdam: North-Holland (1981), p. 101
6. Dover, C. B., *Nucl. Phys.* A374: 359c (1982); in *Proc. Int. Nucl. Phys. Conf.*, Harrogate, Engl., ed. J. L. Durell, J. M. Irvine, G. C. Morrison. Bristol: IOP Publ., Inst. Phys. Conf. Series No. 86 (1987), p. 99; in *Proc. Int. Symp. Medium Energy Phys.*, ed. H.-C. Chian, L.-S. Zheng. Singapore: World Sci. (1987), p. 257
7. Chrien, R. E., ed., *Proc. Int. Symp. on Hypernuclear and Kaon Phys.*, Brookhaven. *Nucl. Phys.* A450: 1c–586c (1986)
8. Bandō, H., Hashimoto, O., Ogawa, K., eds., *Proc. INS Int. Symp. on Hypernuclear Phys.*, Tokyo, Inst. for Nucl. Study, Univ. Tokyo (1986), pp. 1–448
9. Speth, J., ed., *Proc. Int. Symp. on Strangeness in Hadronic Matter*, Bad Honnef, Germany, *Nucl. Phys.* A479: 1c–466c (1988)
10. Bressani, T., ed., *Proc. Int. Symp. on Hypernuclear and Low Energy Kaon Phys.*, Padua, Italy (1988); *Nuovo Cimento* A102 (1989)
11. Gal, A., Klieb, L., *Phys. Rev.* C34: 956 (1986)
12. Morimatsu, O., Yazaki, K., *Nucl. Phys.* A435: 727 (1985), A483: 493 (1988)
13. Dover, C. B., Ludeking, L., Walker, G. E., *Phys. Rev.* C22: 2073 (1980)
14. Bandō, H., Motoba, T., *Prog. Theor. Phys.* 76: 1321 (1986)
15. Enge, H., in *Proc. LAMPF II Workshop*, Los Alamos Rep. LA-9416-C (1982), ed. H. A. Thiessen, p. 34;

Enge, H., Kowalski, S. B., *AIP Conf. Ser.* ed. R. E. Mischke. New York: AIP (1984), 123: 824
16. Holt, J. A., Imazato, J., Nakai, K., eds., *KEK-PS Users Guide Book*, Natl. Lab. for High Energy Phys. (KEK), Tsukuba, Japan (1988); Fukuda, M., Doctoral dissertation, Osaka Univ. (1988), unpublished
17. Tamura, H., et al., *Nucl. Phys.* A479: 161c (1988)
18. Chrien, R. E., et al., *Phys. Lett.* B89: 31 (1979)
19. Dover, C. B., Gal, A., Walker, G. E., Dalitz, R. H., *Phys. Lett.* B89: 26 (1979)
20. Bamberger, A., et al., *Nucl. Phys.* B60: 1 (1973)
21. Bedjidian, M., et al., *Phys. Lett.* B62: 467 (1976)
22. Bedjidian, M., et al., *Phys. Lett.* B83: 252 (1980)
23. May, M., et al., *Phys. Rev. Lett.* 51: 2085 (1983)
24. Gibson, B. F., Lehman, D. R., *Phys. Rev.* C23: 573 (1981); in *Proc. Int. Conf. on Hypernuclear and Kaon Phys.*, Heidelberg, ed. B. Povh, Max Planck Inst. Rep. MPIH-1982-V20 (1982), p. 161; Gibson, B. F., *Nucl. Phys.* A450: 243c (1986)
25. Bodmer, A. R., Usmani, Q. N., *Nucl. Phys.* A477: 621 (1988)
26. Gibson, B. F., Lehman, D. R., *Phys. Rev.* C23: 404 (1981)
27. Bodmer, A. R., Usmani, Q. N., *Phys. Rev.* C31: 1400 (1985)
28. May, M., *Nucl. Phys.* A450: 179c (1986)
29. Chrien, R. E., in *Proc. 4th Int. Symp. on Mesons and Light Nuclei*, Bechyně, Czech., ed. R. Mach; *Czech. J. Phys.* (1989) In press; Chrien, R. E., et al., *Brookhaven preprint* (1989)
30. Gal, A., Soper, J. M., Dalitz, R. H., *Ann. Phys.* 63: 53 (1971); 72: 445 (1972); 113: 79 (1978)
31. Dalitz, R. H., Gal, A., *Ann. Phys.* 116: 167 (1978)
32. Millener, D. J., Gal, A., Dover, C. B.,

Dalitz, R. H., *Phys. Rev.* C31: 499 (1985)

33. Nagels, M. M., Rijken, T. A., de Swart, J. J., *Phys. Rev.* D12: 744 (1975); D15: 2547 (1977)
34. May, M., et al., *Phys. Rev. Lett.* 47: 1106 (1981)
35. Auerbach, E. H., et al., *Ann. Phys.* 148: 381 (1983)
36. Milner, E. C., et al., *Phys. Rev. Lett.* 54: 237 (1985)
37. Chrien, R. E., *Nucl. Phys.* A478: 705c (1988)
38. Pile, P. H., in *Clustering Aspects in Nuclear and Subnuclear Systems*, eds. K. Ikeda, K. Katori, Y. Suzuki. Suppl. to *J. Phys. Soc. Jpn.* 58: 394 (1989)
39. Halderson, D., Yizhang, M., Pingzhi, N., *Phys. Rev. Lett.* 57: 1117 (1986)
40. Bandō, H., Motoba, T., *Prog. Theor. Phys.* 76: 1321 (1986)
41. Motoba, T., *Nucl. Phys.* A479: 227c (1988)
42. Motoba, T., Bandō, H., Wünsch, R., Žofka, J., *Phys. Rev.* 38: 1322 (1988)
43. Hausmann, R., Weise, W., *Nucl. Phys.* A491: 598 (1989)
44. Yamazaki, T., et al., *Phys. Rev. Lett.* 54: 102 (1985)
45. Hayano, R. S., *Nucl. Phys.* A478: 113c (1988)
46. Paul, S., et al., *Nucl. Phys.* A479: 137c (1988)
47. Iwasaki, M., Doctoral dissertation, Tokyo Univ. (1987), unpublished
48. Bertini, R., et al., *Phys. Lett.* B90: 375 (1980); Mayer, B., *Nukleonika* 25: 439 (1980)
49. Bertini, R., et al., *Phys. Lett.* B136: 29 (1984)
50. Bertini, R., et al., *Phys. Lett.* B158: 19 (1985); Walcher, T., *Nucl. Phys.* A479: 63c (1988)
51. Piekarz, H., et al., *Phys. Lett.* B110: 428 (1982)
52. Chrien, R. E., Kishimoto, T., Hungerford, E. V., *Phys. Rev.* C35: 1589 (1987)
53. Wünsch, R., Žofka, J., *Phys. Lett.* B193: 7 (1987)
54. Kohno, M., Hausmann, R., Siegel, P. B., Weise, W., *Nucl. Phys.* A470: 609 (1987)
55. Halderson, D., Philpott, R. J., *Phys. Rev.* C37: 1104 (1988)
56. Tang, L., et al., *Phys. Rev.* C38: 846 (1988)
57. Hayano, R. S., See Ref. 10
58. Roosen, R., et al., *Nuovo Cimento* 49: 217 (1979)
59. Harada, T., Myint, K. S., Akaishi, Y., *Hokkaido Univ. preprint* (1988)
60. Batty, C. J., Gal, A., Toker, G., *Nucl. Phys.* A402: 349 (1983)

61. Dover, C. B., Gal, A., Millener, D. J., *Phys. Lett.* B138: 337 (1984)
62. Gal, A., Dover, C. B., *Phys. Rev. Lett.* 44: 379 (1980); Dabrowski, J., Rozynek, J., *Phys. Rev.* C23: 1706 (1981); Johnstone, J. A., Thomas, A. W., *Nucl. Phys.* A392: 409 (1983); Brockmann, R., Oset, E., *Phys. Lett.* B118: 33 (1982); Dover, C. B., Feshbach, H., *Phys. Rev. Lett.* 59: 2539 (1987)
63. Pirner, H. J., *Phys. Lett.* B85: 190 (1979)
64. Morimatsu, O., Ohta, S., Shimizu, K., Yazaki, K., *Nucl. Phys.* A420: 573 (1984)
65. He, Y., Wang, F., Wong, C. W., *Nucl. Phys.* A448: 652 (1986); A451: 653 (1986); A454: 541 (1986)
66. Brockmann, R., *Phys. Lett.* B104: 256 (1981); Bouyssy, A., *Nucl. Phys.* A381: 445 (1982)
67. Dover, C. B., Gal, A., Millener, D. J., *Phys. Rep.* In press (1989)
68. Dover, C. B., Gal, A., *Prog. Part. Nucl. Phys.* 12: 171–239
69. Yamamoto, Y., Bandō, H., *Prog. Theor. Phys. Suppl.* 81: 9 (1985); *Phys. Lett.* B214: 173 (1988)
70. Nagata, S., Bandō, H., *Prog. Theor. Phys.* 72: 113 (1984)
71. Yamamoto, Y., *Prog. Theor. Phys.* 75: 639 (1986)
72. Kohno, M., *Prog. Theor. Phys.* 78: 123 (1987)
73. Bodmer, A. R., Usmani, Q. N., *Nucl. Phys.* A450: 275c (1986)
74. Bodmer, A. R., Usmani, Q. N., Carlson, J., *Phys. Rev.* C29: 684 (1984)
75. Shoeb, M., Rahman Khan, M. Z., *J. Phys. G: Nucl. Phys.* 10: 1047 (1984)
76. Ansari, H. H., Shoeb, M., Rahman Khan, M. Z., *J. Phys. G: Nucl. Phys.* 12: 1369 (1986)
77. Ahmad, I., Mian, M., Rahman Khan, M. Z., *Phys. Rev.* C31: 1590 (1985)
78. Mian, M., *Phys. Rev.* C35: 1463 (1987)
79. Daskaloyannis, C. B., Grypeos, M. E., Koutroulos, C. G., Massen, S. E., Saloupis, D. S., *Phys. Lett.* B121: 91 (1983)
80. Daskaloyannis, C., Grypeos, M., Koutroulos, C., Saloupis, D., *Lett. Nuovo Cimento* 42: 257 (1985)
81. Grypeos, M. E., Koutroulos, C. G., *Nucl. Phys.* A450: 307c (1986)
82. Rahman Khan, M. Z., Shoeb, M., *J. Phys. Soc. Jpn.* 55: 3008 (1986)
83. Hungerford, E. V., Biedenharn, L. C., *Phys. Lett.* B142: 232 (1984)
84. Takeuchi, S., Shimizu, K., *Phys. Lett.* B179: 197 (1986)
85. Millener, D. J., Dover, C. B., Gal, A., *Phys. Rev.* C38: 2700 (1988)

86. Yamamoto, Y., Bandō, H., Žofka, J., *Prog. Theor. Phys.* 80: 757 (1988)
87. Negele, J. W., Vautherin, D., *Phys. Rev.* C5: 1472 (1972)
88. Rayet, M., *Ann. Phys.* 102: 226 (1976); *Nucl. Phys.* A367: 381 (1981)
89. Vautherin, D., Brink, D. M., *Phys. Rev.* C5: 626 (1972)
90. Goldman, T., in *Intersections Between Part. and Nucl. Phys.*, Steamboat Springs, Colorado, AIP Conf. Proc. No. 123, ed. R. E. Mischke. New York: AIP (1984), p. 799
91. Yamazaki, T., *Nucl. Phys.* A463: 39c (1987)
92. Auerbach, E. H., et al., *Phys. Rev. Lett.* 47: 1110 (1981)
93. Dalitz, R. H., Gal, A., *Phys. Rev. Lett.* 36: 362 (1976)
94. Sunami, Y., Narumi, H., *Prog. Theor. Phys.* 66: 355 (1981)
95. Revai, J., Žofka, J., *Phys. Lett.* 101B: 228 (1981)
96. Portilho, O., Alencar, P. S. C., Coon, S. A., *Few-Body Systems, Suppl.* 2: 417 (1987)
97. Motoba, T., Bandō, H., Ikeda, K., *Prog. Theor. Phys.* 79: 189 (1983)
98. Yamada, T., Ikeda, K., Bandō, H., Motoba, T., *Prog. Theor. Phys.* 73: 397 (1985); *Phys. Lett.* B172: 149 (1986)
99. Mobota, T., Bandō, H., Ikeda, K., Yamada, T., *Prog. Theor. Phys. Suppl.* 81: 42 (1985)
100. Yamada, T., Mobota, T., Ikeda, K., Bandō, H., *Prog. Theor. Phys. Suppl.* 81: 104 (1985)
101. Yamada, T., Ikeda, K., Bandō, H., Motoba, T., *Phys. Rev.* C38: 854 (1988)
102. Bodmer, A. R., Usmani, Q. N., Carlson, J., *Nucl. Phys.* A422: 510 (1984)
103. Bodmer, A. R., Usmani, Q. N., *Nucl. Phys.* A468: 653 (1987)
104. Brückner, W., et al., *Phys. Lett.* B79: 157 (1978)
105. Bertini, R., et al., *Nucl. Phys.* A360: 315 (1981)
106. Bouyssy, A., *Phys. Lett.* B91: 15 (1980)
107. Dalitz, R. H., Gal, A., *Ann. Phys.* 131: 314 (1981)
108. Dalitz, R. H., Herndon, R. C., Tang, Y. C., *Nucl. Phys.* B47: 109 (1972)
109. Bonazzola, G. C., et al., *Phys. Rev. Lett.* 34: 683 (1975)
110. Brückner, W., et al., *Phys. Lett.* B55: 107 (1975); B62: 481 (1976)
111. Bertini, R., et al., *Nucl. Phys.* A368: 365 (1981)
112. Prowse, J., *Phys. Rev. Lett.* 17: 782 (1966)
113. Danysz, M., et al., *Nucl. Phys.* 49: 121 (1963); *Phys. Rev. Lett.* 11: 29 (1963)
114. Dover, C. B., Gal, A., *Ann. Phys.* 146: 309 (1983)
115. Bodmer, A. R., Usmani, Q. N., *Nucl. Phys.* A463: 221c (1987)
116. Miyahara, K., Ikeda, K., Bandō, H., *Prog. Theor. Phys.* 69: 1717 (1983)
117. Bandō, H., *Prog. Theor. Phys.* 69: 1731 (1983)
118. Ikeda, K., Bandō, H., Motoba, T., *Prog. Theor. Phys. Suppl.* 81: 147 (1985)
119. Wang, X. C., Takaki, H., Bandō, H., *Prog. Theor. Phys.* 76: 865 (1986)
120. Baltz, A. J., Dover, C. B., Millener, D. J., *Phys. Lett.* 123B: 9 (1983)
121. May, M., See Ref. 10
122. Shoeb, M., Rahman Khan, M. Z., *J. Phys. G: Nucl. Phys.* 10: 1739 (1984)
123. Yamamoto, Y., *Prog. Theor. Phys.* 75: 639 (1986)
124. Lalazissis, G. A., Grypeos, M. E., Massen, S. E., *J. Phys. G: Nucl. Part. Phys.* 15: 303 (1989)
125. Jaffe, R. L., *Phys. Rev. Lett.* 38: 195 (1977)
126. Mulders, P. J., Aerts, A. T. M., de Swart, J. J., *Phys. Rev.* D19: 2635 (1979); D21: 2653 (1980)
127. Aerts, A. T. M., Dover, C. B., *Phys. Lett.* B146: 95 (1984)
128. Aerts, A. T. M., Dover, C. B., *Nucl. Phys.* B253: 116 (1985)
129. Piekarz, H., *Nucl. Phys.* A479: 263c (1988)
130. Frascaria, R., See Ref. 10
131. Dover, C. B., *Nucl. Phys.* A450: 95c (1986); Locher, M. P., Sainio, M. E., Svarc, A., *Adv. Nucl. Phys.* 17: 47 (1986)
132. Carroll, A. S., et al., *Phys. Rev. Lett.* 41: 777 (1978)
133. Condo, G. T., et al., *Phys. Lett.* B144: 27 (1984)
134. Shahbazian, B. A., Kechechyan, A. O., Tarasov, A. M., Martynov, A. S., *Z. Phys.* C39: 151 (1988)
135. Aoki, S., et al., in *Abstr. 17th INS Int. Symp. on Nucl. Phys.* at *Intermediate Energy*, Tokyo, Japan, ed. T. Nagae, T. Fukuda, Inst. for Nucl. Study, Univ. Tokyo (1988), p. 2
136. Franklin, G. B., *Nucl. Phys.* A450: 117c (1986)
137. Aerts, A. T. M., Dover, C. B., *Phys. Rev. Lett.* 49: 1752 (1982); *Phys. Rev.* D28: 450 (1983)
138. Aerts, A. T. M., Dover, C. R., *Phys. Rev.* D29: 433 (1984)
139. Buras, A. J., *Nucl. Phys.* A479: 399c (1988)
140. Dover, C. B., Walker, G. E., *Phys. Rep.* 89: 1 (1982)
141. Bandō, H., Takaki, H., *Prog. Theor.*

Phys. 72: 106 (1984); *Phys. Lett.* 150B: 409 (1985)

142. Itonaga, K., Motoba, T., Bandō, H., Z. *Phys.* A330: 209 (1988)
143. Motoba, T., Itonaga, K., Bandō, H., *Nucl. Phys.* A489: 683 (1988)
144. Oset, E., Salcedo, L. L., *Nucl. Phys.* A443: 704 (1985)
145. Kurihara, Y., Akaishi, Y., Tanaka, H., *Phys. Rev.* C31: 971 (1985)
146. Barnes, P., *Nucl. Phys.* A478: 127c (1988); A479: 89c (1988); McKellar, B. H. J., See Ref. 8, p. 146
147. Carr, J. A., McManus, H., Stricker-Bauer, K., *Phys. Rev.* C25: 952 (1982)
148. Dubach, J. F., *Nucl. Phys.* A450: 71c (1986); in *Proc. Int. Conf. on Intersections Between Part. and Nucl. Phys.*, Lake Louise, Canada, ed. D. F. Geesaman, AIP Conf. Proc. No. 150. New York: AIP (1986), p. 946
149. Dover, C. B., *Few-Body Systems, Suppl.* 2: 77 (1987)
150. Gibson, B. F., See Ref. 10
151. McKellar, B. H. J., Gibson, B. F., *Phys. Rev.* C30: 322 (1984); Nardulli, G., *Phys. Rev.* C38: 832 (1988)
152. Takeuchi, K., Takai, H., Bandō, H., *Prog. Theor. Phys.* 73: 841 (1985)
153. de la Torre, L., Donaghue, J. F., Dubach, J. F., Holstein, B. R., Cited in Ref. 148
154. Cheung, C.-Y., Heddle, D. P., Kisslinger, L. S., *Phys. Rev.* C27: 335 (1983)
155. Heddle, D. P., Kisslinger, L. S., *Phys. Rev.* C33: 608 (1986)
156. Block, M. M., Dalitz, R. H., *Phys. Rev. Lett.* 11: 96 (1963)
157. Szymanski, J. J., Doctoral dissertation,

Carnegie Mellon Univ. (1987), unpublished
158. Miller, H. G., Holland, M. W., Roalsvig, J. P., Sorenson, R. G., *Phys. Rev.* 167: 922 (1968)
159. Kenyon, I. R., et al., *Nuovo Cimento* 30: 1365 (1963)
160. Grace, R., et al., *Phys. Rev. Lett.* 55: 1055 (1985)
161. Ejiri, H., et al., *Phys. Rev.* C36: 1435 (1987)
162. Bandō, H., Motoba, T., Sotona, M., *Phys. Rev.* C39: 587 (1989)
163. Bocquet, J. P., et al., *Phys. Lett.* B182: 146 (1986)
164. Polikanov, S., *Nucl. Phys.* A478: 805c (1988)
165. Pile, P. H., *Nucl. Phys.* A450: 517c (1986)
166. Yamazaki, T., See Ref. 10
167. Cotanch, S. R., Hsiao, S. S., *Nucl. Phys.* A450: 419c (1986); *Phys. Rev.* C28: 1668 (1983)
168. Donnelly, T. W., Bernstein, A. M., Epstein, G. N., *Nucl. Phys.* A358: 195 (1981)
169. Rosenthal, A. S., Halderson, D., Tabakin, F., *Phys. Lett.* B182: 143 (1986); Rosenthal, A. S., Halderson, D., Hodgkinson, K., Tabakin, F., *Ann. Phys.* 184: 31 (1988)
170. Cohen, J., *Phys. Rev.* C32: 543 (1985); *Int. J. Mod. Phys.* A4: 1–78 (1989)
171. Witten, E., *Phys. Rev.* D30: 272 (1984)
172. Farhi, E., Jaffe, R. L., *Phys. Rev.* D30: 2379 (1984); D32: 2452 (1985)
173. Greiner, C., Rischke, D. H., Stöcker, H., Koch, P., *Phys. Rev.* D38: 2797 (1988)

Annu. Rev. Nucl. Part. Sci. 1989. 39: 151–82

SOME TECHNICAL ISSUES IN ARMS CONTROL

H. L. Lynch

Stanford Linear Accelerator Center, Stanford, California 94309, USA

R. Meunier[1]

CERN, 1211 Geneva 23, Switzerland

D. M. Ritson

Department of Physics, Stanford University, Stanford, California 94305, USA

KEY WORDS: nuclear proliferation, directed energy, missile defense, SLCM.

CONTENTS

[1] Visitor from CEA, Saclay, France.

0163–8998/89/1201–0151$02.00

1. INTRODUCTION

Arms control is not a goal by itself; rather it is a means to reduce the danger of war. The arms control process is primarily a political one, but technical reality must direct and temper those efforts. Physicists are in part responsible for the existence of the awesome weapons that now threaten the world; they must play an important role in providing sound technical judgment to the political leaders and the population as a whole to deal with these inventions. It is the aim of this article to explore just a few technical issues in arms control: nuclear proliferation, directed energy weapons, and nuclear armed sea-launched cruise missiles. Many other areas, not treated by reason of the limited scope of this article, are equally interesting, such as nuclear testing verification measures, the need for nuclear testing for weapon stockpile maintenance, detection of nuclear weapons in space, surveillance techniques to increase confidence, submarine detection techniques, and nuclear winter.

Nuclear weapons are at this time confined to a relatively small number of holders. For a long time, there has been considerable concern that these weapons would spread to many hands, some of which might not be responsible. The first section deals with the problems faced by parties, be they countries or terrorist groups, who wish to acquire nuclear weapons for whatever purpose they see fit. There are technical reasons why this proliferation has been rather slow.

Defense against strategic missiles has taken a new importance since President Reagan announced the inception of the Strategic Defense Initiative (SDI). This is an area of very advanced technology whose potential scope exceeds all previous weapons systems of the past. There are many technological aspects of the SDI, and the second section discusses how directed energy weapons might be used and what difficulties that application faces.

The world is finally ready for major reductions of nuclear weapons by the major powers, the US and USSR. Having concluded the intermediate range nuclear forces (INF) agreement to eliminate an entire class of nuclear weapons in Europe, both parties are trying to go on to reduce the enormous stores of strategic nuclear weapons. Any such agreements depend upon a certain amount of trust between the two parties, but that trust must be based on other sources of confidence that the agreement is being upheld. Sea-launched cruise missiles are new delivery vehicles for nuclear weapons, and their very nature considerably undermines the means used in the past to determine that an adversary adhered to agreed-upon limits of deployment. The last section discusses some of the problems associated with monitoring such missiles.

2. NUCLEAR PROLIFERATION

The official members of the nuclear club—the US, the USSR, France, England, and China—have stockpiled $\sim 25,000$ strategic nuclear warheads and an equivalent number of tactical warheads. Disarmament talks between the big powers naturally center on the control and/or possible elimination of these stockpiles. One is only too aware these days of the ease with which even an individual can obtain access to chemical explosives. They can be obtained via direct sale, illegal acquisition, or by simple manufacture. Similar possibilities exist for the acquisition of nuclear explosives (1, 2), and it has long been of concern that nuclear explosives would proliferate to nations, terrorists, and possibly even to individuals. It is feared that this might lead to the use of nuclear explosives in local wars or for acts of terrorism, and that such usage could conceivably trigger a global nuclear exchange. Proliferation outside the nuclear nations presently appears limited to some nonsignatories to the nuclear proliferation treaty, namely India (actual testing), Israel [strong circumstantial evidence (3)], and possibly South Africa, Pakistan, Brazil, and Argentina (4). Many other nations such as Canada, Germany, Japan, Sweden, and Italy have very sophisticated nuclear production and delivery capabilities. Without question such nations could very quickly produce nuclear weapons should they perceive a need for them.

Even though proliferation to date has been limited and has not led to any disaster, this should not lead to complacency. Nuclear technologies are commercially obtainable, and training in their use is readily available. We examine below the feasibility of possible routes to proliferation that could result in a limited production of the order of one bomb a year.

2.1 *Requirements for Fast-Fission Nuclear Bombs*

The basic physics of bomb construction is well described in the now declassified 1942 Los Alamos lectures of Serber (5). ^{235}U, ^{239}Pu, and ^{233}U are fissile by neutrons with the production of more than one neutron per fission reaction. The fissile core of the bomb is constructed of one of these materials and is surrounded by a tamper of heavy material, typically tungsten or depleted uranium, both to reflect back neutrons that would otherwise be lost from the fissile core and to provide ballast to slow the expansion of the exploding core and thus maintain criticality. The minimum masses required for efficient nuclear explosions are approximately 20 kg for ^{235}U, 6 kg for ^{239}Pu, and 2 kg for ^{233}U (6, 7). ^{233}U meets the requirements of small critical masses but has not been used as a nuclear explosive. Its production by neutron irradiation of ^{232}Th is accompanied by the production of the undesirable uranium isotopes [232]U, a hard gamma emitter, and [234]U, a dilutant.

Nuclear explosions are produced by combining subcritical pieces into a supercritical mass. If the chain reaction is started prematurely by stray neutrons while in a barely critical configuration, the energy release will be low (a fizzle). Complete fission of 1 kg of plutonium releases energy equivalent to 17 kilotons of conventional chemical explosive, but only a fraction of this energy is released before the fissile core becomes subcritical. In modern sophisticated weaponry a small admixture of tritium is added to enhance combustion efficiencies. The tritium admixture enhances fast neutron production by fusion (8). Efficient energy release further requires very low ambient neutron backgrounds and/or very fast assembly to prevent premature ignition while the core is still only marginally critical. The time scale for a chain reaction to develop into an explosion is $\sim 10^{-7}$ s. The time scale for mechanical assembly is set by the maximum velocities that can be imparted by a chemical explosion typically 1000 m s^{-1} for an explosion and 7000 m s^{-1} for sophisticated implosions. The required movements are of the order of centimeters and therefore mechanical assembly requires times of $\sim 10^{-5}$–10^{-6} s. If a stray neutron starts a chain reaction before assembly is completed, i.e. when it is barely critical, the chain reaction is terminated by vaporization and expansion of the assembly back to subcriticality. The bomb therefore must be designed to ensure that no stray neutrons are available to trigger the explosion prior to assembly. At the moment of final assembly the explosion is initiated by a deliberately induced trigger burst of neutrons. Historically the trigger burst of neutrons was obtained through the admixture of an alpha source with a light material by impaction at final assembly. Currently it appears that the preferred trigger is a burst of neutrons produced from a pulse of accelerated deuterons striking a target containing tritium (8, 9).

Ambient backgrounds of neutrons are provided by a number of sources. The most important of these are neutrons produced from light element contaminants (removable by chemical purification) and an irreducible background arising from spontaneous fission.[2] The background of neutrons is sufficiently low in separated [235]U and for pure [239]Pu that relatively conventional explosive assembly will suffice. [239]Pu, however, does not occur naturally; it is produced in reactors from [238]U by neutron capture followed by beta decay. The produced [239]Pu in turn captures neutrons with pro-

---

[2] Values of relevant partial half-lives for spontaneous fission are $\sim 10^{11}$ years for [240]Pu, $\sim 5 \times 10^{15}$ years for [239]Pu, $\sim 1 \times 10^{17}$ years for [233]U, $\sim 2 \times 10^{17}$ years for [235]U, and $\sim 8 \times 10^{15}$ years for [238]U; cf. the *Table of Radioactive Isotopes*, by E. Browne and R. B. Firestone, edited by V. Shirley and published by Wiley in 1986.

duction of ^{240}Pu. The produced plutonium is therefore isotopically impure, and the longer the irradiation the greater the contamination. ^{240}Pu is fissionable but also has a very high rate of spontaneous fission producing large neutron background. To achieve the fast assembly times needed to prevent premature ignition in a plutonium bomb requires implosion assembly. Implosion results from surrounding the pieces to be assembled by an explosive shell. This shell is ignited simultaneously at a large number of points and with suitable design produces a symmetric inwardly moving spherical shock front that provides criticality by an increase of density. Such techniques are highly sophisticated and not easily achievable. The difficulties of assembly increase with increasing isotopic contamination by ^{240}Pu. The isotopic contamination increases with the degree of neutron irradiation of the parent uranium. The original bomb fabricated at Los Alamos was made from irradiated uranium containing one part in four thousand of plutonium requiring chemical processing of approximately forty tons of irradiated uranium. Figure 1 shows a typical rate of production of plutonium isotopes vs irradiation (11). Plutonium produced at these dilutions contains approximately 1–2% isotopic contamination. For obvious reasons a large safety margin was demanded at that time. Plutonium suitable for use in bomb construction is defined by the US government as containing less than 6% isotopic contamination (10). Figure 1 shows a 6% isotopic contamination to be equivalent to a plutonium concentration in the irradiated parent uranium of about one part per thousand. At these concentrations it is necessary to process ~ 10 tons of uranium per bomb.

Herein lies the dilemma faced by the would-be bomb constructor. He cannot use natural uranium because the isotope of interest (^{235}U) is only present at the 0.7% level. Therefore highly enriched ^{235}U ($>90\%$ purity)

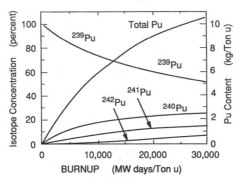

Figure 1   Rate of production of plutonium isotopes as a function of irradiation.

must be produced by large-scale isotopic separation, which is a difficult process. Once highly enriched uranium is obtained, the metallurgy and fabrication into an operational bomb are relatively straightforward. Alternatively, plutonium can be produced in a reactor through neutron irradiation and then separated and purified by conventional and well-known chemical techniques. This avoids the problems inherent in isotopic separation but results in the need for sophisticated implosion bomb assembly. We discuss these alternatives in detail below.

2.2   The Production of Bombs via Isotopic Separation

The original Hiroshima bomb was constructed with a 3 m long gun barrel that fired one subcritical hemisphere of uranium to mate with a second piece (12) to form a supercritical assembly. The rate of spontaneous fission is sufficiently low that efficient energy release was obtained with relatively conventional explosives technology.

The methods used for isotopic separation of uranium are based on small differences in physical properties such as the rates of gaseous diffusion of uranium hexafluoride gas through a porous barrier (13,14), the centrifugal separation of gaseous uranium hexafluoride (15), and more recently, laser separation through small differences in spectral excitation potentials of uranium vapor (16). Except for the latter process the enrichments achieved in a single stage are small and therefore substantial enrichment can only be achieved by the use of multistage cascaded plants. For instance the original US gaseous diffusion plants required several thousand stages. Centrifugal separation can be accomplished with substantially fewer stages, and currently is the preferred production method. Irrespective of the above, the plants for this purpose are very large, have large power consumption, and their operation cannot be disguised. The existing uranium isotope separation plants around the world are summarized in the fuel cycle survey of *Nuclear Engineering International* (17). Among nations listed with separation facilities able to provide usable quantities of weapons grade uranium are India, Pakistan, Brazil, Argentina, and South Africa.

The use of laser separation techniques has recently attracted considerable attention. Theoretically this process has the advantage over the diffusion or centrifugal separations in that it directly separates the wanted isotope and does not simply change slightly the relative isotopic proportions. The US Department of Energy (DOE) announced in 1985 the choice of the Atomic Vapor Laser Isotope Separation, AVLIS, as its preferred future enrichment technology (18). This system will use a "brute force" approach. Uranium will be vaporized into an atomic beam by bombardment with an electron beam. The ^{235}U in the atomic beam will be selectively ionized by precisely tuned dye lasers. The charged isotope will then pass through a magnet and be swept into a collector. The laser system will be relatively standard, consisting of a copper-excimer pump laser, tunable dye lasers, and a second-stage high-powered copper-excimer pump laser and amplifier dye lasers. Such a system would have to process many tons of natural uranium in the form of a low pressure atomic beam to produce even one nuclear warhead. It is both an expensive large-scale operation and one whose existence would be hard to conceal. Further, it has not yet been proven to operate in a production environment.

Present commercial plans envisage large-scale separation of reactor grade plutonium from the spent fuel rods in commercial reactors. It is feared that widespread availability of commercial reactor grade plutonium may be accompanied by clandestine diversions. Laser separation has been proposed to clean up reactor grade plutonium to bomb grade plutonium at the SIS facility at Hanford (10). To purify reactor grade plutonium requires a throughput of two orders of magnitude less material than is needed for uranium to provide one warhead. However, the strong alpha radioactivity of plutonium causes sputtering, and the atomic beam systems will require frequent dismantling and cleanup (18). This is rendered dangerous from the radioactive health hazards associated with plutonium. Neither is such a facility more readily concealed than a reactor. For a nation desirous of recycling reactor grade into weapons grade plutonium, the simplest route would be to use the isotopically contaminated plutonium to fuel a conventional reactor and subsequently to separate out from the reactor clean produced plutonium. Thus laser separation does not appear to offer easy or attractive routes to proliferation.

2.3   *Production of Weapons Grade Plutonium*

Plutonium is produced when ^{238}U is irradiated with neutrons. Natural uranium contains 99.3% ^{238}U and 0.7% ^{235}U. As the ^{238}U is the main component of nonenriched reactors, neutrons are predominantly lost by capture and production of plutonium. Highly enriched reactors can be made substantially more compact. Such compact reactors produce concomitantly smaller quantities of plutonium. Additionally, for the same output power levels the neutron flux levels of a more compact reactor are larger and neutron capture on ^{239}Pu will in turn lead to production of ^{240}Pu and of ^{241}Pu. These isotopes are highly undesirable components in material to be used for weapons production.

The production levels of plutonium can be easily estimated. Fission produces about 2.5 neutrons, of which only one is necessary to sustain the chain reaction, and up to one neutron is available for capture in ^{238}U. This corresponds to up to one atom of produced plutonium per fission or ~ 200 MeV of released energy. This in turn corresponds to ~ 0.35 kg of produced plutonium per MW year of thermal energy. In reality, for normal power reactor operation, the ratio of plutonium produced to fuel consumed is closer to $\sim 60\%$ (19). Typically the fuel is removed only after substantial burnup of the produced plutonium. A typical reactor with a 1-GW electric production rating (or 3-GW thermal rating) operating with a 70% capacity factor will produce ~ 250 kg of reactor grade plutonium annually (20, 21). With approximately three hundred such plants in operation around the world enormous quantities of plutonium are in fact produced in commercial power production, though only a small fraction is now reprocessed. Such plutonium is isotopically heavily contaminated [typically 40% contamination (22)] and luckily, as discussed above, ill suited to bomb construction.

A reactor producing weapons grade plutonium will of necessity have lower fuel burnup. Additionally it will be configured to provide somewhat higher ratios of produced plutonium to consumed uranium. Furthermore, should it use an efficient moderator such as heavy water, the number of excess neutrons available for plutonium production will be increased by up to a factor of two. Accordingly it is expected that efficiencies up to a factor of three higher are possible for dedicated plutonium production than for conventional power production. For instance, the Israeli heavy-water reactor at Dimona (3) is believed to be capable of producing up to 0.25 kg of weapons grade plutonium per MW year of thermal power. At this level of production relatively low power reactors, including research reactors, can produce plutonium sufficient for one or more bombs per year.

2.3.1 NATURAL URANIUM GRAPHITE REACTORS   Historically reactors based on natural uranium and graphite moderators were the main vehicles for production both of power and weapons grade materials. Reactors of this type must be large (~ 10 m cubic size and masses of thousands of tons) to attain criticality because neutrons are captured both in the ^{238}U and in the graphite moderators and cooling system. Such reactors require very pure graphite, and their structural materials and coolants must be selected for low neutron absorption. Despite the large size, technical engineering problems limit the total power output to levels comparable with those from substantially smaller modern power reactors. This of necessity retricts the reactors to operation at neutron flux levels an order of magnitude lower than for the corresponding light-water, modern reactors.

This characteristic of low flux densities makes these reactors very suitable for the production of weapons grade materials in large quantities. The on-load refueling feature of these reactors permits the maintenance of a schedule suitable for efficient military production of plutonium. They are extensively used by the large weapons-producing nations to manu-facture the bulk of their weapons grade materials.

The very large size of these reactors and the difficulties of obtaining large quantities (in the range of thousands of tons) of reactor grade graph-ite make them hard to control and unattractive for illicit plutonium pro-duction. A modern version of such reactors, the high temperature graphite (HTG) reactor (23), is under consideration in Germany and the US, mainly intended for tritium production (24). It would require exotic ceramic fuel technology and very large quantities of helium gas coolant. Even if loading and unloading of fuel for reprocessing proves to be relatively easy, the reprocessing of the ceramic fuel is unconventional, and the technology is speculative. Acquisition of helium gas on the required scale would be very difficult for nations intending to produce plutonium illicitly. In any event, as discussed below, much easier routes are available for medium- or small-scale plutonium production.

2.3.2 LIGHT-WATER POWER REACTORS   Nuclear electric power pro-duction throughout the world is based mainly on light-water reactors. The production of plutonium per year is much larger than that from dedicated weapons production reactors. Such reactors are technically separated into pressurized light-water (PWRs) and boiling water reactors (BWRs). The differences between the two classes are small and are mainly concerned with the mode of converting heat to power. For the purposes of plutonium production, we do not differentiate between the two reactor types.

There are 307 operable units spread through 22 countries in the world (25). With the easy commercial availability of such reactors, it might appear that they would provide the most important and least easily con-trolled avenue for illicit production of nuclear weapons. As we discuss below, this is not the case.

Commercial power reactors are of necessity intended to be economically competitive with conventional power production. To be economically competitive they operate at high temperature in order to achieve good thermal efficiencies. Efficient conventional power plants require boilers at high pressure, and analogously nuclear power production also requires pressurized assemblies. Such reactors operate at comparatively high power density. The fuel elements are typically enriched to 3%. Lower enrichment factors would limit the use of necessary structural elements such as stainless steel tubing and light-water heat transfer elements, both neutron absorb-ing. In addition, to reach criticality the reactor would have to be larger, with attendant problems for the containment structures, particularly the pressure vessel. Enrichment factors for the fuel assemblies beyond the 3% level would lead to substantially higher costs for preparation of fuel elements and would result in a smaller core restricted to lower power levels for a given reactor. Over the last thirty years technical considerations and constraints have resulted in this class of reactors being constructed with almost uniform size and utilizing fuel with 3% enrichment. Typically a 1-GW reactor requires yearly replacement of about one third of its fuel elements and reconfiguring of the core. Replacement of fuel elements requires an extensive shut down. Mechanical access to the pressurized systems is not possible without allowing a substantial interval for thermal and radioactive cool down and is therefore time consuming. Technical development is thus aimed at extending the periods between shutdowns by attaining more complete fuel utilization through higher burnup ratios. In a modern plant, the aim is to achieve up to 40 GW days per ton of fuel before replacement. At this level of operation the produced plutonium contains only $\sim 60\%$ ^{239}Pu, with a 40% admixture of higher isotopes, to be contrasted with the US Department of Energy requirement of 94% isotopic purity for weapons grade plutonium. Using such reactors to produce weapons grade plutonium would require operations schedules and power production outputs quite different from those used in normal operation, and would be easily detectable. Therefore these reactors are not designed for weapons grade plutonium production, and clandestine usage for such purposes is essentially precluded.

2.3.3 HEAVY-WATER MODERATED REACTORS   Heavy water does not appreciably capture slow neutrons. It is thus a very effective moderator that leaves a substantial fraction of the 2.5 neutrons produced in fission available to sustain the chain reaction and to be used for other purposes. Almost one neutron can be made available for production of plutonium, and it can approach a breeder regime. It can easily reach criticality with the natural unenriched uranium. Because of the high cost of enrichment, most commercial reactors using heavy water as a moderator use natural or only very slightly enriched uranium for fuel. For illicit production of plutonium the ready and essentially unrestricted availability of natural uranium for fuel and the possibility of illicit acquisition of heavy water[3] (26) make this class of reactors of particular interest. Easy attainment of criticality permits a very wide range of configurations to be used, from very

[3] For example G. Milhullin reported three known diversions of heavy water produced in Norway almost certainly intended for use in weapons production. In 1959 Norway exported 20 tons of heavy water to the Israeli Dimona reactor and 1 ton in 1970. In 1983 West Germany transshipped 15 tons of heavy water from Norway to Bombay, and subsequent to 1986 Romania diverted 12.5 tons of heavy water to an unknown destination, probably India or Israel (*Herald Tribune*, 8 October 1988).

small to very large. The original German wartime reactor was designed for two tons of natural uranium and two tons of heavy water (27). Providentially this project was stopped from coming to fruition by the destruction of the Norwegian heavy-water plant and supplies of heavy water.

In the commercial context, the savings on uranium enrichment costs are offset by the costs of separating out the heavy water. One of the routes taken by commercial producers of heavy-water reactors (HWRs) is to use large reactors, necessitated partially by the use of unenriched uranium, and to cool the fuel elements individually with pressurized tubes. The commercial advantage of this cooling method is that replacement of individual fuel elements can occur online without shutting down the reactor. From a commercial viewpoint this provides potentially longer times online for these reactors. For a country interested in clandestine production of plutonium the ability to pull out fuel elements, before they are overcooked, without interrupting reactor operations is obviously a desirable feature. HWRs are available in a very wide range of configurations of power output and fuel inventories. Such reactors thus are well suited on all counts for clandestine plutonium production. They appear in many instances to have been acquired as units that can be used for commercial power production but that in case of need could be efficient sources of weapons grade plutonium.

2.3.4 RESEARCH REACTORS   Reactors designed to be used for research are diverse in nature and construction. They share some common characteristics. They must permit flexible operation in variable configurations. Simple access must be provided to permit neutron irradiations and studies. They must provide excess neutrons for research purposes or for material irradiation. They are clearly intended for isotope production. These characteristics make it possible to consider such reactors for production of about one bomb per year. The relatively low neutron flux densities in such a reactor would facilitate the production of weapons grade plutonium. This clearly is not a route that would be used by a major power whose objective was to accumulate a large inventory of weapons.

The most direct route to plutonium production is that purportedly used by the Israelis at the Dimona reactor. The Dimona reactor is a heavy-water "research" reactor using natural or slightly enriched uranium. The research reactor was originally designed to operate at a 30-MW power level but has reputedly been upgraded to over a 100-MW operating level to achieve production levels reportedly in the neighborhood of 40 kg of plutonium per year (3, 28). Such reactors are easy to obtain and easy to operate.

2.4   *Use by Terrorist Groups*

In the modern context there is the overshadowing question of whether nuclear weapons could be obtained or produced by small groups of individuals for the purposes of blackmail, intimidation, or retaliation. (Of course, we must hope that present precautions are adequate to prevent the direct theft of fabricated warheads or weapons grade materials.) At this time, direct production of nuclear weaponry does not seem to be a realistic option for terrorist groups. Terrorist groups could conceivably obtain materials clandestinely. The total plutonium production throughout the world in reactors can be estimated. About 30,000 tons of uranium are produced annually for reactor usage, containing 200 tons of ^{235}U. The consumption of this material will in turn produce of the order of 100 tons of plutonium annually. The nuclear industry plans to reprocess a fraction of this spent reactor fuel to recover the plutonium to be reused in turn as reactor fuel. Diversion of tens of kilograms out of the many hundreds of tons scheduled for eventual commercial production is certainly conceivable.

The problems faced by a terrorist group to turn this material into an operational bomb are daunting. In contrast to uranium, pure ^{239}Pu has a 24,000-year α-decay half-life and is only fifteen times less radioactive than a corresponding quantity of pure radium. In the form of the oxide or nitride there is intense neutron production. It cannot be handled without special confinement. Its dust is radiologically harmful. Isotopically impure plutonium is even more dangerous as the half-life of ^{240}Pu is 6500 years and that of ^{241}Pu is 14.4 years. ^{241}Pu is present typically in reactor grade plutonium at a 12% level. ^{241}Pu also produces ^{241}Am as a daughter element whose decay via K capture produces 60-keV x rays. Sophisticated chemical processing and metallurgy would be needed to provide the bomb assembly. To produce a nuclear explosion with some reasonable degree of probability is possible (6) but would require technology and knowledge of implosion techniques beyond that possessed by the original wartime Manhattan project. However, a terrorist group might still conceivably build such a device in the expectation that even should it "fizzle" it would create widespread radiological contamination (6). While such possibilities cannot be ruled out, they do not appear credible.

A simpler approach would be the construction by a terrorist group of a low yield assembly to contaminate radiologically, but not necessarily destroy, objectives such as airfields, dockyards, or military bases. Creating such a disaster could be accomplished by the sudden assembly of a partially moderated reactor with large supercriticality (28). Various assemblies of small size have been used for the production of neutron bursts. Such assemblies do not require high grades or large quantities of fissionable materials and enrichment >20% is sufficient (28, 29). One reactor was deliberately destroyed by running it into supercriticality to test safety features of boiling water reactors [Borax 1 (28)]. A device deliberately designed for terrorist use would require moderation by a low Z refractory material to prevent early termination of the reaction by vaporization, such as occurred in the previously mentioned experiment, Borax 1, Such reactors could be highly compact and would require neither high quality nor a large quantity of plutonium. Additionally, uranium enriched only up to a 20% level would be usable as the fuel elements. Supercriticality could be induced through explosive removal of control rods or by explosive combination of elements of the reactor. The consequences of illegal diversions of nonweapons grade plutonium or uranium could therefore be grave.

2.5   Inspection and Safeguards

The International Atomic Energy Agency (IAEA) was created on 29 July 1957. It is an independent, intergovernmental organization within the United Nations system. It currently has 112 member states pledged to ensure both the diffusion of peaceful atomic technologies and that facilities are not used for weapons production.

The nonproliferation treaty (NPT) came into being in 1970. Signatories fall into two categories. The first category comprises the US, USSR, France, England, and China, all countries that prior to 1970 had established weapons production facilities. These countries pledge to provide facilities or materials to nonnuclear countries, provided that proper safeguards and accountability exist. The nonnuclear nations in turn pledge that they will permit inspection of nuclear facilities as provided by the regulations of the IAEA plus possibly additional constraints that may be agreed upon with their suppliers.

Nonsignatories to the NPT, however, still comprise one third of the nations of the world. Of particular note among the nonsignatories are Argentina, Brazil, India, Israel, Pakistan, and South Africa, all countries that have either achieved or may well be close to having a nuclear capability (4).

The difficulties faced by the IAEA are exemplified by the difficulties inherent in the inspection for weapons grade plutonium production via heavy-water natural uranium reactors. The online loading capabilities of HWRs and specifically the Canadian "CANDU" class of HWRs make inspection difficult and such reactors were in many instances almost certainly acquired by nations desiring to produce weapons grade materials directly or to have such capabilities in reserve. Only relatively recently have tighter inspection procedures been developed (30) and certainly for many years the procedures could have been circumvented.

Another aspect of problems faced by the IAEA is illustrated by developments in Argentina (31). While not a signatory to the NPT, Argentina has acquired from West Germany production and refining capabilities of 150 tons per year of uranium plus an indigenous pilot for refining ores; it has enrichment capabilities by diffusion and an annual mining production of yellow-cake (uranium oxide) of 185 tons per year (17). It possesses two pressurized heavy-water reactors (PHWRs) supplied by West Germany (650 MWe) and Canada (360 MWe) under IAEA supervision, and one under construction from West Germany (750 MWe) (25) under IAEA supervision. Heavy water originating in the US has been reshipped from West Germany (26). Enriched uranium and heavy water have been supplied by the USSR (32). It has obtained a heavy-water plant from Switzerland (33). It is developing in association with West Germany for commercial sale a small 380-MWe PHWR with relatively low installation costs "that lends itself well to implementation in developing countries" (34; also 2). This constitutes a formidable potential for production and possible export of weapons grade materials.

The IAEA has played an important role in containing and monitoring the spread of nuclear weaponry. However, its powers are limited, and its function is well summarized by Fischer (35), "Within these constraints the Agency's limited but important task is to verify that governments, true to their word, are not secretly making the bomb. The purpose of safeguards is thus to create and maintain confidence, not to enforce compliance."

3.   DIRECTED ENERGY WEAPONS IN MISSILE DEFENSE

In the space available here, one cannot comprehensively treat the possible use and technical challenges of directed energy weapons (DEWs). Only a few topics are treated here. The American Physical Society (APS) report (36) is by far the best single reference available for a comprehensive treatment of the subject. The Office of Technical Assessment (OTA) report (37) provides a more succinct discussion of the issues at the expense of the physics arguments. The book by Velikhov et al (38) gives a succinct discussion of the physics. For the purpose of this discussion, directed energy weapons mean lasers and heavy, neutral-particle beams.

Several kinds of lasers have been considered for ballistic missile defense (BMD) weapons: (*a*) chemical lasers, meaning that the population inversion required for the lasing action (pumping) is achieved by an exothermic chemical reaction, (*b*) free electron lasers in which an electron beam coherently radiates, and (c) x-ray lasers pumped by a nuclear explosion. All these weapons attempt to destroy the target by thermal-mechanical means, i.e. by melting a hole or by mechanical shock. They are primarily useful against the missile booster. A neutral-particle beam consisting of hydrogen atoms could be used to destroy electronics by means of ionizing radiation, to detonate explosives in the warhead, or to identify the presence of fissile material by means of induced fission.

3.1 *Laser Performance Criteria*

Laser damage to a target depends upon several parameters, the most obvious of which is the energy density (J cm^{-2}) absorbed by the target, which in turn depends upon the energy density arriving at the target called the fluence F. The fraction of the fluence absorbed depends upon the characteristics of the surface and the wavelength λ of the light, as well as on the intensity of radiation because of nonlinear effects, such as the change of surface reflectivity due to surface damage and the formation of a plasma at the surface due to vaporized material. These interactions are quite complex (38) and are the subject of extensive calculations and measurements or "lethality experiments." Today's missile boosters are relatively soft targets, and fatal damage may be inflicted at a fluence of 1 kJ cm^{-2} (36). It makes no sense, however, to design a missile defense system for tomorrow (i.e. 10 or 20 years hence) using the missile design criteria of today; adversaries can and will take measures to protect their missiles. By means of spinning the missile and/or adding ablative coatings to the missile it is possible that a fluence of 10 to 100 kJ cm^{-2} may be required to destroy the missile (38; also 36, 39). There is a certain amount of arbitrariness in the choice of a credible fluence needed; the APS report (36) adopted 10 kJ cm^{-2}, and we use that value as representative.

The brightness B or power per unit solid angle of a laser system (W $ster^{-1}$) is a measure of its ability to deliver power density to a distant target. The fluence delivered in a time Δt to a target a distance R from the last optical element of a laser system is $F \approx B\Delta t / R^2$. Of course, the fluence may be decreased below this value if there are transmission losses, such as passing through the atmosphere. The brightness needed to deliver a given fluence can be expressed in this form

$$B \approx \frac{10^{20}}{\Delta t \text{ s}} \left(\frac{F}{10 \text{ kJ cm}^{-2}} \right) \left(\frac{R}{1 \text{ Mm}} \right)^2 (\text{W ster}^{-1}). \qquad 1.$$

Thus for an attack time of 1 s at a range of 1 Mm (1000 km) the brightness needed is 10^{20} W $ster^{-1}$. If longer ranges, greater missile hardness, or shorter attack times are contemplated, larger laser brightness is required.

This relation is not controversial; it is in the choice of parameters that controversies arise in the brightness needed.

The choice of attack time is limited on one hand by the accuracy with which the beam can be trained upon the moving target, which may be rotating, and on the other hand by how quickly targets must be engaged and destroyed. A certain amount of time is required to acquire the target, aim the beam, and keep it steady at the level of 5×10^{-8} radians in order to keep the jitter below the beam divergence set by diffraction. The choice of attack range involves a compromise between the number of laser stations and their power needed to cover the adversary's missile threat adequately. The APS and OTA reports consider two levels of performance, an "entry level" system having a brightness of 10^{20} W ster^{-1} and operational requirements for a "fully responsive threat" of 10^{22} W ster^{-1}. One of the problems here is that no law of physics sets a useful limit to the brightness that may be required to deal with all credible responses by the adversary. The optimists and pessimists can come to very different conclusions on the feasibility of directed energy weapons by different choices of credible sets of parameters.

Diffraction limits the brightness for a system of a given power and aperture. The peak brightness (in the forward direction) of a diffraction-limited circular aperture of diameter D is $B = P(\pi/4)(D/\lambda)^2$ (40). The characteristic angle θ_c is of the order of λ/D; for $\lambda = 1$ μm and $D = 10$ m, $\theta_c \approx 10^{-7}$ radian. Note that while the brightness depends on the diameter of the aperture, the terms $\theta_c D$ and B/D^2 do not. This feature is discussed in more detail below.

The average laser power needed to achieve the desired fluence is

$$P \approx \frac{1}{\eta\rho}\left(\frac{F}{10\text{ kJ cm}^{-2}}\frac{1\text{ s}}{\Delta t}\right)\left(\frac{R}{1\text{ Mm}}\frac{\lambda}{1\text{ }\mu\text{m}}\frac{10\text{ m}}{D}\right)^2 (\text{MW}), \qquad 2.$$

where $\eta \leq 1$ represents the transmission efficiency, and $\rho \leq 1$ is the Strehl ratio. The Strehl ratio describes how closely the system approaches the diffraction limit, because of aberrations, jitter, etc. For example, it is necessary to hold the jitter to $\leq \theta_c/4$ to have $\rho \geq 0.8$. This makes severe demands on the mechanical stability of the system (36). Because fabricating diffraction-limited, large-aperture, optical systems will be a very demanding task, there is a premium on using small wavelengths. The choice of wavelength will be strongly coupled to the choice of laser, a topic discussed below. The same problem of choosing appropriate parameters for Equation 2 appears as in the discussion of the brightness.

It is important to understand that brightness is not an intrinsic characteristic of a laser itself like its total power; rather it characterizes only the system making the link from the final optical element to the target. The final link requires a very slightly converging beam focused at a very large distance from the exit element. The laser itself may have a rather large divergence, which may be converted into a nearly parallel beam by optical elements. A free electron laser is an example of such a laser. There is a Liouville theorem for ideal optics in which the coordinates of the phase space are the transverse dimension and angle of a photon of the beam. The evolution with time of the photons of the beam in phase space is like that of an incompressible fluid, so that the volume of phase space occupied is a constant. The phase-space density or emittance is thus an intrinsic property of a laser system. A simple example of this property is the variation of the angular divergence (from ideal) of a lens with the diameter of the lens; the product of the diameter and the diffraction-limited beam divergence is a constant of the order of λ.

There are two ways that nonideal optical systems degrade performance; they can lose power (photons) and they can increase the volume of phase space. The loss of power is ordinary attenuation. The loss of phase-space density results from aberrations, restricted apertures, or scattering by, for example, the lenses of the system. The object of a practical design is to produce an optical system that approaches the diffraction-limited emittance as nearly as possible.

3.2   Chemical Lasers

It is assumed that the reader is familiar with the principles of lasers. For those desiring some background material, an excellent overview of the basic ideas is contained in the first few chapters of Siegman's book (41).

The lasing medium most often discussed for chemical lasers is either hydrogen fluoride (HF), having a wavelength of about 2.8 μm, or deuterium fluoride (DF), having a wavelength of about 3.8 μm. Such lasers do not produce a monochromatic beam but have numerous different lines covering a range of about 0.3 μm. A major difference between HF and DF lasers is that water vapor in the atmosphere absorbs HF radiation, but the atmosphere is very transparent to DF radiation. HF lasers could only be used down to altitudes of about 30 km (37). This has important consequences for the laser's basing and ability to attack boosters that are still in the upper atmosphere.

The MIRACL laser represents the present state of the art of a demonstrated laser weapon candidate. It is a DF laser that has produced an average power of the order of 1 MW and has an exit mirror 1.5 m in diameter (36, 42). Fuel is forced into a mixing chamber at a speed of about 2000 m s^{-1} by a set of horizontal nozzles. Accurate estimates of the brightness of the MIRACL laser are apparently not available in the unclassified literature (37). If the laser were diffraction limited, its brightness would be of the order of 5×10^{17} W ster^{-1}. The American Physical Society report states that the MIRACL laser has achieved an emittance of about twice the diffraction limit in the vertical plane, but it makes no statement on the horizontal plane, where gas mixing downstream of the nozzles is important. The MIRACL laser is near the limit of its design capability in power output. In order to increase the laser power further, the lasing volume must be increased. The length of the laser is limited by the accumulated inhomogeneities of the index of refraction of the gases; such effects degrade the emittance. The horizontal direction is limited by the fact that the chemical reaction takes place in the first few centimeters, after which there is no new energy input to the system. The only other dimension is vertical, and making the nozzle stack much larger results in an unfavorable aspect ratio for the beam. Reaching substantially higher powers requires a different approach.

The ALPHA laser, being built by TRW, is an HF design that offers the possibility of higher power. Instead of a horizontal row of nozzles, the ALPHA laser has a cylinder of nozzles 1.1 m in diameter and 7 m long directed radially outward. An ingenious beam-combining optical system takes the thin annular beam into a more compact round beam with a hole. The OTA report (37) states that the maximum brightness of a single HF laser is about 10^{20} W ster^{-1}, limited by one-dimensional gas flow dynamics and optical inhomogeneities. Reaching higher brightness requires coupling multiple amplifiers coherently. Ideally, 10 exactly phased units could increase the brightness by a factor of 100; this comes about by compressing the diffraction pattern. Experiments with six CO_2 lasers have shown a power increase of a factor of 23 compared to the maximum possible of 36. Coupling large HF or DF lasers is complicated by the large dimensions and the great stability required to maintain phase coherence. In addition, these lasers normally operate with many spectral lines, but the phasing could only be accomplished with a single line, or perhaps a consistent group of lines.

One particularly troublesome aspect of the ALPHA laser is that a realistic test is mostly cleanly done in space because of the atmospheric absorption of the beam, and the Strategic Defense Initiative Organization (SDIO) proposes to do so in conjunction with a large mirror (LAMP) 4 m in diameter as the Zenith Star experiment, tentatively scheduled for 1993. Such a test would presumably be in direct conflict with the antiballistic missile (ABM) treaty.

Both lasers convert chemical energy to light by a ratio of about 500 J g^{-1} (38). This estimate includes a dilution of $1+3$ needed to prevent detonation, because the chemical reaction is so violent. A weapons grade laser is not unlike a rocket: If all the fuel for a 30-MW laser were ejected in one direction, the thrust would be of the order of 10% of a Minuteman III booster! Such a reaction can produce severe vibration, which could degrade the emittance of the beam, although it is believed that this can be controlled (36). In addition unless the thrust is accurately balanced and isolated from the optical system there can be serious problems with aiming the device with the necessary precision.

3.3   Free Electron Lasers

A free electron laser (FEL) coherently generates light by means of an electron beam that passes through an "undulator" magnet producing a periodic magnetic field transverse to the direction of motion. Marshall's book (43) provides a good discussion of the basic physics of the subject. The wavelength of radiation may be understood by the simple physical argument that the wavelength of the undulator field is Lorentz contracted into a frame moving at the velocity v_z of the electron bunch along the beam line. The radiation produced by the oscillation of the bunch is in turn Doppler shifted back to the laboratory system. Thus the wavelength of the radiation λ is given by

$$\lambda \approx \frac{\lambda_w}{2\gamma_\parallel^2} \approx \frac{\lambda_w}{2\gamma^2}(1+a_w^2), \qquad\qquad 3.$$

where $\gamma_\parallel \gg 1$ is the Lorentz transformation factor $1/\sqrt{1-v_z^2/c^2}$, λ_w is the wavelength of the magnetic field, B is the magnetic field strength, $\gamma = E/mc^2$ is the energy of the electron in units of its mass, and $a_w = eB\lambda_w/(2\pi mc)$ in MKS units. The quantity a_w is usually of the order of 1.

Because the output wavelength is determined by macroscopic parameters, λ_w and E, it can be tuned over a wide range. This allows a free choice of wavelength to satisfy other needs, like an optical window in the atmosphere. On the other hand, the variation of the tune with beam energy means that as energy is extracted from the electron beam the tune will shift, lowering the gain. One way to avoid this problem is to taper the wiggler to keep a constant tune. This can be done by either changing the spatial wavelength or adjusting the magnetic field. This has been shown to be quite successful.

Free electron lasers can be operated either as oscillators or as single-pass amplifiers. There are advantages and disadvantages to each configuration. In both cases the light amplification occurs by virtue of the electron beam. It is important to keep in mind that the gain increases with wavelength, and much of the work to date has been done with relatively long wavelengths, 10.6 μm and longer, but shorter wavelengths are desired.

Oscillators may be tuned to any frequency desired. A resonant cavity causes the light beam to traverse the electron beam many times; in so doing, relatively low gain per pass is needed. Because the laser beam is so intense at the electron beam, the resonator mirrors are placed far away, which allows the beam to spread by diffraction before it is reflected and refocused for another pass. This has the disadvantage that the mechanical tolerances on the reflectors are very stringent to keep the light in phase with the electron beam. For example, present lasers must maintain a tolerance of 10 μm over a distance of 10 m or more, and realistic weapon resonators may be several hundred meters long; these will require automatic feedback systems to maintain the alignment. The reflectors themselves are very vulnerable to damage due to the high laser power. In addition to expanding the beam by diffraction to limit the intensity at the mirror, schemes have been proposed to spread the beam even further by having it strike the reflector only at grazing incidence. Very high reflectivities of mirrors may be obtained by means of dielectric coatings that are tuned to specific wavelengths. These coatings are subject to severe damage if high power at spurious frequencies is present in the beam. Such spurious frequencies can come about by sideband or harmonic generation.

Amplifiers, on the other hand, only amplify an input signal; thus they are limited to the frequency of the source laser. Amplifiers must achieve all their gain in a single pass, so it is essential to achieve high gain. Because there is only one pass there are no stringent demands upon the mechanical alignment of reflectors with respect to the electron beam. There are, however, significant alignment problems of the electron beam with respect to the input and output light beams. In particular, it is essential that the diffraction of the light not cause it to grow beyond the diameter of the electron beam. The photon beam would normally diffract out of the electron beam of diameter d in a distance $L_R = d^2/(1.22\lambda)$ called the Rayleigh length. For example, with a wavelength of 1 μm and a 1 mm diameter electron beam, the Rayleigh length is 0.8 m, but an amplifier might be 100 m long (36). Calculations indicate that at high beam currents the electron beam can provide "optical guiding" in which the electron beam behaves somewhat like a graded index optical fiber, keeping the photon beam contained (44). Recent measurements with the Paladin experiment at the Lawrence Livermore National Laboratory (LLNL) indicate that such optical guiding has been observed (45). Proposed amplifiers are driven by induction linear accelerators (linacs) (see below) and have very high peak power. To expand the beam enough to prevent damage to redirection mirrors, it is proposed that the natural diffraction of the beam after it leaves the wiggler be used over a distance of a few kilometers in an evacuated tube.

A third possibility is a "master oscillator power amplifier" (MOPA), which is a hybrid of the two technologies. This allows splitting the problem into two parts and exploiting the strengths and avoiding some of the weaknesses of each of the other alternatives.

For current work on high powered FELs the source of the electron beam has been either an RF or an induction linac (43). One of the important differences between RF and induction linacs is that the duty cycle (duration of laser pulse divided by time between pulses) of RF linacs is considerably larger than induction linacs; projected RF linacs could have a duty cycle of the order of 1×10^{-3}, while induction linacs could be of the order of 2×10^{-4}. Present FEL weapon candidate duty cycles are much lower. This means that for a given average power, the peak power of the RF linac is much lower than for the induction linac. Very high peak power can damage the optical elements of the laser; as a result RF linacs are more suitable for oscillators than induction linacs. Vigorous work with both kinds of technology is in progress. RF linear accelerator oscillators are being researched by Los Alamos National Laboratory (LANL) and Boeing Aerospace; induction linac amplifier research is being done by Lawrence Livermore National Laboratory (LLNL). Boeing has proposed building a MOPA FEL for a next generation device to be built at White Sands.

The basic question of FEL technology is the ability to produce a weapons grade multi-MW laser. Today's lasers are very far from this goal, much farther than the chemical lasers, although the results are promising. At this time the RF linac oscillator has demonstrated higher average power than the induction linac amplifier. The LANL oscillator has produced 6 kJ per pulse at a wavelength of 10.6 μm with a repetition rate of 1 Hz. SDIO believes that raising the repetition rate to 5 kHz should not be a serious problem; doing so would produce an average power of 30 MW. The OTA report (37) expressed doubts that either technology can be scaled much beyond this level. Reaching higher powers would require combining multiple lasers coherently.

There are advantages and disadvantages to the RF linac and induction linac FELs. One of big challenges of FEL design is achieving a high brightness $B = I_b/(\pi\varepsilon)^2$ beam (43), where I_b is the beam current and ε is the normalized beam emittance (taken to be the same in each dimension in this case). A high brightness electron beam is necessary to produce a high brightness photon beam since the gain is proportional to the beam current and inversely proportional to the spread of $\gamma_\parallel$. Ideally, the normalized emittance is a constant of the motion during acceleration, but in practice there is growth. RF linac oscillators are less sensitive to emittance growth than induction linac amplifiers. Because FELs are notoriously delicate in alignment, they are generally assumed to be ground based to provide easy service access. Nonlinear atmospheric propagation problems, such as thermal blooming and stimulated Raman scattering (see below), can be severe for both kinds of FELs. Thermal blooming is the same for both lasers, but stimulated Raman scattering is more of a problem for induction linac FELs than for RF linac FELs. Power conversion efficiency for induction linacs is generally better than for RF linacs. This is a major consideration for a multi-megawatt weapon. Induction linear accelerators are inherently high current devices, and there is much less uncertainty of their ability to be scaled to high current weapons. There is, however, a concern about possible beam breakup, caused by the interaction of very intense current pulses with the cavity.

Thermal blooming refers to the focusing or defocusing of the beam as a result of heating due to the passage of the beam through the atmosphere (46, 47). Such an effect can seriously degrade the emittance of the beam. For example, a 30-MW laser having a 10-m diameter at a wavelength of 1 μm would suffer an angular spread of the order of 10^{-6} radian in a time of 0.1 s, and growing linearly with time. This is to be compared to the diffraction-limited spread of 10^{-7} radian. This time is needed for a single engagement; the effect of multiple engagements through the same air path will be cumulative. One way to deal with thermal blooming is to use adaptive optics (36) in which the upgoing beam is predistorted by just the right amount so that when it emerges from the atmosphere it has the proper wave front to propagate as if the atmosphere had not interfered. Doing so requires a downgoing reference beam of high quality coming from above the atmosphere. The principle has been demonstrated on a small scale to compensate for atmospheric turbulence, but the thermal blooming problem is much more severe, requiring several thousand variable optical elements controllable on a time scale of 1 ms. If the blooming is severe enough and all of the reference beam cannot be detected, compensation is impossible; the information is lost. Another way to deal with the problem is to direct the beam through different air columns for each shot to avoid the cumulative degradation. Note that thermal blooming depends upon wavelength; for example the thermal blooming is a factor of 10 less severe for $\lambda = 3.8$ μm than for $\lambda = 1.0$ μm.

Stimulated Raman scattering is a lasing action taking place in the atmosphere at a slightly different frequency than the primary beam, and noncollinear with the beam (36, 37, 47, 48); this spoils the emittance of the beam. In contradistinction to thermal blooming, stimulated Raman scattering depends upon the peak intensity of the beam rather than the average intensity. In addition, the phenomenon takes effect rather abruptly with increasing intensity, which creates an upper limit to what can be propagated successfully. For a wide beam, $\lambda = 1$ μm, penetrating one air mass, the maximum intensity is about 2 MW cm^{-2}. The limit is nearly proportional to the wavelength and inversely proportional to the amount of atmosphere penetrated (because of zenith angle of the beam). Because of the low duty cycle of FELs, this limitation can be serious, especially for the induction linac FEL. In estimating the limit we take the optimistic projections for the FEL duty cycle rather than present-day cases. Because of the characteristic times, the effective peak power for stimulated Raman scattering of an RF linear accelerator is only 50 to 100 times the average power, instead of 1000 times for the real peak (47); the microstructure of the beam does not play as large a role as it might. For a 1-μm laser having a 10-m diameter penetrating one air mass, this corresponds to a limit of about 16 GW average power; this should not be a serious problem. On the other hand, an induction linear accelerator has a peak power about 7000 times the average; in this case the upper limit is of the order of 220 MW average power.

3.4 X-Ray Lasers

The greatest amount of controversy over DEW components has surrounded the x-ray laser. Such a laser would be pumped by a nuclear explosion and would deliver a short burst of x rays on multiple targets simultaneously. This device is much closer to the frontier of basic physics than the other DEW devices, and it has also been shrouded in considerable secrecy, which greatly compounds the difficulty of an informed public discussion of the device's potential. Conflicting viewpoints of the level of development of the concept have led to a severe confrontation among weapons experts: The director of the project at LLNL resigned because he was prevented from giving a dissenting view of very optimistic projections of the state of development of the x-ray laser (49, 50).

Conceptually (38, 51, 52) an x-ray laser consists of a large number of rods about 0.1 mm in diameter and about 3.4 m long arranged in a cylinder of radius about 30 cm about a nuclear weapon at the center. A roughly 150-kt-equivalent explosion releases an intense burst of x rays that ionize the rods supplying the inverted population for the lasing action. Normally, x rays would be strongly absorbed in traversing the material because the absorption is primarily due to the photoelectric effect on bound electrons. In the totally ionized state there is almost no absorption, and the medium is said to be "bleached." The lasing action occurs in a single pass; there are no reflectors to make a classical resonator. Depending upon the lasing material chosen, the wavelength of the radiation is of the order of 1 nm. The whole process takes place in a time of the order of 10^{-8} s, which is the length of the burst of pump x rays. At about 10^{-6} s the whole assembly is blown apart by the arrival of the mechanical part of the explosion.

One of the central issues of x-ray laser performance is the angular divergence of the beam. An early paper discussed the minimum divergence angle of a beam coming from a rod of diameter D and length L (53). The minimum obtained comes from the competition between the divergence due to geometry, D/L, and that due to diffraction, λ/D. This turns out to be a fairly large angle, ~ 25 μrad. This limit, however, is not really applicable, as was pointed out in testimony by Wood & Canavan (54). The reason is that it is possible to introduce focusing of the x-ray beam, so that the geometrical divergence constraint is irrelevant (52). It is possible to use the plasma itself to form a lens to focus the beam by properly choosing the magnitude and nonuniformity of the electron density. By such means one can reduce the divergence angle by a factor of the order of 5, and thus increase the brightness by a factor of 25.

Many fundamental questions must be answered before a serviceable x-ray laser can be brought to the engineering stage. Many of these questions can only be answered by actual tests with large-scale nuclear weapons. A comprehensive test ban treaty or a low threshold test ban treaty could effectively stop the development of this device and keep the Genie in the bottle.

3.5   Neutral-Particle Beams

Neutral-particle beams (NPBs) offer a function different from and complementary to lasers in ballistic missile defense. While they may be used against boosters, they can also be useful during the midcourse phase of reentry vehicle (RV) delivery. A beam of neutral hydrogen atoms could be useful for structural damage like a laser, for igniting propellant in a booster or high explosives in the RV, or for disabling electronics. In addition to doing damage, the NPBs may be used for target-vs-decoy discrimination by means of the interaction of the protons and the target.

Since the electron of an atom is rather loosely bound, it is essential that the beam travel in vacuo, or the atom will be ionized. This means that the NPB equipment must be space based. A charged beam will be severely bent by the Earth's magnetic field of about 0.3 gauss; e.g. a 200-MeV beam has a radius of curvature of about 70 km. Since the engagement distance might be of the order of 1000 km, such bending is not tolerable. The residual atmosphere below an altitude of about 120 km will render a NPB useless.

The interaction of a NPB with a target is very different from that of a laser beam, which interacts only at the surface. The NPB penetrates to relatively great depths, 10 to 100 cm, depending upon the material struck and energy of the beam. The rate at which beam energy is lost is given roughly by $dE/dx \approx \zeta(1 + 1/\beta^2)$ [MeV/(g/cm^2)], where x is the distance traveled in the medium times the density of the medium; ζ depends mildly on the medium, ranging from 2 for carbon to 1 for lead; and β is v/c for the projectile particle. The beam energy is most conveniently in the range of 100 to 500 MeV. Note that the energy deposition increases markedly as the particle slows down near the end of its range. If it is known that the beam will stop in the target, for example in a booster, then this high energy deposition may be exploited to assure damage, such as ignition of fuel. Propellant can be ignited by depositions of the order of 200 to 400 J g^{-1} (36). This requires about 5×10^{13} particles per cm^2. On the other hand, if electronic damage is the goal, it is necessary to assume that the beam will pass completely through the target, causing damage on the way. A reasonable criterion is a radiation dose of the order of 10^7 rad, corresponding to about 5×10^{13} particles per cm^2. In either case, if the beam spot on the target is of the order 100 cm in diameter, the particle flux required is of the order of 5×10^{17} particles or 0.1 coulomb of delivered charge. For an engagement time of the order of 1 s, a beam current of the order of 100 mA is needed. If the beam is used for discrimination, then the beam charge can be a factor of 10 to 1000 lower (36).

A beam spot 1 m in diameter at a range of 1 Mm requires that the residual beam divergence be of no more than 1 μrad (the beam must be focused on the target). Stated differently, beams of low emittance are required. This is a stringent requirement but not an impossible one. The present state of the art of generation and acceleration, however, is of the order of a factor of 50 in emittance larger than that needed for this task (37). Low emittance sources have been built at the requisite current of the order of 100 mA.

The technology of particle acceleration is rather well developed, even at the high currents needed. The primary difficulty lies in emittance growth of the beam during the various stages, passing from the source to a preaccelerator and from there to an RF linear accelerator. These difficulties appear to be manageable, although some development of recognized techniques is required. The most cumbersome problem is that the present accelerator technology has given no importance to minimizing the weight, and such technology for a space-based NPB would be inappropriate.

Because the accelerator aperture must be relatively small, of the order of 1 cm, for economic reasons, the beam must be greatly expanded, of the order of 1 m radius, in order to reduce the divergence of the beam to the 1 μrad required. Doing so requires some very high precision magnetic "lenses" in order to preserve the emittance of the beam. There is no physical reason that such lenses cannot be built, but there are at this time no full-scale prototypes. The demands on the field purity to preserve the emittance are very stringent. High energy physics accelerators now struggle with quadrupoles and sextupoles with apertures of the order of a few centimeters to achieve the desired beam quality.

Finally, the negative ion must be stripped of the extra electron to make a neutral atom. This has been done with thin foils (~ 100-nm graphite), low pressure gas, and high power laser beams. The most favored means is the thin foil. The low pressure gas is not suitable in space, and the laser required to do the job would have a power of tens of MW, which is comparable to a laser weapon by itself. The stripping efficiency of a thin foil is of the order of 50%. The stripping itself introduces some divergence to the beam. This is limited by the momentum transfer $\Delta p \approx (2m_e E_i)^{1/2}$ to the loosely bound extra electron making up the ion. Since the binding energy E_i is about 0.7 eV, a nonrelativistic beam of energy E and atomic number A suffers an angular change $\Delta \theta \approx 20 \times 10^{-6}/(EA)^{1/2}$ radians; for a 400-MeV beam of hydrogen, the change is $\Delta \theta \approx 1$ μrad. In practice this number is the extreme, and a root-mean-squared deviation would be somewhat smaller.

Because the NPB has a charged beam before the stripping, it is possible to steer the beam toward the target, over a limited range, by means of bending magnets. The bending, however, depends upon the momentum of the beam, which means that an additional chromatic correction for the bend must be made to preserve the emittance of the beam (which has some energy spread). It is not clear how much the beam can be steered magnetically before such chromatic effects and the required aperture of the magnets and stripping foil become excessive. Going beyond this range would require reorienting the entire accelerator, not a very attractive option.

4.   NUCLEAR ARMED SEA-LAUNCHED CRUISE MISSILES

Sea-launched cruise missiles (SLCMs) create a relatively new problem to the arms control business. They, like MIRVed missiles, are Genies that have escaped the bottle, much to the world's regret. SLCMs are small missiles, about 0.5 m in diameter and 6.5 m long, and may be carried and launched by almost any kind of vessel. SLCMs can be equipped with either conventional or nuclear warheads. Both kinds share the same airframe, and in present designs no obvious, external characteristics distinguish a conventional from a nuclear armed missile. Deployment of several thousand SLCMs, most of which would be conventionally armed, has been proposed by the US (55). While SLCMs are not by themselves suitable for first-strike weapons, there is some diversity of opinion on the danger they present in real, military terms (56, 57). It is clear that a large uncertainty in possible nuclear delivery systems is politically undesirable. Without some means of limiting the number of nuclear SLCMs the strategic arms reduction talks (START talks) will be severely hampered.

The US and the USSR negotiators have agreed that a goal of the START talks is to reduce the number of strategic offensive delivery systems (ICBMs, sea-launched ballistics missiles, and heavy bombers) to 1600 and the number of warheads to 6000. Nuclear armed sea-launched cruise missiles pose a potentially severe threat to such a strategic arms reduction agreement, because the number of such systems could be very large and difficult to detect or verify. The joint communiqué issued by President Reagan and General Secretary Gorbachev at the close of the Washington summit in December of 1987 committed the US and USSR to finding "a mutually acceptable solution to the question of limiting the deployment of long-range, nuclear-armed sea-launched cruise missiles."

The INF agreement is a very attractive model for cooperative measures to inspect and verify compliance with provisions of a treaty by means of on-site inspections. It is a relatively simple matter to determine if a missile presented to an inspector contains a nuclear weapon; the challenge is to find nuclear missiles without such an intrusive inspection or to find them if some attempt is made to conceal them. General Secretary Gorbachev announced at the Washington summit in December 1987 that the USSR had the ability to detect the presence of nuclear weapons on ships at a distance of the order of 100 m. Taken at face value, this would be an extremely valuable tool to assure that no nuclear armed SLCMs were on board a ship being inspected. Upon examination, however, this claim appears to be very optimistic (58).

One proposal for detecting nuclear weapons is to use helicopter-borne gamma-ray and neutron detectors, perhaps in conjunction with a pulsed neutron source. Natural or induced fission in the fissile material creates characteristic photon lines and neutron emission. The cleanest case for passive detection involves detection of the 1.001-MeV photon line resulting from ^{238}Pa decay, a daughter product of ^{238}U decay. This line would be copious for a weapon with a depleted uranium shell, of the order of 10^6 such photons per second. This narrow line is a clear signature of the uranium. In the absence of any shielding, passive detectors could detect nuclear weapons at ranges of 50 to 100 m in a reasonable length of time. If, however, the shell were made of tungsten rather than depleted uranium, the signature would be very much weaker (58).

Another possible signature is a broad, low energy spectrum of photons resulting from fission reactions. These can be detected, although the signature is not as clean as the 1.001-MeV ^{238}Pa photon line. Similarly, the neutrons from spontaneous fission could be detected. While the spontaneous fission rate for ^{235}U, ^{238}U, and ^{239}Pu is low, the rate for ^{240}Pu is rather high. The production of ^{239}Pu is always accompanied by an admixture of ^{240}Pu, so a plutonium-based weapon (the most desirable from a weight standpoint) will have a large photon flux at admixtures of the order of 3% ^{240}Pu, which are normal in weapons grade plutonium. Although it would be expensive, it is possible to separate isotopically the plutonium to get rid of the ^{240}Pu if desired.

One of the problems with passive detection techniques is that the party being inspected may try to conceal the weapon. Simply stowing the warhead deep below deck would provide a substantial amount of steel shielding. Beyond that, it is a relatively simple matter to shield the photons or neutrons if one is willing to dedicate ~ 100 tons of material to this purpose. This is not a serious problem on a ship.

One way to deal with shielding is to use active detection techniques, in which a pulsed neutron source would induce fission in the fissionable warhead. This induced fission signal could be much larger than the natural signal if the source is strong enough. By pulsing the source one has a clear signature to distinguish the nuclear "reply" from background, significantly enhancing the signal-to-noise ratio. Unfortunately, the source required to deal with substantial amounts of potential shielding would result in a rather large radiation dose to the crew of the ship, not to mention the inspecting party.

Finding ways to verify agreed upon limits of SLCM deployment requires new ideas. Some possibilities are discussed in (55).

5. SUMMARY

The examination of potential misuse of reactors to produce weapons grade plutonium clandestinely shows that commercial light-water reactors are both unsuited and unlikely to be used for this purpose. Heavy-water reactors, on the other hand, are very well adapted to produce relatively large quantities of weapons grade plutonium from easily obtainable natural uranium. A number of countries have obtained such reactors, in many cases probably to have such a capability in reserve. Only recently have IAEA inspections been tightened to a point where such activity would be manifestly detectable if inspection was permitted.

Research reactors, particularly those using heavy water, are flexibly configured, and in many cases can be used for weapons production of a few bombs per year. Again these reactors are widely spread around the world.

Nonsignatories to the nonproliferation treaty such as India, Pakistan, and Argentina have been able to circumvent international restrictions and obtain from both the West and USSR substantial supplies of materials and plants required for the production of weapons grade materials. Several nonnuclear nations have created plants for uranium enrichment at a level sufficient to produce a few bombs per year. As it is easy to make bombs from ^{235}U, such nations could probably (if they have not already done so) assemble operational weapons.

On balance we assess the technology and knowledge required for proliferation to be irreversibly diffused throughout the world. However, actual production of weapons still seems confined to members of the nuclear club and a small number of other nations. Proliferation appears to be held in check more by the fact that nations do not see nuclear weapons as an effective means to further national policies rather than by technical considerations. Fortunately, to date proliferation has not occasioned the major confrontation feared in the 1960s.

The possibility that nuclear bombs could be fabricated by terrorist groups appears happily remote. The most credible route for terrorists to obtain nuclear bombs is by the direct theft of warheads or weapons grade materials. It is conceivable that a terrorist group could fabricate from illicitly obtained reactor grade plutonium or moderately enriched uranium the supercritical reactor assemblies capable of creating large-scale radiological contamination. Probably the most credible hazard is dispersal of purloined radioactive material from reactor elements. Radiological contamination in an urban environment could generate major disruption.

The technologies needed for directed energy applications in ballistic missile defense are still a long way from practical application. Chemical lasers are the closest to having weapon potential but are at least two orders of magnitude below the brightness needed to provide an "entry level" capability. Work is under way to take that next step, but the time scale is unknown. The proposed test of the Zenith Star with a very powerful HF laser with large optics may cause serious problems with the ABM treaty. Dealing with a "responsive" threat can probably not be achieved by a simple extrapolation of present technology; rather it will require phase-locking multiple lasers to provide another order of magnitude in brightness.

Free electron lasers show a great deal of promise in principle, but no existing FEL remotely approaches weapon capability. A vigorous research program on two different technologies is being pursued. A major concern of FEL performance is the development and maintenance of low emittance electron beams. Because FELs would probably be ground based, there are serious problems of nonlinear atmospheric propagation of the high power beams. Thermal blooming absolutely must be avoided, or any ground-based laser will be useless as a weapon. The use of adaptive optics to do this has been demonstrated as a principle, but the practice for large, high power beams requires a large extrapolation. Stimulated Raman scattering in the atmosphere may set a limit to the power of FELs powered by induction linear accelerators but a much less restrictive one for RF linear accelerator FELs.

The technology of x-ray lasers is still in its infancy, and much work needs to be done to demonstrate that a weapons grade laser is even possible. The lasing principle has been demonstrated, but more detailed information is not available in the unclassified literature. A crucial part of the development of x-ray lasers is the ability to focus the beam to achieve a smaller divergence than the geometrical limit set by the dimensions of the lasing rods. Because developing x-ray weapon lasers requires nuclear bomb explosions, this technology could be stopped before development by a more restrictive nuclear test ban than that now in force.

Neutral-particle beams (hydrogen atoms) potentially serve a role complementary to lasers in ballistic missile defense. The accelerator technology is well developed, but the existing equipment is too massive to be placed in space, where a NPB must reside. The major problems are the production and acceleration of low emittance beams to be delivered to the target. There is no existence proof for the beam optical elements of the very large apertures required for this role.

Like MIRVs, nuclear armed sea-launched cruise missiles already exist, even though the world would probably be better off without them. Nuclear armed SLCMs pose a serious political threat to major strategic arms reduction agreements because they are small, difficult to detect, and easily proliferated in large quantities to a large class of vessels. New ideas in cooperative measures between the US and USSR as well as detection methods are needed to verify any agreement to limit SLCM deployment.

ACKNOWLEDGMENTS

We are indebted to Frank von Hippel for helpful comments and to Richard Wilson for a careful reading of the section on proliferation.

Literature Cited

1. Spector, L. S., *Bull. At. Sci.*, p. 14 (June/July 1986)
2. *Eng. Tech. Rev.*, p. 17. LLNL (April 1983)
3. Vanunu, M., *Sunday Times* interview (5 Oct. 1986); Cohen, A., Frankel, B., *Bull. At. Sci.*, p. 15 (Mar. 1987); Spector, L. S., *Bull. At. Sci.*, p. 17 (May 1987)
4. Spector, L. S., *Bull. At. Sci.*, p. 16 (Sept. 1985); Khalilzad, Z., *Bull. At. Sci.*, p. 11 (Jan. 1980); *New Scientist*, p. 20 (Oct. 1985); Jackson, A., *Nucl. Eng. Int.*, p. 35 (Aug. 1986)
5. Serber, R., *Lectures, The Los Alamos Primer*. LANL (1943)
6. Mark, J. C., Taylor, T., Fyster, E.,

Maraman, W., Wechsler, J., *Bull. At. Sci.*, p. 42 (Feb. 1983)

7. Cochran, T. B., Arkin, W. M., Norris, R. S., Hoenig, M. M., *Nuclear Weapons Databook.* Cambridge, Mass: Ballinger (1987), 1: 24–25

8. Albright, D., Taylor, T., *Bull. At. Sci.*, p. 39 (Jan./Feb. 1988)

9. Philips Data Handbook, *Electron Tubes, Part 6, Neutron Generator Tubes (Types 18601 18603)* (Sept. 1972)

10. Cochran, T. B., Arkin, W. M., Norris, R. S., Hoenig, M. M., See Ref. 7, 2: 135

11. Cochran, T. B., Arkin, W. M., Norris, R. S., Hoenig, M. M., See Ref. 7, 2: 137

12. Rhodes, R., *The Making of the Atomic Bomb.* New York: Simon & Schuster (1988)

13. Glasstone, S., *Principles of Nuclear Reactor Engineering.* New York: Van Nostrand

14. Delerousse, P., *Nucl. Eng. Int.*, p. 32 (Aug. 1986)

15. Inglis, G. H., *Nucl. Eng. Int.*, p. 41 (Aug. 1986)

16. *Eng. Tech. Rev.*, pp. 34–37. LLNL (July 1987); Update, *Nucl. Eng. Int.*, p. 13 (Feb. 1981); Update, *Nucl. Eng. Int.*, p. 13 (Apr. 1980); Opinion, *Nature* 315: 702 (June 1988)

17. The World's Nuclear Fuel Cycle Facilities, *Nucl. Eng. Int.*, p. 47 (Dec. 1987)

18. Longenecker, J. R., *Nucl. Eng. Int.*, p. 39 (Aug. 1986)

19. Monitor, *Nucl. Eng. Int.*, p. 21 (Apr. 1987)

20. Albright, D., Feiveson, H. A., *Annu. Rev. Energy* 13: 239 (1988)

21. Gambier, G., *Rev. Gen. Nucl.* 6: 475 (Nov./Dec. 1983)

22. Bairot, H., *Nucl. Eng. Int.*, p. 27 (Jan. 1984)

23. Baust, E., *Nucl. Eng. Int.*, p. 50 (Aug. 1986)

24. Crawford, M., *Science* 241: 526 (July 1988)

25. *World Nuclear Industry Handbook.* Sutton, England: Nuclear Engineering International Publication (1988)

26. Budiansky, E., *Nature* 305: 30 (Oct. 1983)

27. Heisenberg, W., Wirtz, K., *FIAT Review of German Science 1939–1945: Nuclear Physics and Cosmic Rays II.* Wiesbaden: Dieterische Verlagsbuchhandlung (1948), pp. 143–46

28. Chastain, J. W., *US Research Reactor Operation and Use.* Reading, Mass: Addison Wesley (1958), pp. 65–182

29. Barbry, F., Leclerc, J., Manaranche, J. C., Maubert, L., *Rev. Gen. Nucl.* 4: 377 (1982)

30. Monitor, *Nucl. Eng. Int.*, p. 19 (Sept. 1984)

31. World Survey, *Nucl. Eng. Int.*, p. 29 (June 1987)

32. News Review, *Nucl. Eng. Int.*, p. 13 (June 1982)

33. *Nucl. Eng. Int.*, p. 32 (Sept. 1982)

34. Frischengruber, K., *Nucl. Eng. Int.*, p. 33 (May 1986)

35. Fischer, D. A. V., *Bull. At. Sci.*, p. 30 (June/July 1986)

36. American Physical Society, *Science and Technology of Directed Energy Weapons* (April 1987)

37. Office of Technical Assessment, *SDI Technology Survivability and Software*, OTA-ISC-353 (May 1988)

38. Velikhov, Y., Sagdeev, R., Kokoshin, A., *Weaponry in Space: The Dilemma of Security.* Moscow: Mir (1986)

39. Office of Technical Assessment, *Ballistic Missile Defense Technologies*, OTA-ISC-254 (Sept. 1985)

40. Jackson, J. D., *Classical Electrodynamics*, New York: Wiley (1975), p. 443. 2nd ed.

41. Siegman, A. E., *Lasers.* Mill Valley, Calif: University Science Books (1986)

42. Schroeer, D., *Direct Energy Weapons and Strategic Defense*, Adelphi Paper 221. London: Int. Inst. for Strategic Studies (July 1987)

43. Marshall, T. C., *Free Electron Lasers.* New York: Macmillan (1985)

44. Scharlemann, E. T., Sessler, A. M., Wurtele, J. S., *Phys. Rev. Lett.* 54: 1925 (1985)

45. Prosnitz, D., et al., presented at Washington APS Meeting (April 1988)

46. Walsh, J. L., Ulrich, P. B., *Laser Beam Propagation in the Atmosphere*, ed. J. W. Strohbehn. Berlin/Heidelberg/New York: Springer-Verlag (1978), p. 223

47. Lynch, H. L., *Technical Evaluation of Offensive Uses of SDI*, Working Paper Cent. for Int. Security and Arms Control, Stanford Univ. (Feb. 1987)

48. Shen, Y. R., *Principles of Nonlinear Optics.* New York: Wiley (1984)

49. Blum, D., *Bull. At. Sci.*, p. 7 (July 1988)

50. Conahan, F. C., *Strategic Defense Initiative Program, Accuracy of Statements Concerning DOE's X-Ray Laser Research Program*, Report to Hon. George E. Brown (30 June 1988)

51. Morrison, D. C., *Lasers and Optronics*, p. 20 (Nov. 1988)

52. Ritson, D. M., *Nature* 328: 487 (1987)

53. Walbridge, E., *Nature* 310: 180 (1984)

54. Wood, L., Canavan, G., *Joint Opening Statement Before the House Republican Research Committee* (19 May 1987)

55. Harvey, J., Ride, S., Co-Chairs. *Poten-*

tial Verification Provisions for Long-Range, Nuclear-Armed Sea-Launched Cruise Missiles, Workshop Rep. Cent. for Int. Security and Arms Control, Stanford Univ. (July 1988)

56. Drell, S. D., Johnson, T. H., *Foreign Affairs* (Summer 1988), p. 1027

57. Postol, T. A., *International Security* 13 3: 191–202 (1988)

58. Sagdeev, R. Z., Prilutskii, O. F., Frolov, V. A., *Problems of Monitoring Sea-Based Cruise Missiles Bearing Nuclear Warheads*, Rep. Committee of Soviet Scientists for Protecting the World from the Nuclear Threat. Presented at Seminar on Verification Problems held jointly with the Federation of American Scientists, Key West Florida (3 Feb. 1988)

Annu. Rev. Nucl. Part. Sci. 1989. 39: 183–230

D MESONS

Rollin J. Morrison and Michael S. Witherell

Physics Department, University of California, Santa Barbara, California 93106

KEY WORDS: charm decays, heavy flavor, semileptonic decay, Cabibbo suppression, meson spectroscopy.

CONTENTS

0163–8998/89/1201–0183$02.00

1. INTRODUCTION

1.1 *Overview and Physics Introduction*

The discovery (1, 2) in 1974 of the J/ψ, a narrow resonance of mass 3.1 GeV/c^2, revolutionized elementary particle physics. This state was thought to be a bound system composed of the charmed quark and its antiparticle, predicted by Glashow, Iliopoulos & Maiani (3) (GIM). Two years later the first D mesons, states composed of the new charmed quark and a u, d, or s antiquark, were observed (4, 5). The discovery of these particles, which explicitly carry the charm quantum number (or flavor) and can only decay to noncharmed states by the weak interaction, definitively confirmed the charm hypothesis and the GIM picture of quark generations.

Since that time, the study of the production, spectroscopy, and decays of the D mesons has been an extremely fruitful area of elementary particle physics. The charmed quark is sufficiently massive that some aspects of perturbative QCD are applicable, both in its production and decay. Because the weak couplings of the charmed quarks are (theoretically) determined in the standard model with three quark generations, charm decays offer a clean laboratory to study strong interaction effects at the boundary between the perturbative and nonperturbative regimes. In addition, the D-meson system has been investigated in great detail experimentally. The knowledge gained from D decays will be useful in understanding B decays, for which the weak couplings are not known and the experimental studies are more limited.

In this review we focus almost exclusively on recent progress in charmed meson physics. We refer the reader to two excellent early reviews (6, 7) and to elementary particle textbooks for discussions of the physics of charm and for reviews of the earlier experimental results. More recent results are covered in detail by Ye & Huang (8).

The decays of quarks are governed by the weak charged current, which couples the upper and lower members of the three quark generations,

$$\begin{pmatrix} u \\ d \end{pmatrix}, \ \begin{pmatrix} c \\ s \end{pmatrix}, \ \begin{pmatrix} t \\ b \end{pmatrix}.$$

Prior to the discovery of the b quark, only the first two quark generations were known and the weak charged current had the form

$$J_\mu = (\bar{u}\bar{c})\gamma_\mu(1-\gamma_5)U\begin{pmatrix} d \\ s \end{pmatrix}, \qquad\qquad 1.$$

where U is the unitary matrix,

$$U = \begin{pmatrix} V_{ud} & V_{us} \\ V_{cd} & V_{cs} \end{pmatrix} = \begin{pmatrix} \cos\theta_c & \sin\theta_c \\ -\sin\theta_c & \cos\theta_c \end{pmatrix}.$$

This matrix is expressed in terms of θ_c, the Cabibbo angle.

The value of $V_{us} = \sin\theta_c$, measured in strange particle decays, is 0.220 ± 0.003. Thus charmed quark decays to strange quarks, with rates proportional to $|V_{cs}|^2$, are favored over decays to down quarks by a factor of $\sim |\cos\theta_c/\sin\theta_c|^2 \approx 20$. Charm decays are therefore dominated by "Cabibbo-favored" decays to strange quarks.

Kobayashi & Maskawa (8a) generalized the mixing matrix to the case of three generations, with a charged weak current

$$J_\mu = (\bar{u}\bar{c}\bar{t})\gamma_\mu(1-\gamma_5)U\begin{pmatrix} d \\ s \\ b \end{pmatrix}. \qquad\qquad 2.$$

The K-M matrix,

$$U = \begin{pmatrix} V_{ud} & V_{us} & V_{ub} \\ V_{cd} & V_{cs} & V_{cb} \\ V_{td} & V_{ts} & V_{tb} \end{pmatrix},$$

can be expressed in terms of four measurable real parameters. In the parameterization favored by the particle data group (8b) U takes the form,

$$\begin{pmatrix} C_{12}C_{13} & S_{12}C_{13} & S_{13}e^{-i\delta} \\ -S_{12}C_{23}-C_{12}S_{23}S_{13}e^{i\delta} & C_{12}C_{23}-S_{12}S_{23}S_{13}e^{i\delta} & S_{23}C_{13} \\ S_{12}S_{23}-C_{12}C_{23}S_{13}e^{i\delta} & -C_{12}S_{23}-S_{12}C_{23}S_{13}e^{i\delta} & C_{23}C_{13} \end{pmatrix},$$

where $S_{ij} = \sin\theta_{ij}$ and $C_{ij} = \cos\theta_{ij}$. The four real parameters are the three angles θ_{ij}, where $\theta_{12} \equiv \theta_c$, and the phase δ. The observation of the very long b-quark lifetime and the low fraction of $b \to u$ transitions indicate that θ_{23} and θ_{13} are very small. As a consequence, the upper left-hand part of the K-M matrix, which controls the charmed- and light-quark decays, is equal to the 2×2 matrix of Equation 1 to a precision of about 0.2%. Charm decays are essentially unaffected by the presence of the third quark family.

1.2 Experimental Background

There are many sources for producing D mesons, each of which has its own advantage. The D mesons result from the fragmentation of c quarks produced in e^+e^- annihilation, or in collisions of neutrinos, real or virtual photons, or hadrons with nuclear targets. D mesons are also very prominent in the decay products of the more massive B mesons and are expected to be produced copiously in Z decays.

A large fraction of the experimental information on D mesons comes from e^+e^- annihilation experiments operating just above $D\bar{D}$ threshold, where backgrounds of noncharm events are relatively small. Electron-positron annihilation at a center-of-mass energy 3.770 GeV is particularly advantageous for the study of the two lowest-mass states, the D^0 and the D^+. Because this energy corresponds to the mass of the ψ'' resonance, which decays almost entirely into $D^0\overline{D^0}$ and D^+D^- states, nearly all of the experimentally produced events contain only a D and a $\bar{D}$ particle, and the charge of one D meson tags the charge of the other D meson. Further, the D mesons have a unique momentum. Since the uncertainty in the beam energy is small, the constraint that the energy of each D must be equal to the beam energy significantly improves the resolution and background rejection. The Mark III collaboration (9) working at the SPEAR storage ring has a detector designed for the study of these D decays. A typical Mark III event is shown in Figure 1. Important recent experimental results have also been produced by the ARGUS (10) and CLEO (11) collaborations using e^+e^- annihilation in the 10-GeV region, above the $\Upsilon(9460)$ resonance.

A major limitation of e^+e^- annihilation as a source of information on the D mesons is the relatively low integrated machine luminosity that is currently available. Much higher rates of charm production are available with high energy photon or hadron beams striking dense fixed targets. The difficulty with these sources has been the low fraction of the interactions producing charm, which is typically only 0.1% and 0.5% for hadro- and photoproduction, respectively. In these charmed events, there are typically also many other particles produced along with the charmed particles. As a result, the charm signals are often overwhelmed by a continuous background mass spectrum.

The solution to this problem makes use of the physical property that the lowest-mass D mesons decay only by the weak interaction. The corresponding lifetimes, typically a fraction of a picosecond, mean that a moving D may travel an observable distance before decaying. With sufficiently precise tracking detectors placed near the production target it is possible to determine which tracks originate from a decay vertex that is measurably separated from the production vertex. By using just these tracks in candidate mass combinations, one can reduce the background by a factor as large as several thousand. Fermilab experiment E691 (12) used an incident photon beam and the high precision vertex detector

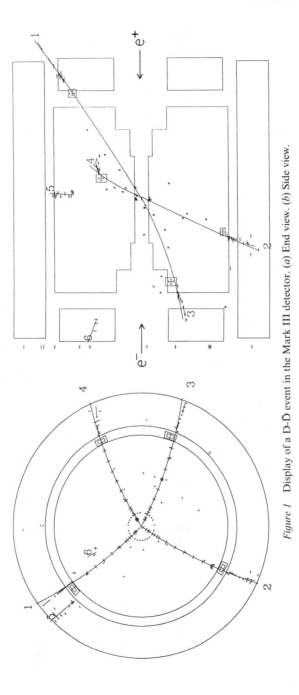

Figure 1 Display of a D-D̄ event in the Mark III detector. (*a*) End view. (*b*) Side view.

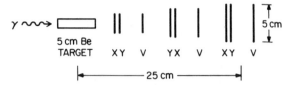

Figure 2 The layout of the target and vertex detector region of E691. The nine silicon microstrip planes contain a total of 7000 strips. The strips of each plane are spaced 50 μm apart and are oriented perpendicular to the X, Y, and V (20.5°) directions as indicated.

arrangement shown in Figure 2, coupled with a large multiparticle spectrometer, to isolate a very large sample of charm particles.

2.   THE SPECTROSCOPY OF CHARMED MESONS

2.1   *Introduction*

The charmed mesons are bound states of a charmed quark and a light antiquark, all of which are expected to belong to the multiplets of the 4-quark SU(4) scheme (6, 7, 13). The charmed mesons of lowest mass, the D^0 (cū), the D^+ (cd̄), and the D_s^+ (cs̄), are quark spin-singlet states with $J^P = 0^-$ (pseudoscalars) and belong to the pion SU(4) multiplet. These states decay weakly with directly measurable lifetimes, as discussed in Section 4. They are identified with charm because their dominant decays satisfy the $\Delta C = \Delta S$ rule of the GIM model and because the decays are weak. The properties of these states are given in Table 1.

The vector D mesons, or D*'s, also have orbital momentum (L) zero but are quark spin triplets. The D^{*0} (cū), the D^{*+} (cd̄), and the D_s^* (cs̄) therefore have $J^P = 1^-$ and are members of the rho multiplet. The D^{*0} and D^{*+} have just enough mass, as seen from Table 1, to decay strongly to Dπ. The extremely limited phase space in this mode makes the electromagnetic decays competitive. In addition, the states are very narrow, with the natural widths known only to be significantly less than the best experimental resolution. Because of the small phase space and the narrow width, the decay $D^{*+} \rightarrow D^0\pi^+$ can be used to "tag" D^0 decays with very little background. The D_s^* cannot decay strongly because of isospin conservation; it decays by photon emission.

With the exception of the lifetimes, the properties of the nonstrange D's and D*'s have been well established for a number of years. The mass and quantum number information for these particles, listed in Table 1, is taken from the Particle Data Group (15), and the lifetime averages are from Section 4 of this review. On the other hand, the experimental information on the D_s^+ and the D_s^*, formerly known as the F and F*, has only recently

Table 1 D and D* properties

	C	S	I	I_3	J	Parity	Mass (MeV/c^2)[a]	Width/Lifetime	Strong/EM decays
D^0	1	0	1/2	$-1/2$	0	$-$	1864.5 ± 0.6[b]	0.432 ± 0.011 ps[c]	—
D^+	1	0	1/2	1/2	0	$-$	1869.3 ± 0.6[b]	1.082 ± 0.032 ps[c]	—
D_s^+	1	1	0	0	0	$-$	1969.1 ± 0.9[d]	0.46 ± 0.04 ps[c]	—
D^{*0}	1	0	1/2	$-1/2$	1	$-$	2007.1 ± 1.4[b]	<5 MeV/c^2[b]	$D^0\pi^0$, $D^0\gamma$
D^{*+}	1	0	1/2	1/2	1	$-$	2010.1 ± 0.6[b]	<2 MeV/c^2[b]	$D^+\pi^0$, $D^+\gamma$, $D^0\pi^+$
D_s^{*}	1	1	0	0	1	$-$	2111.1 ± 1.8[d]	<4.5 MeV/c^2[e]	$D_s^+\gamma$

Mass differences (MeV/c^2)

$M_{D^+} - M_{D^0} = 4.74 \pm 0.28$[b] $M_{D^{*0}} - M_{D^+} = 142.5 \pm 1.3$[b]

$M_{D_s^+} - M_{D^+} = 99.8 \pm 0.8$[d] $M_{D^{*+}} - M_{D^+} = 140.7 \pm 0.3$[b]

$M_{D^{*0}} - M_{D^0} = 145.45 \pm 0.07$[b] $M_{D_s^{*+}} - M_{D_s^+} = 142.0 \pm 1.6$[d]

[a] The mass scale for masses in the region of interest for charm is defined by the precision measurements of the J/ψ and ψ' masses of Zholentz et al (14). This measurement has been used by Trilling (7) to calibrate the SPEAR energy scale, and therefore the mass scale for the D^0 and the D^+, to a precision of 0.5 MeV. The masses of the other charmed particles are determined in terms of these masses.

[b] All properties from Yost et al (15) unless indicated otherwise.

[c] From Section 4 of this review.

[d] All properties for the D_s^+ and D_s^* are from Section 2.2 unless indicated otherwise.

[e] From Albrecht et al (18).

approached a comparable level and so we discuss these states more thoroughly in Section 2.2. In addition, the first experimental evidence for higher-mass charmed mesons, thought to be $L = 1$ states, has just recently been obtained and is described in Section 2.3. In Section 2.4 the theoretical understanding of the mass spectrum of these states is discussed.

2.2   The D_s^+ and D_s^*

The D_s^+ was discovered by CLEO in 1983 (16). The most precise D_s^+ mass measurement is $1968.3 \pm 0.7 \pm 0.7$ MeV/c^2 by E691 (17). This is based on a mass difference of 99.0 ± 0.8 MeV/c^2 between the D_s^+ and the D^+, where the D_s^+ mass is determined from the $D_s^+ \to K^+K^-\pi^+$ spectrum shown in Figure 3, and the D^+ mass is determined from the high statistics $K^-\pi^+\pi^+$ mode. The ARGUS collaboration (18) measured a D_s^+ mass of $1969.3 \pm 0.8 \pm 1.5$ MeV/c^2 based on a mass difference of 104.7 ± 1.4 MeV/c^2

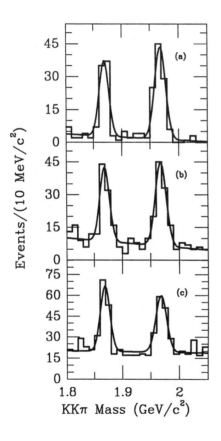

Figure 3   The $K^+K^-\pi^+$ mass spectra of Anjos et al (E691) (17) for the (*a*) $\phi\pi^+$, (*b*) $\overline{K^{*0}}K^+$ and (*c*) nonresonant $K^-K^+\pi^+$ final states. The curve represents a fit with Gaussian peaks for the D^+ and D_s^+ and a linear background.

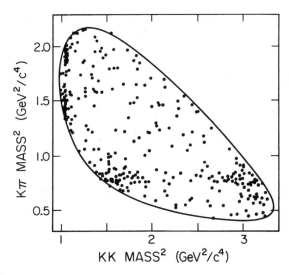

Figure 4   The Dalitz plot, from Anjos et al (17), for the decay $D_s^+ \rightarrow K^+K^-\pi^+$.

between the D_s^+ and the D^0. The other precise measurement is by NA11/32 (19), which gives a value $1972.7 \pm 1.5 \pm 1.0$ but the procedure for defining the mass scale is not clearly specified. From these and other measurements[1] we find an average D_s^+ mass[2] of 1969.1 ± 0.9 MeV/c^2.

The experimental evidence for spin zero for the D_s^+ is based on the angular distribution of the decay products of the ϕ and $\overline{K}^{*0}$, in the decays $D_s^+ \rightarrow \phi\pi^+$ and $\overline{K}^{*0}K^+$. For spin zero, the angular distribution must be $\cos^2\theta$, where θ is the angle between one of the pseudoscalars of the vector particle decay, and the direction of the lone pseudoscalar, in the vector particle center-of-mass frame. The evidence of this distribution is seen from the E691 Dalitz plot of the $D_s^+ \rightarrow K^-K^+\pi^+$ mode shown in Figure 4. The ϕ and K* mass bands are populated predominantly near the kinematic boundaries corresponding to large values of $\cos^2\theta$. This $\cos^2\theta$

---

[1] Other D_s^+ masses, all in MeV/c^2 with references in parentheses, are $1972.4 \pm 3.7 \pm 3.7$ (20), 1980 ± 15 (21), $1963 \pm 3 \pm 3$ (22), $1948 \pm 28 \pm 10$ (23), $1975 \pm 19 \pm 10$ (24), and $1970 \pm 5 \pm 5$ (16).

[2] The average value of the D_s^+ mass has been determined by first removing the common mass scale error of 0.5 MeV/c^2 (see footnote "a" of Table 1) from the systematic error for each measurement, assumed to have been added in quadrature. The remaining systematic and statistical errors were then added in quadrature to determine the weight for each measurement. The error on the average was calculated by adding in quadrature the error determined from the weights and the 0.5-MeV scale error.

effect was first observed at the time of the D_s^+ discovery by CLEO and is the clearest evidence that the D_s^+ has spin zero.

As for strange particles, the relative parity of the D mesons with respect to noncharmed states cannot be determined because of parity non-conservation of the weak decays. The parities of the D mesons are chosen to be negative to be consistent with the other members of the pion multiplet. The parities of the D^+ and D^0 are observed to be the same since the D^{*+} decays to both $D^+\pi^0$ and $D^0\pi^+$. There is no experimental information about the relative parity of the D_s^+ and the nonstrange D's. The isospin of the D_s^+ must be zero if it is the $c\bar{s}$ state expected in the GIM model. The predominance of $s\bar{s}$ quark combinations in the hadronic decays of the D_s^+ (see Section 6) is a clear indication that this identification is correct. In addition no other state has been experimentally observed that could be associated with the D_s^+ in an isotopic multiplet.

As with the D_s^+, the D_s^* mass measurements have recently become more precise. The ARGUS collaboration, which detected D_s^+ decay photons that had converted to electron positron pairs, has the most precise value (18) of the D_s^*-D_s^+ mass difference, $142.5\pm0.8\pm1.5$ MeV/c^2. Including other measurements[3] we compute an average mass difference of 142.0 ± 1.6 MeV/c^2. A striking feature of the D*-D mass differences of Table 1 is the fact that for the three D mesons the mass difference is essentially the same.

The fact that nonstrange D*'s have both photonic and pionic decays implies (7) that the spins of the D*'s are one or greater, and if the spin is one, the parity for the D* is negative. The same argument is not applicable for the D_s^* since hadronic decays are forbidden by isospin conservation. While the spin and parity of the D_s^* have not been unambiguously determined experimentally, assignments different from those of Table 1 would be completely contrary to our present understanding of heavy quark spectroscopy.

2.3   The $L = 1$ D^{**}'s

In addition to the 0^- and 1^- mesons, many higher-mass charm states are expected. Of particular interest are P states with an orbital angular momentum of one. The total quark spin, S, can be either zero or one, which leads to the possible states given in Table 2. The total quark spin may not be a good quantum number so the physical 1^+ states may be admixtures of those given in the table. The parities of all of the states are positive as a result of the odd L. The decays listed are those, to a pion and a D or D*, that are allowed by parity and angular momentum conser-

---

[3] These measurements are $137.9\pm2.1\pm4.3$ by Blaylock et al (20), 143.0 ± 18.0 by Asratyan et al (25), and $139.5\pm8.3\pm9.7$ MeV/c^2 by Aihara et al (23).

Table 2 Decays of possible non-strange P states

	J^P	Decays
$^3P_2$	2^+	$D^*\pi$, $D\pi$
$^3P_1$	1^+	$D^*\pi$
$^1P_1$	1^+	$D^*\pi$
$^3P_0$	0^+	$D\pi$

vation. Each state in the list represents an isodoublet and a related strange charmed isosinglet.

The first observation of such a charmed meson was by the ARGUS collaboration (26). A search was made in the $D^{*+}\pi^-$ mass spectrum, and the peak shown in Figure 5 was found at a mass of about 2420 MeV/c^2. This state has been confirmed by the CLEO collaboration (28) and by E691 (29), and the ARGUS collaboration has improved the statistics (27). The mass and width values from these experiments and the averaged values are given in Table 3. This state is referred to in the literature as the $D^{**0}(2420)$ and by the Particle Data Group as the $D_J^0(2420)$. The fraction of D^{*+}'s originating as decay products of the $D^{**0}(2420)$ has been mea-

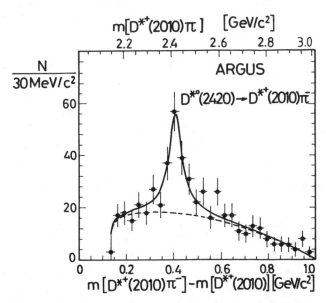

Figure 5 Distribution of the mass difference $M_{D^{*+}\pi^-} - M_{D^{*+}}$ from ARGUS (26). The solid curve represents a fit to the data; the broken line indicates the background.

Table 3 Higher-mass charmed resonances

State	Mass (MeV/c^2)	Width (FWHM)	Ref.
D**^{0}(2420)	2421 ± 5	62 ± 14	27
	2424 ± 6	53^{+30}_{-20}	28
	$2428 \pm 8 \pm 5$	$58 \pm 14 \pm 10$	29
Average mass 2423 ± 4		Average width 60 ± 10	
D**$^+$(2443)	$2443 \pm 7 \pm 5$	$41 \pm 19 \pm 8$	29
D**^{0}(2459)	$2459 \pm 3 \pm 2$	$20 \pm 10 \pm 5$	29

sured to be about 12%. This state could be either of the 1^+ states, the 2^+, or a combination of all three. The statistics and the signal-to-background ratio are not good enough to perform a spin parity analysis. The only spin parity combination excluded is 0^+, but the absence of a signal in the $D^+\pi^-$ spectrum, discussed below, points to the 1^+ assignment.

A state, the D**$^+$(2443), thought to be the isospin companion of the D**^{0}(2420), has been observed by the E691 collaboration in the $D^0\pi^+$ mass spectrum. This peak is due to the decay of the D**$^+$(2443) $\rightarrow$ $D^{*0}\pi^+$, with the D^{*0} decaying to the D^0 and an unobserved photon or π^0. Correcting for the mass displacement and the extra width created by the undetected particles, one finds the mass to be $2443 \pm 7 \pm 5$ GeV/c^2 (Table 3). This state is produced at a rate comparable with that of the D**^{0}(2420).

A more massive and narrower state, the D**^{0}(2459), has been observed by E691 in the $D^+\pi^-$ mass spectrum shown in Figure 6. The peak has a statistical significance of five standard deviations and the mass and width given in Table 3. This state was recently confirmed by the ARGUS collaboration (30). As seen from Table 2, the D**^{0}(2459) could be either the 0^+ or the 2^+, if it is indeed one of the expected P-wave mesons. The decay angular distribution of the ARGUS data indicates that 0^+ is unlikely. Most of the theoretical models predict the 2^+ meson to be the most massive and the narrowest of the P states.

2.4 Model of the Charmed Meson Mass Spectrum

Given the values of the quark masses, the spectrum of the meson masses is in principle calculable on the basis of QCD. It is impressive that the masses of the nonstrange charmed mesons were predicted with a precision of ~ 50 MeV/c^2 after the discovery of the J/ψ (31). The D_s^+ and D_s^* masses were then predicted (32) with an accuracy of a few MeV/c^2 once the nonstrange D masses had been measured.

On the other hand the calculation is nonperturbative and, until lattice gauge calculations are significantly more advanced, will rely on the use of

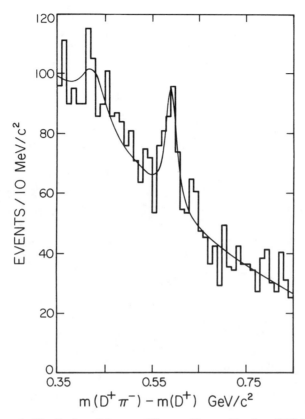

Figure 6 Distribution of the mass difference $M_{D^+\pi^-} - M_{D^+}$ from E691 (29).

a QCD-motivated, phenomenological, nonrelativistic potential (33–35). While different forms of the potential have been used, they have in common a small-distance behavior proportional to $1/r$ due to gluon (vector) exchange and a large-distance, confining, behavior, with an r^{+1} dependence caused by scalar (multigluon) exchange. The potential contains a dominant spin-independent term, and spin-spin, spin-orbit, and tensor terms that cause splittings between states with the same valence quarks. Strange charm states are about 100 MeV/c^2 heavier than the nonstrange particles because of the heavier strange quark mass.

In contrast to charmonium and bottomonium, the charmed mesons contain a light and a heavy quark. The system has a heavy force center, like the hydrogen atom, and probes the form of the potential at larger distance. In this type of model, the splitting between the 0^- and 1^- states of the same quarks is due to the hyperfine (spin-spin) interaction with

$[M(1^-) - M(0^-)]$ proportional to $(|\psi(0)|^2)/(m_i m_j)$, where $\psi(0)$ is the value of the wave function at the origin, and m_i and m_j are the masses of the quarks. The fact that this splitting is the same for strange and nonstrange D's, as seen from Table 1, implies that $|\psi(0)|^2$ must be proportional to the light quark mass. This is the behavior given by the Schrödinger equation for a potential proportional to r, as we expect for the light quark–heavy quark system where the mean value of r is large.

For the $L = 1$ case (31, 34, 36) the spin-spin interaction is unimportant since it is proportional to $|\psi(0)|^2$. As for the hydrogen atom, the mass splitting is expected to be dominated by the spin-orbit interaction involving the spin of the light quark. Instead of the total quark spin, the good quantum number may be j, the total angular momentum of the light quark, and the two 1^+ states of Table 2 may mix, with the mass eigenstates defined by the quantum numbers $J_j^+ = 1_{3/2}^+$ and $1_{1/2}^+$. The sign of the spin-orbit mass shift depends upon the relative importance of the short- and long-distance parts of the potential. The early predictions of De Rújula, Georgi & Glashow (32), made before the first experimental observation of the charmed mesons and renormalized by 60 MeV/c^2 (36) because of the measured D masses, have the 2^+ and $1_{3/2}^+$ nearly degenerate with masses of 2420 and 2390 MeV/c^2, respectively, and the $1_{1/2}^+$ and 0^+ with masses of 2310 and 2300 MeV/c^2. The observed D**(2459) and D**(2420) would clearly be quite consistent with the 2^+ and $1_{3/2}^+$. However, if the long-range part of the interaction dominates the spin-orbit term, the level structure could be inverted. It is clearly very important to determine experimentally the quantum numbers of the D**'s.

3. WEAK DECAYS OF CHARMED MESONS

There are three classes of charmed meson decay: leptonic, semileptonic, and hadronic. For the pseudoscalar D mesons, the decay rate for $D \rightarrow \ell \bar{\nu}$ determines the decay constant f_D. The decay rate is

$$\Gamma(D \rightarrow \ell \bar{\nu}) = \frac{G_F^2}{8\pi} |V_{cq}|^2 f_D^2 m_D m_\ell^2 \left[1 - \frac{m_\ell^2}{m_D^2}\right]^2. \qquad 3.$$

The m_ℓ^2 term represents the familiar helicity suppression of leptonic decays of a pseudoscalar meson. In the case of D mesons, the expected branching fractions are about 10^{-4} for $D^+ \rightarrow \mu^+ \nu_\mu$ and 10^{-3} for $D_s^+ \rightarrow \mu^+ \nu$. Thus direct observation of the purely leptonic decay is very difficult. The only experimental limit comes from the Mark III group (37), which finds $B(D^+ \rightarrow \mu^+ \nu_\mu) < 7.2 \times 10^{-4}$ at the 90% C.L., which translates to a limit of $f_D < 290$ MeV. The best theoretical estimates for f_D lie in the range of

100–250 MeV (38). It is important to improve this limit, because the knowledge of f_D is important for calculations of other modes, and will improve the theoretical estimates for f_B.

The semileptonic decays of charmed mesons occur primarily through the beta decay of the charmed quark. (Figure 7 illustrates the dominant $c \rightarrow s e^+ v_e$ decay.) In this approximation, the inclusive rate can be calculated using the same formula as for muon decay:

$$\Gamma_{SL}(D \rightarrow x\ell^+ \bar{v}) = \Gamma_0 \equiv \frac{G^2}{192\pi^3} m_c^5 f\left(\frac{m_s}{m_c}\right), \qquad\qquad 4.$$

where f is a function that takes into account the finite mass of the strange quark. The error in the charmed quark mass leads to a large error in calculating this rate. To improve the estimate, form factors for particular final states must be used. Section 5 contains a discussion of the exclusive semileptonic decays and measurement of the K-M matrix elements, as well as a more detailed discussion of the semileptonic decay process.

In this picture of semileptonic decays it follows that $\Gamma_{SL}(D^+) = \Gamma_{SL}(D^0)$ $= \Gamma_{SL}(D_s^+)$. Any differences in the total decay rate are due to the hadronic decays. Ignoring strong interactions, there are three diagrams at the valence quark level for hadronic charm decays: spectator decay, W exchange, and W annihilation (Figure 8). The dominant diagram is spectator or flavor decay, in which the light quark is not involved at the weak vertex. Since the D's have spin zero, the W exchange and W annihilation diagrams are suppressed by helicity conservation at the light quark vertex, although the strength of that suppression is a point of some discussion. The valence quark picture, represented by these three diagrams, is clearly naïve but it provides a good starting point to discuss the basic issues in charm decay.

It follows that (a) $\tau_D \approx 0.3$ ps, (b) $\tau_{D^+} = \tau_{D^0} = \tau_{D_s^+}$, and (c) $B(D \rightarrow x\ell^+ v) = 1/(n_\ell + n_c) = 20\%$, where n_c is the number of colors and n_ℓ is the number of light leptons. The average lifetime of the D mesons is about 0.6 ps, but there are large differences in lifetimes of different D mesons, and

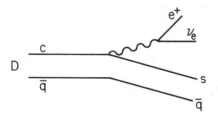

Figure 7   Dominant diagram for Cabibbo-favored semileptonic decays of D mesons.

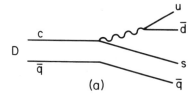

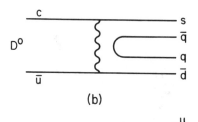

Figure 8 Quark level diagrams for hadronic charm meson decays: (*a*) spectator decay, (*b*) W exchange, and (*c*) W annihilation.

the average semileptonic branching ratio is lower than 20%. These two discrepancies were the experimental clues that led to the somewhat more detailed models of charm decay.

The Hamiltonian for hadronic decay of charmed mesons at the valence quark level has the form

$$H_{\text{had}} = \frac{G}{\sqrt{2}} [V_{\text{cs}}\bar{s}_i\gamma_\mu(1-\gamma_5)c_i + V_{\text{cd}}\bar{d}_i\gamma_\mu(1-\gamma_5 c_i]$$

$$\times [V_{\text{ud}}^*\bar{u}_j\gamma_\mu(1-\gamma_5)d_j + V_{\text{us}}^*\bar{u}_j\gamma_\mu(1-\gamma_5)s_j], \quad 5.$$

where there are summations over color indices. Keeping only the Cabibbo-favored (CF) decays for the moment, this gives $H_{\text{CF}} = (G/\sqrt{2})V_{\text{cs}}V_{\text{ud}}^*(\bar{s}c)_L(\bar{u}d)_L$, where $(\bar{s}c)_L = \bar{s}_i\gamma_\mu(1-\gamma_5)c_i$. The next step is to turn on the strong interactions and add the effects of QCD (39–42). The hadronic decay Hamiltonian has two operators

$$H_{\text{CF}} = \frac{G}{\sqrt{2}} V_{\text{cs}}V_{\text{ud}}^*(c_+O_+ + c_-O_-), \qquad 6.$$

where $O_\pm = (\bar{s}c)_L(\bar{u}d)_L \pm (\bar{s}d)_L(\bar{u}c)_L$. For each operator, the first term is the bare weak Hamiltonian and the second one is the same but with c and d interchanged. (It would be identical were it not for the color indices.) This new term is an effective neutral current induced by hard gluon exchange that rearranges the color such that the $u\bar{d}$ from the virtual W is not a color singlet. The Cabibbo-favored decay rate is

$$\Gamma_{CF}(c \to su\bar{d}) = (2c_+^2 + c_-^2)|V_{cs}|^2|V_{ud}|^2\Gamma_0 \qquad 7.$$

and the total hadronic rate is

$$\Gamma_{had} = (2c_+^2 + c_-^2)\Gamma_0. \qquad 8.$$

Without strong interactions $c_+ = c_- = 1$ and $\Gamma_{had} = 3\Gamma_0$, where 3 is the color factor. These coefficients are calculable in the next-to-leading-log order, and they depend on the momentum scale. We have the relation, $c_+^2 c_- = 1$, and at the charmed quark mass $c_+ \approx 0.7$ and $c_- \approx 2.0$.

Thus if we turn on hard QCD but neglect the soft hadronization process, we find a number of predictions. There is a nonleptonic enhancement, $\Gamma_{had} = (2c_+^2 + c_-^2)\Gamma_0 \approx 5\Gamma_0$ rather than $3\Gamma_0$. This would lower the semileptonic branching ratio from 20% to about 14%. There is an effective neutral current interaction that populates certain "color-suppressed" final states. All of these predictions are verified, which supports the utility of perturbative QCD calculations even at the charmed quark mass.

As discussed in the next section, the three D mesons have different lifetimes; their ratio is $\tau(D^+):\tau(D^0):\tau(D_s^+) = 2.5:1:1$. Although the differences are not as large as the factor of 100 seen in K decays, they are significant. The investigation of these differences has provided a focus for much of the study of hadronic decays of D mesons in the last few years. An understanding of the source of the D-lifetime difference will also be applicable to B mesons, where measurements of individual lifetimes will be available in the future.

First we must test the hypothesis that lifetime differences are due to hadronic decays only. Mark III (43) measures the ratio

$$\frac{B(D^+ \to e^+X)}{B(D^0 \to e^+X)} = 2.3^{+0.5}_{-0.4} \pm 0.1. \qquad 9.$$

If we assume that $\Gamma(D^+ \to e^+X) = \Gamma(D^0 \to e^+X)$, then one can equate the ratio of branching fractions to the ratio of lifetimes, which is determined in Section 4 to be

$$\frac{\tau(D^+)}{\tau(D^0)} = 2.51 \pm 0.10. \qquad 10.$$

Thus, at the level of 20% error, the semileptonic decay rates are equal, and the large differences in lifetime must be explained in the hadronic decay sector.

The most obvious possible source of the lifetime differences is a large contribution from annihilation decays. These include those shown in Figure 8b, for the D^0, and Figure 8c, for the D_s^+. Since the D^+ has no Cabibbo-favored annihilation decay, its lifetime is longer than D^0 or D_s^+ if annihilation decays are important. At the level of simple quark diagrams, the annihilation decays are suppressed by a helicity factor similar to the case of $D \rightarrow \ell v$, shown in Equation 3. It is possible to circumvent this suppression, however, if the effects of strong interactions are sizeable (41, 44).

Even if annihilation decays are small and spectator decays dominate, it is possible to explain large lifetime differences. The Hamiltonian for Cabibbo-favored spectator decays (Equation 6) can be written in the form

$$H_{CF} = \frac{G}{\sqrt{2}} V_{cs} V_{ud}^* \left[\left(\frac{c_+ + c_-}{2} \right) (\bar{s}c)_L (\bar{u}d)_L + \left(\frac{c_+ - c_-}{2} \right) (\bar{s}d)_L (\bar{u}c)_L \right]. \quad 11.$$

This separates the interaction into effective charged and neutral current operators. The diagrams for these two operators are shown in Figure 9. In the case of the D^+, and only the D^+, the two diagrams produce the same final-state quarks and therefore interfere. The rate for the D^+ decay

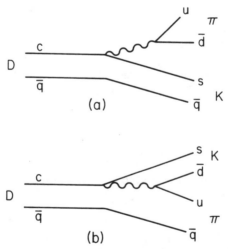

Figure 9   Spectator diagrams corresponding to the effective (a) charged and (b) neutral current (or color-suppressed) operators.

is proportional to c_+^2, whereas the D^0 and D_s^+ rates are proportional to a mixture of c_+^2 and c_-^2. Already at the quark level, there is a large destructive interference that lengthens the D^+ lifetime.

To study D-meson decay modes in detail it is necessary to make some hypothesis about how the quarks become mesons. Bauer, Stech & Wirbel (45) assume factorization and replace the quark currents with effective hadron currents:

$$H_{eff} = \frac{G}{\sqrt{2}} V_{cs}V_{ud}^*[a_1(\bar{s}c)(\bar{u}d)) + a_2(\bar{s}d)(\bar{u}c)], \qquad 12.$$

where

$$2a_1 = (1+\xi)c_1 + (1-\xi)c_2 \qquad 2a_2 = (1+\xi)c_+ - (1-\xi)c_-. \qquad 13.$$

The factor ξ is a color mismatch factor and represents the fraction of the time the quark-antiquark pairs will align the "wrong" way in the process of forming mesons. Naïvely, one expects $\xi = 1/N = 1/3$, from the probability that colors match randomly. However, this is left as a free parameter because this process is not perturbatively calculable. As described below, experimental results prefer $\xi \approx 0$, which follows naturally out of the $1/N$ expansion by Buras et al (46). The QCD parameters c_+ and c_- represent hard gluon effects and are calculable; the factor ξ is connected with hadronic matrix elements. In the simplest version of this scheme, there are two free parameters, a_1 and a_2. The crucial assumption is that the form factors for the final-state mesons can be used to complete the calculation, with no further degree of freedom. In fact there are final-state interactions required and the factorization assumption is not easy to justify. Nonetheless, this scheme gives a good framework for discussing the effects of hadronic matrix elements.

In the case of $D^0 \rightarrow$ pseudoscalar-pseudoscalar (PP) decays, the two diagrams in Figure 9 correspond to $D^0 \rightarrow K^-\pi^+$ and $D^0 \rightarrow \bar{K}^0\pi^0$. For the D^+, they both represent $D^+ \rightarrow \bar{K}^0\pi^+$. This leads to an interference effect in the hadronic D^+ decay rate that is large, as it is at the quark level. More detailed investigation using many exclusive final states (45, 46) leads to the conclusion that it is possible to explain lifetime ratios of about 2 with such an interference effect.

In summary, our picture of hadronic D decays is based on simple valence quark diagrams, hard gluon QCD corrections, form factors for the final-state mesons, and final-state interactions. In the next sections we assemble the experimental facts justifying this picture. We can identify the successes and failures of the simple model and show where more experimental information is needed.

Many theoretical approaches go beyond the one sketched here. A modest extension includes annihilation diagrams at a minor, but not negligible, level (47). Another uses a large number of phenomenological amplitudes, to be determined experimentally (48). One group uses an expansion in $1/N_c$, where N_c is the number of colors, to motivate the small size of annihilation diagrams (46). Finally, there is an attempt to use a somewhat more rigorous model based on QCD sum rules (49).

4. LIFETIMES

The three pseudoscalar D mesons can only decay weakly, and thus have lifetimes of 10^{-13} to 10^{-12} seconds. Proper decay times in this range correspond to decay points, for particles moving at relativistic velocities, measurably displaced from the production point. The mean lifetime, τ, is determined from the distribution[4] of proper decay times,

$$P(t) = \frac{1}{\tau} e^{-(t-t_0)/\tau}, \qquad\qquad 14.$$

where $P(t)$ is the probability that a particle, known to exist at the time t_0, decays at the later time t. The times are related to distances along the flight direction d by $t = d/\gamma v$, where $\gamma = 1/\sqrt{1-v^2/c^2}$ and v is the particle speed.

The error in the lifetime measurement, for a background-free sample of N decays is approximately

$$\delta\tau = \sqrt{\frac{\tau^2 + (\delta t)^2}{N}}, \qquad\qquad 15.$$

where δt is the experimental resolution in proper decay time. The conditions required for a precise lifetime measurement are therefore (a) the particles must be moving at a substantial fraction of the speed of light, (b) δt should be smaller than τ, (c) N should be large, and (d) the background should be small.

A number of techniques have been used to determine charmed particle lifetimes. They all rely on precise measurements of the locations of the vertices of the tracks resulting from the particle decay. The position resolution of ordinary drift chambers is not adequate for this purpose, but that of emulsions (50, 51), silicon detectors (52–56), and high resolution bubble chambers (57, 58) is sufficient. High precision drift chambers (59, 60) typically have a resolution δt comparable to decay times τ. The best measurements have been obtained by the E691 collaboration (53), which

---

[4] For the case of mixing, this distribution is modified as discussed in Section 7.

has a very large statistical sample and used high precision silicon microstrip detectors. The arrangement of the microstrip detectors for this experiment is shown in Figure 2. For this setup the transverse position resolution at the vertex is $\sigma_t = 15$ μm, which results in a proper time resolution $\sigma_t/c = 0.05$ ps, much smaller than the lifetimes of charmed mesons.

The proper time distributions from E691 for the D^0 decays in the channels (a) $D^{*+} \rightarrow \pi^+ D^0$, $D^0 \rightarrow K^- \pi^+$, (b) $D^{*+} \rightarrow \pi^+ D^0$, $D^0 \rightarrow K^- \pi^+ \pi^- \pi^+$, and (c) $D^0 \rightarrow K^- \pi^+$, with events from (a) excluded, are shown in Figure 10a, b, and c, respectively. There are 4200 D^0 decays in the data sample. The measured value of the D^0 lifetime is presented in Table 4, along with the other measurements used in computing the world average. We have adopted the criterion that, to be included in the average, measurements must have been submitted for publication and that the error computed from the statistical and systematic errors taken in quadrature must be less than six times that of the most precise measurement available. The measurements of the D^+ lifetime are also included in Table 4, and D_s^+ results are given in Table 5.

From Tables 4 and 5 we find the ratios of charmed meson lifetimes $\tau_{D^+}/\tau_{D^0} = 2.51 \pm 0.10$, and $\tau_{D_s^+}/\tau_{D^0} = 1.06 \pm 0.10$. As mentioned in the previous section, the large value of τ_{D^+}/τ_{D^0} has motivated a great deal of work in attempting to understand the hadronic D decays. Of the two types of mechanisms mentioned in Section 3, the approximate equality of the D^0 and D_s^+ lifetimes is naturally explained with the destructive interference mechanism. Because of color factors, the D^0 and D_s^+ lifetimes are generally not expected to be equal in models in which the annihilation diagrams are as large as the spectator diagrams.

For the spin-1/2 Λ_c^+ (udc) baryon there is no helicity suppression of the annihilation mechanism. It is therefore not surprising that the Λ_c^+ lifetime measurements of $0.22 \pm 0.03 \pm 0.02$ by E691 (62) and $0.196^{+0.023}_{-0.020}$ ps by NA32 (63) indicate that the Λ_c^+ lifetime is significantly shorter than that of the D meson.

5. SEMILEPTONIC DECAYS

5.1   Introduction

In semileptonic decays the final state contains only two valence quarks. As a consequence, they are particularly amenable to theoretical interpretation and, as with K decays, are the main source of information on the K-M matrix. Semileptonic decays are dominated by spectator diagrams shown in Figures 11a and b, proportional to V_{cs} and V_{cd}, respectively, which to leading order are the only diagrams that can contribute. The only information from the strong interaction needed to calculate the decay rate are

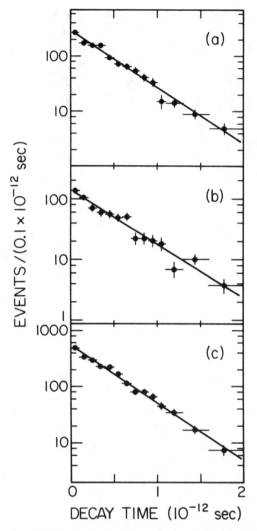

Figure 10 Proper time distribution for the three D^0 decay channels of Raab et al (E691) (53): (*a*) $D^{*+} \to \pi^+ D^0$, $D^0 \to K^- \pi^+$, (*b*) $D^{*+} \to \pi^+ D^0$, $D^0 \to K^- \pi^+ \pi^- \pi^+$, (*c*) $D^0 \to K^- \pi^+$ with events from (*a*) excluded.

the form factors for the final-state mesons. Accurate determinations of V_{cs} and V_{cd} are necessary to test the unitarity of the K-M matrix.

The Cabibbo-favored part of the semileptonic Lagrangian is proportional to $V_{cs}(\bar{s}c)_L(\ell^+ v_\ell)$, which carries zero isospin. These transitions therefore satisfy the rules $|\Delta I| = 0$ and $|\Delta I_3| = 0$, and the D^+ and D^0

Table 4 D^0 and D^+ lifetimes

D^0 lifetime (ps)	D^+ lifetime (ps)	Technique	Ref.
$0.422 \pm 0.008 \pm 0.010$	$1.090 \pm 0.030 \pm 0.025$	Silicon microstrips	53
$0.48 \pm 0.04 \pm 0.03$	$1.05 \pm 0.08 \pm 0.07$	Drift chamber	60
$0.46^{+0.06}_{-0.05}$	$1.12^{+0.14}_{-0.11}$	Bubble chamber	57
0.42 ± 0.05	$1.09^{+0.19}_{-0.15}$	Silicon microstrips	54
$0.50 \pm 0.07 \pm 0.04$	$1.14 \pm 0.16 \pm 0.07$	Drift chamber	61
$0.61 \pm 0.09 \pm 0.03$	$0.86 \pm 0.13^{+0.07}_{-0.03}$	Bubble chamber	58
$0.43^{+0.07}_{-0.05}{}^{+0.01}_{-0.02}$		Emulsion	51
	1.09 ± 0.14	Silicon microstrips	55
0.432 ± 0.011	1.082 ± 0.032	Mean value	

Table 5 D_s^+ lifetimes

Lifetime (ps)	Technique	Ref.
$0.47 \pm 0.04 \pm 0.02$	Silicon microstrips	53
$0.56^{+0.13}_{-0.12} \pm 0.08$	Drift chamber	60
$0.33^{+0.10}_{-0.60}$	Silicon microstrips	56
$0.47 \pm 0.22 \pm 0.05$	Drift chamber	61
0.46 ± 0.04	Mean value	

consequently decay to final states $X\ell^+\nu_\ell$, where X has the $\bar{K}$ quantum numbers: strangeness -1 and isospin $1/2$. It follows that the corresponding exclusive transition rates of the D^+ and D^0 are equal (64): $\Gamma_{\mathrm{CF}}(D^+ \to X^0\ell^+\nu_\ell) = \Gamma_{\mathrm{CF}}(D^0 \to X^-\ell^+\nu_\ell)$. Isospin symmetry does not provide similar rigorous relations relating D_s^+ to D decays, but SU(3) relations can be derived (64). If nonspectator decays are negligible, as expected, the D_s^+ decays should approximately satisfy the relations, $\Gamma_{\mathrm{CF}}(D_s^+ \to \eta\ell^+\nu_\ell) + \Gamma_{\mathrm{CF}}(D_s^+ \to \eta'\ell^+\nu_\ell) = \Gamma_{\mathrm{CF}}(D \to \bar{K}\ell^+\nu_\ell)$, and $\Gamma_{\mathrm{CF}}(D_s^+ \to \phi\ell^+\nu_\ell) = \Gamma_{\mathrm{CF}}(D \to K^*\ell^+\nu_\ell)$. The results to be presented are dominated by semielectronic decays because of the experimental difficulties in identifying low energy muons.

5.2 Inclusive Semileptonic Decays

As mentioned in Section 3, inclusive semileptonic decay rates can be estimated in a free quark model using Equation 4, the expression for muon decay. Using the factor $f(m_s/m_c) = 0.45$ to account for the nonzero mass of the strange quark, this estimate gives a decay rate that is about two thirds the measured value, if we assume a charmed quark mass $m_c = 1.5 \ \mathrm{GeV}/c^2$.

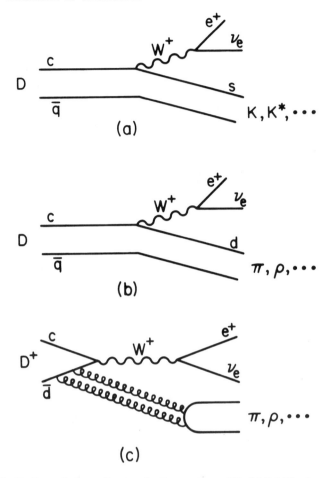

Figure 11 Semileptonic decay diagrams for the nonstrange D's: (*a*) Cabibbo-favored spectator, (*b*) Cabibbo-suppressed spectator, (*c*) Cabibbo-suppressed annihilation. Note that for the D_s^+, diagram (*c*) is Cabibbo favored.

The resonant structure of the hadronic part of the final state is clearly important. Since the mass of the s$\bar{q}$ system is low, the K and K*(892) states are expected to dominate for the Cabibbo-favored decays. Non-resonant Kπ and K$\pi\pi$ states are possible, as well as higher resonances, but these are not expected to comprise a large fraction of the decays. The inclusive rates $\Gamma(D^0 \to x\ell^+\nu)$ and $\Gamma(D^+ \to x\ell^+\nu)$, where just the charged lepton is detected, are the sums of the exclusive rates. They are equal, unless Cabibbo-suppressed nonspectator diagrams such as Figure 11*c* play an important role.

The most precise measurements of inclusive semileptonic decays are by Mark III (65) and are given in Table 6. They detected electrons recoiling against well-defined hadronic D decays produced at the $\psi(3770)$, which decays to D-$\bar{\text{D}}$. The transition rates, computed by dividing the branching fractions by the average lifetimes in Section 4, are also given in Table 6. They are clearly equal within the errors, an indication that the Cabibbo-suppressed nonspectator decays are not important.

5.3   The Exclusive Decay $D \to K\ell v$ and the Determination of $|V_{cd}|$

The calculation of the decay rates for exclusive semileptonic D decays proceeds just as for Ke_3 and Ke_4 decays (66). The amplitude for the decay of charmed meson D to hadron X plus a charged lepton and a neutrino is given by

$$A_{SL}(D \to X) = \frac{G}{\sqrt{2}} V_{ij} L^\mu H_\mu \qquad 16.$$

where G is the Fermi constant, V_{ij} is the relevant K-M matrix element, and L^μ and H_μ are the leptonic and hadronic weak currents, respectively. When X is a pseudoscalar, and in the limit where the mass of the lepton is negligible, we obtain

$$H_\mu = (P_D + P_X)_\mu f^X_+(t), \qquad 17.$$

where $f^X_+(t)$ is a vector form factor depending on the square of the four-momentum transfer,

Table 6   Semileptonic decays

Mode	Branching ratio (%)	Transition rate (10^{10} s^{-1})	Experiment	Ref.
Inclusive				
$D^0 \to e^+x$	$7.5 \pm 1.1 \pm 0.4$	$17.4 \pm 2.5 \pm 0.9$	Mark III	65
$D^+ \to e^+x$	$17.0 \pm 1.9 \pm 0.7$	$15.7 \pm 1.8 \pm 0.6$	Mark III	65
Exclusive				
$D^0 \to K^- e^+ v_e$	$3.8 \pm 0.5 \pm 0.6$	$8.8 \pm 1.2 \pm 1.4$	E691	70
	$3.4 \pm 0.5 \pm 0.4$	$7.8 \pm 1.2 \pm 0.9$	Mark III	71
	3.5 ± 0.5	8.2 ± 1.2	Mean value	
$D^0 \to \pi^- e^+ v_e$	$0.39^{+0.23}_{-0.11} \pm 0.04$	$0.9^{+0.5}_{-0.3} \pm 0.1$	Mark III	71
$D^+ \to \bar{K}^{*0} e^+ v_e$	$4.5 \pm 0.7 \pm 0.5$	$4.2 \pm 0.6 \pm 0.5$	E691	78

$$t \equiv (P_D - P_X)^2 = M_{\ell\nu}^2. \tag{18}$$

In the D center-of-momentum frame the differential decay rate is[5]

$$d^2\Gamma(\tilde{t}, x_\ell) = \frac{G^2 |f_+^X(M_D^2\tilde{t})|^2 |V_{ij}|^2 M_D^5}{64\pi^3}$$

$$\times [(2\eta)^2 - (1 + \eta^2 - \tilde{t})^2 + (1 - \eta^2 + \tilde{t} - 4x_\ell)^2] d\tilde{t} dx_\ell \tag{19}$$

where $\tilde{t} = t/M_D^2$, $x_\ell = E_\ell/M_D$, E_ℓ is the energy of the charged lepton, and $\eta = M_X/M_D$. The only difficult part of the calculation is the evaluation of $f_+^X(t)$. It is reasonable to assume a pole-dominated form,

$$f_+^X(t) = f_+^X(0) \frac{M_y^2}{M_y^2 - t}, \tag{20}$$

where M_y corresponds to the mass of the lowest-mass resonance coupling to D and X. For the vector form factor for D → Keν, the D_s^* has the appropriate $c\bar{s}$ vector quantum numbers. With the mass $M_y = 2.11 \text{ GeV}/c^2$, corresponding to the D_s^*, the integral of Equation 19 gives the transition rate

$$\Gamma(D \to Ke\nu_e) = 1.53 |V_{cs}|^2 |f_+^K(0)|^2 \times 10^{11} \text{ s}^{-1}. \tag{21}$$

The value of $f_+^K(0)$ has been calculated by Wirbel, Stech & Bauer (67) using a model of the hadron wave functions.[6]

Although exclusive semileptonic decays are difficult to isolate because of the undetected neutrino, E691 (70) and Mark III (71) have measured the $D^0 \to K^- e^+ \nu_\ell$ branching ratios, given in Table 6, and E691 has obtained information on the shape of the form factor. A fit to the E691 data using the form factor definition in Equation 20, with the mass M_y as a free parameter, results in $M_y = 2.1^{+0.4}_{-0.2} \text{ GeV}/c^2$, in good agreement with the D_s^* mass. From the mean value of the D → Keν transition rate from Table 6 and the expression in Equation 21 we find, $|V_{cs}|^2 |f_+^K(0)|^2 = 0.54 \pm 0.08$. Assuming that $|V_{cs}| = 0.975$, the value of the cosine of the Cabibbo angle measured in K decays, we find $|f_+^K(0)| = 0.75 \pm 0.05$, in agreement with the theoretical predictions (67, 72) of about 0.78.

The ratio $|V_{cd}|^2/|V_{cs}|^2$ can be determined by the measurement of the branching ratios for both D → Keν$_e$ and D → πeν$_e$. The D → πeν$_e$ decays are described by Equation 19 with M_X taking the value of the pion mass and with $f_+^\pi(t)$ given by Equation 20 with M_y the mass of the nonstrange

[5] This expression can be obtained by appropriate substitution in the equivalent relation for Ke$_3$ decays, for example in Equation 10.72 of Reference 66.

[6] Grinstein et al (68) calculate $f_+^X(t)$ using a quark model approach. A more general discussion of important issues is given by Coffman (69).

D*. The measurement of the ratio of these transition rates then gives the quantity $|V_{cd}|^2 |f_+^\pi(0)|^2/|V_{cs}|^2 |f_+^K(0)|^2$. The Mark III collaboration made a measurement (71) of $D^0 \to \pi^- e^+ \nu_e$ based on seven events. Using their measurements, given in Table 6, and assuming[7] that $f_+^\pi(0)/f_+^K(0)$ is 1.0, the Mark III collaboration finds $|V_{cd}|^2/|V_{cs}|^2 = 0.057^{+0.038}_{-0.015} \pm 0.005$, corresponding to $|V_{cd}|/|V_{cs}| = 0.23^{+0.08}_{-0.03}$.

Obviously this measurement is severely limited by poor statistics. The inverse process can be used with neutrinos and antineutrinos colliding with d, $\bar{\text{d}}$, s, and $\bar{\text{s}}$ quarks in nuclei to form muons and charmed mesons and baryons. The CDHS collaboration (73) at CERN derived $|V_{cd}|$ from the neutrino and antineutrino production of opposite-sign dimuons where one of the muons is from semileptonic charmed meson or baryon decay. This method depends on an analysis that requires a very good understanding of nucleon structure and charm fragmentation and decay. This collaboration finds the value $|V_{cd}| = 0.21 \pm 0.03$, in good agreement with the Mark III value since $|V_{cs}|$ is essentially unity. These measurements are in agreement with $\sin \theta_c$, as expected, within the large errors.

5.4 The Exclusive Decay $D \to K^* \ell \nu$

The semileptonic decays of D's to a vector particle plus a charged lepton and a neutrino are more complicated because the vector is polarized. In the rest frame of the D, the square of the matrix element for the decay to transversely and longitudinally polarized vector mesons respectively[8] are

$$|A_{SL}(D \to V^{(trans)})|^2 = G^2 |V_{ij}|^2 t$$
$$\times [(1 - \cos\theta_e)^2 |H_-(t)|^2 + (1 + \cos\theta_e)^2 |H_+(t)|^2] \times \tfrac{3}{4}\sin^2\theta_v \quad 22.$$

and,

$$|A_{SL}(D \to V^{(long)})|^2 = 2G^2 |V_{ij}|^2 t[(1 - \cos^2\theta_e)|H_0(t)|^2] \times \tfrac{3}{2}\cos^2\theta_v. \quad 23.$$

In this expression the helicity amplitudes,

$$H_\pm(t) = (M_D + M_V)A_1(t) \pm 2\frac{M_D K}{M_D + M_V} V(t) \quad 24.$$

and

---

[7] It is argued that SU(3)-breaking effects should be small (see 67, 68, 72).

[8] The weak decay part of this expression was discussed by Körner & Schuler (74, 75) for B decays and by Bauer & Wirbel (76) for charm decay. Note that different definitions of the form factors, and of $H_+(t)$ and $H_-(t)$, are in use. The additional dependence on the angle between the two decay planes, not given in Equations 22 and 23, can also help in the form factor extraction from the data as discussed by Körner & Schuler (77) for B decays.

$$H_0(t) = \frac{1}{2M_V\sqrt{t}} \left[(M_D^2 - M_V^2 - t)(M_D + M_V)A_1(t) \right.$$

$$\left. -4\frac{M_D^2 K^2}{M_D + M_V}A_2(t) \right], \quad 25.$$

are expressed in terms of the two axial vector form factors $A_1(t)$ and $A_2(t)$ and the vector form factor $V(t)$, which are expected to be dominated by the 1^+ and 1^- $c\bar{s}$ poles, respectively. The strong decay angle θ_V is the angle, in the frame of the vector, between either of the pseudoscalar decay products of the vector and the direction of the D, where the vector is assumed to decay into two pseudoscalars. The weak decay angle θ_e is the angle between the charged lepton and the D in the lepton neutrino (virtual W) frame, and, $K = (1/2M_D)[(M_D^2 - M_V^2 - t)^2 - 4M_V^2 t]^{1/2}$ is the magnitude of the momentum of the K*.

For charm decay, $H_+(t)$ and $H_-(t)$ correspond to negative and positive helicity virtual W's, respectively. For anticharm decays, this association is reversed. In both cases $H_+(t)$ is expected to be larger than $H_-(t)$, exhibiting the parity violation of the V-A structure of the decay. The three helicity components can, in principle, be determined as a function of t from the measured correlation of the angles θ_e and θ_V. The three form factors can be determined from these helicity amplitudes.

The $D^+ \to \overline{K}^{*0}e^+\nu$ decay has been studied by the E691 collaboration (78); results on the branching ratio and corresponding transition rate are given in Table 6. The dominant $\overline{K}^{*0}$ component has been extracted from the mass spectrum of the $K^-\pi^+$ system in the decay $D^+ \to K^-\pi^+e^+\nu$, as shown in Figure 12.

The complete analysis of the data in terms of the amplitudes $|H_+(t)|$, $|H_-(t)|$, and $|H_0(t)|$ has not yet been carried out. The strong decay angular distribution, $w(\theta_v) = 1 + \alpha\cos^2\theta_v$, is obtained from Equations 22 and 23 by integrating over t and θ_e. The coefficient α is given in terms of the ratio of longitudinal to transverse transition rates $\Gamma_L/\Gamma_T = (1+\alpha)/2$. The E691 collaboration finds the $\cos\theta_v$ angular distribution shown in Figure 13.

This distribution, which includes background and is distorted by the experimental acceptance and the effects of the unobserved neutrino, results in a corrected value $\alpha = 3.8^{+3.4}_{-1.8} \pm 0.4$. This is very large compared with the unpolarized case $\alpha = 0.0$. The ratio of longitudinal to transverse polarization is therefore $\Gamma_L/\Gamma_T = 2.4 \pm^{+1.7}_{-0.9} \pm 0.2$, to be compared with the unpolarized value of 0.5 and with the expectation of 0.9 from the Bauer, Stech & Wirbel model (67, 76).

We can estimate the extent to which the K and K* saturate the inclusive semileptonic decay. The average of the D inclusive transition rates measured

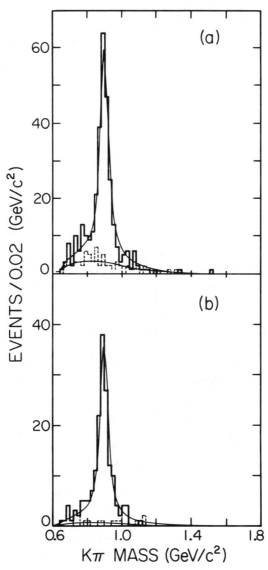

Figure 12   The Kπ mass spectrum in the decay $D^+ \rightarrow K^-\pi^+e^+\nu_e$ from Anjos et al (E691) (78). This figure shows right-sign (*solid*) and wrong sign (*dashed*) combinations for (*a*) loose and (*b*) tight selection criteria. The curves represent a fit to the data.

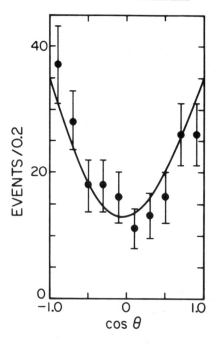

Figure 13 The cos θ_v distribution for $\overline{K}^{*0} \to K^- \pi^+$ in the $D^+ \to \overline{K}^{*0} e^+ v_e$ decay of Anjos et al (78). This distribution includes background and the effects of the unobserved neutrino. The fit shown is for $\alpha = 3.8$.

by Mark III is $(16.3 \pm 1.6) \times 10^{10}$ s^{-1}. The sum of the K and K* transition rates from Table 6 is $(12.6 \pm 1.4) \times 10^{10}$ s^{-1}. This leaves a transition rate of $(3.7 \pm 1.2) \times 10^{10}$ s^{-1} for decays other than K and K*. Approximately one third of this is expected to be in Cabibbo-suppressed $D \to \pi e v$ and $D \to \rho e v$ decays (67). The data are clearly consistent with very small contributions from other semileptonic final states.

5.5   The Status of Semileptonic Decays

There appears to be good agreement between theory and experiment for the $D \to K e v$ and $D \to \pi e v$ decays. On the other hand, both the theoretical and experimental situation must be improved in order to make possible meaningful measurements of $|V_{cs}|$ and $|V_{cd}|$. The present values are consistent with $\cos \theta_c$ and $\sin \theta_c$ as expected. A test of the model for computing the form factor is to predict the more complicated behavior of the semileptonic vector decays. The comparison of the $D^+ \to \overline{K}^{*0} e^+ v$ measurement with theoretical predictions shows that the theory has overestimated the transition rate by a factor of two or more (67, 79) and underestimated the longitudinal polarization (76), also by about a factor of two. A model to account for these discrepancies has been proposed (76), but the dynamics behind this model is not yet well understood.

The inclusive nonstrange D transition rates are consistent with the sum of K, K*, and Cabibbo-suppressed modes but it is important to search more carefully for nonresonant $K\pi$ and $K\pi\pi$ semileptonic modes. Finally, we note that semileptonic D_s^+ decays have not yet been observed. The ratio between D and D_s^+ decay rates for similar modes can probably be calculated with sufficient reliability so that a measurement of a semileptonic D_s^+ branching ratio would establish the scale for all D_s^+ branching ratios.

6. HADRONIC WEAK DECAYS OF D MESONS

6.1 *Introduction*

A full understanding of the weak decay mechanism of D mesons requires detailed study of exclusive decay modes. There is now a wealth of information on exclusive hadronic decays for the D^0, D^+, and even the D_s^+ mesons. A number of theoretical models have also been tested and revised as more modes are measured. A brief review of the major features of these models was given in Section 3. The general level of understanding of charm decays is probably better than the understanding of decays of strange particles.

There are several theoretical models predicting the decays into the two-body final states: pseudoscalar-pseudoscalar (PP), pseudoscalar-vector (PV), or vector-vector (VV). For Cabibbo-favored decays, the quark level diagrams for spectator decays are shown in Figure 9. The color-aligned diagram (*a*) has the coefficient $c_1 = (c_+ + c_-)/2$ and the effective neutral current diagram (*b*) has the coefficient $c_2 = (c_+ - c_-)/2$. Almost all theoretical treatments employ these diagrams at the quark level with coefficients, calculable from QCD, of $c_1 \approx 1.3$ and $c_2 \approx -0.6$. The hadronization process is handled differently in the various models, however. In the scheme of Stech and coworkers (45), one assumes factorization and needs only the form factors describing the transition of the quark-antiquark pair to a final-state pseudoscalar or vector meson. They also ignore nonspectator diagrams in the simplest version of the model.

Theoretical models predict partial decay rates. To determine a decay rate experimentally, one needs to measure two quantities: the lifetimes and an absolute branching fraction. The Mark III group has made use of the unique advantages of the $\psi(3770)$ resonance to measure absolute branching fractions for a set of D^0 and D^+ decay modes (80). The number of events with a single reconstructed D hadronic decay (single tag) is compared with the number of events with both D decays reconstructed (double tags). By using this method, one can extract the individual branching fractions without reference to the production cross section. Figure 14 shows the mass spectra for this study. A global fit is done to three D^0 and four D^+ branching ratios, and the results are given in Table 7. Unfortunately, it has not been possible to collect a sufficiently large sample of doubly tagged $D_s^+ D_s^-$ events to measure the absolute D_s branching fractions in this way.

6.2 Cabibbo-Favored D Decays

The decay that has been analyzed in the greatest detail is $D \to \bar{K}\pi$. In the Stech picture, there are three types of transitions: class I, $D^0 \to K^-\pi^+$ described by Figure 9a; class II, $D^0 \to \bar{K}^0\pi^0$ (Figure 9b); and class III, $D^+ \to \bar{K}^0\pi^+$, which has contributions from both diagrams. The amplitudes for the three classes, ignoring differences in form factors and phase space, are:

(I) $D^0 \to K^-\pi^+$ a_1

(II) $D^0 \to \bar{K}^0\pi^0$ a_2

(III) $D^+ \to \bar{K}^0\pi^+$ $a_1 + xa_2$

where x is about 1 for $\bar{K}^0\pi^+$, but varies in other D^+ decay modes. As shown in Section 3, $a_1 \approx c_1 \approx 1.2$ and $a_2 \approx c_2 \approx -0.5$. This, together with the form factors for pseudoscalar mesons, is all that would be needed to calculate $D \to PP$ decays if final-state interactions were negligible. As we will see, they are significant.

Besides the $D \to K\pi$ decay, the most detailed information on hadronic charm decay comes from $D \to K\pi\pi$ decays and the resonant PV sub-components. Mark III has analyzed four modes (81): $D^0 \to K^-\pi^+\pi^0$ and $\bar{K}^0\pi^+\pi^-$ plus $D^+ \to \bar{K}^0\pi^+\pi^0$ and $K^-\pi^+\pi^+$. The functions used in fitting the Dalitz plots include Breit-Wigner functions for the vector mesons, nonresonant continuum, and background, and they allow for complex phases. Table 8 shows the results of the fits for all four modes. The two D^0 modes and $D^+ \to \bar{K}^0\pi^+\pi^0$ are seen to be dominated by PV channels. The $D^+ \to K^-\pi^+\pi^+$ channel requires a nonresonant contribution that is not even approximately distributed according to phase space.

Final-state interactions in the well-measured $D \to PP$ and $D \to PV$ modes have been analyzed by Chau & Cheng (48), Kamal (82), and Bauer, Stech & Wirbel (45). The isospin decomposition for $K\pi$ modes yields

$$A(D^0 \to K^-\pi^+) = \frac{1}{\sqrt{3}}(\sqrt{2}A_{1/2} + A_{3/2})$$

$$A(D^0 \to \bar{K}^0\pi^0) = \frac{1}{\sqrt{3}}(-A_{1/2} + \sqrt{2}A_{3/2})$$

$$A(D^+ \to \bar{K}^0\pi^+) = \sqrt{3}A_{3/2}. \qquad\qquad 26.$$

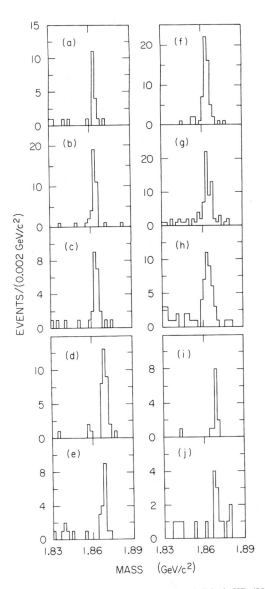

Figure 14 The mass M_X for double tags from Adler et al (Mark III) (80): (*a*) $K^-\pi^+$ vs $K^+\pi^-$, (*b*) $K^-\pi^+$ vs $K^+\pi^-\pi^-\pi^+$, (*c*) $K^-\pi^+\pi^+\pi^-$ vs $K^+\pi^-\pi^-\pi^+$, (*d*) $K^-\pi^+\pi^+$ vs $K^+\pi^-\pi^-$, (*e*) $K^-\pi^+\pi^+$ vs $K^0\pi^-\pi^0$, (*f*) $K^-\pi^+$ vs $K^+\pi^-\pi^0$, (*g*) $K^-\pi^+\pi^+\pi^-$ vs $K^+\pi^-\pi^0$, (*h*) $K^-\pi^+\pi^0$ vs $K^+\pi^-\pi^0$, (*i*) $K^-\pi^+\pi^+$ vs $K^0\pi^-$, (*j*) $K^-\pi^+\pi^+$ vs $K^0\pi^-\pi^-\pi^+$.

Table 7 D^0 and D^+ branching fractions from the global fit

Decay mode	Branching fraction (%)
$D^0 \to K^-\pi^+$	$4.2 \pm 0.4 \pm 0.4$
$D^0 \to K^-\pi^+\pi^-\pi^+$	$9.1 \pm 0.8 \pm 0.8$
$D^0 \to K^-\pi^+\pi^0$	$13.3 \pm 1.2 \pm 1.3$
$D^+ \to K^-\pi^+\pi^+$	$9.1 \pm 1.3 \pm 0.4$
$D^+ \to \overline{K^0}\pi^+$	$3.2 \pm 0.5 \pm 0.2$
$D^+ \to \overline{K^0}\pi^+\pi^0$	$10.2 \pm 2.5 \pm 1.6$
$D^+ \to \overline{K^0}\pi^+\pi^-\pi^+$	$6.6 \pm 1.5 \pm 0.5$

Table 8 Results of the Dalitz plot analysis of the four $D \to K\pi\pi$ decay modes

	Mode	Fit fraction (%)	Branching fraction (%)
$K^-\pi^+\pi^0$	$K^-\rho^+$	$81 \pm 3 \pm 6$	$10.8 \pm 0.4 \pm 1.7$
	$K^{*-}\pi^+$	$12 \pm 2 \pm 3$	$4.9 \pm 0.7 \pm 1.5$
	$\overline{K^{*0}}\pi^0$	$13 \pm 2 \pm 3$	$2.6 \pm 0.3 \pm 0.7$
	nonresonant	$9 \pm 2 \pm 4$	$1.2 \pm 0.2 \pm 0.6$
$\overline{K^0}\pi^+\pi^-$	$\overline{K^0}\rho^0$	$12 \pm 1 \pm 7$	$0.8 \pm 0.1 \pm 0.5$
	$K^{*-}\pi^+$	$56 \pm 4 \pm 5$	$5.3 \pm 0.4 \pm 1.0$
	nonresonant	$33 \pm 5 \pm 10$	$2.1 \pm 0.3 \pm 0.7$
$\overline{K^0}\pi^+\pi^0$	$\overline{K^0}\rho^+$	$68 \pm 8 \pm 12$	$6.9 \pm 0.8 \pm 2.3$
	$\overline{K^{*0}}\pi^+$	$19 \pm 6 \pm 6$	$5.9 \pm 1.9 \pm 2.5$
	nonresonant	$13 \pm 7 \pm 8$	$1.3 \pm 0.7 \pm 0.9$
$\overline{K^-}\pi^+\pi^+$	$\overline{K^{*0}}\pi^+$	$13 \pm 1 \pm 7$	$1.8 \pm 0.2 \pm 1.0$
	nonresonant	$79 \pm 7 \pm 15$	$7.2 \pm 0.6 \pm 1.8$

The amplitudes obey the rule

$$A(D^0 \to K^-\pi^+) + \sqrt{2}A(D^0 \to \overline{K^0}\pi^0) = A(D^+ \to \overline{K^0}\pi^+). \qquad 27.$$

Similar isospin relations hold for the decays $D \to K^*\pi$ and $D \to K\rho$. No solutions are compatible with the data for relatively real amplitudes, which implies that final-state interactions play an important role. The Mark III data can be fit with complex ratios of the isospin 1/2 and 3/2 amplitudes, as shown in Table 9 (42).

The branching ratios for the Cabibbo-favored $D \to PP$ and $D \to PV$ decays are summarized in Table 10. Also listed are the predicted branching ratios from the Bauer, Stech & Wirbel model before annihilation contributions are included. The agreement of the data with this "bare-bones" model is quite good. There are, however, rather large discrepancies in

Table 9 Isospin amplitudes and phase shifts derived from $D \to PP$ and $D \to PV$

Charmed	$|A_{1/2}/A_{3/2}|$	$\delta_{1/2} - \delta_{3/2}$
$D \to \overline{K}\pi$	3.67 ± 0.27	$77 \pm 11°$
$D \to \overline{K}{}^*\pi$	3.22 ± 0.97	$84 \pm 13°$
$D \to \overline{K}\rho$	3.12 ± 0.40	$0 \pm 26°$

Table 10 Cabibbo-favored $D \to PP$ and PV branching ratios

Decay mode	Branching ratio (%)[a]	Theory[b]
$D^0 \to K^-\pi^+$	$4.2 \pm 0.4 \pm 0.4$	5.0
$\overline{K}{}^0\pi^0$	$1.9 \pm 0.4 \pm 0.2$	2.2
$\overline{K}{}^0\eta^0$	$1.5 \pm 0.7 \pm 0.2$	0.3
$\overline{K}{}^0\omega$	$3.2 \pm 1.3 \pm 0.8$	0.3
$\overline{K}{}^0\rho^0$	$0.75 \pm 0.09 \pm 0.47$	0.3
$K^-\rho^+$	$10.8 \pm 0.4 \pm 1.7$	11
$\overline{K}{}^0\phi$	$0.83^{+0.18}_{-0.16}$	0
$K^{*0}\pi^0$	$2.6 \pm 0.3 \pm 0.7$	1.4
$K^{*-}\pi^+$	$5.2 \pm 0.3 \pm 1.5$	2.8
$D^+ \to \overline{K}{}^0\pi^+$	$3.2 \pm 0.5 \pm 0.2$	3.8
$\overline{K}{}^0\rho^+$	$6.9 \pm 0.8 \pm 2.3$	15
$\overline{K}{}^{*0}\pi^+$	$5.9 \pm 1.9 \pm 2.5$	0.4

[a] Experimental results from Mark III, except $\overline{K}{}^0\phi$, which is an average of three experiments as discussed in Section 6.2.
[b] Theoretical results from Bauer, Stech & Wirbel (45), before nonspectator decays are added.

modes such as $D^0 \to \overline{K}{}^0\omega$ and $\overline{K}{}^0\phi$, where the bare amplitudes are particularly small. These might be fed from stronger amplitudes through channel mixing.

The decay $D^0 \to \overline{K}{}^0\phi$ was originally proposed as a test of the existence of nonspectator process (83). As mentioned in Section 3, the contributions from annihilation (Figures 8b,c) enhance the decay rate for the D^0 and D_s^+, but not for D^+. If final-state rescattering is ignored, spectator decays can produce the decay mode $D^0 \to \overline{K}{}^0\phi$ only through OZI violation.

The $\overline{K}{}^0\phi$ mode has been measured by ARGUS (84), Mark III (85), and CLEO (86), and the average value for the branching ratio is $0.83^{+0.18}_{-0.16}\%$. This rate is large, considering the need to produce an additional $s\bar{s}$ quark pair. This was originally interpreted as evidence of a large contribution from annihilation decays, as large as from the spectator decays. Donoghue (87) and Hussain & Kamal (88), however, ascribed the large rate to rescattering effects in final states produced by spectator decay. In the scheme of Chau & Cheng (48), the weak annihilation and strong rescattering are described by the same amplitude. Shifman (88a) also questions the ability to draw a clear distinction between the two sources. The wide range of these interpretations emphasizes the difficulty of drawing conclusions about the underlying decay mechanism on the basis of a single decay mode. To see whether this is a broad pattern of annihilation decays, one must look at D_s^+ decays.

6.3 Cabibbo-Suppressed D Decays

The simple expectation for Cabibbo-suppressed decays is that the decay rate should be equal to $\tan^2 \theta_c \approx 0.05$ compared to related favored decays. Variations by factors of two can be expected in particular modes, but not as a systematic effect in several channels. Large variations from this value can give added information about the mechanism of charm decay.

The Cabibbo-suppressed decays of the D^+ provide a test of the interference effect in charm decay. As described above, the Cabibbo-favored decays such as $D^+ \to \overline{K}^{*0}\pi^+$ are class III decays with contributions from both decay diagrams. The two amplitudes are of opposite sign, so the interference suppresses the decay rate. For Cabibbo-suppressed decays into pions, such as $D^+ \to \rho^0\pi^+$, this interference still occurs, but not for modes including $s\overline{s}$ quark pairs, such as $D^+ \to \phi\pi^+$ or $\overline{K}^{*0}K^+$. Thus the rate $\Gamma(D^+ \to \overline{K}^{*0}K^+)$ should be larger than $\Gamma(D^+ \to \rho^0\pi^+)$. E691 measures, after correcting for phase space, $B(D^+ \to \rho^0\pi^+)/B(D^+ \to \phi\pi^+) = 0.10\pm0.06$ (17, 90), which is much smaller than one. Mark III measures $B(D^+ \to \pi^+\pi^0)/B(D^+ \to \overline{K}^0\pi^+) < 0.15$ at 90% C.L., but $B(D^+ \to \overline{K}^0K^+)/B(D^+ \to \overline{K}^0\pi^+) = 0.317\pm0.08\pm0.048$ (89). Thus the ratio $B(D^+ \to \pi^+\pi^0)/B(D^+ \to \overline{K}^0K^+)$ is less than 0.5, as expected if the $\overline{K}^0K$ rate is not suppressed by destructive interference. There are two relevant cases, one PP and the other PV. Thus in both cases there is evidence that decay modes in which destructive interference is possible are suppressed relative to those without interference.

The ratio $B(D^0 \to K^+K^-)/B(D^0 \to \pi^+\pi^-)$ is measured to be 3.7 ± 1.1 (85). This is larger than the value of 1.4 expected from the fact that $f_K > f_\pi$. It is not clear yet whether this difference is due to final-state interactions. More precise measurement of these and other related decays need to be performed before one can decide whether this is an indication of interesting physics, such as Penguin diagrams (41), or of final-state interactions.

6.4 D_s^+ Decays

In the absence of a resonance similar to the $\psi(3770)$, the study of D_s^+ branching ratios is much more difficult and data are scarce. There have been measurements of the D_s^+ in e^+e^- experiments just above threshold and above the $\Upsilon(1S)$ resonance and in photoproduction. Although the results are not as extensive as for the D^0 and D^+, they are already providing new information for our developing picture of charm decay.

The benchmark decay mode for the D_s^+ is $D_s^+ \to \phi\pi^+$. Figure 3 shows the $K^+K^-\pi^+$ mass spectra from E691 (17), separated into $\phi\pi^+$, $\overline{K}^{*0}K^+$, and nonresonant components. Table 11 gives the results on a number of D_s^+ decay branching ratios; all values quoted are relative to the $\phi\pi^+$ branching ratio. The smallness of the nonresonant $KK\pi$ decays relative to PP and PV decays agrees well with the patterns seen in D^0 and D^+ decay.

The D_s^+ is a particularly good system for studying the effects of the weak annihilation diagram, Figure 8c. The process $c\bar{s} \to u\bar{d}(q\bar{q})$ will only rarely produce a final state with an $s\bar{s}$ pair, while the spectator decay process $c\bar{s} \to s\bar{d}u\bar{s}$ will always have strange quarks in the final state. Thus decays such as $D_s \to \rho\pi$, $\omega\pi$, and 3π should be small unless the annihilation diagram is important. On the other hand, one would expect $\Gamma(D_s^+ \to \rho^0\pi^+) \approx \Gamma(D_s^+ \to \omega\pi^+) \approx \Gamma(D_s^+ \to \phi\pi^+)$ if the two processes contribute equally (41). These predictions may be modified somewhat by interference, which may change the relative rate of $\rho^0\pi^+$ and $\omega\pi^+$ decays.

The decay $D_s^+ \to \pi^+\pi^-\pi^+$ has only been observed by E691 (90). The 3π mass spectrum is shown in Figure 15. The Dalitz plot was analyzed to extract the resonant contributions, and the results are shown in Table 11. The very small upper limit on $\rho^0\pi^+$ supports the picture that the weak annihilation decay is suppressed relative to spectator decay. A significant fraction of the 3π final state is due to $D_s^+ \to f_0\pi^+$, $f_0 \to \pi^+\pi^-$. The $f_0(975)$, previously called $S(975)$, is an $s\bar{s}$ resonance below $K\bar{K}$ threshold that decays primarily to pions. The $D_s^+ \to f_0\pi^+$ decay can probably be attributed to the spectator diagram, so a measure of the relative size of annihil-

Table 11 D_s^+ branching ratios (relative to $\phi\pi^+$)

Decay mode	Branching ratio $B(D_s^+ \to \phi\pi^+)$	Group	Ref.
$\overline{K}^{*0}K^+$	$0.87 \pm 0.13 \pm 0.05$	E691	17
	1.44 ± 0.37	ARGUS	93
$(K^+K^-\pi^+)_{NR}$	$0.25 \pm 0.07 \pm 0.05$	E691	17
$\overline{K}^0K^+$	$1.1 \pm 0.3 \pm 0.5$	Mark III	20
$\rho^0\pi^+$	<0.08	E691	90
$\pi^+\pi^-\pi^+$	$0.44 \pm 0.10 \pm 0.04$	E691	90
$f_0\pi^+$	$0.28 \pm 0.10 \pm 0.03$	E691	90
$(\pi^+\pi^-\pi^+)_{NR}$	$0.29 \pm 0.09 \pm 0.03$	E691	90
$\omega^0\pi^+$	<0.5	E691	91

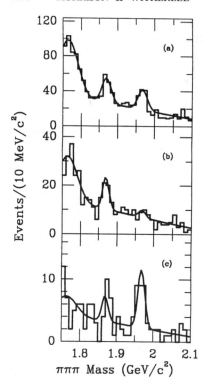

Figure 15 The $\pi^+\pi^-\pi^+$ mass spectrum for Anjos et al (E691) (90): (*a*) inclusive, (*b*) with ρ enhancing $\pi^+\pi^-$ mass cuts, (*c*) with f_0 enhancing $\pi^+\pi^-$ mass cuts.

ation to spectator decays is the ratio $B(D_s^+ \to \pi^+\pi^-\pi^+)_{NR}/B(D_s^+ \to K^+K^-\pi^+) = 0.14 \pm 0.04$. These results confirm that annihilation decays are small, and are not the dominant cause of the D_s^+/D^+ lifetime difference.

There are actually two annihilation diagrams, depending on whether the produced $q\bar{q}$ pair is $u\bar{u}$ or $d\bar{d}$. It is the interference of these two amplitudes that can modify the $\rho^0\pi^+$ rate relative to the $\omega\pi^+$. If the annihilation amplitude was comparable to the spectator, but this interference effect suppressed the $\rho^0\pi^+$ decay completely, the $\omega\pi^+$ rate should be at least as large as $\phi\pi$. E691 has also looked for this decay, and sets an upper limit of $\Gamma(D_s^+ \to \omega^0\pi^+)/\Gamma(D_s^+ \to \phi\pi^+) < 0.5$ (91).

The reason that the D_s^+ branching ratios are normalized to the $D_s^+ \to \phi\pi^+$ decay is that absolute branching ratio measurements are very difficult. A number of experiments have estimated the $D_s^+ \to \phi\pi^+$ branching ratio using charm production cross sections in e^+e^- annihilation far above threshold. It is assumed that $c\bar{c}$ production accounts for 4/10 of the cross section and that the fraction of all charm that produces a D_s^+ is 15%. The results are $B(D_s^+ \rightarrow \phi\pi^+) = 4.4 \pm 1.1\%$ (CLEO) (92), $3.2 \pm 0.7 \pm 0.5\%$ (ARGUS) (93), and $3.3 \pm 1.0\%$ (HRS) (94). Thus 3.5% is usually used as the best estimate. One should keep in mind that the assumed D_s^+ fraction could be off by a factor of as much as 2. As noted in Section 5.5, the absolute D_s branching ratios may also be determined by measuring the semileptonic decays, which are calculable.

6.5 Summary

From all of these experimental facts, a fairly consistent picture of hadronic charm decays emerges. The spectator model, taking into account the interference between the two diagrams where relevant, describes the general features well. Most, but not all, of the decays are due to PP, PV, or VV final states. Final-state interactions are important, but do not obscure completely the predictive power of simple models. The nonspectator diagrams could easily contribute 20% of the total decay rate, but there is convincing evidence that the large difference in the lifetimes is primarily due to interference, and not due to large nonspectator amplitudes.

7. D^0-$\overline{D}^0$ MIXING

The process of particle-antiparticle mixing is a sensitive probe of the weak interaction in the neutral K-, D-, and B-meson systems. Such mixing arises because the eigenstates of the strong interaction, which are flavor eigenstates, differ from the eigenstates of the weak interaction by which the particle decays. In the standard model, the mixing is expected to be small for D^0. Because of this, and because mixing is very sensitive to phenomena such as a new quark generation, it is a good place to look for new physics.

The amount of mixing depends on the mass and width differences of the weak eigenstates. The signature of mixing is the decay of a particle produced as a D^0 into a final state characteristic of a $\overline{D}^0$, i.e. a state containing a "wrong-sign" kaon or lepton. Wrong-sign hadronic final states can be produced either by mixing or by doubly Cabibbo-suppressed decays (DCSD). As for neutral kaons the D^0 system can be characterized by two CP eigenstates (even or odd) with mass difference $\Delta M = M_{odd} - M_{even}$ and width difference $\Delta\Gamma = \Gamma_{odd} - \Gamma_{even}$. The most general expression describing the number of $D^0 \rightarrow K^+\pi^-$ decays as a function of time is, in the limit of small mixing (95),

$$I(D^0 \to K^+\pi^-) = e^{-\Gamma t} \left\{ \frac{t^2}{4} \left[(\Delta M)^2 + \left(\frac{1}{2}\Delta\Gamma\right)^2 \right] + |\rho|^2 \right.$$
$$\left. + t \left[\frac{1}{2}\Delta\Gamma \, \mathrm{Re}\left(\frac{1-\varepsilon}{1+\varepsilon}\rho\right) - \Delta M \, \mathrm{Im}\left(\frac{1-\varepsilon}{1+\varepsilon}\rho\right) \right] \right\}. \quad 28.$$

Here ε is the CP parameter familiar from K^0 decay, and ρ is the ratio of the DCSD amplitudes to the Cabibbo-favored amplitude, $\rho = A(D^0 \to K^+\pi^-)/A(D^0 \to K^-\pi^+)$, which is roughly equal to $\tan^2\theta_c$. The terms proportional to $t^2 e^{-\Gamma t}$ are due to mixing, and the DCSD term is proportional to $e^{-\Gamma t}$. The two terms proportional to $te^{-\Gamma t}$ take into account possible interference between the mixing and DCSD amplitudes. The second term is explicitly CP nonconserving; it changes sign for $\overline{D^0} \to K^-\pi^+$. Since it is small, we neglect it in what follows.

In the simplest picture, neglecting interferences, the time dependence of wrong sign events can be written $e^{-\Gamma t}[\frac{1}{2}\Gamma^2 t^2 r_M + |\rho|^2]$, where $r_M = \frac{1}{2}(x^2+y^2)$, $x = \Delta M/\Gamma$ and $y = \Delta\Gamma/2\Gamma$. Integrated over time, the total fraction of wrong-sign to right-sign decays is then $r_t = r_M + |\rho|^2$. If the DCSD term is small, or zero as in the case of semileptonic decays, then $r_t = r_M$; thus r_M is the measure of mixing.

The two parameters x and y determine the amount of mixing. Of these, y is generally expected to be small for heavy quarks because the phase-space differences for the decays of the two CP eigenstates are small. The short-distance contribution to ΔM mixing is calculable from the diagrams in Figure 16. The dominant contribution is $\Delta M/M = G^2/4\pi^2(B_D f_D^2)(m_s^2 - m_d^2)V_{cs}^2 V_{ud}^{*2}$ (96), where B_D is the "Bag constant" and f_D is the decay constant. Rough estimates give $B_D \approx 1$, and as we have discussed $f_D \approx 200$ MeV. Reasonable estimates lead to values of $r_M = \frac{1}{2}x^2 \approx 10^{-7}$. There are, however, important long-distance contribu-

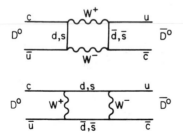

Figure 16 Box diagram short-distance contributions to D^0-$\overline{D^0}$ mixing.

tions, arising from intermediate states such as KK and $\pi\pi$, that couple to both D^0 and $\overline{D}^0$ (97). These cannot be reliably calculated in perturbation theory (98, 99), but might cause r_M to be as large as 10^{-3}.

One example of physics beyond the standard model that could increase the level of D^0-$\overline{D}^0$ mixing is a fourth-generation b′ quark. Direct flavor-changing neutral currents through either Z^0 or Higgs exchange could also cause large mixing. If mixing is not seen, limits can be set on such new effects. If large, however, it would have to be shown to be larger than any possible long-distance effect before new physics could be claimed.

Mark III (100) studied mixing by using double-tag events in which both the D^0 and $\overline{D}^0$ are reconstructed. In the double-tag sample there are 224 events with total strangeness 0 and 3 events with strangeness ± 2. The expected background is 0.4 ± 0.2 events, primarily from misidentified decays. The D^0-$\overline{D}^0$ wave function must be properly symmetrized according to whether the relative angular momentum of the system l is even or odd (101). This leads to a ratio of wrong-sign to right-sign events equal to r_M ($3r_M$) for states of odd (even) l. At the $\psi(3770)$ l must be odd. Naïvely one would expect one wrong-sign event due to DCSD, but quantum statistics suppress DCSD-produced final states with identical two-body decay modes. Two of the three wrong-sign events are consistent with such identical modes. If the three events are attributed to mixing, the mixing rate is $r_M = 1.2 \pm 0.6\%$. Because of the low statistics, and the uncertainty in estimating the DCSD rate, no definite conclusion can be reached.

FNAL experiment E615 studied same-sign pairs of muons in the reaction $\pi N \to \mu\mu X$ at 255 GeV (102). Of about 4000 same-sign pairs, they conclude that less than 63 could be due to D^0-$\overline{D}^0$ mixing. Assuming $\sigma(D^0) + \sigma(\overline{D}^0) = 7.7\ \mu b$ per nucleon for $x_F > 0$ leads to a limit $r_M < 0.0056$ at the 90% confidence level. This result assumes a linear A dependence for the charm cross section. Since a tungsten target was used, the limit could grow by a factor of two if instead the cross section dependence was $A^{0.9}$. Since there is experimental evidence that the cross section only grows as $A^{0.75}$, the limit may be considerably weaker.

FNAL experiment E691 made use of its large charm sample and excellent vertex resolution to study the time dependence (103) of D^0 decays. Pions from the decay $D^{*+} \to \pi^+ D^0$ tag the flavor of the D at production, and the decay modes $K^+\pi^-$ and $K^+\pi^-\pi^+\pi^-$ were used to search for mixing. Figure 17 shows the scatter plots of Q value ($Q = M_{K\pi\pi} - M_{K\pi} - M_\pi$) versus $K\pi$-invariant mass for right-sign ($D \to K^-$) and wrong-sign ($D^0 \to K^+$) events. There is a large cluster of events in the right-sign plot, with small background. Wrong-sign events from DCSD have a standard exponential decay time dependence while mixing evolves as $t^2 e^{-\Gamma t}$. Thus for $t > 2\tau$, 68% of the mixed events survive but only

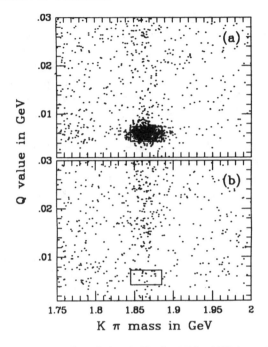

Figure 17 (*a*) The scatter plot of $Q = M(K\pi\pi) - M(\pi) - M(K\pi)$ vs $M(K\pi)$ for the $(K^-\pi^+)\pi^+$ sample of Anjos et al (E691) (103). There is a requirement that $t > 0.22$ ps. (*b*) The same plot for $(K^-\pi^+)\pi^-$ events.

14% of the DCSD events. Figure 18 shows the similar scatter plot for $t > 0.88$ ps; there is only one event in the signal region of the wrong-sign plot, with an expected background of 2.7 events. A fit to the entire sample, with separate terms for mixing and DCSD, gives a total of 909 right-sign events, with 1.2 ± 3.6 from mixing. Including the $K\pi\pi\pi$ mode and correcting for efficiency, which is higher for mixing because of the long decay times, one obtains an upper limit on mixing of $r_M < 0.0037$ at 90% C.L. and a limit on DCSD of $|\rho|^2 < 1.5\%$ for the $K\pi$ and 1.8% for the $K\pi\pi\pi$ modes. If the worst case of interference is assumed, the mixing limit becomes somewhat larger. This case would require the unexpected combination of a very large DCSD amplitude in both modes, a large negative $\Delta\Gamma$, and a small value of ΔM.

If the Mark III events were due to mixing at the 1% level, it would have been very hard to accommodate within the standard model. The limit from E691 rules out this interpretation, however, and sets the limit on mixing

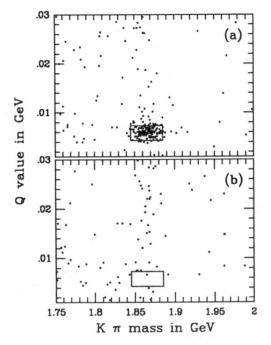

Figure 18   (*a*) The scatter plot of Q vs $M(K\pi)$ for $(K^-\pi^+)\pi^+$ events of Anjos et al (103), with the requirement $t > 0.88$ ps. (*b*) The plot for $(K^-\pi^+)\pi^-$ events with $t > 0.88$ ps.

of $r_M < 0.4\%$. If mixing is seen much below this level, it could be due to the conventional long-range effects. It will be interesting to extend the sensitivity to mixing effects to the 10^{-3} level, and to measure directly the level of DCSD.

8.   RARE DECAYS

In contrast to the strange and bottom quark, most charm quark decays are K-M favored. As a consequence, limits on charm branching fractions to rare modes must be significantly lower than those for the strange or bottom quarks in order to have comparable sensitivity to processes not allowed in the standard model. In addition, it is difficult to compete with the very stringent upper limits that have been measured for rare K decays. There are a number of models, however, such as extended technicolor (104), leptoquarks (105), and some superstring-inspired models (106), in which large flavor-changing neutral current effects would be present in charm, but not in kaon, decays.

The standard model forbids flavor-changing neutral currents (FCNC) in lowest order. Some decays such as $D^0 \to \mu^\pm e^\mp$ and $D^+ \to \mu^\pm e^\mp \pi^+$ also require lepton flavor violation (LFV) and are strictly forbidden in the standard model. The sensitivity of the two-body decays is reduced, as a result of helicity suppression, if the mechanism is a vector or scalar interaction, so that the search for the three-body decays is important. The best experimental limits are presented in Table 12 (102, 107–111). A more complete listing is given in (112). These limits can be used to constrain models; for example in the model of Buchmüller & Wyler (105) the $D^0 \to \mu^\pm e^\mp$ limit implies that the leptoquark mass scale must be greater than 1 TeV.

9.   SUMMARY AND OUTLOOK

During the last few years, there has been a great deal of progress in all areas of charm physics. Important results have come from e^+e^- experiments at threshold and in the Υ region, as well as from photoproduction. The basic properties of the charmed meson have been experimentally determined, often with great precision. The spectroscopy of the pseudoscalar and vector

Table 12   Limits on rare D decays

Mode	Type[a]	Limit (90% C.L.)	Experiment	Ref.
$D^0 \to e^\pm e^\mp$	FCNC	$<3.1 \times 10^{-5}$	E691	107
		$<1.7 \times 10^{-4}$	ARGUS	108
		$<1.3 \times 10^{-4}$	Mark III	111
$D^0 \to \mu^\pm \mu^\mp$	FCNC	$<1.1 \times 10^{-5}$	E615	102
		$<7.0 \times 10^{-5}$	ARGUS	108
$D^0 \to \mu^\pm e^\mp$	FCNC and LFV	$<4.1 \times 10^{-5}$	E691	107
		$<1.0 \times 10^{-4}$	ARGUS	108
		$<1.2 \times 10^{-4}$	Mark III	109
$D^+ \to \pi^+ e^+ e^-$	FCNC	$<5.0 \times 10^{-4}$	E691	107
		$<2.6 \times 10^{-3}$	CLEO	110
$D^+ \to \pi^+ \mu^+ \mu^-$	FCNC	$<1.3 \times 10^{-3}$	E691	107
		$<2.9 \times 10^{-3}$	CLEO	110
$D^+ \to \pi^+ \mu^\pm e^\mp$	FCNC and LFV	$<2.6 \times 10^{-4}$	E691	107
		$<3.8 \times 10^{-3}$	CLEO	110
$D^0 \to \overline{K}^0 e^+ e^-$	FCNC	$<1.7 \times 10^{-3}$	Mark III	111
$D^0 \to \rho^0 e^+ e^-$	FCNC	$<4.5 \times 10^{-4}$	CLEO	110
$D^0 \to \rho^0 \mu^+ \mu^-$	FCNC	$<8.1 \times 10^{-4}$	CLEO	110
$c \to e^+ e^- x$	FCNC	$<2.2 \times 10^{-3}$	CLEO	110
$c \to \mu^+ \mu^- x$	FCNC	$<1.8 \times 10^{-2}$	CLEO	110

[a] FCNC = flavor-charging neutral currents; LFV = lepton flavor violation.

D mesons is complete, and the masses are well measured. Three of the excited charmed mesons have been observed, although it will take further experiments to extract completely the individual particle properties.

A remarkably detailed picture of the charm decay mechanism exists, based on a large number of measured branching ratios. We now know the lifetimes to a few percent, and absolute branching fractions have errors of about 8%. There are accurate measurements of the major exclusive semileptonic decays, which permit detailed comparisons with theoretical models. Of the major hadronic decay modes, only those with high multiplicity remain to be measured.

Through the last few years the cause of the lifetime differences has been intensively studied. A simple picture dominated by spectator decays, with an expected interference that increases the D^+ lifetime, agrees well with the data. Annihilation diagrams are present, probably at a 20% level, but are not as large as spectator diagrams and are not the primary source of lifetime differences. This picture draws on measurements of the lifetimes and many branching ratios. Final-state interactions are important in all modes, although we can only measure them in relatively few cases. They do not, however, completely obscure the pattern of decay rates that comes from the quark diagrams and simple hadronization schemes.

There are a number of measurements that would advance our picture of charm decays. The measurement of $D^+ \to \mu^+ \nu_\mu$ (and $D_s^+ \to \mu^+ \nu_\mu$, $\tau^+ \nu_\tau$) would pin down the D-meson decay constant, f_D. The absolute branching fractions for the D_s^+ need to be normalized properly. To complete our picture of semileptonic decays we need to measure or set tight limits on modes with high hadron multiplicity, and we need to observe the D_s^+ semileptonic decays. We also need measurements of four- and five-body modes to complete our picture of hadronic decays. There is a natural level for mixing and doubly Cabibbo-suppressed decays that is only just beyond the present limits. The sensitivity of searches for these two effects should be increased to that level.

The photoproduction experiment E691 obtained charm samples with a few thousand events for the simplest decay modes, with little background even in high multiplicity modes. Future experiments using hadro- and photoproduction should obtain another factor of 10. For such measurements as $D^+ \to \mu^+ \nu_\mu$, it is necessary to use the double-tagged and kinematically constrained sample of $D\bar{D}$ pairs at the $\psi(3770)$. The new Beijing Electron Positron Collider (BEPC) should reach a luminosity that is a factor of almost 10 beyond SPEAR and could surpass the remarkable results from Mark III. With the increased luminosity at CESR, operating in the Υ region, much better charm spectroscopy results will become possible.

Studies of charmed mesons have provided a way of testing our ability to calculate strong interactions just beyond the region of high momentum transfer in which perturbative effects dominate. As a result, we are able to extrapolate the nonperturbative effects with reasonable reliability to the area of bottom mesons. Such measurements as lifetime differences and form factors in semileptonic decays are crucial to this program. In the future this improved knowledge of the strong interaction effects will help us to extract the unknown weak interaction quantities in B-meson decays. The hope is that, in a period of about ten years, it will be possible to write a similarly detailed review on B-meson physics.

ACKNOWLEDGMENTS

We want to thank Ms. Debbie Alspaugh for her help and care in preparing this manuscript. This work was supported by the US Department of Energy, and R.J.M. would like to acknowledge the hospitality of the Rutherford Appleton Laboratory, where part of this report was prepared. We profited greatly from discussions with Prof. B. Stech, Prof. J. Körner, Dr. M. Wirbel, Dr. M. Bauer, and especially our colleagues from the E691 collaboration. We also relied on earlier reviews by Professors G. Trilling, F. Gilman, D. Hitlin, and I. Bigi.

Literature Cited

1. Auburt, J. J., et al., *Phys. Rev. Lett.* 33: 1404 (1974)
2. Augustin, J. E., et al., *Phys. Rev. Lett.* 33: 1406 (1974)
3. Glashow, S. L., Iliopoulos, J., Maiani, L., *Phys. Rev.* D2: 1285 (1970)
4. Goldhaber, G., et al., *Phys. Rev. Lett.* 37: 255 (1976)
5. Peruzzi, I., et al., *Phys. Rev. Lett.* 37: 569 (1976)
6. Goldhaber, G., Wiss, J. E., *Annu. Rev. Nucl. Part. Sci.* 30: 337 (1980)
7. Trilling, G. H., *Phys. Rep.* 75: 57 (1981)
8. Ye, M.-H., Huang, T., eds., *Charm Physics*, London: Gordon & Breach (1988)
8a. Kobayashi, K., Maskawa, T., *Prog. Theor. Phys.* 49: 652 (1973)
8b. Yost, G. P., et al. (Particle Data Group), *Phys. Lett.* B204: 107 (1988)
9. Bernstein, D., et al. (Mark III), *Nucl. Instrum. Methods* 226: 301 (1984)
10. Albrecht, H., et al. (ARGUS), *Phys. Lett.* B134: 137 (1984)
11. Andrews, D., et al. (CLEO), *Nucl. Instrum. Methods* 211: 47 (1983)
12. Raab, J. R., et al. (E691), *Phys. Rev.* D37: 2391 (1988)
13. Gaillard, M. K., Lee, B. W., Rosner, J. L., *Rev. Mod. Phys.* 47: 277 (1975)
14. Zholentz, A. A., et al., *Phys. Lett.* B96: 214 (1980)
15. Yost, G. P., et al. (Particle Data Group), *Phys. Lett.* B204: 1 (1988)
16. Chen, A., et al., *Phys. Rev. Lett.* 51: 634 (1983)
17. Anjos, J. C., et al., *Phys. Rev. Lett.* 60: 897 (1988)
18. Albrecht, H., et al., *Phys. Lett.* B207: 349 (1988)
19. Becker, H., et al., *Phys. Lett.* B184: 277 (1987)
20. Blaylock, G. T., et al., *Phys. Rev. Lett.* 58: 2171 (1987)
21. Ushida, N., et al., *Phys. Rev. Lett.* 56: 1767 (1986)
22. Derrick, M., et al., *Phys. Rev. Lett.* 54: 2568 (1985)
23. Aihara, H., et al., *Phys. Rev. Lett.* 53: 2465 (1984)
24. Althoff, M., et al., *Phys. Lett.* B136: 130 (1984)
25. Asratyan, A. E., et al., *Phys. Lett.* B156: 441 (1985)
26. Albrecht, H., et al., *Phys. Rev. Lett.* 56: 549 (1986)

27. Prentice, J., et al., in *Proc. Europhysics Conf.*, ed. O. Botner, Uppsala Univ., Sweden (1987), p. 910
28. Bebek, C., et al., see Ref. 27, p. 916
29. Anjos, J. C., et al., *Phys. Rev. Lett.* 62: 1717 (1989)
30. Albrecht, H., et al., *DESY 88-179*, submitted to *Phys. Lett. B*
31. De Rújula, A., Georgi, H., Glashow, S. L., *Phys. Rev.* D12: 147 (1975)
32. De Rújula, A., Georgi, H., Glashow, S. L., *Phys. Rev. Lett.* 37: 785(c) (1976)
33. Kwong, W., Rosner, J. L., Quigg, C., *Annu. Rev. Nucl. Part. Sci.* 37: 325 (1987)
34. Gilman, F. J., see Ref. 8, p. 37
35. Martin, A., *Comments Nucl. Part. Phys.* 16: 249 (1986)
36. Rosner, J. L., *Comments Nucl. Part. Phys.* 16: 109 (1986)
37. Adler, J., et al., *Phys. Rev. Lett.* 60: 1375 (1988)
38. Novikov, V. A., et al., *Phys. Rev. Lett.* 38: 626 (1977); Golowich, E., *Phys. Lett.* B91: 271 (1980); Krasemann, H., *Phys. Lett.* B96: 397 (1980); Mathur, V. S., et al., *Phys. Lett.* B107: 127 (1981); Shuryak, E., *Nucl. Phys.* B198: 83 (1982); Ward, B. F. L., *Phys. Rev.* D28: 1215 (1983); Reinders, L., *Phys. Rep.* 127: 1 (1985); Godfrey, S., Isgur, N., *Phys. Rev.* D32: 189 (1985); Suzuki, M., *Phys. Lett.* B162: 392 (1985); Voloshyn, R. M., et al., *TRI-PP-87-62* (1987); De Grand, E. A., Loft, R., *Phys. Rev.* D38: 954 (1988)
39. Rückl, R., Habilitationschrift, Univ. Munich (1983)
40. Gilman, F., see Ref. 8, p. 1
41. Bigi, I., see Ref. 8, p. 339
42. Hitlin, D., see Ref. 8, p. 219
43. Baltrusaitis, R. M., et al., *Phys. Rev. Lett.* 54: 1976 (1985)
44. Rosen, S. P., *Phys. Rev. Lett.* 44: 4 (1980); Fritzsch, H., Minkowski, P., *Phys. Lett.* B90: 455 (1980); Bernreuther, W., et al., *Z. Phys.* C4: 252 (1980); Bigi, I., *Z. Phys.* C5: 313 (1980); Bander, M., et al., *Phys. Rev. Lett.* 44: 7 (1980); Erratum, *Phys. Rev. Lett.* 44: 962 (E) (1980)
45. Bauer, M., Stech, B., *Phys. Lett.* B152: 380 (1985); Bauer, M., Stech, B., Wirbel, M., *Z. Phys.* C34: 103 (1987)
46. Buras, A. J., Gerard, J. M., Rückl, R., *Nucl. Phys.* B268: 16 (1986)
47. Kamal, A. N., Verma, R. C., *Phys. Rev.* D35: 3515 (1987)
48. Chau, L. L., Cheng, H. Y., *Phys. Rev.* D36: 137 (1987)
49. Blok, B., Shifman, M. A., *Sov. J. Nucl. Phys.* 45: 135, 301, 522 (1987)
50. Fuchi, H., et al., *Nuovo Cimento Lett.* 31: 199 (1981)
51. Ushida, N., et al., *Phys. Rev. Lett.* 56: 1771 (1986)
52. Albini, E., et al., *Phys. Lett.* B110: 339 (1982)
53. See Ref. 12
54. Barlag, S., et al., *Z. Phys.* C37: 17 (1987)
55. Palka, H., et al., *Z. Phys.* C35: 151 (1987)
56. See Ref. 19
57. Aguilar-Benitez, M., et al., *Phys. Lett.* B193: 1401 (1988)
58. Abe, K., et al., *Phys. Rev.* D33: 1 (1986)
59. Gladney, L., et al., *Phys. Rev.* D34: 2601 (1986)
60. Albrecht, H., et al., *Phys. Lett.* B210: 267 (1988)
61. Csorna, S. E., et al., *Phys. Lett.* B191: 318 (1987)
62. Anjos, J. C., et al., *Phys. Rev. Lett.* 60: 1379 (1988)
63. Barlag, S., et al., *Phys. Lett.* B218: 374 (1989)
64. Pais, A., Treiman, S. B., *Phys. Rev.* D15: 2529 (1977)
65. Baltrusaitis, R. M., et al., see Ref. 43
66. Commins, E. D., *Weak Interactions*. New York: McGraw-Hill (1977)
67. Wirbel, M., Stech, B., Bauer, M., *Z. Phys.* C29: 637 (1985)
68. Grinstein, B., Wise, M. B., Isgur, N., *Cal Tech Preprint CALT-68-1311* (1985)
69. Coffman, D. M., PhD thesis, Calif. Inst. Technol. (unpublished) (1986)
70. Anjos, J. C., et al., *Phys. Rev. Lett.* 62: 1587 (1989)
71. Adler, J., et al., *Phys. Rev. Lett.* 62: 1821 (1989)
72. Dominguez, C. A., Daver, N., *Phys. Lett.* B207: 499 (1988); erratum B211: 500 (1988)
73. Kleinknecht, K., Renk, B., *Z. Phys.* C34: 209 (1987)
74. Körner, J. G., Schuler, G. A., *Z. Phys.* C38: 511 (1988); erratum *Z. Phys.* C41: 690 (1989)
75. Körner, J. G., *Mainz Preprint MZ-TH/88-12* (1988)
76. Bauer, M., Wirbel, M., *Heidelberg Preprint HD-THEP-88-22* (1988); *Dortmund Preprint DO-TH 88-19* (1988)
77. Körner, J. G., Schuler, G. A., *Mainz Preprint MZ-TH/89-01* (1989)
78. Anjos, J. C., et al., *Phys. Rev. Lett.* 62: 722 (1989)
79. Suzuki, M., *Phys. Lett.* B155: 112 (1985)
80. Adler, J., et al., *Phys. Rev. Lett.* 60: 89 (1988)

81. Adler, J., et al., *Phys. Lett.* B196: 107 (1987)
82. Kamal, A. N., *Phys. Rev.* D33: 1344 (1986)
83. Bigi, I. I., Fukugita, M., *Phys. Lett.* B91: 121 (1980)
84. Albrecht, H., et al., *Phys. Lett.* B158: 525 (1985)
85. Baltrusaitis, R. M., et al., *Phys. Rev. Lett.* 56: 2136 (1986)
86. Bebek, C., et al., *Phys. Rev. Lett.* 56: 1983 (1986)
87. Donoghue, J., *Phys. Rev.* D33: 1516 (1986)
88. Hussain, F., Kamal, A. N., *Alberta Preprint Alberta-Thy-1-86* (1986)
88a. Shifman, M. A., in *Proc. 1987 Int. Symp. on Lepton and Photon Interactions at High Energies*, ed. W. Bartel, R. Rückl. Amsterdam: North Holland (1987)
89. Baltrusaitis, R. M., et al., *Phys. Rev. Lett.* 55: 150 (1985); erratum 55: 639 (1985)
90. Anjos, J. C., et al., *Phys. Rev. Lett.* 62: 125 (1989)
91. Anjos, J. C., et al., *Fermilab Preprint Fermilab-PUB-89/23-E* (1989)
92. Chen, C., et al., *Phys. Rev. Lett.* 51: 634 (1983)
93. Albrecht, H., et al., *Phys. Lett.* B153: 343 (1985); B179: 398 (1986); B195: 102 (1987)
94. Derrick, M., et al., *Phys. Rev. Lett.* 54: 2568 (1985)
95. Bigi, I., Sanda, A. F., *Phys. Lett.* B171: 320 (1986)
96. Commins, E. D., Bucksbaum, P. H., *Weak Interactions of Leptons and Quarks.* Cambridge Univ. Press (1983)
97. Datta, A., *Phys. Lett.* B154: 287 (1985)
98. Wolfenstein, L., *Phys. Lett.* B164: 170 (1985)
99. Donoghue, J. F., et al., *Phys. Rev.* D33: 179 (1986)
100. Gladding, G., in *Proc. 5th Int. Conf. on Physics in Collision*, ed. B. Augert. Gif-sur-Yvette: Editions Frontieres (1986); Gladding, G., *Ann. NY Acad. Sci.* 535: 178 (1988)
101. Bigi, I. I., Sanda, A. I., *Nucl. Phys.* B281: 41 (1987)
102. Louis, W. C., et al., *Phys. Rev. Lett.* 56: 1027 (1986)
103. Anjos, J. C., et al., *Phys. Rev. Lett.* 60: 1239 (1988)
104. Farhi, E., Susskind, L., *Phys. Rep.* 74: 277 (1981); Eichten, E., et al., *Phys. Rev.* D34: 1547 (1986)
105. Buchmüller, W., Wyler, D., *Phys. Lett.* B177: 377 (1986)
106. Campbell, B., et al., *Int. J. Mod. Phys.* A2: 831 (1987)
107. McHugh, S., in *Proc. 24th Conf. on High Energy Physics*, Munich, ed. R. Kotthaus, J. H. Kühn. Berlin: Springer-Verlag (1989), p. 557
108. Albrecht, H., et al., *Phys. Lett.* B209: 380 (1988)
109. Becker, J., et al., *Phys. Lett.* B193: 147 (1987); B198: 590 (1987)
110. Haas, P., et al., *Phys. Rev. Lett.* 60: 1614 (1988)
111. Adler, J., et al., *SLAC-PUB-4671* (1989); *Phys. Ref. Brief. Rep.* (1989), in press
112. Ammar, R., *Proc. DPF-88 Conf.*, Storrs, Conn. (1988)

Annu. Rev. Nucl. Part. Sci. 1989. 39: 231–58

SUPERNOVAE AND THE HADRONIC EQUATION OF STATE[1]

S. H. Kahana

Physics Department, Brookhaven National Laboratory, Upton, New York 11973.

KEY WORDS: hydrodynamics, general relativity, neutrinos, neutron stars, nuclear physics.

CONTENTS

[1] This chapter has been authored under contract DE-AC02-76CH00016 with the US Department of Energy. Accordingly, the US Government retains a nonexclusive, royalty-free license to publish or reproduce the published form of this chapter, or allow others to do so, for US Government purposes.

1. INTRODUCTION

There have been several articles recently reviewing this subject. Note quite a decade ago, Lattimer (1) wrote a review of the equation of state of hot dense matter. Very recently, Cooperstein & Baron (2, 3) summarized the direct mechanism and equation of state. In Volume 38 of this series (4), Hans Bethe discussed the physics of supernovae. I myself contributed a chapter on the role of the equation of state in supernovae to Gerry Brown's festschrift (5). It is not at all clear that another review can shed further light on this subject, especially when one recalls that these authors are or have been colleagues and collaborators in common work.

Given all of this, I narrow my perspective here, concentrating on those hadronic aspects of the equation of state (EOS) most at question in gravitational collapse, namely the high density behavior. Gravitational collapse of the inner core of a massive star, inevitable after the exhaustion of its nuclear fuel, produces close to a solar mass ($M_\odot$) of matter at or above normal nuclear density. The result of this collapse can be an explosion, i.e. a supernova, produced by bounce of the infalling core, or else a black hole, should the infall not be halted. The direct mechanism (prompt explosion from the shock wave generated at core bounce) is sensitive to the properties of dense hadronic matter. The previous reviews have dealt with the EOS at all densities with, in most cases, only a minimal treatment of the most compressed matter. This makes some sense. Most of the progress attained in dealing with the direct mechanism has come from use of a phenomenological equation of state for high densities introduced (3) precisely to separate the problem into two parts: the hydrodynamic simulation of core collapse and rebound, and the much more difficut problem of derivation of the equation of state. Some tentative progress has been made on this latter problem, and Hans Bethe in his usual encyclopedic fashion considered this topic in his review (4). I wish to expand on his comments.

The most noteworthy event in this subject in recent times is undoubtedly SN1987a (6), certainly the closest supernova in several centuries and the occasion for greatly increased research activity. The detection of neutrinos from SN1987a (7, 8) is arguably the most important observation, serving at one and the same time to fix SN1987a as being induced by gravitational collapse and to confirm in broad terms the validity of existing theoretical scenarios. Unfortunately, a hope dear to the hearts of my collaborators and myself (9) that SN1987a would differentiate between the prompt (direct) and delayed (10) shock mechanisms has not yet been realized. The detection of hydrogen lines in the spectra from SN1987a identifies it as

Type II. The correlation in SN1987a between Type II and gravitational collapse is expected. The progenitors of Type II supernovae are believed to be the most massive stars, above 10 solar masses, in whose short life of 1–10 million years little time has been available for loss of their hydrogen atmosphere. The other broad class of supernovae, Type I, whose spectra exhibit no hydrogen lines, is not dealt with here.

Matter at several times normal densities, taken for symmetric nuclear matter to be $\rho_0 = 0.16$ nucleons/fm^3, is difficult to create in the laboratory, certainly under equilibrium conditions. Heavy-ion collisions at relativistic energies do produce high densities; we are still debating the nature of the constraints imposed on the EOS by such collisions. Nonrelativistic theories of nuclear matter can be extrapolated to high densities, and these theories have achieved a high degree of sophistication in the intervening years since Brueckner (11) and Bethe & Goldstone (12) first expounded them. Nevertheless, since the essential algorithm in their construction involves the counting of hole lines and amounts to an expansion in powers of density, it is difficult to take such theories seriously at high density. With two-body potentials fitted to free-space nucleon-nucleon data, calculations are unable even to account for the correct saturation density, which in our context is a low density. More recent work by Pandharipande and collaborators (13, 14) has been able to address dense matter by invoking a variational approach.

Eventually one wishes to understand the EOS at a deeper level. Some attempts to obtain such understanding (5, 15, 16), for either gravitational collapse or for ion collisions, are based on the mean field theory outlined by Serot & Walecka (17) and chiral variants of this theory (15). The conceptual basis for this approach is simple enough, one is tempted to add even simpleminded: interactions between nucleons are included only insofar as they generate average fields. Be this as it may, the method possesses one overriding virtue: it is formally Lorentz covariant and can in principle describe high Fermi momentum, i.e. high density. Super-luminality is excluded, and asymmetry properties arise naturally from the introduction of isovector-vector mesons. Some uneasiness must attach, however, to the extreme density dependences seen in the unmodified theory (17). The nucleon effective (Dirac) mass, for example, is reduced even at saturation by a factor of close to two.

Where then is one to turn for a theoretical basis for the EOS at high density? The highest relevant densities correspond at most to a Fermi momentum just less than 2.5 times that of normal nuclear matter, i.e. $2.5 \times (260 \text{ MeV}/c)$; this is hardly high enough to justify the use of asymptotic freedom. One is dealing with strongly interacting matter, the ultimate description of which derives from quantum chromodynamics. Unfortunately, Monte Carlo lattice gauge theory, the one tool applicable to the nonperturbative treatment of strong interactions, fails for deep-seated technical reasons in the case of finite density (18).

To place the subject in context, I outline in Section 2 the evolution leading to collapse in massive stars, with particular attention to the role of high density properties. I refer for the most part to simulations of the direct mechanism using the phenomenological EOS developed by Baron, Cooperstein & Kahana (3, 19), which is also oversimplified. For details on the delayed mechanism, which has had its ups and downs but which may yet prove necessary, I refer the reader to the work of Wilson, Bethe & Wilson and Mayle (10). Both mechanisms must make use of the same properties of highly condensed matter. Neutron stars and the even higher densities found within them are also touched upon in Section 2. Of interest is the possibility that these compact objects may consist in part of two- or three-flavor quark-gluon plasma. Some small evidence for a phase transition from hadronic to quark matter might be discernible in the few late neutrinos detected at Kamiokande (7).

In Section 3 I discuss the historical derivation of the equation of state, and in Section 4 the more recent mean field treatment of the subject and variations on this theme. In Section 5, I summarize briefly, attempt to draw some conclusions, and indicate likely future directions.

2.   ASTROPHYSICAL SETTING

2.1   Stellar Evolution Toward Supernova

Bulk nuclear matter at supersaturation density is found only in gravitationally collapsed objects or in the early universe. The equation of state enters into the collapse in two ways, first as the critical element in halting collapse and in determining the shock energy, and second in defining the stable configuration for the surviving neutron star. The evolution can be divided into several phases; as follows.

QUASI-STATIC BURNING   The precollapse period involves the staged fusion of light elements in the central core of the star and takes less than 10^7 years, short in terms of the age of solar-like stars but very long indeed in terms of the future evolution of the massive star. The result of this slow burning phase is a concentric structure of shells of elements, the lightest outside (H, He) and the heaviest inside (C, O, Si, Fe, etc). The final burning Si to Fe takes only a few days. The central core eventually contains elements near iron in atomic number, i.e. near the maximum in the binding energy. Fusion is therefore no longer possible, and collapse inevitable. The precollapse evolution (20, 21; S. E. Woosley & T. A. Weaver and K. Nomoto & M. Hashimoto, private communication) provides the initial conditions for the ensuing hydrodynamical collapse, and in particular the "iron" core mass is crucial. To a first approximation, this is given by the Chandrasekhar mass

$$M_{Ch} \approx 5.8 Y_e^2 M_\odot, \hspace{2cm} 1.$$

where $Y_e \approx Z/A$ is the fraction of electrons present per baryon. Thus the β-capture taking place in this initial phase is highly important. Uncertain is the treatment of convection in the later more rapid burning of O, C, and Si (20, 21).

COLLAPSE In some calculations collapse is initiated by photo-disintegration of iron nuclei into α particles, but eventually is driven by the pressure drop from rapid β-capture. For the most part this capture is on free protons, converting them into neutrons:

$$e^- + p \to n + v_e. \hspace{2cm} 2.$$

As the density rises in the collapsing core, nuclei melt and, at densities approaching normal saturation density $\rho_0(Z/A)$ and above, only nuclear matter remains. Eventually neutrons outnumber protons at the core center by approximately two to one. Neutrino trapping (22), which takes place within the core at densities above 10^{12} g cm^{-3}, halts the β-capture. It is the properties of highly asymmetric matter that are then of concern. This second phase and the ensuing evolution up to possible explosion takes only about 500 milliseconds. Collapse is halted only if nuclear matter under compression can stiffen sufficiently. Otherwise black holes result. This is then the first point at which the properties of nuclear matter at supersaturation density enter decisively. The collapsing core divides into two regions: an inner core, homologously collapsing, subsonic core (infall velocities proportional to radius) and an outer supersonic shell.

BOUNCE AND SHOCK FORMATION Core rebound and shock formation occur closely in time, just about when the density at the edge of the homologous core reaches that of saturated nuclear matter. The shock's initial energy is a critical factor in determining its eventual fate. The core center contains material at considerably higher densities, with ρ_{core} at maximum compression providing a good indication of the likelihood of explosion. Softer equations of state lead to higher ρ_{core}^{max} and to explosion.

SHOCK PROPAGATION: EXPLOSION Sufficiently energetic shocks at formation will traverse the high density core and impart enough kinetic energy to the spectator material in the stellar mantle and envelope to unbind them and produce explosion. Energy losses to the shock are from two sources, dissociation of heavy nuclei into lighter components and production and escape of some neutrinos; both follow from the heat and entropy generated by the energetic shock. The complete dissociation of $0.1M_\odot$ of bound material costs 1.6×10^{51} ergs. This is to be compared with the $(0.3–0.4) \times 10^{51}$ ergs binding of the material outside the core and with the $(1–2) \times 10^{51}$ ergs of energy in a healthy shock.

The precollapse mass of the iron core M_c and the final homologous core mass $\sim 0.6–0.7M_\odot$ then become very important parameters in the efficacy of the direct mechanism. Should the shock be required to traverse too much dense material it will not escape the core. This was found in the earliest adequate simulations (23–26). Shocks generated in these calculations, using a core mass of $1.55M_\odot$ (27), stalled and degenerated into accretion shocks; black holes seemed inevitable.

Subsequently, explosion by a prompt shock, i.e. the direct mechanism, has been advocated by a BNL-SUNY-Cornell collaboration (3, 9, 19) and a delayed mechanism by Wilson and coworkers (10). In the latter, the stalled shock is resuscitated by neutrino energy deposition in the material just behind the shock. Neutrinos are trapped over the dynamical times associated with rapid collapse but diffuse outwardly during any appreciable stalling of the shock. Both mechanisms are gravitationally driven, and both depend on the properties of dense matter. Thermal pressure is appreciable, with temperatures in the core rising to somewhat less than 20 MeV near the shock. However, for theoretical purposes, temperatures less than the Fermi energy can be considered as low. This is in contrast to relativistic ion collisions where, although equilibrium is not a foregone result, $T \gtrsim 100$ MeV has already been attained and $T \gtrsim 200$ MeV can be anticipated.

2.2   Energetics and the Direct Mechanism

The bounce mechanism in supernova collapse is balanced on a knife edge. It is driven by a highly inefficient process, with small changes able to tip the scales toward explosion or black hole. A useful energy accounting is provided by a nonrelativistic virial theorem just before collapse giving the total gravitational plus internal energy as

$$U_G + U_I = \frac{(\Gamma - 4/3)}{\Gamma - 1} U_G \qquad\qquad 3.$$

for a gas whose adiabatic changes are governed by

$$P \propto \rho^\Gamma. \qquad\qquad 4.$$

Since the pressure is dominated at this point by the degenerate, relativistic electron component, the adiabatic index Γ is close to $4/3$ and the total energy close to zero. Neutrino escape during collapse is slight, and the cancellation between U_G and U_I persists. The drop in radius from 3000 km to perhaps 15–20 km drives the gravitational energy to several times 10^{53} ergs, well above the few times 10^{51} ergs needed for explosion. Positive energy can be transferred to the shock only if the core is left gravitationally bound. Some moderation in the stiffness of nuclear matter found in non-relativistic calculations allows collapse to proceed to higher central density and consequently leads to an increase in shock formation energy. True to the mentioned inefficiency, much of this increase goes into further neutrino production and loss and into more dissociation. There is some enhanced possibility, however, for explosion (3, 19).

2.3 Hydrodynamics and the Equation of State

The hydrodynamic scheme employed by Baron, Cooperstein & Kahana (3, 28, 29) (BCK) is spherically symmetric, fully general relativistic, energy conserving, and in various stages has treated neutrino transport by trapping and free streaming, by leakage, and most recently as a two-fluid model (2). The neutrinos represent the only nonequilibrium aspect of the evolution, and since one knows the basic interactions between neutrinos and matter it is in principle possible to treat their transport completely. However, the space-time and energy dependences complicate the situation, and some simplification is necessary. A two-fluid approach, assigning different temperatures to matter and to the neutrinos is one such simplification. Although the preliminary results (2) for neutrino production and transport are similar to those of Bruenn (30) using a different approach, nagging differences in, for example, neutrino-electron scattering persist. Undoubtedly, the last word on neutrino diffusion in the supernova core has not been uttered.

The inclusion of general relativity is not a luxury. The delicate energy balance dictates this, as does the comparison between the 2-km Schwarzschild radius and 15–20-km collapsed radius of the homologous core. The full hydrodynamic equations need not be set down here (28, 29); they represent the conservation laws in the presence of gravitation. It is useful, however, to keep in mind the simpler Newtonian force equation

$$\rho \frac{dU}{dt} = -\frac{GM(R)\rho}{R^2} - \frac{dP_T}{dR},$$ 5.

where the Lagrangian coordinate $M(R)$ is the mass within some radius R, while $U(R)$ is the velocity field and $P_T = P_N + P_\ell$ is the sum of nuclear and leptonic pressures. One should note that the leptonic pressure dominates the initial stage of collapse and is significant to quite high hadronic density, perhaps to $3\rho_0$.

Two-dimensional simulations have been considered, but the calculations are complex (31, 32) and the results not definitive. A simple estimate of the energy of a rigidly rotating pulsar yields

$$E_{rot} = \frac{1}{2} I_{rigid} \omega^2 = \frac{15 \times 10^{51}}{\tau_{ms}^2} ergs \qquad 6.$$

for a neutron star of mass $1.5 M_\odot$, radius 8 km, and period $\tau = \tau_{ms} \times 10^{-3}$ seconds. There do exist millisecond pulsars (33, 34) and for these the above rotational energy, though a small fraction of the final gravitational or internal energies (few $\times 10^{53}$ ergs), could still have some definite influence on the collapse phase (31, 32). Accretion just after collapse or during the ensuing history of the neutron star can lead to appreciable spin-up (33–35), while nonrigid rotation can cut down what we imagine to be the total rotational energy. It is also not clear whether pulsars have similar or greatly differing periods at birth. In the latter case, rotation will not affect supernovae uniformly, and for example will not help or hurt the slowest pulsars, which still must be exploded by some mechanism. For this and for other reasons we anxiously await direct observation of the pulsar in SN1987a.[2]

To solve the hydrodynamic equations, one clearly requires a relationship between pressure and density, i.e. an equation of state. In the absence of a good theoretical basis for the EOS, it is advisable to use an easily adjusted analytic form. The BCK EOS was devised to elicit from the simulations those properties of matter most strongly sensed in the collapse. The cold hardonic pressure has the form

$$P_N(\rho) = \frac{K_0(\Delta)}{9\gamma} \rho_0(\Delta) \left[\left(\frac{\rho}{\rho_0(\Delta)} \right)^\gamma - 1 \right] \qquad 7.$$

with an asymmetry parameter $\Delta = (N - Z)/A$, and where $\rho_0(\Delta)$ and $K_0(\Delta)$ are the saturation density and incompressibility modulus determined from

$$P_N[\rho_0(\Delta)] = 0 \qquad 8.$$

---

[2] Just as this article was being completed, preliminary findings by Pennypacker and co-workers in Chile (private communication) suggested the rather startling value of $\frac{1}{2}$ millisecond for the pulsar period. Repeated attempts by several groups to confirm this sighting have been unsuccessful. It is very difficult to accept such a rapid rotation rate; for example, the edge of a 10-km neutron star would possess a linear velocity $0.4c$. Stability then would require a very soft equation of state indeed. More definite sightings of the pulsar, anticipated soon, will clarify this situation.

$$K_0(\Delta) = K_0[\rho_0(\Delta)] = 9\left(\frac{\partial P_N}{\partial \rho}\right)_0.$$
9.

High density properties are controlled by the parameter γ, which is the value the hadronic adiabatic index

$$\Gamma = \frac{d \ln P_N}{d \ln \rho}$$
10.

assumes in the limit $\rho \gg \rho_0$. Finite temperature behavior is not unimportant, but as long as the temperature T remains appreciably less than the Fermi energy, thermal properties at high density may be described simply by introducing an effective mass m^* (k_F) at the Fermi surface. The hadronic energy per baryon is

$$\varepsilon(\rho, T) = \frac{E(\rho, T)}{A}$$

$$= M + (-16 + W_s\Delta^2) + \varepsilon_\rho + \frac{\pi^2}{2} \frac{(m^*)T^2}{\hbar k_F^2},$$
11.

where ε_ρ is the compressional energy per baryon

$$\varepsilon_\rho = \int_{\rho_0}^{\rho} \frac{d\rho}{\rho^2} P_N(\rho),$$
12.

and $W_s(\rho)$ is the symmetry energy coefficient. The BCK EOS involves four parameters, with $K_0(0)$, W_s, and m^* more or less determined by properties of equilibrium symmetric matter, but with γ not very well understood. The effective mass, taken to be in the range found in Brueckner theories of nuclear matter $m^*/m \approx 0.7$, determines the energy density at the Fermi surface. A large value of m^* would inhibit explosion. We have laboratory evidence concerning W_s from semiempirical mass formulae and concerning $K_0(0)$ from the breathing mode in heavy nuclei. The latter is a 0^+ state, occurring in all nuclei above, say, oxygen in mass, presumably associated with a radial density fluctuation. Because of the saturation properties of nuclear matter the excursion from normal density, in this fluctuation, is no more than a few percent. Thus although the position of this state in energy serves to fix the compressibility of nuclear matter near saturation, one learns very little about high density. Reasonable values for symmetric matter, $\Delta = 0$, are

$$K_0(0) = K_0 = 210 \pm 30 \text{ MeV}$$
13.

$$W_s[\rho_0(0)] = 34\text{–}38 \text{ MeV}.$$
14.

The incompressibility is determined from the analysis by Blaizot (36) of the breathing mode in Zr and Pb. Other authors (37) reexamining this problem recently, but with a more schematic theory than that used by Blaizot, claim a significantly higher value. There is, I believe, no reason to doubt the analysis leading to Equation 13.

One should, of course, mention the relevance of heavy-ion collisions to the equation of state. Attempts to use meson production in such collisions (38) to probe the EOS can, for a variety of reasons, be discounted. A more serious effort is the extraction from the fragment momentum distributions of a flow angle (39–41). A kinetic sphericity tensor

$$F_{ij} = \sum_{\text{fragment } v} \frac{p_i(v)p_j(v)}{2m(v)} \qquad 15.$$

is diagonalized and its transverse and longitudinal eigenvalues used to define an aspect ratio and flow angle. In a hydrodynamical situation this angle vanishes for peripheral collisions and reaches its maximum for central collisions. Stiffer equations of state lead to higher flow angles for a given impact parameter. Theoretical analysis (41, 42) is by way of a classical Boltzmann equation incorporating both a mean field and a collision term. Some controversy attaches to the interpretation (41, 42), but the inclusion of momentum dependence in the average field makes it possible to explain the appreciable flow observed with a relatively soft equation of state (43).

2.4   Asymmetric Matter

As indicated previously, it is the compressibility of asymmetric matter, $\Delta \approx 1/3$, that enters most critically into the supernova calculations; at the instant of bounce the homologous core is highly neutron rich. This circumstance has been exploited in the calculations referred to here. A considerable softening in the matter in the core obtains simply because one must extrapolate from $\Delta = 0$ to $1/3$ to deduce the saturation compressibility appropriate to supernova material. I emphasize this point because I believe some misconception has risen about the exact degree of softening required for the direct mechanism to work. Any liberty that has been taken with respect to nuclear properties has been vested in the parameter γ in Equation 7, which fixes the truly high density behavior and about which we know little. Dependence on the asymmetry in the range $\Delta = 0$ to $1/3$ is crudely given (3, 5, 19) by

$$K_0(\Delta) = K_0(0)(1 - a\Delta^2) \qquad 16.$$

$$\rho_0(\Delta) = \rho_0(1 - b\Delta^2). \qquad 17.$$

Kohlemainen et al (44) find $a = 2.0$, $b = 0.75$ in a calculation of nuclear matter energies using Skyrme forces. The general behavior in Equation 16 is a straightforward consequence of a smooth, monotonic extrapolation toward neutron matter, which is not expected to be bound nor to possess a saturation minimum in energy. At some value of Δ between 0 and 1, $K_0(\Delta)$ must then vanish, i.e. at a point of inflection in the energy curve.

The simple forms in Equations 16 and 17 should be improved upon, although our knowledge of highly asymmetric matter is somewhat limited.

2.5   Early Results and Recent Worries

Representative results from simulations producing explosions are displayed in Table 1. Equation of state parameters $[K_0(0), \gamma]$ are the incompressibility at saturation for symmetric matter ($\Delta = 0$) and the high density adiabatic index, as given in Equation 7. All calculations take full account of general relativity except for model 38, which is Newtonian. The maximum central density reached in the calculation ρ^c_{max} is in units of the saturation density appropriate to the asymmetric matter ($\Delta = 1/3$) relevant to bounce, $[\rho_0(1/3) = 2.4 \times 10^{14} \text{ g cm}^{-3}]$. The precollapse models for 40–45 are the main sequence $M = 12$, $15 M_\odot$ models of Woosley & Weaver (20), while 61–63 are from the $13 M_\odot$ model of Nomoto & Hashimoto (21). The explosion energy E_{expl} was obtained from the estimated shock energy by correcting for oxygen burning in the mantle and gravitational binding of mantle and envelope. Models are further distinguished by the symmetry energy W_s, which we believe is experimentally closer to the higher values in the table and by a trapping density that is set at $0.4 \times 10^{12} \text{ g cm}^{-3}$ in the first six models in the table, and at the more realistic $1 \times 10^{12} \text{ g cm}^{-3}$ for models 62 and 63.

These results have been reproduced and greatly expanded upon in the modeling of Bruenn (30), who has explored the parameter space (K_0, γ, W_s)

Table 1   Explosion energies from the prompt mechanism

Model number	Mass $(M_\odot)$	$K_0(0)$ (MeV)	γ	W_s (MeV)	$\dfrac{\rho^c_{max}}{\rho_0(0.33)}$	E_{expl} (10^{51} ergs)
38	12	180	2	29.3	2.3	0.1
40	12	180	2	29.3	12.0	3.2
41	12	180	3	29.3	3.1	0.8
43	15	180	2.5	29.3	4.1	1.7
45	15	90	3	29.3	4.0	0.8
61	13	180	2.5	29.3	4.1	2.4
62	13	180	2.5	36.0	4.1	2.6
63	13	180	2.5	34.0	4.1	1.9

thoroughly. The standard parameter set in Table 1, $\gamma = 2.5$, $K_0 = 180$ MeV, $W_s = 29.3$–36.0 MeV (model numbers 43, 61, 62, 63) yields an EOS soft enough to lead to explosion, but not so soft that a reasonable extrapolation to higher density fails to describe known masses of neutron stars. It must be emphasized that the BCK EOS is for use in a limited range of density (ρ_0 to $4\rho_0$) and of Δ. The explosion energies for the realistic models 43, 61, 62, and 63 are large and could still describe observed supernovae after some reduction. Interestingly, these all lead to a maximum central density near $4\rho_0$. The Newtonian model 38, which barely leads to explosion despite the very soft behavior at high density, has a considerably lower central density.

It is the felicitous combination of relativistic gravitation with softer equations of state that is behind these explosions. There is a region of parameter space (W_s, K_0, γ) permitting prompt explosions (30) that will probably persist somewhat changed through updating of the theory, i.e. for improved neutrino transport and changing initial conditions. Suffice it to say that large increases in K_0 and/or γ will stall the outward progress of the shock.

Figure 1 shows the so-called sonic point at bounce for both Newtonian and general relativistic simulation on the same initial model. Any dis-

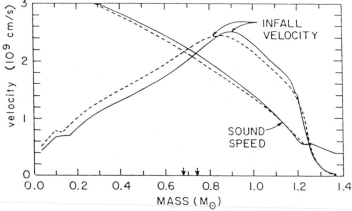

Figure 1   Sonic points near bounce. The velocity of sound and the infall velocity are plotted as a function of the Lagrangian coordinate $M(R)$ (the mass within the radius R) for Newtonian (*solid curves*) and Einsteinian (*dashed curves*) simulations on the same initial model. The crossover of these curves defines the sonic point (*arrows*). Disturbances propagating outwardly from bounce become supersonic in the outside region, generating the shock wave. The general relativistic sonic point (the first arrow on the abscissa) occurs at a smaller mass point than the Newtonian source point (second arrow), $0.67 M_\odot$, and the shock in this case must traverse an extra $0.06 M_\odot$ compared to the Newtonian.

turbance passing out of the subsonic core must go supersonic at this point where infall and sound velocities are equal. The shock forms near this sonic point. The smaller mass inside the general relativistic sonic point suggests that the stronger, relativistic gravitation is not unambiguously helpful to the shock. However, in the presence of sufficiently soft nuclear matter for densities in the range $\rho = \rho_0$ to $4\rho_0$, the increase in formation energy overwhelms other effects.

Presupernova simulations are performed by Woosley & Weaver (20; and private communication) and by Nomoto & Hashimoto (21; and private communication). This modeling perhaps is an even more lively battle-ground. Over the past decade the crucial "Fe" core mass has edged down-ward from $1.55M_\odot$ (27) to $1.12M_\odot$ (21), and reexamination of convection and β-decay may reduce it more.

The symmetry energy in Table 1, $W_s = 29.3$ MeV, is rather low. Just as the initial lepton fraction $Y_\ell = Y_e$ sets the initial core mass (Equation 1), so does the lepton fraction at bounce $Y_\ell^f = Y_e^f + Y_\nu^f$ determine the homologous core mass. Too small a value for Y_e^f results in a smallish, homologous core and kills the shock. The β-decay occurring after the initiation of collapse is on free protons and will be suppressed by a large symmetry energy W_s. Increasing the symmetry energy to an empirically more favored value $W_s \approx 36$ MeV prevents Y_ℓ^f from dropping excessively. Values of Y_ℓ^f below perhaps 0.375, down from the initial value close to 0.42, will prevent explosion by a prompt shock.

The story is by no means complete at this juncture. Preliminary results of the latest simulations (2) devised to treat the long time evolution up to a few seconds after bounce, and including improved neutrino transport, show the shock to be appreciably weakened. A final assessment must await proper completion of these calculations, and perhaps also future developments in the precollapse evolution. One thing is certain: smaller initial core masses, the use of general relativity, and reasonably soft equations of state produce an environment favorable to prompt explosion.

2.6 SN1987a

The recent supernova SN1987a has taught us more about the mis-conceptions in the precollapse modeling (45, 46), unfortunately, than about the explosion mechanism. This situation may be somewhat rectified when we are able to view the surface of the neutron star left behind. Observation of iron x-ray lines strongly redshifted in the gravitational field of the star would place an interesting constraint on the equation of state, albeit at higher densities than seen during the supernova phase. This shift is given by $\Delta\lambda \propto 2M_{ns}/R_{ns}$. With SN1987a one had a unique opportunity to cor-relate the supernova with its progenitor, in this case the blue giant Sanduleak-69 202. Red giants, not blue, were widely expected, though not by all (47), to explode. The variations in stellar atmospheric dynamics that distinguish blue and red giants (45, 46) have little effect on the core beneath. Indeed, the evolution of the star is controlled not by its main sequence mass but rather by the mass of its helium core (20, 21). Analyses of the expanding supernova cloud have placed reasonably serious bounds on the explosion energy between 0.5 and 1.5×10^{51} ergs (45, 46). Had the lower bound been higher, the rather weaker delayed explosions might have been questioned. The total explosion energy includes the shock energy, some 0.8×10^{51} ergs from burning of fuels in the mantle, and the small gravitational binding (0.2–0.3) of the material outside the core, that is, $E_{explosion} = E_{shock} + E_{burning} - B$.

One is left with the very definite message of neutrino emission from SN1987a (7, 8). Apart from the interesting limits placed on the electron-neutrino mass, some information on the mass of the proto-neutron star has also been obtained. The total emission energy, estimated from either or both IMB and Kamiokande (KII) detection, is essentially the star's binding energy. In one analysis (48) this is found to be $(2 \pm 0.5) \times 10^{53}$ ergs; in others (49) it is somewhat larger. For an acceptable range of equations of state, this implies a neutron star mass in the range $1.2 M_\odot$ to $1.7 M_\odot$. Table 2 lists the times of arrival of the KII events. The late time neutrinos, perhaps associated with reheating of the proto-neutron star, are

Table 2 Time sequence for Kamiokande neutrinos

Event number	$t_{detection}$ (sec)	Energy (MeV)	$(t - t_1)$ source (sec) $m_{\bar{v}_e} = 10$ eV	$m_{\bar{v}_e} = 15$ eV
1	0.0	20.0	0.0	0.0
2	0.107	13.5	−0.63	−1.55
3	0.303	7.5	−3.50	−8.27
4	0.324	9.2	−2.00	−4.90
5	0.507	12.8	−0.38	−1.49
6 (rejected)	0.686	6.5		
7	1.541	35.4	1.97	2.51
8	1.728	21.0	1.80	1.89
9	1.915	19.8	1.92	1.89
10	9.2	8.6	6.45	7.39
11	10.4	13.0	9.55	8.49
12	12.4	8.9	9.87	6.71
Pulse width at source (first 8 events)	1.92	—	5.42	10.8

10, 11, and 12. Event 6 was rejected by Hirata et al (7) because of its low energy, below background. The last two columns can be used to determine a limit to the electron antineutrino mass. This is done by mapping the KII event sequence back in time to the source as a function of neutrino mass. The pulse width for the first eight events becomes increasingly long as the neutrino mass $m_{\bar{\nu}_e}$ increases, and the spectrum becomes increasingly hardened. Knowledge of the supernova and neutron star cooling strongly suggest a mass of 15 eV as an upper limit.

Curiously, a gap of more than seven seconds exists between events 9 and 10. If IMB and KII events are viewed as starting together, then this gap is filled in somewhat. The final three neutrinos in Table 2 are by no means very low energy; the probability that all three are background is $\sim 10^{-4}$ (7). This five- or seven-second hiatus is viewed by some authors to be statistically acceptable (49). While it is difficult to build a case one way or another on such sparse data, it is equally difficult to refrain from speculation. An appreciable reheating of the core material is suggested. The hydrodynamic time scale is controlled essentially by gravitational infall rates, with the entire collapse taking less than half a second. Material falling back into the core after explosion is a possible source for the late neutrinos, but this is unlikely to last more than a second or two, unless of course the proto-neutron star is rotating very rapidly. Phase changes due to pion condensation or the formation of a two-flavor quark plasma could generate more binding energy, but should take place without much time delay. A conversion of the hadronic material into a three-flavor plasma (50) does, on the other hand, require time for the weak conversion of ordinary into strange quarks, with a time constant perhaps not far from ten seconds (51; M. Soyeur, private communication).

2.7   Neutron Stars

Masses have been determined for several pulsars (34; J. H. Taylor, unpublished results), all of course members of binaries. The best known values are for the remarkable system PSR 1913+16, thought to consist of two neutron stars, one a pulsar, with masses $1.444M_\odot$ and $1.384M_\odot$ (34). From these known systems there is no good evidence that any neutron star possesses a mass greater than $1.5M_\odot$. Since accretion is always a possibility for a neutron star with a companion, one might conclude that the clustering about masses $\sim 1.4M_\odot$ implies the maximum stable mass is close to this value. The baryonic mass of these neutron stars is some 7 to 10% higher, since the measured gravitational masses include some binding.

The implications of neutron star masses for supernova modeling is unfortunately not very direct. The explicit choice of high density adiabatic index $\gamma = 2.5$ in Table 1 was made to acknowledge this constraint on the

EOS. Too soft an EOS cannot support the requisite neutron star, and a literal extrapolation to higher density with the parameter set $\gamma = 2.5$ and $K_0 = 180$ MeV yields $M_{ns}^{max} = 1.29 M_{\odot}$, somewhat low. However, the central density of the heaviest stable star is above $10\rho_0$, and the average density is $(7-8)\rho_0$, whereas changes in the EOS at densities above $(3-4)\rho_0$ have little impact on supernovae.

I have made an illustrative calculation for present purposes, dividing the density range in the neutron star arbitrarily into three regions in each of which the EOS is taken to be

$$P_i(\rho) = \frac{K_N^i \rho_{0i}}{9\gamma_i} \left(\frac{\rho}{\rho_{0i}}\right)^{\gamma_i}, \qquad i = 1, 2, 3. \qquad \qquad 18.$$

In keeping with the unsaturated nature of neutron matter, the second term in Equation 7 is dropped. A single EOS fitting the totality of ρ, Δ behavior more smoothly would be preferable, but the results would be little altered. One must not take the analytic form (Equation 7) too seriously, especially in regions to which it does not apply. Indeed, we have committed at least one transgression (3): the symmetry energy

$$W_s(\rho, \Delta) = \varepsilon(\rho, \Delta) - \varepsilon(\rho, 0) \qquad \qquad 19.$$

taken straightforwardly from Equations 11 and 12 becomes negative for sufficiently high density.

With the choices $\gamma_1 = 2.3$, $\gamma_2 = 4.0$, $\gamma_3 = 2.0$, $\rho_{01} = \rho_0$, $\rho_{02} = 3.4\rho_0$, $\rho_{03} = 7.0\rho_0$, $K_N^1 = 180$ MeV, and the constraint that the mock phase changes between regions are second order, one finds $M_{ns}^{max} \approx 1.58 M_{\odot}$. This very schematic calculation yields results similar to those of Prakash, Ainsworth & Lattimer (53). It is worth noting the outer region $u = 1$ to 3.4 in Equation 18 is actually softer than that in Table 1. This could be an important factor in the eventual fate of the direct mechanism.

3.   THEORETICAL EQUATIONS OF STATE

3.1   *More Phenomenology: Skyrme Forces*

For the purposes of supernova and neutron star modeling, analytic equations of state fitted to known properties of nuclear matter are highly useful if somewhat arbitrary. Neither the nonrelativistic theories of nuclear matter nor the mean field theory (MFT) are free of phenomenology, both requiring adjustments to describe correctly the saturation properties of symmetric nuclear matter. In addition, MFT makes no pretense of connecting matter properties with those of the elementary two-body system. These, perhaps, represent the best that can be expected from a theory of strong interactions, yielding parametrized structures within which symmetries and fundamental interactions can play a role in determining the relation between energy and density.

The phenomenological form (Equation 11), with a compression energy per baryon

$$\varepsilon_c(\rho, \Delta) = \frac{K_0(\Delta)}{9\gamma(\gamma-1)}\left(u^{\gamma-1} + \frac{\gamma-1}{u} - \gamma\right)$$ 20.

$$u = \frac{\rho}{\rho_0(\Delta)},$$ 21.

is reminiscent of the density expansions of Blaizot (36) and Zamick (54)

$$\rho_0\varepsilon(\rho, 0) = M + a'u^{2/3} + b'u + c'u^{d+1},$$ 22.

resulting from a Skyrme-like effective interaction

$$\langle k|V|k'\rangle = \frac{1}{S}(t_0 + t_3\rho^d).$$ 23.

The density parameter d in Equation 23 plays the role of a range for this force, with larger d associated with shorter range. Applying the saturation conditions

$$P(\rho_0) = \rho^2 \frac{\partial\varepsilon}{\partial\rho} = 0, \qquad \varepsilon(\rho_0, \Delta) = M - B(\Delta)$$ 24.

and identifying the kinetic energy with

$$a' = \frac{3k_{F_0}^2}{10M},$$ 25.

one obtains (36) for $\Delta = 0$

$$K_0 = \frac{a'}{\rho_0} + 9B(0) + \left(9B(0) + \frac{3a'}{\rho_0}\right)d = 166 + 210d \text{ MeV}$$ 26.

for $B(0) = 16$ MeV and $\rho_0 = 0.16$ fm^{-3} ($k_{F_0} = 1.33$ fm^{-1}). Thus incompressibilities from 376 MeV for $d = 1$ to 201 MeV for $d = 1/6$ are compatible with symmetric matter saturating at the empirical density and energy. An even longer-range force $d = -1/3$ is possible, and then $K_0 = 96$ MeV. More generally, Equations 24, together with the expansion about symmetric matter

$$\varepsilon(\rho,\Delta) = \varepsilon_0(\rho) + \Delta^2 W_s(\rho) + \ldots, \qquad\qquad 27.$$

lead directly to Equations 16 and 17 (5).

The high density adiabatic index in Equation 22 is in fact (see Equation 10)

$$\Gamma_{\text{h.d.}} = d+2, \qquad\qquad 28.$$

and hence for $d > 0$ and $\gamma > 2$ both Equations 20 and 22 exhibit super-luminality above some density. For the canonical choice $\gamma = 2.5$, $K_0(0) = 180$ MeV, this occurs for the BCK EOS at $u \gg 10$, high enough not to perturb the simulations described in Section 2. The stiffer equations of state produced in all of the nonrelativistic theories in this section contain sound speeds above the speed of light at disturbingly low densities. The Brueckner-based theories that evaluate the energy only up to some power $(\rho)^N$ arbitrarily fix the high density index at $\Gamma_{\text{h.d.}} = N+1$. If Equation 22, for example, were used to represent such a theory, the saturation constraints together with $\Gamma_{\text{h.d.}}$ would predetermine the compressibility in an ad hoc fashion.

3.2   Nuclear Matter

I wish to limit my observations on the nonrelativistic theories to the work of Pandharipande and collaborators (13, 14). The variational technique of these authors makes their work applicable to dense matter. Since they base their interactions partly on the free space two-body interaction, relativity may enter through the back door, though the basic structure cannot be correct. As indicated above though, the two- and three-nuclear interactions are parametrized so as to describe correctly not just these free few-body systems but also properties of the many-body system, namely the binding energy, density, and compressibility at saturation.

The calculational procedure is based on work by Jastrow (55). The strong short-range nucleon-nucleon correlations are built into the variational wave function

$$\Psi_{\text{J}} = \sum_{ij} f_j(r_{ij})\Phi \qquad\qquad 29.$$

through the pair correlation functions f_j; Φ is the Fermi gas wave function. This is in contrast to the work of Brueckner-Bethe-Goldstone (11, 12), who employed selective summation of perturbation theory to mute these strong effects. The Jastrow theory, however, has difficulty coping with the spin complexities in the two-body force. This problem is handled to some extent by generalizing the correlation factors f_J to operators

$$\mathscr{F}_{ij} = \sum_{v=1}^{14} f^v(r_{ij}) 0_{ij}^v.$$

30.

The summation in Equation 30 is over the eight operators familiar from nucleon-nucleon scattering and an additional six quadratics in the orbital angular momentum or spin-orbit operator, producing the v_{14} model (14). Hypernetted chain methods developed earlier (56, 57) were applied to the variational energy

$$E = \frac{\langle \Psi_J | H | \Psi_J \rangle}{\langle \Psi_J | \Psi_J \rangle}.$$

31.

This v_{14} model of Lagaris & Pandharipande (13) fits NN scattering up to 425 MeV, overbinds nuclear matter a bit, but errs seriously in the position of the minimum in the bulk energy. The predictions are $B(0) = 17.5$ MeV and $k_{F0} = 1.7$ fm^{-1}, compared to the empirical $B(0) = 16$ MeV, $k_{F0} = 1.33$ fm^{-1}. Differences are then wholly ascribed to a repulsive three-body interaction, TNI, which is fitted to give $B(0)$, k_{F0}, and $K_0(0)$. The semimicroscopic EOS of Friedman & Pandharipande (FP) (14) can reasonably be applied to asymmetric nuclear matter, in particular neutron matter, with a caveat on the sound speed. In the unmodified v_{14} model the speed of light is exceeded at $\rho \approx 2$ fm^{-3}, whereas in the $v_{14}+$TNI phenomenologically fixed theory, this happens at a lower density $\rho \approx 1$ fm^{-3}, i.e. at 1.6×10^{15} g cm^{-3}. Central densities in the maximum mass neutron stars of Section 2 exceed the latter value by $\sim 80\%$. These two FP models both give $M_{ns}^{max} \approx 2.0 M_{\odot}$ but differ in radii: $R_{ns}(v_{14}) \approx 7$ km, $R_{ns}(v_{14}+$TNI$) = 9.6$ km.

To my knowledge, no supernova simulation has used the FP equation of state. In any case, putting in the incompressibility $K_0(0) = 240$ MeV by hand begs the issue. Given the much lower onset of superluminality for FP, $u \approx 6$ vs $u > 10$ in BCK, one can guess the FP matter is too stiff to permit explosion by prompt shock.

4. RELATIVISTIC EQUATIONS OF STATE

4.1 *Mean Field Theory*

The mean field theory (MFT) of Serot & Walecka (17) and of Celenza & Shakin (58) can very easily be turned to the problem of matter at high density. The basic Lagrangian for this theory

$$\mathscr{L} = \bar{\psi}\left(i\gamma_\mu\partial_\mu + M + g_w\gamma_\mu V_w^\mu + \frac{g_\rho\gamma^\mu V_\rho^{a\mu}\tau^a}{2} - g_s\phi\right)\psi$$

$$+ \frac{m_w^2}{2}(V_w^\mu)^2 + \frac{m_\rho^2}{2}(V_\rho^{a\mu})^2 - \frac{m_s^2}{2}\phi^2 + [\text{boson kinetic energy}] \quad 32.$$

describes nucleons of mass M interacting with vector mesons of masses m_w, m_ρ, and with a scalar meson of mass m_s. In uniform matter, scalar and vector fields are space-time independent, and the meson equations of motion reduce to

$$m_s^2\phi = g_s\langle\bar{\psi}\psi\rangle = g_s\rho_s \quad\quad\quad 33a.$$

$$m_w^2 V_w = g_w\langle\psi^\dagger\psi\rangle = g_w\rho \quad\quad\quad 33b.$$

$$m_\rho^2 V_\rho^{30} = g_\rho\langle\psi^\dagger 1/2\tau^3\psi\rangle = g_\rho(1/2\Delta)\rho. \quad\quad 33c.$$

The π meson, for example, does not contribute since it cannot possess a finite mean value in the normal state of nuclear matter.

The nucleon is described by the Dirac equation

$$[i\gamma_\mu\partial^\mu - (g_w V_w^0 + 1/2\tau^3 g_\rho V_\rho^{30})\gamma^0 + M^*]\psi = 0 \quad\quad 34.$$

with the effective mass

$$M^* = M - \frac{g_s^2}{m_s^2}\frac{\nu}{(2\pi)^3}\int d^3k \frac{M^*}{\sqrt{M^{*2}+k^2}}, \quad\quad 35.$$

where $\nu = 4$ (2) for nuclear (neutron) matter. The equation of state for the pressure as a function of energy density $\epsilon = \rho\varepsilon(\rho, \Delta)$ can be obtained from the energy momentum tensor $T_{\mu\nu}$ for cold nuclear matter:

$$\epsilon = \langle T_{00}\rangle = \frac{g_w^2}{2m_w^2}\rho^2 + \frac{g_\rho^2}{8m_\rho^2}\Delta^2\rho^2 + \frac{g_s^2}{2m_s^2}\rho_s^2 + \frac{\nu}{(2\pi)^3}\int d^3 k E^*(k) \quad 36.$$

$$P = \frac{1}{3}\langle T_{ii}\rangle = \frac{g_w^2}{2m_w^2}\rho^2 + \frac{g_\rho^2}{8m_\rho^2}\Delta^2\rho^2 - \frac{g_s^2}{2m_s}\rho_s^2 + \frac{\nu}{(2\pi)^3}\int d^3 k \frac{k^2}{E^*(k)},$$

$$E^*(k) = \sqrt{M^{*2}+k^2}. \quad\quad 37.$$

At high density the scalar density ρ_s falls off more rapidly than the baryon density ρ, and the vector meson terms in Equations 36 and 37 dominate the free Fermi gas integrals, leaving

$$P = \epsilon \propto \rho^2$$

and a sound speed $c_s^2 = \partial P/\partial\epsilon = 1$.

Phenomenology enters the theory by way of the coupling constant and mass factors

$$C_w^2 = \frac{g_m^2}{m_w^2} = \frac{196}{M^2} \qquad C_s^2 = \frac{g_s^2}{m_s^2} = \frac{267}{M^2}, \qquad 38.$$

with the indicated values selected to yield $B(\Delta = 0) = 15.75$ MeV and $k_{F0} = 1.42$ fm^{-1} (somewhat high).

Strong density dependences in the theory evident in Figure 2 dictate a surprisingly low effective mass, $M^*/M \approx 0.55$, for symmetric matter at saturation. Very stiff saturated matter, $K_0(0) = 560$ MeV, is one consequence. A simple expression illuminating the cancellations in MFT that produce this large incompressibility, and derived (5) from the conditions in Equations 24, is

$$\frac{K_0(0)}{9} = 2\left[M - B(0) - \frac{g_w^2}{2m_w^2}\rho_0 - E^*(k_F)\right] + \frac{k_F}{3}\frac{\partial E^*}{\partial k_F}(k_F)$$

$$= \{2[(938 - 15.75) - 165.2 - 587.7] + [44.5 - 323]\} \text{ MeV}$$

$$\approx 61.4 \text{ MeV}. \qquad 39.$$

Clearly, MFT taken literally would not benefit the direct mechanism. Another result is the symmetry energy deduced from the Δ dependence in the free gas and ρ-meson terms in Equation 36. Using the empirical $C_\rho^2 = g_\rho^2/2m_\rho^2 \approx 100/M^2$ one finds $W_s(\rho_0) = 37$ MeV, in agreement with the higher values necessary for suppressing electron capture after collapse and before bounce.

Independent attempts by ter Haar & Malfliet (59) to include relativity through solution of a Bethe-Salpeter equation are more closely related to the two-body problem, but they yield similar results. MFT provides a useful relativistic framework on which to hang the theory of high density matter. It is relatively simple to incorporate other hadronic species (16). In a variant of the mean field theory, Ainsworth et al (15) obtained an even softer EOS than BCK for $\gamma = 2.0$, $K_0 = 232$ (see Figure 3). They replace the scalar contribution in Equation 36 by that for a chiral σ model

$$\varepsilon_\phi \approx \frac{\lambda^2}{\rho_s}\left[\left(\frac{\phi}{\phi_0}\right)^2 + \left(\frac{\phi}{\phi_0}\right)^3 + \frac{1}{4}\left(\frac{\phi}{\phi_0}\right)^4\right], \qquad 40.$$

and add opposite-sign vacuum fluctuation contributions from both boson (ϕ) and fermion (nucleon) fields

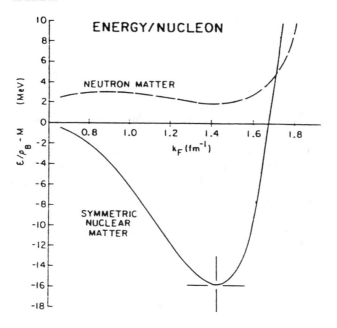

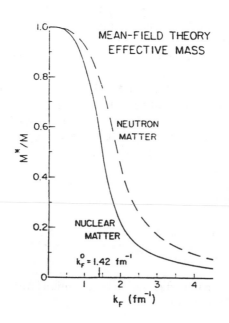

$$\varepsilon_{\text{nucleon}}^{\text{vac}} + \varepsilon_{\phi}^{\text{vac}} = -\frac{(M^*)^4}{8\pi^2\rho_s}F(\eta) + \frac{m_s^4}{64\pi^2\rho_s}F\left(\frac{3}{2}\eta\right) \tag{41.}$$

with

$$\eta = \frac{2\phi}{\phi_0} + \left(\frac{\phi}{\phi_0}\right)^2$$

$$F(\eta) = (1+\eta)^2\ln(1+\eta) - \eta - \left(\frac{3}{2}\right)\eta^2 \tag{42.}$$

and with ϕ_0 the chiral-breaking mean value of the scalar field in terms of which $M = g_s\phi_0$.

The presence of vacuum fluctuation energy in the calculations of these authors is interesting; it permits them to use a considerably higher effective mass $M^* \approx 0.85M$. They also suggest that strong velocity dependence appears in the mean field and is capable of explaining the very stiff but hot EOS extracted from relativistic heavy-ion collisions (43).

4.2   Quarks: Phase Changes

There is little guidance from quantum chromodynamics with regard to the equation of state at finite density (18), i.e. for finite chemical potential μ. Monte Carlo lattice gauge theory for finite temperature has been successful (60) in describing the phase transition to quark-gluon matter for vanishing baryonic density, but for technical reasons cannot perform the same task for finite density (18). Simply put, the energy density with, say, $T = 0$, $\mu \neq 0$, can be calculated analytically on the lattice to be

$$\varepsilon = c_2\mu^4 + c_1\mu^2a^{-2} + O(a^2). \tag{43.}$$

The second term can be anticipated on dimensional grounds, but creates a clear problem in the limit of vanishing lattice constant. The same problem is present in the continuum energy for free fermions:

$$\varepsilon = \int d^4p \frac{p_0^2}{(p_0+i\mu)^2+p^2+m^2} - \int d^4p \frac{p_0^2}{p_0^2+p^2+m^2}. \tag{44.}$$

Figure 2   Energies and effective masses for MFT. The relativistic mean field energy per nucleon and effective mass, M^*, are plotted as functions of the Fermi momentum (from 17). The sharp rise in the energy curve for $k_F > k_{F0}$ heralds the high incompressibility in this model. The low value of $M^*(k_{F0}) \approx 0.55M$ is perhaps a disturbing feature of this model. The vanishing of effective mass for large k_F signals a restoration of chiral symmetry at high density.

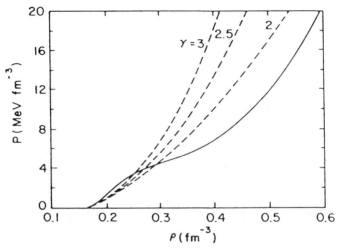

Figure 3 Equations of state. The BCK equation of state is plotted (*dashed lines*) for $K_0(0) = 180$ MeV and $\gamma = 3.0, 2.5, 2.0$ alongside that from Ainsworth et al (15) (*solid line*). For $\gamma = 3.0$, BCK is reasonably stiff, but that resulting from the mean field theory of Ainsworth et al (15) is considerably softer than any of the BCK equations of state for densities greater than $2\rho_0$.

Curing the disease within the Euclidean metric necessary for the lattice calculations renders the action complex and therefore not amenable to a Monte Carlo algorithm (18). Solutions have been sought within a strong coupling limit (61) or using the quenched approximation for fermions (62), but so far without satisfactory results.

A recent promising approach to finite density by Bernard, Meissner & Zahed (63) exploits the dynamical symmetry-breaking theory of Nambu & Jona-Lasinio (NJL) (64). Mass is generated in a fashion closely akin to the production of the gap in superconductivity. Following a recent exposition by D. E. Kahana (65), one can write the NJL Lagrangian as

$$\mathscr{L} = \bar{\psi}i\gamma_\mu\partial_\mu\psi + 1/2\mathrm{tr}\left[-m(\phi_+ + \phi_-) + g^2\phi_+\phi_-\right] \qquad 45.$$

where

$$\phi_\pm = \left[\bar{\psi}\lambda^a\frac{(1\pm\gamma_5)}{2}\psi\right]\lambda^a. \qquad 46.$$

The trace is over flavor indices only, while the λ^a are generators of the $U(N_f)$ flavor group satisfying

$$\mathrm{tr}\,[\lambda^a \lambda^b] = 2\delta^{ab} \qquad (\lambda^0)_{ij} = \sqrt{\frac{2}{N_\mathrm{f}}}\,\delta_{ij}. \qquad\qquad 47.$$

Explicit chiral symmetry breaking is introduced via the current masses $(m)_{ij} = m_i \delta_{ij}$. The field operators represent quarks (not baryons) with N_c colors and N_f flavors.

A gap equation (63, 64) describing the generation of constituent quark masses M can then be derived from an effective potential treatment (65) that also yields the masses generated for scalar and pseudoscalar meson nonets

$$M - m = 4g^2 N_c \int \frac{\mathrm{d}^4 k}{(k^2 + M^2)} M. \qquad\qquad 48.$$

Here M and m are diagonal flavor matrices. The theory is clearly highly divergent, but in the recent work (63, 65) the introduction of a cutoff and of a few other parameters (current masses and the π-meson mass) produces a remarkably consistent picture of low energy properties of hadrons. A very nice feature is the ease with which both finite temperature *and density* are accommodated in the model.

Chiral phase transitions result from the vanishing of the constituent mass in the modified gap equation (63)

$$M - m = \frac{4g^2 N_c (4\pi)}{(2\pi)^3} \int_0^\Lambda k^2\, \mathrm{d}k\, \frac{1}{4E}\left[\tanh\frac{\beta(E-\mu)}{2} + \tanh\frac{\beta(E+\mu)}{2}\right] M$$

$$E = E(k) = \sqrt{k^2 + M^2}$$

$$M = M(\beta, \mu), \qquad \beta = (k_B T)^{-1}. \qquad\qquad 49.$$

Remarkably, with parameters taken from D. E. Kahana's work (65), an approximate solution of Equation 49 yields at zero density a chiral restoration transition with $T_c \approx 185$ MeV and for vanishing temperature a transition at a density $\rho_c \approx 3\rho_0$. Using a somewhat different parametrization, Bernard, Meissner & Zahed (63) find higher values of both T_c and ρ_c.

Kahana (65) described the nucleon as a soliton of quarks loosely bound together. It should be possible to see under what circumstances of T and μ this binding melts, i.e. to depict a deconfining transition perhaps distinct from the chiral restoration. In any case, use of Equation 49 together with knowledge of the density dependences in vector and scalar meson masses should lead to an equation of state for highly compressed matter.

5. COMMENTS

I hesitate to draw definitive conclusions from this sketch of a broad subject still very much in flux. The state of hydrodynamic simulation of collapse has not yet settled down enough to establish a mechanism for explosion, though the direct mechanism still seems a likely candidate. The most uncertain parts of the theory are the quasi-static early evolution and neutrino transport during and after collapse. In both of these further progress can be expected. Nuclear structure and reaction studies can play an important role in the precollapse phase, clarifying, for example, the extent of electron capture and evaluating critical reaction rates, such as for $^{12}C(\alpha, \gamma)^{16}O$. The thornier problems of convection in the late stages of burning may be finally resolved only by three-dimensional simulation, not an attractive eventuality.

Theoretical derivations of the high density equation of state leave much to be desired, but interesting avenues have been opened. The possibility that quark-gluon degrees of freedom are relevant is worth pursuing. Meanwhile phenomenological approaches, constrained by general principles and by laboratory studies of relativistic ion collisions, are a powerful tool.

Evidence from SN1987a has generated confidence in the gravitationally driven nature of Type II supernovae, but has not settled details of the mechanism. The two mechanisms, prompt or delayed, that have existed in the literature in one form or another for several decades (66, 67) are still with us. The analysis of SN1987a continues and will undoubtedly further refine our knowledge. Unfortunately, there is no way of knowing when another such nearby event will occur.

ACKNOWLEDGMENTS

I would like to thank my colleagues E. Baron, H. A. Bethe, G. E. Brown, J. Cooperstein, and J. Lattimer for contributing considerably to my knowledge of supernovae and for use of their work and ideas in this review. I am particularly indebted to J. Cooperstein for adjusting his neutron star code for application to the three-region equation of state.

Literature Cited

1. Lattimer, J., *Annu. Rev. Nucl. Part. Sci.* 31: 337–74 (1981)
2. Cooperstein, J., Baron, E., Supernovae: The Direct Mechanism and the Equation of State, to be published in *Supernovae*, ed. A. Petschek. New York: Springer-Verlag (1989)
3. Baron, E., Cooperstein, J., Kahana, S., *Nucl. Phys.* A440: 744 (1985); *Phys. Rev. Lett.* 55: 126 (1985)
4. Bethe, H., *Annu. Rev. Nucl. Part. Sci.* 38: 1–28 (1988)
5. Kahana, S., *Windsurfing the Fermi Sea*, Proc. Int. Conf. Symp. on Unified Concepts of Many-Body Problems, Sept. 1986, ed. T. T. S. Kuo, J. Speth. Amster-

dam: Elsevier Science (1987), I: 13–29

6. Shelton, I., *International Astronomical Circular No. 4316.* 24 Feb. 1987

7. Hirata, K., et al., *Phys. Rev. Lett.* 58: 1490 (1987)

8. Bionta, R. M., et al., *Phys. Rev. Lett.* 58: 1494 (1987)

9. Baron, E., Bethe, H. A., Brown, G. E., Cooperstein, J., Kahana, S. H., *Phys. Rev. Lett.* 59: 726 (1987)

10. Wilson, J. R., In *Numerical Astrophysics,* ed. J. Centrella, J. Leblanc, R. Bowers. Boston: Jones & Bartlett (1985); Bethe, H. A., Wilson, J. R., *Astrophys. J.* 295: 14 (1985); Mayle, R., PhD Thesis, Univ. Calif., Berkeley. Unpublished (1984)

11. Brueckner, K. A., *Phys. Rev.* 96: 908 (1954)

12. Bethe, H. A., Goldstone, J., *Proc. R. Soc.* A238: 551 (1957); Goldstone, J., *Proc. R. Soc.* A239: 267 (1957)

13. Pandharipande, V. R., Wiringa, R. B., *Rev. Mod. Phys.* 51: 821 (1979); Lagaris, I. E., Pandharipande, V. R., *Nucl. Phys.* A359: 349–64 (1981)

14. Friedman, B., Pandharipande, V. R., *Nucl. Phys.* A361: 502 (1981)

15. Ainsworth, T. L., Baron, E., Brown, G. E., Cooperstein, J., Prakash, M., *Nucl. Phys.* A464: 740 (1987)

16. Glendenning, N., *Phys. Rev. Lett.* 51: 1120 (1986)

17. Serot, B. D., Walecka, J. D., *Adv. Nucl. Phys.* 16: 1–321 (1985)

18. Cleymans, J., Gavai, R. V., Suhonen, E., *Phys. Rep.* 130: 217–92 (1986)

19. Kahana, S., Baron, E., Cooperstein, J., In *Problems of Collapse and Numerical Relativity,* ed. D. Bancel, M. Signore. Dordrecht: Reidel (1984), pp. 163–82

20. Woosley, S. E., Weaver, T. A., *Bull. Am. Astron. Soc.* 16: 971 (1984); *Annu. Rev. Astron. Astrophys.* 24: 205 (1986); *Phys. Rep.* 163: 79 (1988)

21. Nomoto, K., Hashimoto, M., *Phys. Rep.* 163: 13 (1988); *Prog. Part. Nucl. Phys.* 17: 2670 (1986)

22. Mazurek, T., *Astrophys. Space Sci.* 35: 117 (1975)

23. Mazurek, T., Cooperstein, J., Kahana, S., In *DUMAND '80,* ed. V. J. Stenger. Honolulu: Dumand Center (1981); Mazurek, T., Cooperstein, J., Kahana, S., In *Supernovae: A Survey of Current Research,* ed. M. Rees, R. J. Stoneham. Dordrecht: Reidel (1982)

24. Wilson, J. R., *Ann. NY Acad. Sci.* 336: 358 (1980)

25. Hillebrandt, W., In *Supernovae,* see Ref. 23

26. Arnett, W. D., In *Supernovae,* see Ref. 23

27. Weaver, T. A., Zimmerman, B., Woosley, S. E., *Astrophys. J.* 225: 1021 (1978)

28. Cooperstein, J. PhD Thesis, State Univ. New York, Stony Brook. Unpublished (1983)

29. Baron, E. PhD Thesis, State Univ. New York, Stony Brook. Unpublished (1985)

30. Bruenn, S., *Phys. Rev. Lett.* 59: 938 (1987)

31. Monchmeyer, R., Müller, E., In *Timing in Neutron Stars,* NATO ASI Ser. Dordrecht: Reidel (1989); Müller, E., Hillebrandt, W., *Astron. Astrophys.* 103: 358 (1981)

32. Bodenheimer, P., Woosley, S. E., *Astrophys. J.* 269: 381 (1983)

33. van der Heuvel, P. T., *J. Astrophys. Astron.* 5: 209 (1984)

34. Weisberg, J. M., Taylor, J. H., *Phys. Rev. Lett.* 52: 1348 (1984)

35. Shapiro, S. L., Teukolsky, S. A., *Black Holes, White Dwarfs, and Neutron Stars.* New York: Wiley (1983)

36. Blaizot, J. P., *Phys. Rep.* 64: 171 (1980)

37. Sharma, M. M., Borghols, W. T. A., Brandenburg, S., Crona, S., van der Woude, A., Harakeh, M. N., *Phys. Rev.* C38: 2562 (1988)

38. Harris, J. W., et al., *Phys. Lett.* 153: 377 (1982)

39. Gustafsson, H. A., et al., *Phys. Rev. Lett.* 52: 1590 (1984)

40. Stock, R., *Phys. Rep.* 135: 259 (1986)

41. Stöcker, H., Greiner, W., *Phys. Rep.* 137: 277 (1986)

42. Aichelin, J., Rosenhauer, A., Peilert, G., Stöcker, H., Greiner, W., *Phys. Rev. Lett.* 58: 1926 (1987)

43. Gale, C., Bertsch, G., Das Gupta, S., *Phys. Rev.* C35: 1666 (1987)

44. Kohlemainen, K., Prakash, M., Lattimer, J., Treiner, J., *Nucl. Phys.* A439: 535 (1985)

45. Woosley, S. E., Pinto, P. A., Ensman, L., *Astrophys. J.* 324: 466 (1988)

46. Saio, H., Nomoto, K., Kato, M., *Nature* 334: 508 (1988)

47. Brunish, W. M., Truran, J., *Astrophys. J. Suppl.* 49: 447 (1982)

48. Kahana, S., Cooperstein, J., Baron, E., *Phys. Lett.* B196: 259 (1987)

49. Lattimer, J., *Nucl. Phys.* A478: 199c (1988)

50. Baym, G., Jaffe, R., Kolb, E. W., McLerran, L., Walker, G. H., *Phys. Lett.* B160: 181 (1985)

51. Olinto, A., *Phys. Lett.* 192: 71 (1987)

52. Deleted in proof

53. Prakash, M., Ainsworth, T. L., Lattimer, J. M., *Phys. Rev. Lett.* 61: 2518 (1988)

54. Zamick, L., *Phys. Lett.* B45: 313 (1973)
55. Jastrow, R., *Phys. Rev.* 98: 1479 (1955)
56. Fantoni, S., Rosati, S., *Nuovo Cimento* A10: 145 (1972)
57. Ripka, G., *Phys. Rep.* 56: 1 (1979)
58. Celenza, L. S., Shakin, C. M., *Relativistic Nuclear Physics: Theories of Structure and Scattering.* Singapore: World-Scientific (1985)
59. ter Haar, B., Malfliet, R., *Phys. Rev. Lett.* 56: 1237 (1986)
60. Engels, J., Karsch, F., Montray, J., Satz, H., *Phys. Lett.* B101: 819 (1987); Engels, J., Karsch, F., Satz, H., *Nucl. Phys.* B205: 239 (1982)
61. Karsch, F., Mütter, K. U., *CERN preprint* (1988)
62. Kogut, J., Matsuaka, M., Stone, M., Wyld, H. W., Shenker, J. H., et al., *Nucl. Phys.* B225: 93 (1983)
63. Bernard, V., Meissner, U., Zahed, I., *Phys. Rev.* D36: 819 (1987)
64. Nambu, Y., Jona-Lasinio, G., *Phys. Rev.* 122: 345 (1961)
65. Kahana, D. E., *Phys. Lett.* In press (1989)
66. Burbidge, E. M., Burbidge, G. R., Fowler, W. A., Hoyle, F., *Rev. Mod. Phys.* 29: 547 (1957)
67. Colgate, S., White, R. H., *Astrophys. J.* 143: 626 (1966)

Annu. Rev. Nucl. Part. Sci. 1989. 39: 259–310

DEEP INELASTIC LEPTON-NUCLEON SCATTERING

Sanjib R. Mishra and Frank Sciulli

Nevis Laboratory, Columbia University, Irvington, New York 10533

KEY WORDS: structure functions, quantum chromodynamics, quark-parton model sum rules, universality, scaling violations.

CONTENTS

0163–8998/89/1201–0259$02.00

1. INTRODUCTION

The elastic scattering of electrons by neutrons and protons provided direct experimental evidence for the compositeness of hadrons (1). Appropriately, subsequent deep inelastic experiments using lepton beams revealed that hadrons are composed of point-like constituents whose properties are identical with those of quarks, the fundamental hadronic states representing the SU(3) algebra. The quark model had been successful in describing masses and magnetic moments of hadrons (2). The use of high energy leptons as probes of nucleon structure is unique because the leptons do not interact strongly, so that they are able to penetrate the nuclear surface; their short wavelength implies that the leptons collide with individual charged or weakly interacting constituents.

About twenty years ago, systematic measurements (3) were begun of the inelastic reaction in which the nucleon, N, does not remain intact:

$$e^- + N \rightarrow e^- + X, \qquad\qquad 1.$$

where X is some ensemble of hadrons. These experiments gave larger cross sections than had been anticipated and manifested some surprising regularities that were dubbed "scaling" (4). A consistent and attractive interpretation emerged in which the scattering occurred between the lepton and one of the fundamental constituents, or "partons," of the nucleon (5). A short time later, charged-current neutrino experiments (6),

$$\nu_\mu + N \rightarrow \mu^- + X, \qquad\qquad 2.$$

were interpreted in the same way.

This model interpretation of deep inelastic scattering has led to two decades of experimentation using all available leptonic probes: electrons (e^-), muons (μ^-), and neutrino (ν), along with their antiparticles. (Several experimental methods are described in Section 2.) The scattering of leptons from nucleon targets cleanly delineates the properties of those parton constituents containing electric or weak charge. In the last decade, with the advent of an attractive theory of the strong interactions, quantum chromodynamics (QCD) (7), deep inelastic data have provided important contact with experiment; in the process, measurements have been made of the dimensionless coupling (α_s) between quarks and gluons, a fundamental parameter of QCD.

Understanding nucleon structure has been important; however, deep inelastic scattering has contributed to our understanding in other ways. Neutral-current phenomena, which provided the first demonstration of the SU(2) × U(1) unification of weak and electromagnetic interactions

(8), were discovered (9) and corroborated (10) in neutrino-nucleon experiments; interference between neutral weak and electromagnetic propagators was demonstrated first in an electron-nucleon inelastic scattering experiment (11). To date, the most precise measurement of the "weak mixing" or "Weinberg" angle, θ_W, parametrizing weak neutral-current coupling, comes from neutrino-nucleon data. (The advent of precise measurements at Z^0 factories should soon change this.) The electroweak theory, together with the quark hypothesis and QCD, forms the basis for the "standard model" of elementary particles. For reasons of brevity and because good reviews exist (12a,b,c), we do not discuss neutral-current measurements in this review except insofar as they relate to nucleon structure.

Deep inelastic scattering has been a fertile ground for searching out new phenomena. For example, there had been indications of anomalous behavior in the neutrino production of like-sign dimuon events, which we address below. In addition, experiments have ruled out important regions of parameter space for existence of neutral massive leptons. A short summary of the status of these and other searches for exotic phenomena is included in Section 5.

1.1 *Formalism of Deep Inelastic Scattering*

The description of events in deep inelastic scattering relies on three important event parameters: the square of the center-of-mass energy, s, and the scaling parameters, x and y. The scaling parameters provide a simple intuitive picture of the scattering between lepton and parton. The meaning of the x parameter is pictured in Figure 1: in a Lorentz frame in which the

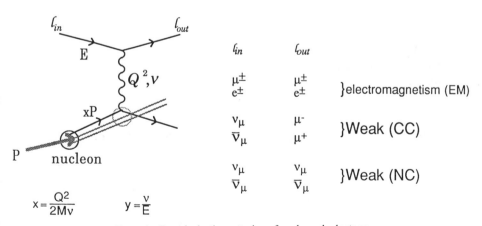

Figure 1 Deep inelastic scattering of nucleons by leptons.

nucleon is traveling at high momentum, P, the struck parton carries momentum, xP. The y variable is directly related to the two-body scattering angle, θ^*, of the outgoing lepton in the center-of-mass frame of the initial-state lepton and the struck parton: $\cos\theta^* = (1+y)/2$. The interesting point is that these and other parameters of the event can be calculated from a few measurable laboratory quantities:

E = incident beam lepton energy,

E' = outgoing lepton energy,

θ = outgoing lepton angle with respect to the beam direction, and

v = outgoing hadronic energy.

For a discussion of the relationship between the scaling parameters and experimentally measured quantities, see for example (13) and references quoted there. The parameters noted here are relevant to the fixed-target experiments. For the ep collider case, see (16).

Specifically, the deep inelastic parameters depend on the energy, but not on other details of the hadronic final state. Conservation of energy constrains E, E', and v. Unless otherwise stated, all masses are assumed small, including the nucleon mass (M), in comparison to the momentum transfer between the lepton and participating parton. The momentum transfers in present fixed-target experiments are small in comparison to the mediating weak boson masses, so effects of those masses are ignored in the formulae that follow.

The overall center-of-mass energy between lepton and nucleon is

$$s = 2ME.$$

The four-momentum transfer between lepton and nucleon vertices, q_μ, has a magnitude determined directly from measurements of the initial and final leptons:

$$-q^2 = Q^2 = 2EE'(1-\cos\theta) \simeq EE'\theta^2.$$

The dimensionless scaling variables, x and y, are calculated from the measured quantities as follows:

$$x = Q^2/2Mv \qquad y = v/E.$$

These follow (13) from the more general Lorentz-invariant relations,

$$x = Q^2/2P\cdot q \qquad y = 2P\cdot q/s.$$

1.2   Structure Functions and Cross Sections

Individual events are characterized by unique values of x, y, and Q^2; accumulation of many such events permits measurement of the differential cross sections with respect to two of these variables. Cross sections are commonly parametrized in terms of structure functions, which are related by the quark model to the densities of partons within the nucleon (see Section 1.3). We illustrate this for the fixed-target scattering of leptons with nucleons.

Consider the differential cross section for the νN charged-current scattering (Reaction 2, above),

$$\frac{d^2\sigma^{\nu N}}{dx\,dy} = \frac{G^2 ME}{\pi}\{F_2^{\nu N}(x, Q^2)[1-y] + 2xF_1^{\nu N}(x, Q^2)[y^2/2]$$

$$+ xF_3^{\nu N}(x, Q^2)y[1-(y/2)]\}, \quad 3.$$

in terms of the three structure functions: F_2, $2xF_1$, and xF_3. The y dependence (or $\cos\theta^*$ dependence) is a consequence of the helicity structure of the weak current. The function, xF_3, measures the parity-violating part of the cross section. The analogous expression for muon or electron scattering from nucleons (Reaction 1) may be obtained by making the following formal replacements in Equation 3:

$$G^2/\pi \rightarrow 8\pi\alpha^2/Q^4;$$

$$F_2^{\nu N}(x, Q^2) \rightarrow F_2^{\mu N}(x, Q^2);$$

$$xF_1^{\nu N}(x, Q^2) \rightarrow xF_1^{\mu N}(x, Q^2);$$

$$xF_3^{\nu N}(x, Q^2) \rightarrow 0. \qquad\qquad 4.$$

The function xF_3 is absent since muon scattering is dominated by parity-conserving photon exchange at momentum transfers typical of fixed-target experiments.

The historical fact is that the structure functions, while strongly dependent on x, were seen to be largely independent (at the $\pm 10\%$ level) of the momentum transfer variable, Q^2, when x is fixed. This independence of Q^2 is the quantitative statement of scaling (4). It led to the physical interpretation that the nucleon consisted of nearly free partons and that the structure functions were the momentum densities (or squared wave functions) of partons within a nucleon (5). The deviations from scaling could be attributed to the potential and kinetic energies of these partons: scaling deviations are now expected to be understandable as consequences of the strong QCD binding forces. It follows that certain features of the structure functions are predictable within the context of perturbative QCD (see Section 2.5).

Another important historical point, evident from the earliest experiments, was that $F_2(x) \approx 2xF_1(x)$ (the Callan-Gross relation). The small value (less than about 0.2) of

$$R = (F_2 - 2xF_1)/2xF_1 \qquad \qquad 5.$$

was recognized (14) to be a consequence of the nearly free spin-1/2 constituents dominating the scattering process. Again, deviations from zero are expected and predictable at high Q^2 by perturbative QCD. We return to these important questions of deviations from exact scaling and finite R values below.

1.3   Parton Densities and Structure Functions

Within the context of the quark model, the compositeness of any hadron can be described by the densities of its constituent quarks. For example, the function $u(x)\,\mathrm{d}x$ specifies the differential probability for finding a u quark in a proton carrying a fraction between x and $x+\mathrm{d}x$ of the proton momentum. Knowledge of the quark and antiquark densities [denoted generically $q(x)$ and $\bar{q}(x)$, respectively] is of considerable importance in making reliable estimates of parton luminosities for collider experiments; indeed, it is the limiting factor in constraints on the number of neutrino generations and on the mass of the top quark that can be obtained from present measurements using hadron colliders (15). Furthermore, quark densities are important in predicting the standard model behavior of the cross section at HERA, where the Q^2 and x range will be extended by two orders of magnitude (16).

A complete theory of nucleon constituents and of the strong interactions would permit us, at least in principle, to predict parton densities for any hadron. This ability eludes us at present. Nevertheless, QCD allows us to evolve densities from one energy to another. Furthermore, the quark model does permit us to calculate structure functions for any deep inelastic process directly from parton densities. [For a detailed recipe of structure functions in terms of parton densities, see for example the review by Fisk & Sciulli (13).] Conversely, measured structure functions permit extraction of parton densities. In Section 3 we discuss the extracted densities for u and d quarks inside the nucleon from proton and neutron structure functions. The overall isoscalar quark and antiquark densities, which are more precisely known, are also discussed. Comparison of structure functions measured with differing nuclear targets allows us to determine how the nuclear environment influences the parton densities.

We also discuss in Section 3 measurements of the density of strange quarks from v_μ-induced events with two opposite-sign muons in the final state. These events arise mainly from the charged-current production and semileptonic decay of charm, which is preferentially produced from strange quarks.

The use of polarized muons with polarized hydrogen targets permits measurements of the cross sections with muon and proton spin aligned and opposite. Such asymmetries measure the "spin-dependent structure function," $xg_1^p(x)$ (see Equation 10).

One important qualitative distinction between the weak and electromagnetic interactions in Reactions 2 and 1, respectively, is that the scattering by neutrinos depends on x times the density of quarks inside the nucleon, while the scattering by electrons or muons depends on the square of the quark electric charge as well. The sum of the neutrino and antineutrino total cross sections is proportional to the integrals of the structure functions and provides a measure of the fraction of the proton momentum carried by struck quarks. Early measurements of these cross sections, at both low (18) and high (19) energies, demonstrated that only about half the nucleon momentum is carried by quarks. This provided a direct indication that hadrons carried fundamental quanta in addition to quarks; the missing momentum is known to be carried by gluons (20). The ratio of muon to neutrino scattering, on the other hand, measures the mean-square charge of quarks, as discussed below.

The quark model inherently provides beautiful relationships among structure functions measured by different inelastic processes; it further permits predictions about certain properties of those structure functions. Some of the simplest predictions and best experimental tests involve isoscalar nucleon targets, containing equal numbers of u and d quarks. The measurements, accomplished with heavy nuclear targets to provide high rate, sometimes require small corrections for neutron excess. The following quark model predictions for isoscalar targets can be tested:

1. The ratio of the F_2 structure functions for muon (or electron) to neutrino scattering is the mean-square quark charge in units of the square of the electron charge. That is,

$$\frac{F_2^{\mu N}(x)}{F_2^{\nu N}(x)} = \frac{5}{18}\left(1 - \frac{3}{5}\frac{s+\bar{s}}{q+\bar{q}}\right), \qquad 6.$$

where the small x-dependent correction is due to the asymmetric number of strange and charm quarks in the nucleon.

2. Structure functions obtained with neutral-current interactions ($v_\mu +$ N $\rightarrow v_\mu + X$) are directly predictable from those obtained with charged-

current interactions. Measurements of neutral-current structure functions are much more difficult since both the initial- and final-state leptons are neutral. This is discussed in Section 3.1.

3. Certain of the sum rule predictions, described in the next section, can be tested.

1.4   Sum Rules

Sum rule predictions were originally obtained using rigorous current algebra formulations, but they can also be obtained from a naive non-relativistic quark model, which we describe here for pedagogical reasons. The relevant quantities are obtained by integrating appropriate measured structure functions weighted by $1/x$. These integrals are then directly related to

N_u = number of valence u quarks inside the proton;

N_d = number of valence d quarks inside the proton;

e_u = charge of u quark; and

e_d = charge of d quark.

The quark model of the proton predicts $N_u = 2$ and $N_d = 1$; $e_u = +2/3$ and $e_d = -1/3$. The comparison with data is discussed in Section 3. We mention here several specific sum rules.

1. The Gross–Llewellyn Smith (GLS) sum rule is the best known because it is the best tested [21]. It states [22]

$$S_{GLS} \equiv \int_0^1 \frac{xF_3^{\nu N}}{x} \, dx = (N_u + N_d)\left(1 - \frac{\alpha_s}{\pi}\right). \qquad 7.$$

This relation is a consequence of the fact that the parity-violating structure function, in the quark model, is equal to the difference between the quark and antiquark densities, which is just the valence quark density. An experimental value for S_{GLS} near 3 corroborates, then, that the partons defining the nucleon quantum numbers carry baryon number $1/3$. The presence of the term with α_s illustrates that some sum rules should have perturbative QCD corrections, like GLS, while others, like the Adler sum rule, do not.

2. Adler [23] sum rule: This integral can be obtained from neutrino data using hydrogen and deuterium targets:

$$S_A \equiv \frac{1}{2} \int_0^1 \frac{dx}{x} (F_2^{\nu n} - F_2^{\nu p}) = (N_u - N_d), \qquad 8.$$

which is unity for quark constituents.

3. Gottfried (24) sum rule: This integral is obtained using muon or electron data from hydrogen and deuterium targets:

$$S_G \equiv \int_0^1 \frac{dx}{x} (F_2^{\mu n} - F_2^{\mu p}) = (N_u - N_d)(e_u^2 - e_d^2), \qquad 9.$$

which gives $1/3$ for quark constituents.

4. Bjorken (25) sum rule: The spin structure functions from proton and neutron targets are expected to satisfy

$$\int_0^1 [g_1^n(x) - g_1^p(x)] \, dx = \frac{1}{2} [(N_u^\uparrow - N_u^\downarrow) - (N_d^\uparrow - N_d^\downarrow)](e_u^2 - e_d^2)\left(1 - \frac{\alpha_s}{\pi}\right), \qquad 10.$$

where $N^\uparrow$ ($N^\downarrow$) refers to quarks with their spins parallel (antiparallel) to the spin of the nucleon. In the nonrelativistic quark model, the asymmetry of spin-oriented up and down quarks, which appears in brackets on the right-hand side of Equation 10, is equal to the expectation value of the longitudinal spinor and this is just the ratio of axial vector to vector weak couplings. That is $\langle \sigma_z \rangle = G_A/G_V$ (26; see also 25).

The predictions of these important sum rules (Equations 7–10) are discussed at length, together with comparisons with contemporary experimental results, in Section 4. To reiterate, the specific predictions quoted there are consequences of current algebra, and as such are firm predictions of the standard model.

1.5 *Quantum Chromodynamics*

QCD has been described as a "radically conservative" theory: it results from extrapolating general principles like locality, causality, and renormalizability, reconciling these with experimental fact, and accepting the conclusions that "fall short of actual contradictions" (27). It serves as a paradigm for the description of strong interactions in the same way that electromagnetic theory does for electricity. But, unlike the case for quantum electrodynamics (QED), we are severely limited in testing the predictions of QCD. One important difference is that QED becomes simple where experiments are easiest: at low momentum transfers; for QCD, the simpler predictions occur at the highest momentum transfers.

The "best" way of checking QCD, and measuring the coupling strength of quarks to gluons, is a subject of some controversy because of uncalculable contributions from "nonperturbative" effects. While there are important predictions for other processes (28), elegant and unambiguous predictions do exist for the behavior of deep inelastic structure functions. The dependence (29) of structure functions on Q^2 at fixed x and the dependence (30) of R on x and Q^2 are the most important.

Perturbative QCD only describes the evolution of structure functions (i.e. the Q^2 dependence) at collision energies where the bound kinetic and potential energies of the struck quark may be neglected. Such "higher twist" effects may be estimated but they cannot be reliably calculated because they are related to unknown bound-state quark wave functions. From simple dimensional considerations, such terms must behave as

$$\text{higher twist terms} \approx \left(\frac{A(x)}{Q^2}\right)^n, \qquad n \geq 1, \tag{11.}$$

where the numerator, while a function of x, is presumably of the order of the size of the bound state ($A \leq 1 \text{ GeV}^2$).

The diagrams contributing to the leading-order perturbative QCD corrections are shown in Figure 2, in which effects arise either because the bound quark can emit a gluon (Figure 2, *left*) or because a gluon within the nucleon can produce a quark-antiquark pair (Figure 2, *right*). In both cases, the magnitude of the effect depends on the quark-gluon coupling constant, α_s, which in leading order may be written

$$\alpha_s(Q^2) = \frac{12\pi}{(33 - 2N_f) \ln(Q^2/\Lambda^2)}, \tag{12.}$$

where N_f is the number of quark flavors.

R-PARAMETER   If a beam lepton were to interact with a free fermion, at rest or moving collinear with the beam, the R parameter would be identically zero. As described above, effects due to the wave functions of bound quarks would create a functional dependence like that of Equation 11. If this were the complete description, the value of R would fall to zero very soon after we pass momentum transfers characteristic of soft hadronic processes. QCD, on the other hand, predicts (30) that diagrams like those in Figure 2 will contribute with the logarithmic Q^2 dependence characteristic of α_s:

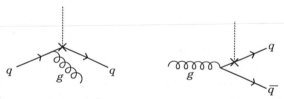

Figure 2   (*Left*) Gluon bremsstrahlung by a quark. (*Right*) Quark-antiquark pair production by a gluon. The dotted line represents the intermediate boson ($\gamma/W^\pm/Z^0$) coupling to the interacting quark. The coupling strength at the quark-gluon vertex is α_s.

$$R(x, Q^2) = \frac{\alpha_s(Q^2)}{2\pi} x_2 \int_x^1 \frac{dz}{z^3} \left[\frac{8}{3} F_2(z, Q^2) + 4f\left(1 - \frac{x}{z}\right) zG(z, Q^2) \right] \Bigg/ 2xF_1.$$

13.

Here f is the number of flavors (N_f) for the neutrino case, and the sum over the quark charges (Σe_i^2) for the muon or electron case. (The resulting predicted value of R is the same.) The first and second terms inside the integral correspond to contributions represented in Figures 2a and 2b, respectively. Since α_s (Equation 12) falls only logarithmically with Q^2, the value of R at fixed x should fall slowly with Q^2, so long as the momentum transfers are high enough.

STRUCTURE FUNCTION EVOLUTION   Structure functions are of two types:

1. The leading-order evolution of a "nonsinglet" structure function, $F^{NS}(x, Q^2)$, is determined completely by the interaction represented in Figure 2a. Here the rate of change at fixed x depends directly on an integral over the same structure function (29)

$$\frac{d F^{NS}(x, Q^2)}{d \ln(Q^2)} = \frac{\alpha_s(Q^2)}{2\pi} \int_x^1 F^{NS}(z, Q^2) P_{qq}\left(\frac{x}{z}\right) dz$$

14.

and is therefore particularly simple. Here, P_{qq} is a "splitting function" determined in perturbative QCD. The xF_3 structure function and the function $F_2^n - F_2^p$ are examples of nonsinglet structure functions.

2. Singlet structure functions satisfy considerably more complicated evolution equations (29) because the Q^2 evolution has additional contributions from the process represented in Figure 2b. Equation 14 is modified by an additional integral on the right-hand side that depends on the gluon structure function. The gluon structure function itself satisfies a separate evolution equation dependent on two integrals. We discuss these further in Section 4.

2. DEEP INELASTIC SCATTERING EXPERIMENTS

2.1 SLAC eN Experiments

The program of fixed-target electron-nucleon inelastic scattering at SLAC (31) has been in operation for over twenty years. The Stanford Linear Accelerator has delivered a beam within the range 4.5–20 GeV to experiments with various targets. The scattered electron is momentum reconstructed in one of the two magnetic spectrometers; these devices, a powerful complex of detection equipment and computers, were a standard for later large detector systems.

The upstream of the two spectrometers can measure particle momenta up to 8 GeV; the downstream device operates, with smaller solid-angle acceptance, up to 20 GeV. The energy (E) of the incident electron, as well as the energy (E') and the angle (θ) of the scattered electron, is measured with good resolution. The spectrometers employ scintillation counter hodoscopes, gas Čerenkov counters, and electromagnetic shower detectors. Charged pions and electrons arising from neutral pions are backgrounds for the scattered electron, particularly at low E'. Techniques for flux determination using Faraday cups and redundant monitors are well established; measurements to better than the 1% level are standard. The limit in normalization precision is typically dependent on knowledge of the spectrometer acceptance.

2.2   Experiments with μ Beams

Experiments utilizing muons[1] extend the eN program to higher energies ($20 < E < 300$ GeV) using beams of (polarized) muons from pion decay. The experiments enjoy much less beam flux than that at SLAC, but compensate by using detectors with large acceptance. Normalization is obtained by counting the beam particles. Programs have been operated at Fermi National Accelerator Laboratory (FNAL) (BFP collaboration) and at CERN (EMC and BCDMS collaborations). For measurements on iron, the BFP and EMC collaborations use a calorimeter as target, which provides a measure of the hadron energy (ν), while the BCDMS experiment does not. We further compare differences in technique in Section 4.

2.3   Experiments with ν Beams

Neutrinos, because they permit measurement of the nonsinglet (parity-violating) structure functions (xF_3), are unique probes. Studies have been carried out by collaborations using bubble chambers[2] and electronic detectors.[3] The best statistical precision on heavy materials (isoscalar targets) is available from the electronic detectors, but the bubble chamber experiments have carried out more extensive studies on hydrogen and deuterium targets. In a neutrino-induced charged-current event, the directly measured quantities are the energy transferred to the hadronic system (ν), the final-state muon energy (E'), and the muon scattering angle (θ).

---

[1] The major collaborations are Berkeley-Fermilab-Princeton (BFP) (32a), European Muon Collaboration (EMC) (32b,c), and Bologna-CERN-Dubna-Munich-Saclay (BCDMS) (33a,b,c).

[2] The CERN Gargamelle and BEBC programs, and the FNAL 15-foot bubble chamber programs, are described in (13).

[3] The CERN-Dortmund-Heidelberg-Saclay (CDHS) collaboration (34a), Chicago-Columbia-Fermilab-Rochester (CCFR) collaboration, the FMMF group (34b), and the CHARM collaboration (35).

Two types of high energy ν beams are commonly employed: narrow band beams and wide band beams [for a review of neutrino beam types, see (13)]. The narrow band beam provides an independent measurement of the incident neutrino energy, E. Furthermore, the beam is either almost purely neutrino or almost purely antineutrino, with little cross contamination. Normalized cross sections are most easily obtained because the flux of the parent pions and kaons is directly measured (36). Wide band beams, created by focusing with pulsed horns or with quadrupole triplets, provide an order of magnitude more neutrinos; however, there is no direct check on the event neutrino energy and no direct measure of the flux.

Since neutrino flux cannot be measured directly in wide band beams, alternative techniques for normalization must be used. Such techniques have been developed (37); they allow measures of relative normalization among the several neutrino energies, as well as the relative flux between neutrinos and antineutrinos. (The overall cross-section normalization must still be obtained from narrow band or other measurements.) These relative flux measurements, calculated directly from the events, are more precise than those that can presently be measured in more direct ways.

2.4   HERA Collider eN Experiments

The technology of colliding beams permits the extension of deep inelastic scattering experiments to much higher momentum transfers. This will indeed happen in the early 1990s with the commissioning of the HERA electron-nucleon collider at the DESY laboratory in Hamburg, West Germany (38). The machine utilizes counter-rotating beams of 30-GeV electrons and 820-GeV protons to achieve a center-of-mass energy (314 GeV) an order of magnitude higher than achievable with present fixed-target beams. Figure 3 shows the range in $1/x$ versus Q^2 probed by several accelerators; leptonic probes at high Q^2 and very small x will uniquely characterize the "soft" region of the nucleon (39; also W.-K. Tung, private communication).

The HERA facility will provide substantial capability for finding exotic particles and interactions (e.g. 40). Some, like certain kinds of massive, strongly interacting particles with lepton number, will be uniquely accessible. Others, such as supersymmetric particles, will have complementary searches performed at other colliding facilities and HERA. The facility will permit measurements of nucleon structure to be extended by about two orders of magnitude in Q^2 and in $1/x$ (see Figure 3). The typical momentum transfers at HERA can be comparable to the masses of the exchanged weak bosons.

HERA will be a unique laboratory for certain tests of the standard

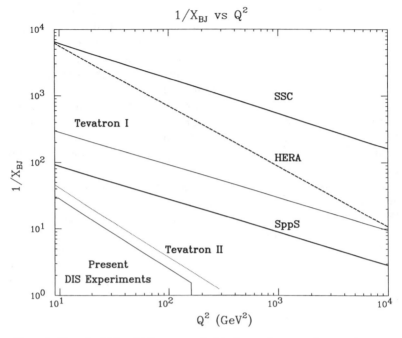

Figure 3   The $x (= X_{BJ})$ and Q^2 ranges probed by interactions at various accelerators.

model, particularly sensitive to the existence of nucleon and/or lepton substructure. Typically such structure is parametrized as a point-like four-fermion interaction conserving helicity and flavor with effective weak coupling reduced by a scale parameter, $\Lambda_{\pm}$; this contribution is assumed to interfere either positively or negatively, respectively, with the ordinary electroweak coupling. With unpolarized beams, the anticipated sensitivities extend to about $\Lambda_{\pm} \approx 4\text{--}7$ TeV, depending on the form of the coupling (41).

The accelerator is expected to provide a transversely polarized electron beam whose spin can be rotated either parallel or antiparallel to the beam direction in the intersection regions. Polarization will provide an important tool for understanding the spin dependence of known or new interactions; its use should extend the sensitivity to $\Lambda_{\pm}$ by approximately 1 to 1.5 TeV (41). In general, the ability to vary the flavor ($e^{\pm}$) and the polarization of the lepton will provide a unique parameter space for exploiting the lepton-nucleon interaction.

HERA is designed to provide a luminosity of 1.5×10^{31} cm^2 s^{-1}, or about one inverse picobarn per day. This high value, necessary so that structure functions can be measured with adequate event rates, will necessitate high average proton beam currents contained in 200 stored bunches. The short time interval between beam bunches (96 ns) creates new problems for the readout and triggering electronics in an adverse radiation environment. Such technical problems are the forerunners of even more severe problems to be faced with hadron colliders (like SSC and LHC) planned for even higher energies and luminosities.

For neutral-current reactions, the direction and energy of the outgoing electron will provide measurements of the appropriate event parameters (Q^2, x, y). However, in the charged-current process

$$e + p \rightarrow \nu_e + X,$$

only the system of hadronic debris (X) is observable. The jet of hadrons that follows the direction of the outgoing parton must be used to reconstruct the event parameters. For this reason, experiments at HERA have paid close attention to designing apparatus that can reconstruct the global properties (energy and angle) of hadronic jets.

Of crucial importance in reconstructing the event parameters is the jet energy measurement. The ZEUS experiment, in response to this challenge, has opted to build a 4π compensating calorimeter, using depleted uranium and scintillator. The detector, now under construction, has a measured hadronic resolution of $\delta E/E \approx 0.35/\sqrt{E(\text{GeV})}$.

3. QUARK-PARTON MODEL: TESTS AND SUM RULES

3.1 Universality on Isoscalar Targets by $\gamma/W^\pm/Z^0$ Exchange

The structure functions that parametrize the deep inelastic cross sections are uniquely related to the nucleon quark densities. An essential element of the quark-parton picture is the universality of quark densities whether they are measured in electromagnetic, ν charged-current, or ν neutral-current interactions. For example, in μ (or e) scattering from an isoscalar target, the structure function $F_2^{\mu N}(x, Q^2)$ is proportional to the square of the electric charges and momentum densities of the interacting partons, and can be expressed (neglecting R) as

$$F_2^{\mu N}(x, Q^2) = \sum_i e_i^2 [xq_i(x, Q^2) + x\bar{q}_i(x, Q^2)], \qquad 15.$$

where the sum extends over the densities of all quark (q) and antiquark ($\bar{q}$) types and e_i is the electric charge (in units of the electron charge).

For charged-current ν scattering from an isoscalar target, the structure function $F_2^{\nu N}(x, Q^2)$ is obtained by the replacement $e_i \rightarrow 1$; while

$xF_3^{\nu N}(x, Q^2)$ is simply the appropriate difference between the quark and antiquark densities $(xq_i - x\bar{q}_i)$.

The corresponding expressions for neutral-current structure functions are obtained by the formal replacement of e_i^2 by a (known) combination of left- and right-handed coupling constants (42). These, in turn, are parametrized in terms of the usual standard model neutral-current couplings, $\sin^2 \theta_W$ and ρ, where θ_W is the Weinberg angle (12a,b,c).

NEUTRAL-CURRENT STRUCTURE FUNCTIONS   Measurements of the total neutrino and antineutrino neutral-current cross sections are used to measure θ_W and ρ. The standard model, with appropriate electroweak radiative corrections, predicts the value of ρ and requires that θ_W be independent of the process. The ratio of the total neutral-current cross section to the total charged-current cross section in deep inelastic experiments provides a sensitive measure of both ρ and $\sin^2 \theta_W$. These measurements have been reviewed by others (12a,b,c), who have concluded that the deep inelastic value for ρ is in good agreement with the standard model prediction. An average from several deep inelastic measurements gives a value for $\sin^2 \theta_W$ in good agreement with the average of measurements from the vector boson masses. Other methods for measuring these parameters are not as precise, but are generally in agreement. It should be noted that the principal limitation in the deep inelastic measurement of the Weinberg angle is the uncertainty in the threshold behavior of the production of charm from down and strange quarks.

A more directed test of the quark model comprises measuring (with neutrinos) the x-dependent neutral-current structure functions and comparing them to the structure functions predicted from quark densities measured in electromagnetic and charged-current deep inelastic experiments. The measurement of these structure functions is inherently difficult because the measurement of the value of x in an individual event is imprecise: neutral-current interactions have a neutrino in the final state, and so the event parameters must come from measurements of the hadronic jet. The FMM (43a, 43b) and CHARM (44) collaborations have reported neutral-current structure functions with a typical resolution on x of about 20%. These data demonstrate good agreement between quark densities extracted from neutral-current interactions and those extracted from charged-current and electromagnetic interactions.

MEAN-SQUARE CHARGE TEST   The probe independence of the parton densities requires that the ratio $F_2^{\mu N}/F_2^{\nu N}$ be nearly equal to the mean-square charge $(5/18)$ of the quarks, as expressed in Equation 6. This is most accurately tested with high statistics data from experiments employing heavy isoscalar targets. Figure 4 shows a comparison of four data sets

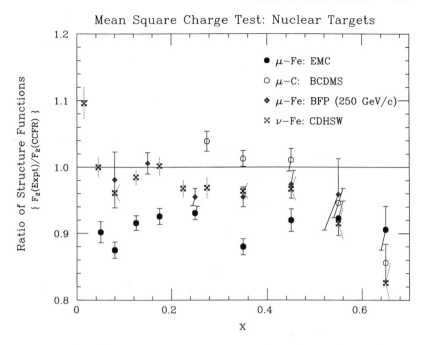

Figure 4 Ratios of F_2 measurements among two sets of neutrino experiments [CDHSW (46) and CCFR (45)] and three sets of muon experiments [BCDMS (33b), BFP (32a), and EMC (32c)]. The muon experiments have been corrected by the quark charges, so that a ratio of unity implies a mean-square charge of 5/18. All ratios are shown with respect to the CCFR data. The assumption $R = R_{QCD}$ is made with the exception of the BFP data, where $R = 0$. Small corrections for the neutron-proton excess and for the strange-charm asymmetry are included in the comparison.

[CDHSW (46), BCDMS (33b), EMC (32c), and BFP (32a)] relative to the CCFR (45) neutrino data.[4] The reported values of F_2 were interpolated or extrapolated for each x value to the mean Q^2 of the CCFR data (32a, 32c, 33b, 45, 46). The comparison of the ratios, averaged over x, of F_2 measured by several experiments relative to the CCFR experiment is presented in Table 1.

The two sets of neutrino data are in good overall agreement, although the CCFR data are somewhat different from those of CHDSW at extreme values of x where the comparison is particularly sensitive to the Q^2 extra-

---

[4] The data were corrected using $R = R_{QCD}$, except for BFP where $R = 0$ was assumed by the authors. Other small corrections, including the effects of neutron-proton excess in the iron target data and the effect of the strange-charm asymmetry have also been applied. See (45) for details.

Table 1   Ratio, r^{exp}, of F_2^{exp} with respect to F_2^{CCFR}

r^{CDHSW}	r^{EMC}	r^{BFP}	r^{BCDMS}
0.988	0.905	0.958	0.989

polation. These data agree within the quoted systematic normalization error of each experiment ($\sim 3\%$). There now exists a consistency among neutrino experiments that was lacking a few years ago (47) when newer CDHSW measurements came into agreement with those of CCFRR.

The EMC (BFP) data are 10% (4%) lower than those of the CCFR on average, but do not exhibit any appreciable x dependence. The disagreement of the EMC data average with both neutrino and other muon data is somewhat outside the quoted 5% systematic errors. While the average level of the BCDMS data shown in Table 1 is in agreement with that of the neutrino experiments, the figure does exhibit some x dependence. It should be noted that the BCDMS data are reported for $Q^2 > 30 \text{ GeV}^2$, which results in a higher mean Q^2 than other experiments and consequently these data undergo larger extrapolation; this seems, however, inadequate to explain the discrepancy among the muon experiments. (A similar conclusion is reached in the comparison of the EMC hydrogen data with that of BCDMS discussed in Section 3.3.)

Thus, the mean-square charge of the quarks is measured to be 5/18, within the 5–10% differences appearing among the muon experiments. This disagreement of normalization and shape among muon experiments should be considered a problem.

3.2   Sum Rules

GROSS–LLEWELLYN SMITH SUM RULE   The Gross–Llewellyn Smith sum rule, the most accurately tested of the sum rules, measures the number of valence quarks in the nucleon. Equation 7, with corrections due to finite Q^2, may be written

$$S_{\text{GLS}} \equiv \int_0^1 \frac{dx}{x} xF_3(x, Q^2) = 3\left[1 - \frac{\alpha_s(Q^2)}{\pi} - \frac{4}{27}\frac{\langle G_1 \rangle}{Q^2} + O(Q^{-4})\right], \qquad 16.$$

where the second term in the brackets corresponds to the known leading-order perturbative QCD correction and the third term, an estimate of higher twist contribution (48), is expected to be small here. The sum rule therefore predicts, using $\Lambda = 186 \pm 60$ MeV and including only the first-order perturbative QCD correction, at $Q^2 = 3 \text{ GeV}^2$

$S_{GLS}^{pred} = 2.74 \pm 0.06.$

Because of the $1/x$ weighting, the experimental value of the integral is dominated by low-x events. We illustrate the extraction of the experimental value from data following the technique of the CCFR collaboration (49). The data are binned in fine x bins (good low-x resolution is imperative for this) and within each bin interpolated to $Q^2 = 3$ GeV^2. Figure 5 shows the values of xF_3 (right scale and squares) and the distribution of the weighted integral (left scale and diamonds) as a function of x. A fit of the form,

$$xF_3(x, Q^2 = 3) = ax^b(1 - x)^c,$$

is used to calculate the unsampled, very low-x contribution to the integral. The best such fit is shown superimposed on the data in the figure. Fits are made to various functional forms (always constrained to vanish as $x \to 0$), and the integration limits are varied to estimate the systematic biases of the procedure. The overall systematic error, determined by applying the

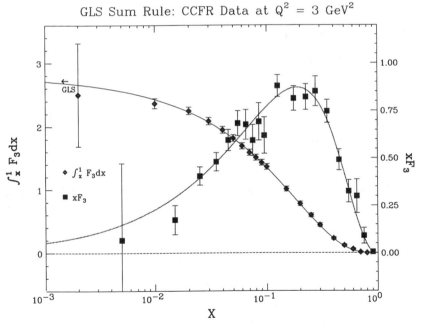

Figure 5 The xF_3 function and the GLS sum rule: The CCFR (45) measurements of xF_3 (solid squares and right scale) and the corresponding integral (diamonds and left scale) as a function of x. The dotted curve and the solid curve represent the best fit to the data as outlined in the text.

same procedure in a high statistics Monte Carlo calculation, is about a third of the statistical error.

Table 2 summarizes the published measurements (45, 49–51) of this integral. The average, excluding the CCFRR (1984) value [which is superseded by that of CCFR (1988)] is

$$\langle S_{GLS}^{expt} \rangle = 2.79 \pm 0.13,$$

which gives a difference between experiment and theory of 0.05 ± 0.14. We conclude that the GLS sum rule is consistent with the expectations of the quark-parton model and QCD at the level of the 5% experimental error.

ADLER SUM RULE   The Adler sum rule, shown in Equation 8, predicts the integrated difference between neutrino-neutron and neutrino-proton structure functions. Unlike the previous case, this sum rule is expected to be valid to all orders of perturbative QCD.

The neutrino scattering measurements performed on deuterium by the WA25 (BEBC) collaboration (51) are consistent with the quark-parton model prediction,

$$S_A^{pred} = 1.$$

The experimental measurement is

$$S_A^{expt} = 0.01 \pm 0.20,$$

where we have averaged the quoted statistical and systematic errors.

GOTTFRIED SUM RULE   The Gottfried sum rule, shown in Equation 9, predicts that the integrated difference between electromagnetic structure functions for neutrons and protons should be

$$S_G^{pred} = 0.33,$$

independent of Q^2. The EMC muon measurements on deuterium have

Table 2 Measurements of the Gross–Llewellyn Smith sum rule

Group (Ref.)	S_{GLS}
CDHS (1979) (50a)	3.2 ± 0.5
CHARM (1983) (50b)	2.56 ± 0.42
CCFRR (1984) (49)	2.83 ± 0.20
WA25 (1985) (51)	2.70 ± 0.40
CCFR (1988) (45)	2.79 ± 0.16

been used to test this (52a, 52b). The data, while not very precise, are in agreement with the prediction:

$$S_G^{\text{expt}} = 0.24 \pm 0.11.$$

BJORKEN SUM RULE AND SPIN STRUCTURE FUNCTION   Measurements of the spin structure functions are important: they provide unique information about nucleon structure that could lead to understanding the nucleon spin in terms of the quark and gluon constituents. The Bjorken sum rule shown in Equation 10 predicts the value of the integral over the difference between appropriately weighted neutron and proton spin structure functions

$$S_B = \frac{1}{6}\frac{G_A}{G_V}\left(1 - \frac{\alpha_s}{\pi}\right)$$ 17.

to have the values $S_B^{\text{pred}} = (0.191 \pm 0.002)$ for $G_A/G_V = (1.254 \pm 0.006)$ and $\alpha_s = (0.27 \pm 0.02)$, as quoted by the EMC group (53). Unfortunately, there are no measurements at present that permit a direct check of this sum rule.

The EMC collaboration has, however, reported (53) a measurement of the scattering asymmetry for a longitudinally polarized muon beam from a polarized proton target. These measurements are consistent with, but more precise than, earlier measurements (54); they extend the data into the small-x region, which is critical for the evaluation of the sum rule integral.

The asymmetry data permit evaluation of the spin structure function, $xg_1^p(x)$, characterizing the proton. Information is lacking on xg_1^n, the corresponding structure function for the neutron, which would permit a direct test of the sum rule. Even so, the measurement has generated considerable interest.

Figure 6 shows xg_1^p (right scale and squares) and the corresponding integrated value of $\int g_1^p(x)\,dx$ (left scale and diamonds) as a function of x. While there is some uncertainty in extrapolating the integral to very low x, this is not a problem unless there is some anomalous discontinuity for $x < 0.01$. With this caveat, the measured data cover about 98% of the integral. The value of the integral, evaluated at $\langle Q^2 \rangle = 10.7$ GeV^2, is

$$\int g_1^p(x)\,dx = 0.114 \pm 0.012 \pm 0.026,$$ 18.

where the first error is statistical and the second is systematic.

One reason for the surprise engendered (55) by this result is that there is a sum rule by Ellis & Jaffe (56) predicting that the integral (Equation 18) should equal 0.19, as estimated by the EMC group [although estimates (55) as low as 0.17 have been quoted]. This sum rule, however, assumes

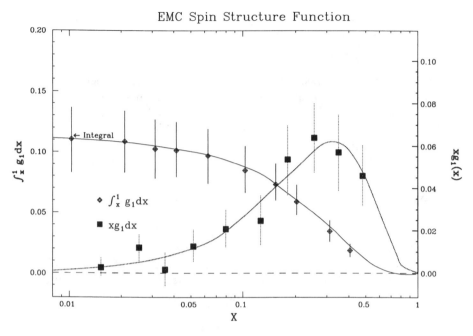

Figure 6   The EMC (53) spin structure function, $xg_1^p(x)$ (solid squares and right scale), and the corresponding integral, $\int g_1^p \, dx$ (diamonds and left scale) as a function of x.

flavor SU(3) and ignores contributions of strange quarks to the net polarization; the Bjorken sum rule, in contrast, depends only on isospin symmetry.

Assuming the validity of the Bjorken sum rule, the EMC collaboration calculates a negative value for the integrated neutron spin structure function:

$$\int g_1^n(x) \, dx = -0.077 \pm 0.012 \pm 0.026. \qquad 19.$$

Summing these two integrals and ignoring the effect of strange quarks, one infers that the net fractional spin carried on average by the valence quarks in the proton is only $14 \pm 9 \pm 21\%$. If we assume the violation of the Ellis-Jaffe sum rule is due to polarized strange quarks, the net fractional spin carried by quarks is estimated to be $1 \pm 12 \pm 24\%$. Note that for both cases the quarks carry a rather small fraction of the proton spin. This result is intuitively surprising because the nonrelativistic model of constituent quarks has been so successful in describing the masses and magnetic moments of baryons.

There exists some controversy as to whether the violation of the Ellis-Jaffe sum rule and the apparently small fraction of the proton spin carried by quarks represent serious problems. Several authors point out that there are potentially large common contributions to the proton and neutron integrals from gluon spins (57), expected from estimates of higher order terms in the perturbative expansion. An alternative (nonperturbative) explanation, based on chiral symmetry and the $1/N$ expansion as expressed in the Skyrme model, suggests that the average quark spin and gluon spin contributions should be separately small; that is, the proton spin is principally a consequence of orbital angular momentum (58).

We note that new data with polarized leptons and polarized target nucleons would be important in clarifying these questions. Such data could be used in checking the Bjorken sum rule directly, as well as helping to understand the relationship of gluons and up, down, and strange quarks to the spin structure of the nucleon. Information from many sources may well be brought to bear on this question, including measurements of the form factors describing elastic eN scattering and pseudoelastic νN scattering (59).

3.3 *Quark Densities*

Measurements of nucleon quark densities are important for the several reasons discussed in Section 1.3. Deep inelastic scattering from heavy isoscalar nuclei provides only combinations of proton quark densities, so that measurements from light targets are required. Neutrino and anti-neutrino data with hydrogen (alone) provide the input for a direct method of extraction of the individual quark densities. The cross sections involve the following densities and y dependences:

$$\nu p: \qquad x[d(x)+s(x)]+x[\bar{u}(x)+\bar{c}(x)](1-y)^2 \qquad\qquad 20a.$$

$$\bar{\nu} p: \qquad x[u(x)+c(x)](1-y)^2+x[\bar{d}(x)+\bar{s}(x)], \qquad\qquad 20b.$$

where, for simplicity, we have suppressed the Q^2 dependence of the proton quark densities. The charm component is expected to be small in the present Q^2 domain, and hence is ignored. The strange component is accessible from the subset of events with leptonic charm decays, as discussed below. Equations 20 indicate, then, that observation of the y dependence of the cross section from neutrino and antineutrino scattering in principle permits extraction of individual u, d, $\bar{u}$, and $\bar{d}$ components of the proton.

Unfortunately, the available neutrino data are statistically limited. Electron and muon experiments have much larger event samples, but data from both hydrogen and deuterium are necessary, and additional assumptions must be made to extract flavor-separated quark densities. For muon scattering on hydrogen, the dependence is

$$\mu p: \qquad x[(d+\bar{d}+s+\bar{s})+4(u+\bar{u}+c+\bar{c})][1+(1-y)^2]/9. \qquad 21a.$$

Assuming isospin symmetry for the nucleon doublet (e.g. $u_n = d_p \equiv d$), one obtains a dependence in scattering from neutrons of

$$\mu n: \qquad x[(u+\bar{u}+c+\bar{c})+4(d+\bar{d}+s+\bar{s})][1+(1-y)^2]/9. \qquad 21b.$$

Since antiquark components enter with the same y dependence as quark components, information on the antiquark part must be assumed from neutrino measurements.

Two experiments, CDHS and WA21 (BEBC), have studied $\nu/\bar{\nu}$ scattering (60a,b; see also 47) from hydrogen. Figure 7 shows the ratio of quark and antiquark components as measured by the two groups. The measurements agree at the statistical level of the experiments, which is about 15%. [For these and other comparisons, early CDHS measurements have been

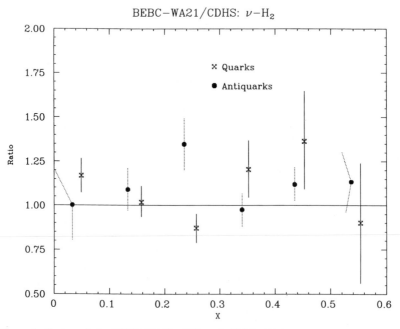

Figure 7 Ratios of the BEBC-WA21 (60b) and CDHS (60a) measurements of quark (*crosses*) and antiquark (*solid circles*) densities from ν-H$_2$ interactions.

adjusted in normalization to agree with their more recent measurements of the total νN cross section (61).]

There are three high statistics structure function measurements using electrons and muons on hydrogen targets (33b, 62a,b). The two muon experiments, EMC (62b) and BCDMS (33b), employing incident beam energies up to 280 GeV, reach the highest Q^2. These two measurements, however, disagree outside the statistical errors. Figure 8 shows the structure function, F_2, as a function of Q^2 in four representative x bins; $x = 0.125$, 0.250, 0.350, and 0.650 for the EMC and BCDMS data. We note that there is a discrepancy in normalization: the EMC data are lower than those of BCDMS by about 8–9% over most of the range. In addition, the two data sets exhibit a pronounced x-dependent difference. The solid curves in these figures come from an illustrative parametrization obtained from a best fit to all existing data on individual quark densities (63; W.-K. Tung, private communication).

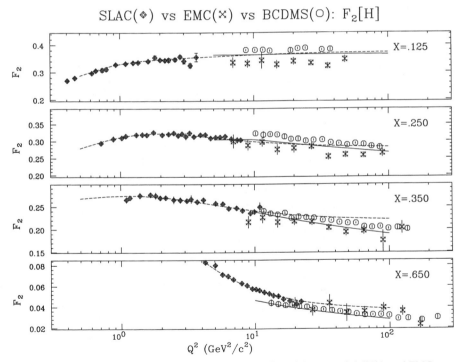

Figure 8   A comparison of SLAC (*diamonds*) (62a), BCDMS (*open circles*) (33b), and EMC (*crosses*) (62b) measurements of F_2 as a function of Q^2 in four x bins. The dotted curve shows a parametrization of the SLAC data extrapolated to high Q^2; the solid curve is an illustrative parametrization by Tung (63).

While comparing the BCDMS and EMC data to the SLAC data (shown as solid points), which admittedly are at a lower Q^2, it was noted (64) that the extrapolated fits to SLAC data, shown as dotted curves in the figure, do not fit either muon data set well. The SLAC fits are typically 9% higher than the EMC data, but generally agree with the EMC x dependence. The converse is true when the fits are compared to the BCDMS data: while the average levels agree, there is a discrepancy in the x dependence (64). Similar disagreements between the EMC and the BCDMS data are found for comparisons of F_2 measured on nuclear targets (iron and carbon respectively), as discussed at the end of this section. This glaring discrepancy between two deep inelastic muon experiments is an outstanding experimental problem and needs to be reconciled.

There are also deuterium data from SLAC (62a) and EMC (52b) evincing similar concerns: the relative x dependence of the two measurements is similar, but the EMC data consistently lie lower than extrapolated fits to SLAC data. Neutrino-deuterium data from WA25 (51) agree with EMC data (52b) within the precision of the neutrino data, except near $x \approx 0.2$.

VALENCE QUARK DENSITIES   Figure 9 summarizes the available information on the separate valence quark components of the proton. Three of the four measurements are extracted from hydrogen-deuterium data with neutrinos; the fourth measurement is from the EMC muon experiment. The smooth curve is the parametrization of Tung et al (63; W.-K. Tung, private communication). The EMC valence u-quark measurements, though slightly higher than those of CDHS and WA25, agree within the statistical precision of the experiments. The valence d-quark measurements from neutrinos agree with each other; the EMC measurement, however, disagrees with those from the neutrino experiments by nearly a factor of two at $x \approx 0.2$. This discrepancy represents a problem.

The overall valence quark density for isoscalar nucleons (i.e. neutron-proton average) is essentially the xF_3 structure function measured in the high statistics neutrino experiments on isoscalar targets. The various measurements are in good agreement. This structure function was discussed in Section 3.2 and an example of one measurement is shown in Figure 5.

SEA-QUARK DENSITIES   Figure 10 shows the density of one of the proton antiquark components, as obtained from the $\bar{\nu}_\mu$ hydrogen data of CDHS (60a) and WA21 (60b) and from the deuterium data (51) of WA25. The smooth curve is the Tung parametrization (63; W.-K. Tung, private communication). The statistical precision of these measurements is not very high.

The overall nucleon antiquark density $[x(\bar{u}+\bar{d})/2]$ obtained from CHDSW (46) and CCFR (45) data, shown on a log scale in Figure 11, is measured relatively precisely. There is reasonable agreement between the

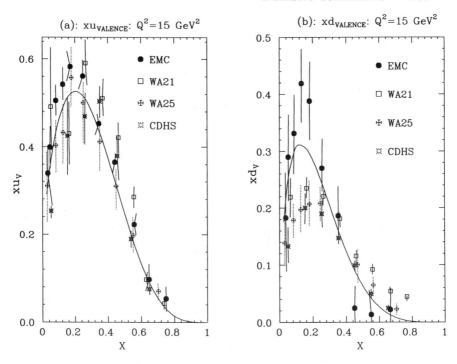

Figure 9   Valence quark densities in a proton: (*a*) xu_V; (*b*) xd_V as a function of x. The data are from EMC (52b), WA21 (60b), WA25 (51), and CDHS (60a). The solid curve is the Tung et al (63) parametrization.

CDHSW and CCFR data over most of the x range. (The lowest x values disagree because the trends of the two data sets in Q^2 are somewhat different, so that the extrapolation to $Q^2 = 5$ GeV^2 results in the differing values shown.) These antiquark distributions can be important in determinations of the gluon distribution.

MEASUREMENT OF THE STRANGE SEA   Neutrino production of opposite-sign dimuons ($\mu^- \mu^+$) offers a unique probe to measure the strange component of the nucleon sea. Dimuon events originate from the production of a single charmed particle:

$$\nu_\mu + (d, s) \rightarrow \mu^- + c + X \qquad\qquad 22a.$$

$$\bar{\nu}_\mu + (\bar{d}, \bar{s}) \rightarrow \mu^+ + \bar{c} + X. \qquad\qquad 22b.$$

Here the charmed quark emerges as a charmed particle (typically D^0 and D^+), which subsequently decays, for example $D^0 \rightarrow \mu^+ + \nu_\mu + X$. The $d(\bar{d}) \rightarrow c(\bar{c})$ conversions in Reactions 22 are Cabibbo suppressed. Hence,

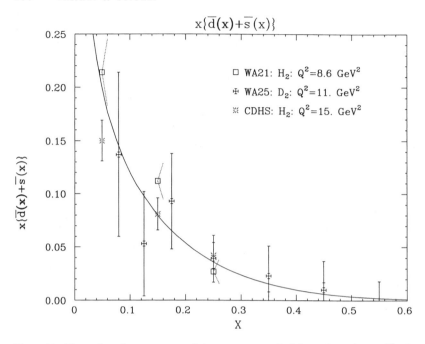

Figure 10   The antiquark component of the proton as probed by antineutrinos: $x(\bar{d}+\bar{s})$. The three sets of antineutrino data are by WA21 (60b), WA25 (51), and CDHS (60a). The curve is the Tung et al (63) parametrization.

the $\bar{\nu}$-induced $\mu^+\mu^-$ events arise predominantly from production from strange quarks ($\sim 90\%$), and so provide a direct measurement of the strange sea-quark density. For the ν_μ-induced events, production from s quarks competes favorably with production from valence d quarks. The ratio of the $\mu^+\mu^-$ events relative to single-muon events provides a measure of the momentum fraction, κ, carried by the strange sea quarks relative to that carried by the nonstrange sea:

$$\kappa = 2s/(\bar{u}+\bar{d}). \qquad\qquad 23.$$

If there were no strange quarks, $\kappa = 0$; if the sea were flavor-SU(3) symmetric, $\kappa = 1$. Certain model assumptions are required to extract κ from the dimuon data: (*a*) a model of the threshold dependence for charm production; and (*b*) the function describing the charm fragmentation into D mesons. In addition, the product of the parameters $B|V_{cd}|^2$ enters. Here, B is the semileptonic D-decay branching ratio for the appropriate admixture of D^+ and D^0, and V_{cd} is the Kobayashi-Maskawa matrix element connecting the charm and down quarks. The usual choice for (*a*)

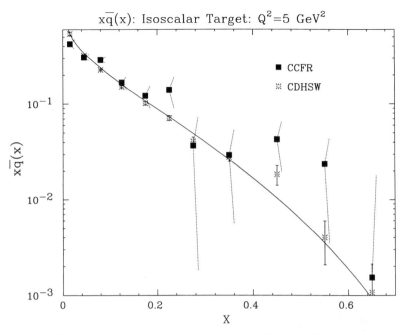

Figure 11 The antiquark density of an isoscalar target. The data shown are from the CDHSW (46) and CCFR (45) collaborations. The curves are from the Tung et al (63) parametrization.

is the "slow-rescaling" formulation, which has a single parameter, m_c, an effective charm-quark mass for this process (65–67).

The parameter, κ, is extracted from the x distributions of the dimuon events. (Some resolution effects must be unfolded since the neutrino from the charm decay is unobserved.) Figure 12*a* (12*b*) shows recent, high statistics CCFR (68) x distributions of ν_μ- ($\bar{\nu}_\mu$)-induced $\mu^+\mu^-$ events after correcting for resolutions and backgrounds. The individual contributions from valence and sea components from the fits are shown as smooth curves. The recent results on κ and $B|V_{cd}|^2$ from the CCFR collaboration's wide band data, along with their earlier CCFRR narrow band results (69) and those of the CDHS group (70), are presented in Table 3. The quoted errors include contributions from statistical and systematic sources.

These measurements of κ are in good agreement; they demonstrate that the nucleon sea has half as many strange quarks as nonstrange quarks [or is roughly one half SU(3) symmetric]. The average value for the strange sea fraction from CDHS and CCFR is

$$\kappa = 0.52 \pm 0.07. \qquad 24.$$

The value of B can also be independently calculated (69) using emulsion

x–Distribution of Neutrino Dimuons

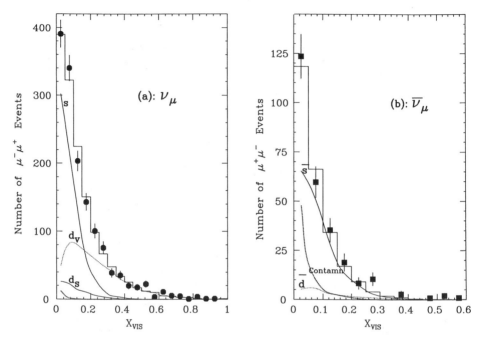

Figure 12 CCFR (65–67) x_{VIS} distribution of $\mu^-\mu^+$ events: (*a*) ν_μ-induced dimuon data (*solid circles*) after background subtraction (from π/K decays); the histogram is the charm Monte Carlo prediction, which is a sum of contributions from strange quarks (*upper solid curve*), valence d quarks (*upper dotted curve*), sea d quarks (*lower solid curve*), and contamination from antineutrino-induced events (*lower dotted curve*). (*b*) $\bar\nu_\mu$-induced dimuon data (*solid squares*) and corresponding Monte Carlo (*histograms*). The contributions to the Monte Carlo are from the antistrange quarks (*solid curve*), $\bar{d}$ (*lower dotted curve*), and contamination (*upper dotted curve*).

data (71) on neutrino production of charmed particles and the muonic branching fractions (72) for charmed hadrons, giving $B_{calc} = 0.110 \pm 0.009$. Evaluation (73) of $|V_{cd}|$ follows by using B_{calc}. This technique is currently the only direct measure of this parameter. The last column in Table 3 shows the resulting values of $|V_{cd}|$. The average is

$$|V_{cd}| = 0.21 \pm 0.014. \qquad\qquad 25.$$

In the limiting case that two of the three angles in the Kobayashi-Maskawa matrix approach zero, this value would approach $\sin\theta_c$, where θ_c is the Cabbibo angle.

NUCLEAR TARGET EFFECTS ON STRUCTURE FUNCTIONS   Structure functions

Table 3   Values of κ and B from neutrino-induced $\mu^+\mu^-$ events

Group (Ref.)	κ	$B\lvert V_{cd}\rvert^2$	$\lvert V_{cd}\rvert$
CDHS (70)	0.52 ± 0.09	$(0.41 \pm 0.07) \times 10^{-2}$	0.19 ± 0.02
CCFRR (69)	$0.52^{+0.17}_{-0.15}$	assumed	
CCFR (68)	$0.53^{+0.13}_{-0.09}$	$(0.51 \pm 0.09) \times 10^{-2}$	0.22 ± 0.02

in a free nucleon are different from those measured in a heavy nuclear target. This was first demonstrated by the EMC collaboration (74) and corroborated by the Rochester-SLAC-MIT group (75). The "EMC effect," which states that the structure function is dependent at the 10% level on the details of the nuclear atomic number, A, is now established. It has engendered many theoretical conjectures [for a review of nuclear effects in deep inelastic scattering, see for example (76)]. Among these conjectures are (a) "Q^2 rescaling," which asserts that the effective momentum transfer in heavy nuclei is different from that in a free nucleon; and (b) that virtual pions and other hadrons within the nucleus give a larger antiquark component than for free nucleons.

The salient features of the EMC effect are illustrated by various data sets in Figure 13 (75, 77a,b). The tendency of the ratio of structure functions for a nucleus of atomic weight A to that for the deuteron, $F_2(A)/F_2(D)$, to decrease for $x < 0.15$ is presumably caused by nuclear shadowing (78), a well-known effect that is responsible for the A dependence of photon ($Q^2 = 0$) cross sections. In the intermediate x range ($0.15 < x < 0.60$), there is a clearly decreasing ratio that can be understood by a variety of mechanisms such as those mentioned above. For $x > 0.60$, the ratio increases as a result of Fermi motion.

The agreement in the crucial intermediate x region among these various experiments performed at widely different energies is nontrivial. The SLAC data cover the range $2 \le Q^2 \le 15$ GeV^2, while the muon scattering data cover $14 \le Q^2 \le 200$ GeV^2, a factor of about ten higher. It seems likely, therefore, that this effect is largely independent of Q^2. A recent measurement of $R = \sigma_L/\sigma_T$ for various nuclear targets demonstrated that this quantity does not differ for light and heavy nuclei (79).

Neutrino interactions can distinguish quark flavors and hence would be ideally suited to study the EMC effect. Neutrino measurements (80) of $F_2(A)/F_2(D)$ are consistent with those made with charged leptons. The statistical precision of these experiments, however, falls short of shedding light on nuclear shadowing (low-x behavior), or of precisely measuring antiquark composition at larger x values where the competing mechanisms (a) and (b) may differ.

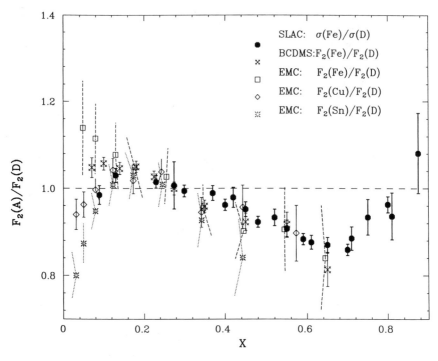

Figure 13   The EMC effect illustrating dependence of $F_2(A)/F_2(D)$ as a function of x. The data shown are those of SLAC (75), BCDMS (77b), and EMC (77a).

4.   QCD TESTS AND MEASUREMENTS

4.1   *Measurements of* $R = \sigma_1/\sigma_t$

It is expected from the arguments of Section 1.5 that the value of R should be small ($R \ll 1$). Furthermore, for x fixed and low Q^2, R should fall as a power of $1/Q^2$; at higher Q^2, R should fall as the logarithm of Q^2. The earliest low-Q^2 measurement from SLAC (3) did obtain a small value ($R \approx 0.18$) as anticipated from the Callan-Gross prediction; this was the first direct demonstration of spin-1/2 constituents for hadrons. On the other hand, subsequent measurements (81; earlier articles include 3, 62a, 82) of R at SLAC seemed nearly independent of Q^2, although the systematic uncertainties were large. Measurements with neutrino and muon beams at higher Q^2, as discussed below, gave smaller values of R. The situation has been perplexing for two reasons: (*a*) the continuity of all these measurements was questionable; and (*b*) a flat Q^2 dependence is difficult to reconcile with any reasonable model of spin-1/2 constituents.

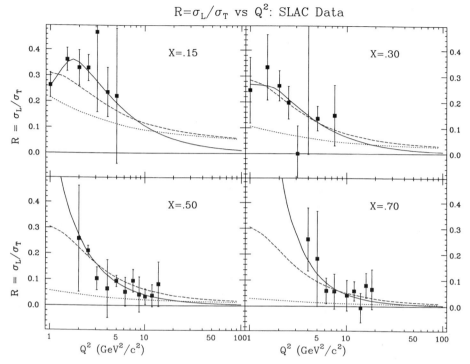

Figure 14   The SLAC (83) data showing $R = \sigma_1/\sigma_t$ as a function of Q^2 in four representative *x* bins. The lowest dotted curves represent predictions from the QCD calculation (see footnote 5), the dashed curves represent QCD prediction with a higher twist contribution; the solid curves represent a higher twist fit of the type shown in Equation 26 without the perturbative QCD term.

Recent measurements taken at SLAC have provided a new perspective to this historical problem. The new data (83), shown for a subset of *x* values in Figure 14, qualitatively show the expected behavior at low Q^2. The lower dotted curves are the predictions of a QCD calculation (see Equation 13) used by the experimenters. As expected, nonperturbative effects contribute heavily to the experimental data in this low-Q^2 regime. The upper dashed curve includes a higher twist term fitted to the data.[5] The authors

[5] The QCD prediction and the empirical fitted term used by Dasu et al (83) is

$$R(x, Q^2) = \frac{1.11(1-x)^{3.34}}{\ln(Q^2/0.04)} + \frac{0.11(1-x)^{-1.94}}{Q^2},$$

where the first term is an approximate parametrization of Equation 13 and the second term comes from an empirical fit to the SLAC data.

indicate, and the figure illustrates, that this hypothesis fits the data well. On the other hand, these data do not demonstrate the predicted perturbative QCD behavior (lower dotted curves). The measured values of R at these low Q^2 are evidently due principally to higher twist effects.

We have used this SLAC data to predict the value of R with the hypothesis that all the Q^2 dependence in the SLAC data is described by higher twist (85). This is useful in judging whether other existing data at higher Q^2 require a QCD explanation. We adopt the parametrization for the higher twist

$$R_{HT} = \frac{\alpha(x)}{Q^2} + \frac{\beta(x)}{Q^4}. \qquad\qquad 26.$$

The resulting fitted values of α and β at each x value were then used to calculate the solid curves shown in Figure 14. These fits are quite acceptable; furthermore, the values of α and β are typically in a reasonable range for the higher twist hypothesis; i.e. $\alpha < 1$ GeV^2 and $\beta < 1$ GeV^4. The fitted values of $\alpha(x)$ and $\beta(x)$ provide an estimate of the maximum effect that higher twist could have at the higher Q^2 of other experiments. Typically, this hypothesis predicts a small value of R_{HT} when $Q^2 > 10$ GeV^2, as expected.

Averaged values of the high-Q^2 data on R are shown as a function of x in Figure 15 (32c, 33b, 86–88). The data point at a given x usually consists of averaged points from many different values of Q^2. The data cover the range $0.015 < x < 0.65$ and $1.4 < Q^2 < 70$ GeV^2. Not surprisingly, the data with the largest positive excursions from zero are at the lowest Q^2. Overall, detailed comparison of the data contained in Figure 15 with R_{HT} gives an acceptable χ^2. For a fit to the QCD hypothesis with inclusion of the higher twist effects described previously,[5] and χ^2 is no better.

We conclude that the qualitative behavior of data on R agrees well with quark model expectations that R should be small and falling with Q^2. The behavior at higher Q^2 should follow the quantitative expectations of QCD, but the data are not yet precise enough to verify this prediction.

4.2   Scaling Violations in Nonsinglet Structure Functions

In the Q^2 domain where perturbative QCD is expected to apply, Equation 14 states that, for nonsinglet structure functions, the logarithmic derivative with respect to Q^2, at fixed x, should be a predictable function:

$$\frac{d \ln x F_3(x, Q^2)}{d \ln (Q^2)} = \psi(x, Q^2). \qquad\qquad 27.$$

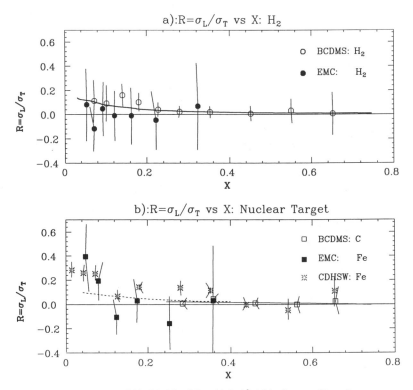

Figure 15 Measurements (33b, 86–88) of R at high Q^2: (*a*) hydrogen, (*b*) nuclear targets as a function of x. Data at many differing Q^2 values have been averaged at each value of x.

The term $\psi(x, Q^2)$, shown up to leading order in Equation 14, involves an integral of the $\xi F_3(\xi, Q^2)$ structure function for all values $\xi > x$. This integral is calculated, either directly or using a fit, from the same data used for the logarithmic derivative on the left-hand side. In principle, the only unknown in $\psi(x, Q^2)$ is α_s, which in turn is known as a function of Λ (Equation 12). Figure 16 illustrates this for a particular parametrization of xF_3. Note that a specific curve exists for a specific value of Λ.

Figure 16 also shows the slopes extracted from the CDHSW wide band data (46) for xF_3 as a function of x for $Q^2 > 6$ GeV^2. These data, which come from the largest statistical sample published, are superimposed on the same family of curves. Note that no curve agrees well with the data. The figure also shows the narrow band data of CCFR (45). Here, disagreement is not obvious, but the statistical precision is not as high as the wide band CDHSW data. New wide band data at very high Q^2 from CCFR should help in addressing this question.

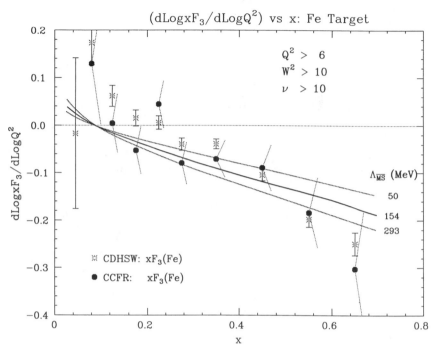

Figure 16   The measured logarithmic derivative of xF_3 with respect to Q^2 as a function of x, reported by the CDHSW (46a) and the CCFR (45) collaborations. QCD predictions, for specific values of $\Lambda_{\overline{MS}}$, are shown by the solid and dotted curves.

It should be noted that the curves in Figure 16 are calculated in next-to-leading order of the perturbative expansion. (The general features are the same in leading order.) However, no corrections for higher twist have been applied because such effects are not well understood. Such higher twist corrections, if they exist, must fall with a power of $1/Q^2$. Their contribution would be much reduced if the data were restricted to higher momentum transfers. When such a cut, say $Q^2 > 20$ GeV^2, is applied, the statistical precision of the CDHSW data is such that, while consistent with the requirements of QCD, the test is not very stringent.

4.3   Scaling Violations in Singlet Structure Functions

The evolution of singlet structure functions (e.g. F_2) follows equations more convoluted than Equation 14, because of the presence on the right-hand side of an additional integral over the gluon distribution. This can be viewed as either a problem or a bonus. On the one hand, it complicates any test of QCD behavior; on the other hand, under the assumption that QCD behavior is valid and well understood, it permits extraction of the gluon distribution. Techniques have been developed to test QCD with singlet structure functions for two reasons: (a) because the statistical precision is so much better in neutrino F_2 measurements relative to those of xF_3, and (b) because xF_3 is not available from muon scattering experiments.

The procedure followed by the BCDMS group (33a,b,c) is typical. First, note that in regions of large x, we anticipate that $R \to 0$, $\bar{q} \to 0$, and $G \to 0$. In such a limit, $F_2 \to xF_3$, and the evolution equation should look identical to that for a nonsinglet. The BCDMS group uses data for $x > 0.275$ and $Q^2 > 20$ GeV^2. The comparison of d $(\ln F_2)$/d $(\ln Q^2)$ for the BCDMS data on hydrogen and on carbon is shown in Figure 17 with the corresponding QCD predictions. The agreement is extremely good; the

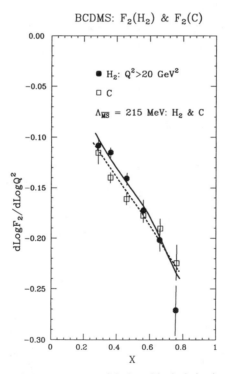

Figure 17   BCDMS (33b,c) measurement of the logarithmic derivative of F_2 with respect to Q^2 vs X with H$_2$ (*solid*) and C (*open*) targets. The corresponding QCD predictions are shown by solid and dotted curves, respectively.

small predicted differences (due to the differences in the F_2 shape) between hydrogen and carbon seem to be followed by the data.

This procedure, applied to the EMC data (32c) with $Q^2 > 10$ GeV^2, yields the comparison between data and prediction shown in Figure 18a. Note that the entire trend of the data is unlike the predicted curve. A similar comparison occurs for the F_2 structure function of CDHSW shown in Figure 18b. At face value, the EMC and CDHSW data question the predictive capability of perturbative QCD, at least in this Q^2 regime. We discuss the conflicts among these data below.

For hydrogen data, the BCDMS measurements extend down to low x. In Figure 19, we see the logarithmic slope data plotted into the low-x region. The lower curve is the prediction for the nonsinglet case; the difference between the lower curve and the data is attributed to the effect of gluons at low x. It should be noted that the procedure has assumed that

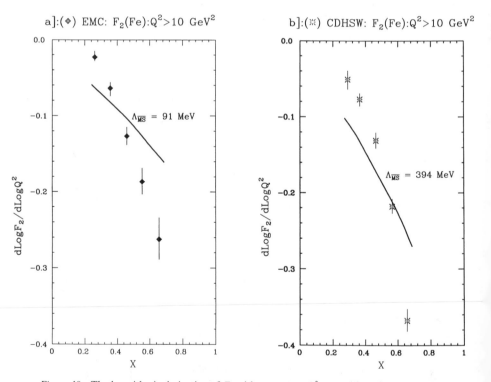

Figure 18 The logarithmic derivative of F_2 with respect to Q^2 vs x with an iron target as measured by (a) EMC (32b) and by (b) CDHSW (46a) collaborations. The QCD curves with typical values of $\Lambda_{\overline{MS}}$, as analyzed by the BCDMS collaboration (33c), are also shown.

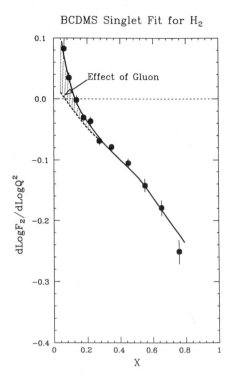

Figure 19   The BCDMS (33b,c) measurements of the logarithmic slope of F_2 for H_2 target. The perceived effect of the gluons, prominent at low x, is shown by the hatched region between the singlet (*solid*) and the nonsinglet (*dotted*) QCD predictions.

$G = 0$ for $x > 0.275$ and so the calculated gluon distribution must be very small in that region; the assumption cannot be verified from the same data.

The fits to the BCDMS data, treated as nonsinglets in the large-x region, in next-to-leading order, give the results

$$\Lambda_{\overline{MS}} = 230 \pm 20 \pm 60 \text{ MeV for carbon;}$$

$$\Lambda_{\overline{MS}} = 200 \pm 22 \pm 60 \text{ MeV for hydrogen.}$$

4.4   *Experimental Conflict on Structure Function Evolution*

The preceding discussion indicates a clear conflict between conclusions from the BCDMS muon scattering experimental results in Figure 17, on the one hand, and from EMC (μ scattering) and CDHSW (ν scattering) data in Figure 18, on the other. (The narrow band data from CCFR do not have enough statistical precision to permit a definite conclusion; CCFR Tevatron wide band data are still being analyzed.) While the BCDMS data show a logarithmic dependence that is consistent with the predictions of QCD, the EMC and CDHSW data clearly do not. The data in Figure 18 show a slope at large x $(x > 0.55)$ that is much more negative than predicted by QCD, and a slope at intermediate x $(x < 0.45)$ that is nearer zero than predicted.

Attempting to ascribe a precise cause for this conflict would be mere speculation. It is interesting, however, to detail some differences among the experiments, which is done in Table 4. Note that other experiments (like CCFR, CHARM, and BFP), which are more similar to the first column than the second, do not at present lend substantial weight to either side of the controversy.

The use of carbon rather than iron should be inconsequential, as discussed below. However, the energy measurement is very important. Because of the geometry of the BCDMS experiment, the final-state hadronic energy must be inferred from the difference between the beam and outgoing muon energies. This imposes a very strong requirement on the precision of the muon energy measurements, particularly in regions where v is small (larger x and smaller Q^2). The toroidal magnetic field used for the measurement is calibrated to a precision of 2×10^{-3}, which is not adequate. So data from differing muon energies are used to make a final adjustment. If the R parameter were known and the calibration correct, differing incident muon energies should give precisely the same structure functions at the same values of x and Q^2. An incorrect value of R would give a pattern of structure function differences distinct from the pattern of differences resulting from an incorrect calibration. In the case of the carbon data, a final adjustment of 1.0013 was made in the magnetic field value, based on these considerations (33b). The BCDMS experimenters make a convincing argument that this final calibration adjustment can be performed without bias.

Table 4 Experimental differences among experiments

Experimental feature	CDHSW and EMC	BCDMS
Target material	Iron	Carbon
Energy measurement	E_μ magnet	E_μ magnet
	v calorimetry	$v = E_\mu - E_{\mu'}$ with $E_{\mu'}$ measured to about 10^{-3}
Q^2 range	Q^2 down to 10 GeV^2 or lower	$Q^2 > 25$ GeV^2

The validity of perturbative QCD in deep inelastic scattering is not settled. There are differing conclusions from the experiments. Several resolutions of the experimental conflict have been proposed:

1. "Iron is different from Carbon." We recall that the test (Equation 14) has only to do with the validity of the evolution equation. In principle, this equation should be valid for any "bag of quarks and gluons," so long as the evaluation of the integral on the right-hand side is characteristic of the same bag as the data on the left-hand side, and the integration is being done correctly. It should not matter which bag of quarks is used, provided higher twist effects can be ignored. It is difficult to understand how a specifically nuclear effect could create higher twist terms that persist beyond $Q^2 \approx 10 \text{ GeV}^2$. Suggestions have been made (43b) that there might be a difference between carbon and iron that has a logarithmic Q^2 dependence. Such an effect must be QCD in origin and would imply that the evolution equations for carbon and iron are different. It is difficult to conceive how conventional quantum chromodynamics could accommodate this. In any case, comparison among low-Q^2 and high-Q^2 experiments, as mentioned in Section 3.3, provides no evidence that such an effect exists.

2. "Uncorrected systematic errors" in one or more experiments might produce this difference. Such difficulties, if they exist, are unlikely to be simple. There are substantial systematic differences among these experiments, like the normalization problems between the EMC experiment and the neutrino experiments or in the x dependence and normalization conflicts between the high statistics muon experiments on hydrogen described in the previous section. In addition, different QCD evolution programs and parametrizations are often used.

3. "Different ranges of Q^2 are being compared." As noted earlier, when the higher Q^2 cut of BCDMS is imposed on the EMC or CDHSW data, the statistical precision is not really adequate to make a definitive statement; that is, the conflict largely disappears. The BCDMS data, on the other hand, do not extend to lower Q^2, where a direct numerical comparison between experimental data could be made. More data at high Q^2, from both muon and neutrino experiments, are very important.

4.5 Measurements of the QCD Parameter, Λ

Even though there are strong disagreements among experiments about the presence or absence of QCD-like behavior, there are no strong disagreements among the various published values of $\Lambda_{\overline{MS}}$, the value of Λ in the next-to-leading order $\overline{MS}$ scheme. Table 5 shows the values for this parameter obtained from data on the true nonsinglet

Table 5 Nonsinglet fits

Group	$\Lambda_{\overline{MS}}$ (MeV)
CDHS (1983)	200 ± 100
CHARM (1983–84)	$310 \pm 140 \pm 70$
CCFR (1984)	155^{+230}_{-130}

xF_3 structure function. The corresponding values from F_2 are shown in Table 6.

An average of values (47) from true nonsinglet fits in Table 5 is

$$\Lambda_{\overline{MS}} = 186 \pm 60 \text{ MeV},$$

which is consistent with the BCDMS average of $215 \pm 15 \pm 60$ MeV. Even the EMC values are near (within their quoted systematic errors) the BCDMS values. There is a controversy as to whether perturbative QCD describes these data; however, the values of the quark-gluon coupling do not seem to be controversial. The use of data that are marginally compatible with the stated QCD hypothesis to provide a value for $\Lambda_{\overline{MS}}$ is, of course, suspect.

5. SEARCHES FOR EXOTIC PROCESSES

5.1 *Like-Sign Dimuons*

Early measurements (69, 89–92) of neutrino-induced like-sign dimuons events,

$$\nu_\mu + N \rightarrow \mu^- + \mu^- + X,$$

with $E_\nu < 300$ GeV, posed an intriguing problem for the standard model. These measurements indicated an excess of such events over the expected backgrounds at a statistical level of two to three standard deviations. (The backgrounds were expected principally from decays of π and K mesons in

Table 6 Singlet fits

Group	$\Lambda_{\overline{MS}}$ (MeV)
EMC–Fe (1986)	$115 \pm 20^{+90}_{-45}$
EMC–H$_2$ (1985)	110^{+60+90}_{-50-50}
EMC–D$_2$ (1987)	$65^{+95+155}_{-50-45}$
BCDMS–C (1987)	$230 \pm 20 \pm 60$
BCDMS–H$_2$ (1987)	$200 \pm 22 \pm 60$

the hadronic showers of charged-current events.) Furthermore, the rate exhibited a tendency to increase with neutrino energy.

A signal of like-sign dimuons is expected because of associated production and decay of pairs of charmed particles. However, the perturbative QCD prediction, assuming gluon bremsstrahlung as the mechanism, fell short by almost an order of magnitude (93). It was noted (94), however, that the systematic uncertainty in the theoretical prediction could be as large as 60%, in large part because of the uncertainty in the threshold behavior of the process.

Data from the recent high statistics CCFR experiment (95), conducted at the Tevatron with neutrino energies up to 600 GeV, obviate the need for any new threshold and are consistent with, but do not require, the expected charm contribution at higher energies.

Figure 20 presents a compilation (69, 89–92, 95) of the world data along with the CCFR 90% confidence limit (CL) on the rate of production of

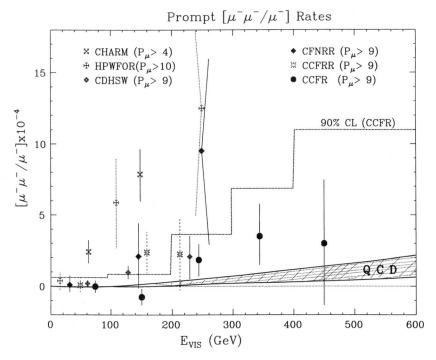

Figure 20   A compilation of the world data (69, 89–92, 95) on the rate of like-sign dimuon production with respect to single muon as a function of neutrino energy. The leading-order QCD prediction and the subsequent uncertainties are also shown (94).

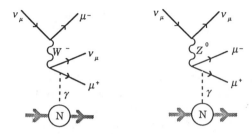

Figure 21 Feynman diagrams showing production of trileptons in neutrino scattering.

like-sign dimuons with respect to charged-current events.[6] The theoretical prediction is shown as a band signifying the uncertainty in the calculation. While like-sign dimuons no longer offer a threat to the standard model, it would be of interest to test whether the QCD prediction is correct at high energy.

5.2 *Coherent-Trilepton Production*

Neutrino-induced trilepton events are expected to be created by the coherent interaction of neutrinos in the coulomb field of the target nucleus, as illustrated in the diagrams of Figure 21. Such events would be characterized by a charged lepton pair ($\mu^- \mu^+$) and no visible hadron energy in the final state (recoilless dimuon). This old favorite of neutrino physics has been discussed in a series of theoretical papers (97, 98). An accurate measurement of the rate for this process could corroborate the destructive interference between the $W^\pm$ and the Z^0 diagrams shown in Figure 21, which is expected to create a 40% suppression relative to the rate predicted by the older V-A theory (W exchange alone). Confirmation of a related effect has been quoted in $v_e e^-$ scattering at the 2.7-standard-deviation level (44).

Two collaborations, CHARM and CCFR, have reported observation of recoilless dimuons (99). The CCFR data, shown in Figure 22, exhibit an excess of 13 dimuon events for $E_{had} < 2$ GeV, with an estimated background of 2.3 events from incoherent sources. The standard model prediction for coherent trileptons is 5.0 ± 1.5 events. The CHARM experiment

---

[6] The only data that show a serious statistical conflict with QCD are those of the CHARM group, with more than a three-standard-deviation difference from the prediction. While this might be attributed to differences in experimental cuts, like the lower muon threshold accepted in CHARM, it is unlikely that this is the source of the discrepancy. It has been pointed out (92) that the background subtraction made by the CHARM group for meson decays is substantially smaller than that made by other groups. Background sources that emanate directly from the event vertex have not yet been addressed in published results from the CHARM data.

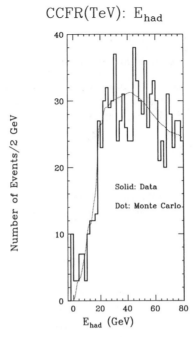

CCFR(TeV): E_{had}

Figure 22 Distribution of hadron energy CCFR (95) of $\mu^-\mu^+$ events. The excess data (*histograms*) over conventional Monte Carlo (*dotted curve*) below 2 GeV represent the recoilless dimuons.

reports similar statistical precision. Thus the available data corroborate the existence of coherent trileptons but do not yet confirm the standard model prediction of destructive interference. Both collaborations have in hand substantially increased statistical samples to address this important question.

5.3 *Search for Neutral Heavy Leptons*

Deep inelastic neutrino scattering provides a unique mechanism for producing and detecting specific types of neutral heavy leptons (η) (100), like those motivated (101) by grand unified theories, string phenomenology, or certain left-right symmetric theories. The coupling of such leptons to the ordinary neutrino is assumed to be suppressed relative to that of the ordinary weak interactions by a factor U; this suppression could be visualized as due to the mismatch of the two different lepton "types" (e.g. differing weak-isospin charges or helicities). The effect of $U < 1$ would be to reduce the production cross section and to increase the lifetime. Searches for leptons of lower mass are best conducted in hadron decay experiments:

(a) $\pi \rightarrow \eta + \mu$ at SIN (102),

(b) $K \rightarrow \eta + X$ at KEK (103a) and LBL (103b) and

(c) $D \rightarrow \eta + X$ in a beam dump experiment at CERN (104a,b).

These searches are principally limited by the mass of the parent meson, so they provide the most sensitive limits for masses below about 1.8 GeV. High energy v_μ-N experiments are most sensitive for high mass NHL searches. In this article, we concentrate upon these. The search is conducted in two channels (105):

(A) $\eta \rightarrow \mu^- + \mu^+ + X$,

with an assumed branching ratio (105, 106) of 10% and kinematics distinct from events due to conventional hadronic sources; and

(B) $\eta \rightarrow \mu^- + X$ and $\eta \rightarrow v + X$,

with assumed branching ratios of 55% and 21%, respectively (105, 106). The sensitivity occurs for those cases in which the lifetimes are long enough so that a decay vertex would appear far enough downstream of the production vertex to result in two clearly distinguished vertices.

Figure 23 shows 90% C.L. upper limits on neutral lepton production from several sources. The CCFR limit from (A) excludes masses up 18 GeV for full coupling, while the sensitivity to the coupling strength (U^2) extends below 10^{-3} for a mass of 2 GeV. The null result of the double vertex search conducted by the CHARM and CCFR collaborations are also shown in the figure.[7] New data should permit these searches to extend to masses of 22 GeV at full coupling and provide coupling limits as low as $U^2 \approx 10^{-6}$ for lower masses.

6.   SUMMARY AND CONCLUSIONS

Deep inelastic scattering has been and continues to be a fertile process for seeking out new phenomena, a vital testing ground for the predictions of the standard model, and an important laboratory for measuring many of the important parameters of the prevalent theory. Beginning with the pioneering SLAC experiments and continuing with experiments of higher precision and higher energy conducted at SLAC, FNAL, and CERN,

---

[7] These limit curves should be closed figures in this plane. The double valued nature is due to two competing effects. For a given mass, if the coupling (U) were large, η would decay too quickly to leave a distinct second vertex; alternatively, if the coupling were too small, the lepton would escape the detector without decaying. We have shown the incomplete curves quoted by the authors.

NHL Limits

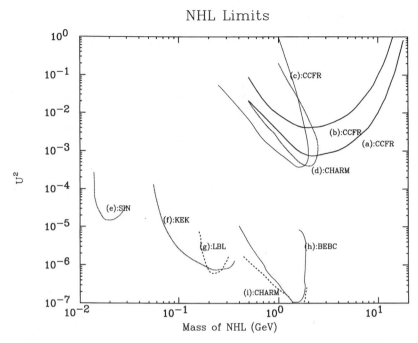

Figure 23 Upper limits (at 90% C.L.) on the neutral heavy lepton coupling, U^2, as a function of its mass by CCFR (100, 105) ν_μ-N interactions (curves *a, b, c*); CHARM (104b) neutrino interactions and beam dump (curves *d, i*); BEBC (104a) beam dump (curve *h*); LBL (103b) K decay (curve *g*); KEK (103a) K decay (curve *f*); and SIN (102) π decay (curve *e*).

these studies have demonstrated the validity of the electroweak theory and the quark structure of nucleons. Furthermore, experiments are in the process of testing the predictions of QCD. A short synopsis of the status of the various topics follows.

The universality of nucleon quark densities is verified by the data. The spin of the scattered quarks is 1/2 ($R \ll 1$). The mean-square charge of the quarks, derived from neutrino and muon scattering, is 5/18 at the 10% level. The number of valence quarks (GLS sum rule) is three to high precision. Two other sum rules, those of Adler and Gottfried, are consistent with measurements. The fundamental Bjorken sum rule, which quantifies the "spin states" of a nucleon invoking only the isospin symmetry, has not yet been tested, but the recent measurement of the proton spin structure function by EMC may well lead to a new understanding of the distribution of nucleon spin among the constituents. New experiments to measure the proton and neutron spin structure are eagerly awaited.

The several neutrino measurements of quark densities agree within their present statistical precision. The more abundant muon data, however, exhibit two outstanding inconsistencies: (a) the lack of agreement in the normalization and shape of $F_2(x, Q^2)$ between BCDMS and EMC data; and (b) the discrepancy between the valence d-quark measurement by the EMC on one hand, and by neutrino experiments (WA21, CDHS, and WA25) on the other. Even though there are systematic problems at the statistical level of these experiments, the agreement among them is still comparable to the statistical precision of the several neutrino experiments. These problems may be resolved with improved analysis and new data.

The nuclear dependence of nucleon structure function, the EMC effect, is now firmly established from various measurements using muons and electrons. Neutrino experiments will have to reach a new level of precision to distinguish between competing theoretical models. The structure functions at low and intermediate values of x scale not quite linearly with atomic weight. Also, measurements ranging more than a factor of 10 in Q^2 show no clear dependence on this parameter.

Measurements of $R = \sigma_L/\sigma_T$ are consistent with, but do not demonstrate, the predictions of perturbative QCD. The recent measurements at SLAC have clarified the low-Q^2 behavior of the parameter, as well as demonstrating its nuclear independence. More precise data at high Q^2 are needed to test QCD predictions unequivocally.

QCD predicts the x dependence of the slope of the structure function evolution with Q^2 $\{d\,[\ln F(x, Q^2)]/d\,[\ln Q^2]\}$. The simplest predictions, from xF_3, do not yet have enough statistical precision at high Q^2 to warrant a strong conclusion. The present experimental data on F_2 permit the following observations: (a) the BCDMS data from hydrogen and carbon agree with theory; (b) the CDHSW neutrino data, as well as the EMC and BFP muon data on F_2, exhibit slopes that are much steeper (at high x) than predicted. The reason for this discrepancy is not understood at present; more data at higher Q^2 may clarify this important question.

The world average of the QCD parameter from nonsinglet fits (xF_3) is $\Lambda_{QCD} = 180 \pm 60$ MeV, in agreement with the corresponding BCDMS value from singlet fits, $215 \pm 15 \pm 60$ MeV.

Study of like-sign dimuons at the Tevatron has obviated any need for invoking mechanisms beyond the standard model. Recoilless dimuons have been observed and are interpreted as trilepton production. The data are not adequately precise to demonstrate the standard model prediction of W-Z destructive interference. Limits on neutral heavy leptons have improved, and searches will continue with new data.

As described in the text and summarized above, some unsettled issues of consistency with theory and some apparent contradictions among the experiments remain to be addressed. New high statistics deep inelastic data from fixed-target experiments should substantially aid this effort, as well as exploring new regions for searching out new phenomena. Finally, with the advent of HERA, an expansion by an order of magnitude in center-of-mass energy, in momentum transfer, and in the x parameter will open. We hope that future contributions of deep inelastic scattering physics to our understanding of particles and forces will match the exciting discoveries and measurements of the past.

ACKNOWLEDGMENTS

We are pleased to thank our colleagues in the CCFR collaboration for much assistance and many helpful conversations. Special help was forthcoming from A. Bodek, and E. Oltman. Communications with F. Close, J. Ellis, P. Herczeg, W. Marciano, A. Mueller, and W.-K. Tung are also gratefully acknowledged. We especially would like to thank A. Caldwell, E. Hyatt, and S. Ritz for critical reading of the draft manuscript and many helpful suggestions. The authors are supported by funds from the National Science foundation.

Literature Cited

1. Hofstadter, R., *Electron Scattering & Nuclear and Nucleon Scattering*. New York: Benjamin (1963)
2. Gell-Mann, M., *Phys. Lett.* 8: 214 (1964); Zweig, G., *CERN Reps. 8182/TH 401, 8419/TH 412* (1964) (unpublished)
3. Miller, G., et al., *Phys. Rev.* D5: 528 (1972); Bodek, A., et al., *Phys. Rev. Lett.* 30: 1087 (1973)
4. Bjorken, J. D., Paschos, E. A., *Phys. Rev.* 185: 1975 (1969)
5. Feynman, R. P., *Phys. Rev. Lett.* 23: 1415 (1969); *Photon-Hadron Interactions.* Reading, Mass: Benjamin (1972)
6. Eichten, T., et al., *Phys. Lett.* 46B: 274 (1973)
7. Bardeen, W. A., Fritzsch, H., Gell-Mann, M., in *Scale and Conformal Symmetry in Hadron Physics*, ed. R. Gatto. New York: Wiley (1973), p. 130; Gross, D. J., Wilczek, F., *Phys. Rev.* D8: 3633 (1973); Weinberg, S., *Phys. Rev. Lett.* 31: 494 (1973)
8. Weinberg, S., *Phys. Rev. Lett.* 19: 1264 (1967); Salam, A., In *Elementary Particle Theory: Relativistic Groups and Analyticity (Nobel Symp. No. 8)*, ed.

N. Svartholm. Stockholm: Almqvist & Wiksell (1968), p. 367
9. Hasert, F. J., et al., *Phys. Lett.* 46B: 138 (1973)
10. Barish, B. C., et al., *Proc. 17th Int. Conf. on High Energy Physics*, London (July 1974), p. 111; Barish, B. C., et al., *Phys. Rev. Lett.* 34: 538 (1975)
11. Prescott, C. Y., et al., *Phys. Lett.* 84B: 524 (1979)
12a. Fogli, G. L., *Riv. Nuovo Cimento* 9(8): 1–67 (1986)
12b. Sirlin, A., in *Proc. Int. Symp. on Lepton-Photon Interactions at High Energy*, Hamburg (1987)
12c. Amaldi, U., et al., *Phys. Rev.* D36: 1385 (1987)
13. Perkins, D. H., *Introduction to High Energy Physics*. Reading, Mass: Addison-Wesley (1982), pp. 287–305; Fisk, H. E., Sciulli, F., *Annu. Rev. Nucl. Part. Sci.* 32: 499–573 (1982)
14. Callan, C. G., Gross, D. J., *Phys. Rev. Lett.* 22: 156 (1969)
15. Eichten, E., et al., *Rev. Mod. Phys.* 56: 579 (1984); Altarelli, G., et al., *Nucl. Phys.* B246: 12 (1984); Halzen, F., "Top Search." *Phys. in Collision* 1987: 275 [*QCD* 161: 1542 (1987)]; Albajar,

C., et al., *Phys. Lett.* B198: 271 (1987); Ansari, R., et al., *Phys. Lett.* B194: 158 (1987)

16. Stirling, W. J., Summary of the Results from Study Group 2—QCD at HERA, *Proc. of the HERA Workshop*, ed. R. D. Peccei. DESY, Hamburg (Oct. 1987), 1: 185–207

17. Deleted in proof

18. Eichten, T., et al., *Phys. Lett.* 40B: 593 (1972)

19. Barish, B. C., et al., see Ref. 10, p. 105; Barish, B. C., et al., *Phys. Rev. Lett.* 35: 1316 (1975)

20. Gross, D. J., Wilczek, F., *Phys. Rev.* D9: 980 (1974); Georgi, H., Politzer, H. D., *Phys. Rev.* D9: 416 (1974)

21. Gross, D. J., Llewellyn Smith, C. H., *Nucl. Phys.* B14: 337 (1969)

22. Beg, M. A. B., *Phys. Rev.* D11: 1165 (1975)

23. Adler, S. L., *Phys. Rev.* 143: 1144 (1966)

24. Gottfried, K., *Phys. Rev. Lett.* 18: 1154 (1967)

25. Bjorken, J. D., *Phys. Rev.* 148: 1467 (1966); *Phys. Rev.* D1: 1376 (1970)

26. Lee, T. D., *Particle Physics and Introduction to Field Theory.* New York: Harwood Academic (1981), pp. 544–73

27. Wilczek, F., *Annu. Rev. Nucl. Part. Sci.* 32: 177 (1982)

28. Söding, P., Wolf, G., *Annu. Rev. Nucl. Part. Sci.* 31: 231 (1981); Wu, S. L. See Ref. 12b, p. 39

29. Altarelli, G., Parisi, G., *Nucl. Phys.* B26: 298 (1977)

30. Altarelli, G., Martinelli, G., *Phys. Lett.* 76B: 89 (1978); Glück, M., Reya, E., *Nucl. Phys.* B145: 24 (1978)

31. Friedman, J. I., Kendall, H., *Annu. Rev. Nucl. Part. Sci.* 22: 203 (1972)

32a. Meyers, P. D., et al., *Phys. Rev.* D34: 1265 (1986)

32b. Allkofer, O. C., et al., *Nucl. Instrum. Methods* 179: 445 (1981)

32c. Aubert, J. J., et al., *Nucl. Phys.* B272: 158 (1986)

33a. Bollini, D., et al., *Nucl. Instrum. Methods* 204: 333 (1983)

33b. Benvenuti, A. C., et al., *Phys. Lett.* B195: 91 (1987)

33c. Benvenuti, A. C., et al., *Phys. Lett.* B195: 97 (1987)

34a. Holder, M., et al., *Nucl. Instrum. Methods* 148: 235 (1978); Merritt, F. S., et al., *Nucl. Instrum. Methods* A245: 27 (1986)

34b. Womersley, W. J., et al., *Nucl. Instrum. Methods* A267: 49 (1988)

35. De Winter, K., et al., *CERN-EP/88-87* (July 1988)

36. Blair, R., et al., *Nucl. Instrum. Methods* 226: 281 (1984)

37. Blair, R., *Proc. 12th Int. Conf. on Neutrinos and Astrophysics*, Sendai, ed. T. Kitagaki, H. Yuta, Singapore (1986), p. 351; Blair, R., PhD Thesis, Calif. Inst. Technol. (Apr. 1982); Auchincloss, P. S., PhD Thesis, Columbia Univ. (1987); Belusevic, R., Reine, D., *Phys. Rev.* D38: 2753 (1988)

38. ZEUS Collaboration, *The ZEUS Detector Technical Proposal*, DESY (Mar. 1986); H1 Collaboration, *Technical Proposal for the H1 Detector*, DESY (Mar. 1986)

39. Tung, W.-K., et al., *Proc. 1988 Snowmass Summer Study on HEP in the 1990's*, Snowmass, CO (to be published)

40. Bartels, J., et al., see Ref. 16, 2: 795–903; Smith, W. H., et al., see Ref. 39

41. Martyn, H., see Ref. 16, 2: 801–12

42. Aitchison, I., Hey, A., *Gauge Theories in Particle Physics.* Bristol: Hilger (1980), p. 238

43a. Mattison, T. S. PhD Thesis, MIT (Dec. 1986)

43b. Taylor, F., in *Proc. Neutrino—88*, Boston (1988), to be published

44. Winter, K., see Ref. 43b; Allaby, J. V., et al., *Phys. Lett.* B213: 554 (1988)

45. Oltman, E., et al., PhD thesis, Columbia Univ. (Mar. 1989) presented at DPF88, Storrs, CN, to be published

46. Vallage, B. PhD Thesis, Saclay CEA-IV-2513 (Jan. 1987); to be published

47. Sciulli, F., in *Proc. Int. Symp. on Lepton and Photon Interactions at High Energies*, Kyoto, Japan (Aug. 1985), pp. 8–48

48. Iijima, B. A., *MIT Preprint CTP993* (1983)

49. MacFarlane, D. B., et al., *Z. Phys.* C26: 1 (1984)

50. deGroot, J. G. H., et al., *Phys. Lett.* B82: 292 (1979); Bergsma, F., et al., *Phys. Lett.* B123: 269 (1983)

51. Allasia, D., et al., *Phys. Lett.* B135: 231 (1984); *Z. Phys.* C28: 321 (1985)

52a. Sloan, T., in *Proc. Int. Europhysics Conf. on HEP*, Uppsala, Sweden (1987)

52b. Aubert, J. J., et al., *Nucl. Phys.* B293: 740 (1987)

53. Ashman, J., et al., *Phys. Lett.* B206: 364 (1988)

54. Alguard, M. J., et al., *Phys. Rev. Lett.* 37: 1261 (1976); 41: 70 (1978); Baum, G., et al., *Phys. Rev. Lett.* 51: 1135 (1983)

55. Close, F. E., Roberts, R. G., *Phys. Rev. Lett.* 60: 1471 (1988); Close, F. E., in *Proc. Spin 88*, Univ. Minn., Minneapolis (Sept. 1988)

56. Ellis, J., Jaffe, R. L., *Phys. Rev.* D9: 1444 (1984)
57. Anselmino, M., Leader, E., *Preprint NSF-ITP-88-142* (1988); Anselmino, M., Ioffe, B. L., Leader, E., *Preprint NSF-ITP-88-94* (1988); Efremov, A. V., Teryaev, O. V., *JINR Rep. E2-88-287* (1988); Altarelli, G., Ross, G. G., *Phys. Lett.* B212: 391 (1988); Carlitz, R. D., Collins, J. C., Mueller, A. H., *Phys. Lett.* B214: 229 (1988)
58. Brodsky, S., Ellis, J., Karliner, M., *Phys. Lett.* 206B: 309 (1988); Ellis, J., Karliner, M., *Phys. Lett.* 213: 73 (1988)
59. Kaplan, D. B., Manohar, A., *Harvard Preprint HUTP-88/A024* (May 1988)
60a. Abramowicz, H., et al., *Z. Phys.* C25: 29 (1984)
60b. Jones, G. T., et al., *Preprint 87*, Rec. Jul. (1987)
61. Berge, J. P., et al., *Z. Phys.* C35: 443 (1987)
62a. Bodek, A., et al., *Phys. Rev.* D20: 1471 (1979)
62b. Aubert, J. J., et al., *Nucl. Phys.* B259: 189 (1985)
63. Tung, W.-K., et al., see Ref. 39
64. Whitlow, L., et al., *Univ. Rochester Preprint UR1102* (Jan. 1989), submitted to *Phys. Rev. Lett.*
65. DeRujula, A., et al., *Rev. Mod. Phys.* 46: 391 (1974)
66. Barnett, R. M., *Phys. Rev. Lett.* 36: 1163 (1976); *Phys. Rev.* D14: 70 (1976)
67. Georgi, H., Politzer, H. D., *Phys. Rev.* D14: 1829 (1976)
68. Foudas, F., et al., presented at DPF88, Storrs, CN (1988), to be published; Mishra, S., et al., invited talk presented at 14th Rencontres de Moriond, Les Arc (Mar. 1989), to be published
69. Lang, K., et al., *Z. Phys.* C33: 483 (1987)
70. Abramowicz, H., et al., *Z. Phys.* C17: 19 (1982)
71. Ushida, N., et al., *Phys. Lett.* 121B: 292 (1983)
72. Baltrusaitis, R. M., et al., *Phys. Rev. Lett.* 54: 1976 (1985)
73. Kleinknecht, K., Renk, B., *Z. Phys.* C34: 209 (1987)
74. Aubert, J. J., et al., *Phys. Lett.* B123: 275 (1983)
75. Bodek, A., et al., *Phys. Rev. Lett.* 50: 1431 (1983); 51: 534 (1983)
76. Berger, E. L., Coester, F., *Annu. Rev. Nucl. Part. Sci.* 37: 463 (1987)
77a. Ashman, J., et al., *Phys. Lett.* B202: 603 (1988)
77b. Benvenuti, A. C., et al., *Phys. Lett.* B189: 483 (1987)
78. Mueller, A. H., Qiu, J., *Nucl. Phys.* B268: 427 (1986); Qiu, J., *Nucl. Phys.*

B291: 46 (1987); Berger, E. L., Qiu, J., *Phys. Lett.* B206: 141 (1988)
79. Dasu, S., et al., *Phys. Rev. Lett.* 60: 2591 (1988)
80. Parker, M. A., et al., *Nucl. Phys.* B232: 1 (1984); Cooper, A. M., et al., *Phys. Lett.* B141: 133 (1984); Guy, J., et al., *Z. Phys.* C36 (1987); Abramowicz, H., et al., *Z. Phys.* C25: 20 (1984); Hanlon, J., et al., *Phys. Rev.* D32: 2441 (1985)
81. Mestayer, M. D., et al., *Phys. Rev.* D27: 285 (1983)
82. Poucher, J. S., et al., *Phys. Rev. Lett.* 32: 118 (1974); Riordan, E. M., et al., *Phys. Rev. Lett.* 33: 561 (1974); Stein, S., et al., *Phys. Rev.* D12: 1884 (1975); Atwood, W. B., et al., *Phys. Lett.* 64B: 497 (1976)
83. Dasu, S., et al., *Phys. Rev. Lett.* 61: 1061 (1988)
84. Deleted in proof
85. Mishra, S., Sciulli, F., *Columbia Univ. Preprint* "Do present data demonstrate $R = R_{QCD}$?" (Feb. 1989)
86. Buchholz, P., in *Proc. Int. Europhysics Conf. on High Energy Physics, Bari*, ed. L. Nitti, G. Preparata, Eur. Phys. Soc., Switzerland (1986)
87. Aubert, J. J., et al., *Nucl. Phys.* B259: 189 (1985)
88. Benvenuti, A. C., et al., *CERN-EP/88*, submitted to *Phys. Lett.*
89. Jonker, M., et al., *Phys. Lett.* B107: 241 (1981)
90. Trinko, T., et al., *Phys. Rev.* D23: 1889 (1981)
91. Nishikawa, K., et al., *Phys. Rev. Lett.* 46: 1555 (1981); 54: 1336 (1985)
92. Burkhardt, H., et al., *Z. Phys.* C31: 39 (1986)
93. Barger, V., et al., *Phys. Rev.* D17: 2284 (1977); Barnett, M., Chang, L. M., *Phys. Lett.* B72: 223 (1977); Smith, J., Vermaseren, J. A. M., *Phys. Rev.* D17: 2288 (1977)
94. Cudell, J. R., et al., *Phys. Lett.* B175: 227 (1986)
95. Schumm, B., et al., *Phys. Rev. Lett.* 60: 1618 (1988)
96. Deleted in proof
97. Kozhushner, M. A., Shabalin, E. P., *Sov. Phys. JETP* 14: 676 (1962); Czyz, W., et al., *Nuovo Cimento* 34: 404 (1964); Fujikawa, F., *Ann. Phys.* 68: 102 (1971)
98. Lovseth, J., Radomski, M., *Phys. Rev.* D3: 2686 (1971); Brown, R. W., et al., *Phys. Rev.* D6: 3273 (1972); Belusevic, R., Smith, J., *Phys. Rev.* D37: 2419 (1988)
99. Bergsma, F., et al., *Phys. Lett.* B122: 185 (1983); Mishra, S. R., et al., *Proc. 24th Int. Conf. on HEP*, Munich (Aug.

1988), Springer-Verlag, presented by M. Oreglia, p. 924

100. Mishra, S. R., see Ref. 43b
101. Pati, J. C., Salam, A., *Phys. Rev.* D10: 275 (1974); Mohapatra, R. N., Pati, J. C., *Phys. Rev.* D11: 566 (1975), D11: 2558 (1975); Fritzsch, J., et al., *Phys. Lett.* B59: 256 (1975); Mohapatra, R. N., Senjanovic, G., *Phys. Rev. Lett.* 44: 912 (1980); Gronau, M., Leung, C. N., Rosner, J., *Phys. Rev.* D29: 2539 (1984); Witten, E., *Phys. Lett.* B91: 81 (1980); Bjorken, J. D., Llewellyn Smith, C. H., *Phys. Rev.* D7: 887 (1973)
102. Abela, R., et al., *Phys. Lett.* B105: 263 (1981)
103a. Yamazaki, T., *Proc. of 22nd Int. Conf. on High Energy Physics*, Leipzig (1984), 1: 262
103b. Pang, C. Y., et al., *Phys. Rev.* D8: 1989 (1973)
104a. Cooper-Sarkar, A. M., et al., *Phys. Lett.* B160: 207 (1985)
104b. Dorenbosch, J., et al., *Phys. Lett.* B166: 473 (1986)
105. Mishra, S. R., et al., *Phys. Rev. Lett.* 59: 1397 (1987)
106. Gronau, M., et al., *Phys. Rev.* D29: 2539 (1984); Bartel, W., et al., *Phys. Lett.* 123B: 353 (1983)

Annu. Rev. Nucl. Part. Sci. 1989. 39: 311–56

MUON-CATALYZED FUSION

W. H. Breunlich and P. Kammel

Institute for Medium Energy Physics, Austrian Academy of Sciences, Boltzmanngasse 3, A-1090 Vienna, Austria

J. S. Cohen and M. Leon

Los Alamos National Laboratory, Los Alamos, New Mexico 87545, USA

KEY WORDS: mesomolecules, mesic atoms, muonic molecules, muonic hydrogen, exotic atoms, cold fusion.

CONTENTS

1. INTRODUCTION

While examining alternatives to the existence of two kinds of mesons demonstrated by Lattes, Occhialini, and Powell in their emulsion experiment, Frank (1) arrived at the rather fantastic idea that negative muons could catalyze the fusion of hydrogen isotopes. Shortly afterward, Sakharov (2) discussed catalysis and energy release in deuterium. The same idea of muon-catalyzed fusion (μCF) was then proposed and amplified

311

0163–8998/89/1201–0311$02.00

upon by Zeldovich (3) a few years later. These theoretical extravagances were, however, completely unknown to Alvarez and his coworkers (4) when in 1956 they discovered, and correctly interpreted, pd fusions being catalyzed by negative muons in the Berkeley hydrogen bubble chamber. This accidental discovery triggered a burst of experimental and theoretical activity (5, 6) examining the physics of muon-catalyzed pd and dd fusion in some detail.

After the early 1960s, interest in this area was primarily motivated by weak interaction studies. In order to measure and interpret unambiguously the basic nuclear capture process ($\mu^- p \to n\nu$) in hydrogen and deuterium, the different atomic and molecular processes induced by μ^- in hydrogen mixtures had to be understood (7). Surprisingly, one of these "background" reactions showed a puzzling behavior: the rate for forming the $dd\mu$ "mesomolecule" (muonic molecular ion) that precedes the fusion was found experimentally to be large and to have a strong temperature dependence (8). The previously calculated Auger process (6), in which the energy released upon binding (typically tens to hundreds of eV) is carried off by an ejected electron, is more than an order of magnitude slower and is not strongly energy dependent. To resolve this discrepancy, Vesman (9) in 1967 suggested that the $dd\mu$ muonic molecule possesses a very loosely bound state, so that the released energy is small enough to go into vibration and rotation of the resulting "compound-molecule" (one "nucleus" of which is the $dd\mu$ mesomolecule). This is necessarily a resonant process, and implies significant temperature dependence coming from the overlap of the Maxwell distribution of initial $d\mu + D_2$ energies with the resonance energy. Once the mesomolecule is formed, its small size leads to rapid fusion (to zeroth order, the linear dimension scales as m_e/m_μ).

A decade of theoretical labor by Ponomarev and his collaborators (10, 11) established that such a loosely bound state does indeed exist for the $dd\mu$ and also for the $dt\mu$ system. Then in 1977 Gershtein & Ponomarev (12) predicted that because of the resonant molecular formation mechanism of Vesman, the μ^- could catalyze $\sim 10^2$ dt fusions in a dense deuterium-tritium target! This startling prediction did much to trigger the present upsurge of interest in muon-catalyzed fusion; experimental programs were started first at JINR,[1] then at PSI,[2] LNPI,[3] and LAMPF.[4] By now experi-

[1] The Joint Institute for Nuclear Research, Dubna.

[2] Austrian Academy of Sciences (OeAW)–Paul Scherrer Institute (PSI), presently OeAW–PSI–Lawrence Berkeley Laboratory–Los Alamos National Laboratory (LANL)–Technical University of Munich collaboration.

[3] Leningrad Nuclear Physics Institute.

[4] Idaho National Engineering Laboratory (INEL)–LANL, presently Brigham Young University–California State University at Los Angeles–INEL–LANL–University of Mississippi collaboration.

ments are being carried out at seven laboratories around the world, and theoretical efforts are even more widespread.

There are two basic motives for this research: First, the resonance process allows the study of some beautiful phenomena spanning several disciplines (atomic, molecular, nuclear, and particle physics) united in μCF. In particular, many facets of the Coulomb three-body problem involving muons and hydrogen isotopes can be explored with unprecedented accuracy. Second, the large number of fusions catalyzed by a single muon has revived old dreams of the practical application of this remarkable process.

For ddμ, an unanticipated and strong dependence of the resonant molecular formation rate on the dμ hyperfine state was found (8a); this allows the precise determination of the rate for dμ hyperfine transitions, important for weak interaction experiments (13, 14). Measurements of the temperature dependence of molecular formation have rigorously tested most elements that enter the calculation of resonant formation; in particular, the ddμ binding energy was determined experimentally with extreme accuracy. Another surprising feature of ddμ fusion is that the branching ratio of the fusion channel (n + ^{3}He) to the (p + t) channel is not one, as would seem to be required by charge symmetry, but 1.4.

For dtμ, the prediction of $\sim 10^2$ fusions per μ^- has been abundantly verified, and some new and surprising phenomena have been uncovered. For example, a totally unexpected three-body contribution to resonant dtμ formation has been found. Furthermore, the loss of muons due to sticking to the alpha particles produced in the fusion reaction has turned out to be significantly less than predicted just a few years ago, and it remains less than even the most recent precise calculations. So far, as many as 150 fusions per muon have been observed.

This review summarizes our current knowledge of the basic physics of μCF, which has advanced so dramatically in the last few years (for other reviews, see 15). The basic aspects are clearly visible in ddμ fusion, where the combination of scrupulous experimental and theoretical work has come close to yielding a complete understanding. In the dtμ system, we are still in an exploratory stage and vigorous research will be needed for some years to come.

2. OVERVIEW OF THE CATALYSIS CYCLE

The steps of the catalysis cycle are sketched in Figure 1 for a dt target. The slowing down, stopping, and initial capture into an atomic bound state, and the subsequent deexcitation of that atomic system are all very fast in a dense hydrogen target ($\sim 10^{-11}$ s) and proceed mainly by ejecting

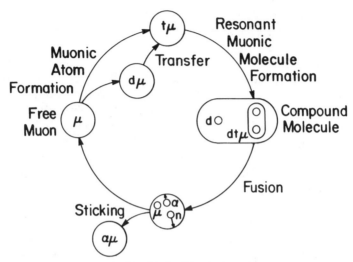

Figure 1   Simplified μCF cycle for a dt target.

electrons from target molecules (16). The μ^- then finds itself in the ground state of a dμ or tμ atom. For the dμ's, collision with a triton inside a DT or T_2 molecule will eventually result in transfer of the muon to the more tightly bound tμ ground state (Section 3). Then in a collision with a D_2 or DT molecule, the tμ can form the compound molecule, [(dtμ)dee]* or [(dtμ)tee]*, via the Vesman resonant molecular formation mechanism (9). Since these last two are the slowest steps of the cycle, they determine the cycling rates (Section 4).

The compound molecule is initially formed with the $(dt\mu)_{Jv}$ in the angular momentum $J = 1$ and vibrational quantum number $v = 1$ state, but de-excitation via one or more Auger transitions to a $J = 0$ state is rapid ($\sim 10^{-12}$ s). In this state fusion occurs in $\sim 10^{-12}$ s (17), because the negative muon with its large mass ($m_\mu \approx 207 m_e$) screens most of the repulsive Coulomb barrier between the two nuclei. Then because the recoil velocity of the α particle produced in the fusion reaction d+t $\rightarrow$ α+n is so large (~ 6 au), the μ^- is usually freed (with $\gtrsim 99\%$ probability) to go around the cycle again (Section 5). The rates[5] and probabilities that characterize these steps are displayed in Table 1. Competing with these processes is the decay of the muon, with rate $\lambda_0 = 0.455 \times 10^6$ s^{-1}.

---

[5] As is conventional, in this article all collisional rates and times are normalized to liquid hydrogen density $N_0 = 4.25 \times 10^{22}$ atoms/cm^3, and density ϕ is expressed relative to this value.

3. MUONIC ATOM PROCESSES

3.1 *Muonic Atom Formation and Deexcitation*

The μCF process starts with a free muon, injected into a mixture of hydrogen isotopes, being stopped to form a muonic atom. The slowing and capture occur primarily by ionization, e.g.

$$\mu + D \rightarrow \begin{cases} \mu + d + e \\ d\mu(n) + e \end{cases} \qquad\qquad 1.$$

and it is now known that the muon is slowed to an energy of ~ 10 eV, comparable to the target ionization potential, and then captured into an orbital with $n \approx \sqrt{m_\mu/m_e} \simeq 14$, which has about the same size and energy as that of the displaced electron (18).

Experimental information on the atomic capture process is very limited; generally only the last few x rays in the cascade are observed. The time for capture of a μ^- starting with a kinetic energy of ~ 2 keV ($v \approx 1$ au) has been observed in a low-density target (19) and is in reasonable agreement with the modern theoretical calculations (18, 20, 21). The capture time (reduced to liquid hydrogen density) is ~ 0.3 ps, which is quite negligible compared with the μCF cycle time.

Following the initial Coulomb capture, the muon cascades to lower levels. The cascade, which depends on the initial capture orbital and the target density, occurs via a variety of Stark mixing, Auger, inelastic, and radiative processes. Though the cascade is complicated and not well understood, it is known to be rapid. It takes only about 14 ps in liquid hydrogen, mostly for the final transitions $n = 3 \rightarrow 2$ and $2 \rightarrow 1$ (16), and is still much shorter than the μCF cycle time. However, this does not necessarily mean that the atomic capture and cascade processes are irrelevant to the quantitative understanding of the main μCF cycle. We mention three effects that require attention: (*a*) Some of the possible steps in the cascade yield hot muonic atoms. (*b*) Though the initial capture is expected to be independent of the isotope to an excellent approximation, the molecular dissociation that occurs early in the cascade is not. Pion experiments have shown that capture in HD has a significantly different consequence than that in H_2/D_2 mixtures (22). A similar situation can be expected in the $D_2/DT/T_2$ mixtures used for dt μCF. (*c*) Even though the excited state cascade is very rapid, the rates of muon transfer from d to t in excited states are competitive.

3.2 *Scattering and Thermalization*

Until recently the dμ atom was expected to reach its ground state and be thermalized before the μ was transferred to t,

Table 1   Parameters of muon-catalyzed fusion in d–t mixtures at liquid hydrogen density [a]

Process	Symbol	30K	300K	Source [b]
Atomic capture rates (same for dμ and tμ)				
$\mu^- + D_2 \rightarrow d\mu^* + \cdots$	λ_a	4×10^{12} s^{-1}	c	ET
Atomic cascade rates (same for dμ and tμ)				
$d\mu^* \rightarrow \cdots \rightarrow d\mu(1s)$	λ'_a	7×10^{10} s^{-1}	c	T
Isotopic exchange rates				
$d\mu(1s) + t \rightarrow t\mu(1s) + d$	λ_{dt}	2.7×10^{8} s^{-1}	2.8×10^{8} s^{-1}	ET
Probability of dμ reaching 1s state	q_{1s}	See text		ET?
Hyperfine transition rates				
$d\mu(\uparrow\uparrow) + d \rightarrow d\mu(\uparrow\downarrow) + d$	$\lambda_{d\mu}^{3/2,1/2}$	3×10^{7} s^{-1}	4×10^{7} s^{-1}	ET [d]
$t\mu(\uparrow\uparrow) + t \rightarrow t\mu(\uparrow\downarrow) + t$	$\lambda_{t\mu}^{1,0}$	1.3×10^{9} s^{-1}	1.3×10^{9} s^{-1}	T
Resonant hyperfine transition rates				
$d\mu(\uparrow\uparrow) + D_2 \rightarrow (dd\mu)dee \rightarrow d\mu(\uparrow\downarrow) + D_2$	$\lambda_{d\mu(res)}^{3/2,1/2}$	1×10^{7} s^{-1}	$<1\times10^{7}$ s^{-1}	ET [d]
Muonic molecule formation rates [e]				
$d\mu(\uparrow\downarrow) + D_2 \rightarrow (dd\mu)dee$	$\lambda_{(dd\mu)d}^{1/2}$	5×10^{4} s^{-1}	3×10^{6} s^{-1}	ET
$d\mu(\uparrow\uparrow) + D_2 \rightarrow (dd\mu)dee$	$\lambda_{(dd\mu)d}^{3/2}$	4×10^{6} s^{-1}	4×10^{6} s^{-1}	ET
$t\mu(\uparrow\downarrow) + D_2 \rightarrow (dt\mu)d$	$\lambda_{(dt\mu)d}^{0}$	4×10^{8} s^{-1}	4×10^{8} s^{-1}	E

$t\mu(\uparrow\uparrow) + D_2 \rightarrow (dt\mu)dee$	$\lambda^1_{(dt\mu)d}$	3×10^4 s⁻¹ f	2×10^6 s⁻¹	T
$t\mu(\uparrow\downarrow) + DT \rightarrow (dt\mu)tee$	$\lambda^0_{(dt\mu)t}$	$\sim10^7$ s⁻¹	1×10^8 s⁻¹	E
$t\mu(\uparrow\uparrow) + DT \rightarrow (dt\mu)tee$	$\lambda^1_{(dt\mu)t}$	3×10^4 s⁻¹ f	1×10^8 s⁻¹	T
$t\mu + T_2 \rightarrow (tt\mu)te + e$	$\lambda_{tt\mu}$	2×10^6 s⁻¹	3×10^6 s⁻¹	E g

Fusion rates

$dd\mu(J=1) \rightarrow \begin{cases} n + (^3He + \mu) \\ p + t + \mu \end{cases}$ n/p branching ratio	$\lambda^f_{dd\mu}$ Y_n/Y_p	4×10^8 s⁻¹ 1.4	c c	ET ET
$dt\mu(J=0) \rightarrow n + (\alpha + \mu)$	$\lambda^f_{dt\mu}$	1.3×10^{12} s⁻¹	c	T
$tt\mu(J=1) \rightarrow n + n + (\alpha + \mu)$	$\lambda^f_{tt\mu}$	1.5×10^7 s⁻¹	c	E

Sticking fractions

$dd\mu \rightarrow \mu^3He + n$	ω_d	0.12	c	ET
$dt\mu \rightarrow \alpha\mu + n$	ω_s	0.0043	c	ET?
$tt\mu \rightarrow \alpha\mu + n + n$	ω_t	0.14	c	E

a Numerical values are authors' *estimates* based on published values.
b E≡experimental, T≡theoretical, ?≡substantial disagreement between E and T.
c No significant temperature dependence.
d The experimental *total* hyperfine transition rate for $d\mu(\uparrow\uparrow)$ at 30K is 3.7×10^7 s⁻¹.
e As conventional, the physical $dt\mu$ formation rates are given by $c_d c_x \phi$ times the rates given (x=d or t).
f Nonresonant.
g Applies to DT as well.

$$d\mu(1s)+t \rightarrow t\mu(1s)+d+48 \text{ eV}. \qquad\qquad 2.$$

Theory and experiment appear to be in agreement that the rate for this reaction is $\sim 2.8 \times 10^8 \text{ s}^{-1}$ (however, see discussion in Section 4.3) and has only a weak temperature dependence for $T < 1000$ K (for references, see Table 5). Likewise the resulting $t\mu$ atom, though it would be given 19 eV by the transfer, was expected to be thermalized before forming $dt\mu$. Menshikov & Ponomarev (24) pointed out that the rates for transfer in excited states

$$d\mu(n)+t \rightarrow t\mu(n)+d+\frac{48}{n^2} \text{ eV} \qquad\qquad 3.$$

are large enough for $n = 2$–5 to compete with the deexcitation rates for $d\mu$, and consequently there is only some probability $q_{1s} < 1$ for $d\mu$ to reach the 1s state. The resulting value of q_{1s} was found to have a strong dependence on c_t and ϕ (see Figure 2). In further theoretical work (25) they predicted that three-body collisions would reduce the already small energy defect and enhance the transfer rates. Existing μCF experiments (26, 27), though they do not determine q_{1s} directly, indicate a much weaker dependence on ϕ and c_t. The above theoretical values of q_{1s} were calculated assuming that the $d\mu(n)$ atoms are thermalized. If the $d\mu(n)$ atoms are

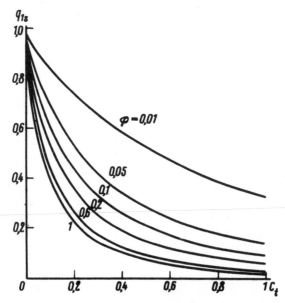

Figure 2   Theoretical $q_{1s}(c_t)$ values for fixed ϕ (from 24). The quantitative behavior of q_{1s} is still uncertain since experiments suggest a much weaker dependence of q_{1s} on ϕ and c_t.

actually hot ($\gg 0.1$ eV), then the transfer rates, especially those corresponding to the three-body mechanism, may be significantly smaller (28). The cascade and transfer may also be complicated by the $d\mu$ sometimes being bound to a normal molecule (29).

Experimental evidence for muonic atom energy distributions is mostly indirect; it is important to calculate them theoretically. These distributions depend on the initial energy, which is a product of the cascade, and the elastic scattering cross sections. For many years the elastic scattering of muonic atoms by hydrogenic nuclei has been calculated in the adiabatic representation of the three-body problem (e.g. 30). The actual targets, however, are molecules and it is now known that the effects of electronic screening and molecular binding can be quite large at low collision energies (31). Calculations of the elastic cross sections for ground- and excited-state $d\mu$ and $t\mu$ with all the required isotopic molecules are just under way. So far, the time dependence of the $t\mu(1s)$ energy distributions has been calculated using the available cross sections for bare nuclear targets (32). It is found that the energy distributions after a very short time depend only weakly on the initial energy as long as this is $\gtrsim 1$ eV. This condition appears to be verified by two recent experiments: One measuring $d\mu$ diffusion at low pressures (47–750 mbar) indicates a mean initial energy of 1.8 eV (33). Another measuring the neutron energy distribution from the $\pi^- p \rightarrow \pi^0 n$ reaction in liquid hydrogen finds surprisingly high mean kinetic energy, 16.2 ± 1.3 eV for the $p\pi$ $n = 3, 4$ states where this reaction usually occurs (34); some of the corresponding kinetic energy for $d\mu$ and $t\mu$ might remain when the ground state is reached.

Another relevant muonic atom cross section is that for hyperfine quenching. Theory suggests that the spin flip is accomplished by charge exchange,

$$d\mu(\uparrow\uparrow) + d \rightarrow d + d\mu(\uparrow\downarrow) + 0.049 \text{ eV} \qquad \text{4a.}$$

$$t\mu(\uparrow\uparrow) + t \rightarrow t + t\mu(\uparrow\downarrow) + 0.24 \text{ eV}. \qquad \text{4b.}$$

For $d\mu$ quenching, experiments and theory have yielded rates in the range $(3–5) \times 10^7$ s^{-1} for $T \lesssim 300$ K (8a, 35). Theory gives 1.3×10^9 s^{-1} for $t\mu$ quenching (37).

4.   MUONIC MOLECULE PROCESSES AND μCF CYCLES

4.1   Basic Theory of Resonant Molecular Formation

4.1.1   ENERGY LEVELS   The calculated Coulomb binding energies of all the muonic molecules are shown in Table 2. Early calculations were unable to demonstrate the existence of the crucial weakly bound states of $dd\mu$ and

Table 2 Coulomb molecular binding energies[a] (in eV)

J, v	ppμ	pdμ	ptμ	ddμ	dtμ	ttμ
0, 0	253.15	221.55	213.84	325.07	319.14	362.91
0, 1	—	—	—	35.84	34.83	83.77
1, 0	107.27	97.50	99.13	226.68	232.47	289.14
1, 1	—	—	—	1.97[b]	0.66[c]	45.21
2, 0	—	—	—	86.45	102.65[d]	172.65
3, 0	—	—	—	—	—	48.70

[a] The $J = 0$ and 1 energies come from Alexander & Monkhorst (40a), the $J = 2$ energies from Frolov (139), and the $J = 3$ energy from Vinitsky (138). These are not necessarily the only publications giving the listed value. For an extensive compilation, see (38).
[b] The accurate energy is 1.9749 eV.
[c] Also (40b). The accurate energy is 0.6603 eV.
[d] Also (140).

dtμ with $J = 1$, $v = 1$. For example, in the Born-Oppenheimer (fixed-nuclei) approximation the state is much too bound, but if adiabatic corrections are included it is not bound at all! The first successful calculations were made by Vinitsky et al (for a review, see 38) by expanding the wave function in a Born-Oppenheimer basis and solving the coupled differential equations; that type of calculation has become known as the "adiabatic representation" though it is truly a nonadiabatic treatment (sometimes called the method of "perturbed stationary states"). Bhatia & Drachman (39) proved variationally that the $J = 1$, $v = 1$ state of dtμ is bound, and Hu (40) shortly afterward demonstrated that very accurate variational calculations were feasible. These calculations use interparticle coordinates (Hylleraas-type wave functions). Subsequently the Coulomb binding energies have been variationally determined with an accuracy better than 0.1 meV, far surpassing the coupled-equation method. The accurate values for the $J = 1$, $v = 1$ states are 1.9749 eV for ddμ and 0.6603 eV for dtμ (40a,b). Beyond this level of accuracy, the energies may be affected by uncertainties in fundamental constants (see 40b).

However, there are corrections to these Coulomb energies that are very important (41). These include relativistic and QED effects, particle spin (hyperfine) effects, nuclear electromagnetic structure effects, and energy shifts caused by the host molecule. Table 3 shows calculated values of the various corrections. For dtμ, the corrections diminish the binding of the lower hyperfine state by ~ 60 meV. The present uncertainty in this correction is thought to be a few meV; an accuracy of ~ 1 meV is needed for many resonant molecular formation calculations. The uncertainty due to the Coulombic wave function should soon be eliminated, but there is also significant uncertainty due to imprecise knowledge of the potential

Table 3 Corrections (in meV) to the energies of the $J = 1$, $v = 1$ states of dtμ and ddμ

	dtμ[a]	dtμ[b]	dtμ[c]	ddμ[a]	ddμ[b]
Nuclear charge distribution	+13.3	+13.3	+10.4[d]	−1.5	−2.1
Darwin-type corrections	−2.5	−2.4	−1.8	−0.9	←[e]
Relativistic mass corrections	−0.5	←		+0.4	←
Recoil corrections	+3.8	←	+2.7	+1.9	←
Vacuum polarization	+16.6	+16.61	+17.1	+8.7	+8.66
Deuteron polarizability	−2.2	←	←	−0.1	←
Finite size of muonic molecule	+1.2	+0.29[f]	←	+1.0	+0.24[g]
Nuclear strong interaction	$\lesssim 10^{-4}$	←	$\sim 10^{-4}$	$\lesssim 10^{-4}$	←
Lower (para) hyperfine state	+35.9	+35.9	←	+16.2	←
Total $\Delta\varepsilon$	+65.6	+64.8	+62.4	+25.7	+24.3

[a] Most recent values of Bakalov and colleagues (41, 127, 135).
[b] G. Aissing and H. J. Monkhorst (unpublished).
[c] Kamimura (93) and M. Kamimura et al (unpublished).
[d] Using the triton charge form factor of Juster et al (136); with the triton form factor of Collard et al (137) used in the other calculations, the calculated value is + 13.3 (M. Kamimura, private communication).
[e] Arrows indicate which value is included in the sum if not calculated.
[f] Scrinzi & Szalewicz (141).
[g] Estimated.

arising from the deuteron polarizability (41) and of the triton charge form factor (M. Kamimura, private communication).

4.1.2 MOLECULAR FORMATION AND BACK DECAY The resonant mechanism of Vesman (9) can result in molecular formation rates very much larger than those resulting from the Auger process. The reactions, for example, are

$$t\mu + (D_2)_{v_i K_i} \rightarrow [(dt\mu)_{11} dee]^*_{v_f K_f}, \qquad 5.$$

etc. K_i and K_f are the initial D_2 and final compound-molecule rotational quantum numbers, and v_i and v_f are the vibrational quantum numbers. Normally $v_i = 0$; henceforth we omit v_i and write v for v_f.

The (normalized) molecular formation rate $\lambda_{dt\mu}$ is found from

$$\lambda_{dt\mu} = N_0 \int_0^\infty \sum_{i,f} W_{K_i} |M_{fi}|^2 \, 2\pi\delta(\varepsilon - \varepsilon_{r_i}) f(\varepsilon, T) \, d\varepsilon, \qquad 6.$$

where W_{K_i} gives the T-dependent probability of the initial D_2 rotational states K_i, ε_{r_i} are the resonance energies (which depend on K_i, K_f, and v), and $f(\varepsilon, T)$ is the distribution of initial kinetic energy ε in the center-of-mass system (the Maxwell distribution if the tμ atom is thermalized). The temperature dependence of $\lambda_{dt\mu}$ comes from the $W_{K_i}(T)$ and, especially, the

Maxwell factor at the resonance energies $f(\varepsilon_{r_i}, T)$. M_{fi} is the transition matrix element.

The pioneering calculation of this process was that of Vinitsky et al (42); important modifications were introduced by Leon (43), Cohen & Martin (44), and Menshikov & Faifman (45). In common with other rearrangement collisions, either the post (bound-state) or prior (scattering-state) interaction Hamiltonian can be used; if evaluated with exact wave functions, the matrix elements M_{fi} are equal (45, 46). Except in the work of Lane (46), the bound-state form is normally used, where the interaction is dipole:

$$H' = e^2 \mathbf{d} \cdot \mathbf{E}; \qquad 7.$$

here $\mathbf{d}$ is the dipole operator of the dtμ system, and $\mathbf{E}$ is the electric field from the spectator nucleus plus electrons at the dtμ center of mass. This interaction takes the initial tμ+d s state to the final $J = 1$ (and $v = 1$) bound state. The tμ-D$_2$ relative motion can introduce an orbital angular momentum L (43), so that the conservation of angular momentum reads

$$\mathbf{L} + \mathbf{K}_i = \mathbf{J} + \mathbf{K}_f \qquad 8.$$

(with $J = 1$). Usually the $L = 0$ contribution is dominant, so that the most probable transitions have $K_f = K_i \pm 1$; however, in one very important case (singlet tμ colliding with D$_2$), $L = 0$ transitions would evidently fall below threshold, so $L > 0$ must be considered.

Once the highly excited mesomolecular complex has been formed, back decay into the entrance channel (the inverse of Equation 5) competes with fusion and Auger deexcitation (47). Consequently, the effective molecular formation rate (i.e. that leading to fusion)[6] is (48)

$$\tilde{\lambda}_{dt\mu} = \lambda_{dt\mu} \frac{\tilde{\lambda}_f}{\tilde{\lambda}_f + \Gamma}, \qquad 9.$$

where $\tilde{\lambda}_f$ is the sum of the fusion rate λ_f and the rate for Auger transitions λ_e. The back-decay rate Γ is related to $\lambda_{dt\mu}$ by detailed balance, but can also be influenced by collisions with target molecules that change K_f. While the asymmetric molecule (dtμ)$_{11}$ rapidly deexcites by E1 Auger transitions, $J = 1 \rightarrow 0$ transitions are strongly suppressed in the symmetric ddμ. Hence, for ddμ back decay significantly reduces the effective molecular formation rates and also leads to resonant hyperfine transitions (48, 49) but is not expected to play as important a role for dtμ (47).

---

[6] In other fields, back decay is referred to as resonant elastic (or inelastic) scattering.

4.2 ddµ Cycle

The study of µCF in pure deuterium has led to a detailed quantitative verification of the resonance formation mechanism, an important step toward understanding the more complex dtµ cycle. This progress for ddµ was possible because recent experiments have separated clearly the basic reaction steps, and *ab initio* calculations of the molecular formation rates have been carried out.

4.2.1 BASIC PROCESSES AND KINETICS The dµ atoms are formed in two hyperfine states ($F = \frac{3}{2}, \frac{1}{2}$) separated in energy by only 48.5 meV. After fast thermalization (30), hyperfine transitions take place by inelastic scattering processes with rates $\lambda_{d\mu}^{FF}$ (37). Furthermore, ddµ molecules are formed both by nonresonant formation processes (rate $\lambda_{dd\mu}^{nr}$) and by resonant ones (rates $\lambda_{dd\mu}^{FS}$, F and S being the total spin of the initial dµ and the final ddµ, respectively). The $J = 1$, $v = 1$ muonic molecule produced in the [(ddµ)dee]* complex by resonant formation has an extraordinarily long lifetime. For this state, back-decay rates $\Gamma \approx 1.5 \times 10^9$ s^{-1} (48) are faster than the fusion rate $\lambda_f = 0.44 \times 10^9$ s^{-1} (50) and the very slow Auger rate $\lambda_e = 0.03 \times 10^9$ s^{-1} (51), and only about one fourth of the ddµ's formed actually undergo fusion. Thus, back decay strongly influences the effective (i.e. experimentally observable) rates shown in the kinetic scheme of Figure 3. The effective molecular formation rates are (48)

$$\tilde{\lambda}_{dd\mu}^F = \lambda_{dd\mu}^{nr} + \sum_S \lambda_{dd\mu}^{FS} \frac{\tilde{\lambda}_f}{\tilde{\lambda}_f + \Sigma_{F'} \Gamma_{SF'}}, \qquad 10.$$

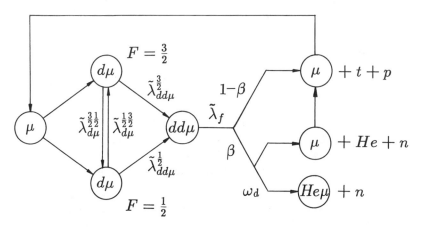

Figure 3 Simplified scheme of the µCF cycle in D$_2$.

where Γ_{SF} are the back-decay rates for $(dd\mu)_S \to (d\mu)_F + d$. In a similar way, total hyperfine transition rates contain a second, resonant term arising from the intermediate formation and subsequent decay of $(dd\mu)_S$ molecules into $(d\mu)_{F'}$ atoms (48, 49):

$$\tilde{\lambda}_{d\mu}^{FF'} = \lambda_{d\mu}^{FF'} + \sum_S \lambda_{dd\mu}^{FS} \frac{\Gamma_{SF'}}{\tilde{\lambda}_f + \Sigma_{F''}\Gamma_{SF''}}. \qquad 11.$$

4.2.2 EXPERIMENTAL RESULTS AND DISCUSSION   Apart from very early bubble chamber studies, experiments use counter techniques to detect the 2.5-MeV neutrons or the charged reaction products from dd fusion. Major achievements during the last decade are the experimental demonstration of the resonant temperature dependence of $dd\mu$ formation at Dubna (52), the separation of the hyperfine components of $dd\mu$ formation at PSI (8a), and the direct measurement of the different fusion channels at LNPI (53). The PSI experiment demonstrates how hyperfine quenching is directly visible in the observed time distributions (Figure 4). At low temperature (~ 30 K), the resonant transition $(d\mu)_{F=3/2} \to (dd\mu)_{S=1/2}$ increases the formation rate $\tilde{\lambda}_{dd\mu}^{3/2}$ by two orders of magnitude over the nonresonant rate. Clearly, the description of $dd\mu$ formation by a single rate as used before

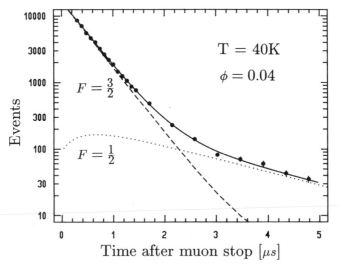

Figure 4 Time spectrum of $dd\mu$ fusion neutrons (54). The intense component (*dashed line*) arising from the large formation rate $\tilde{\lambda}_{dd\mu}^{3/2}$ from the $d\mu$ quartet state rapidly disappears because of hyperfine quenching, while the contribution from the doublet state (*dotted line*) decays with λ_0.

1980 is an oversimplification; experiments must be analyzed in terms of the separate rates $\tilde{\lambda}^F_{dd\mu}$.

The results of counter experiments on ddμ formation rates are collected in Figure 5. Note that two distinct classes of observed rates are displayed, corresponding to different contributions from the two dμ hyperfine states: (a) Experiments (8a, 14, 54, 55) separate the two hyperfine molecular formation rates $\tilde{\lambda}^F_{dd\mu}$. (b) Although hyperfine populations are not always known exactly for experiments (26, 52, 53, 56, 57), they mainly correspond to the steady-state situation and measure an average molecular formation rate $\tilde{\lambda}_{dd\mu}$ that is a superposition of molecular formation rates $\tilde{\lambda}^F_{dd\mu}$. Within this data set there seems to be a normalization problem in the pioneering Dubna experiment (52), which obtains a similar T dependence but has a rate lower by a factor of ~ 4 than all subsequent experiments. Bubble chamber results (58) are not shown in Figure 5, but if interpreted properly

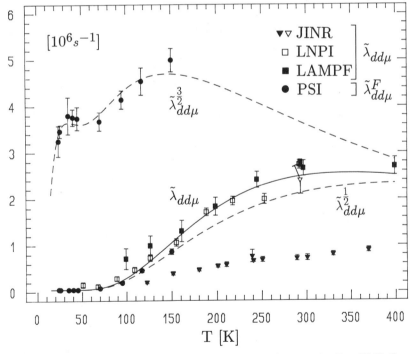

Figure 5 Resonance behavior of the observed ddμ formation rates for $T \leq 400$ K. Two classes of experiments are shown: measurements of hyperfine formation rates $\tilde{\lambda}^F_{dd\mu}$ and measurements of steady-state molecular formation rate $\tilde{\lambda}_{dd\mu}$. Apart from the absolute normalization by Bystritski et al (52), all experiments are explained by current theory with slight adjustments of $\tilde{\lambda}_f$ and ε_{11} (Table 4).

as the total yield from both hyperfine states, they are consistent with recent counter experiments (55).

The most recent theoretical analysis of ddμ formation rates uses the data of Zmeskal et al (54), where hyperfine effects have been separated experimentally. Effective formation rates $\tilde{\lambda}^F_{dd\mu}$ using the formalism and formation matrix elements $|M_{fi}|$ of Menshikov et al (48) are fitted to the experimental data. The results of this analysis (59) are compared with the best theoretical values in Table 4. As seen in Figure 6a, the temperature behavior of $\tilde{\lambda}^{3/2}_{dd\mu}$ is highly sensitive to the resonance transition energies and therefore to ε_{11}. Apart from $\tilde{\lambda}_f$, the fitted results are nearly independent of uncertainties in the magnitudes of the theoretical matrix elements, so that ε_{11} is determined experimentally with extreme accuracy. The agreement between experiment and theory is impressive. The remaining discrepancy should be used in the near future to check improved calculations of the energy corrections discussed in Section 4.1.1. As shown in Figure 5, the results of this analysis are also consistent with the other measurements of the steady-state ddμ formation rates. In particular within absolute calibration errors agreement is found with the accurate measurement of $\tilde{\lambda}_{dd\mu} = (2.76 \pm 0.08) \times 10^6$ s^{-1} at 300 K (53), although this result suggests a somewhat ($\sim 10\%$) larger value of $\tilde{\lambda}_f$.

Once the parameters of ddμ formation are determined, the back-decay contribution to the total hyperfine transition rate $\tilde{\lambda}^{3/2,1/2}_{d\mu}$ (Equation 11) can be calculated without any free parameters. Data on the temperature dependence of $\tilde{\lambda}^{3/2,1/2}_{d\mu}$ just recently became available (59) and are compared with theoretical predictions in Figure 6b. The shape of the data provides some experimental evidence for the effect of back decay in hyperfine transitions. The scattering contribution (37) alone is the right order of magnitude, but its flat temperature dependence does not reproduce the observed structure very well. When the back-decay component is added the theoretical curve substantially exceeds the experimental values. Improved calculations of

Table 4 Parameters of ddμ formation

Parameter	Experiment[a]	Theory	Reference
ε_{11}[b] (meV)	-1965.9 ± 0.3	$-1965.4, -1966.8$[c]	
λ_{nr} (s^{-1})	$(4.60 \pm 0.44) \times 10^4$	4×10^4	89
$\tilde{\lambda}_f$ (s^{-1})	$(3.21 \pm 0.34) \times 10^8$	4.7×10^8	50

[a] Fit of experimental $\lambda^F_{dd\mu}$ (54, 59) with theoretical ddμ formation rates (48).
[b] Total energy excluding hyperfine splitting.
[c] Nonrelativistic energy -1974.9 meV, corrections (excluding hyperfine) 9.5 meV, 8.1 meV (Table 3).

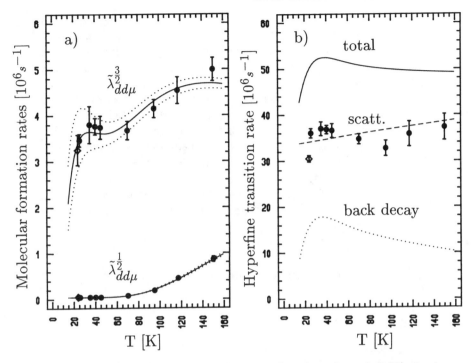

Figure 6 Results of PSI experiments on hyperfine effects (*open diamonds*, liquid; *closed circles*, gaseous D_2). (*a*) Molecular formation rate: the region between dotted curves corresponds to a variation of $\varepsilon_{11} \pm 0.5$ meV from best fit value. (*b*) Hyperfine transition rate: the total calculated rate (*solid line*), which consists of contributions from scattering [*dashed line*; calculation from (37)] and back decay (*dotted line*), disagrees with the experimental points in magnitude while having similar T dependence.

molecular and screening effects on the scattering rates are now in progress to clarify this problem.

In contrast to the dtμ case (see below), experimental results for $\tilde{\lambda}_{dd\mu}$ limit (nonlinear) density effects to $<8\%$ at 24 K (comparison of $\tilde{\lambda}_{dd\mu}^{3/2}$ for $\phi = 0.02$ and 1.15) (54, 55) and to $<10\%$ at 300 K (comparison of $\tilde{\lambda}_{dd\mu}^{3/2}$ for densities from $\phi \approx 0.06$–0.90) (53, 57). The somewhat surprising decrease of the hyperfine transition rate $\tilde{\lambda}_{d\mu}^{3/2,1/2}$ at liquid density (see Figure 6*b*) is also consistent with a density effect in the back-decay term of $\sim 10\%$ (59). Thus the possibility of a significant ϕ dependence of ddμ formation suggested by some theoretical models (discussed in 51) is not supported by experiment.

In summary, *ab initio* calculations of resonant ddμ formation (48) successfully describe all observed resonant formation rates in pure deuterium with an accuracy of a few percent. The $dd\mu$ binding energy ε_{11} determined from experiment differs by only ~ 1 meV from the calculated values. While this excellent agreement is reassuring, some points remain: (a) There exists no stringent experimental test of one basic and theoretically challenging ingredient, namely the magnitude of the resonant formation matrix elements $|M_{\text{fi}}|$, since the magnitude of observable $dd\mu$ formation rates is determined primarily by $\tilde{\lambda}_{\text{f}}$ (see Equation 10). (b) Such a test could be provided by back decay, since its contribution to the hyperfine transition rate is much more sensitive to these matrix elements (59). However, because of the lack of agreement between experiment and theory for the total hyperfine transition rate, an unambiguous separation of the back-decay term is difficult. (c) There is some disagreement between theory and the first experiments in HD mixtures, which measure the $dd\mu$ formation rate $\lambda_{(dd\mu)p}$ on HD molecules (56, 60).

4.3 $dt\mu$ Cycle

The observation of many fusion neutrons per μ^- from high density, high tritium-fraction targets greatly stimulated interest in μCF. While much has been learned, there remain some very important theoretical and experimental questions.

4.3.1 KINETICS AND STEADY-STATE RATES   The $dt\mu$ cycle is obviously more complicated than $dd\mu$, since there are twice as many states and more processes to be considered (61). The competing $dd\mu$ and $tt\mu$ molecular formations can enter, but except for extremely small tritium fraction c_t (or deuterium fraction c_d) these do not significantly influence the cycling rate because the $dt\mu$ formation rate is $\sim 10^2$ times faster. (These competing channels do, however, contribute significantly to muon loss, since they have much higher sticking probabilities than $dt\mu$.) Once the muon arrives in the target, the populations of important states ($d\mu$, $t\mu$-singlet, and $t\mu$-triplet) relax from their initial values according to the kinetics in Figure 7. In contrast to $dd\mu$, however, the much larger rates for $dt\mu$ imply that a steady state usually is reached in only a few nanoseconds; thereafter the fusion yield is given by a single exponential.

The expression for the steady-state cycling rate λ_c is easily obtained by considering the times spent by each part of the cycle (43):

$$\lambda_c^{-1} = \frac{q_{1s}c_d}{\lambda_{dt}c_t} + \frac{3/4}{\lambda_{t\mu}^{10}c_t + \lambda_{dt\mu}^1 c_d} + \frac{1/4 + (3/4)\chi}{\lambda_{dt\mu}^0 c_d}. \qquad 12.$$

The first term is the time in the $d\mu$ ground state, with q_{1s} being the conditional probability for $(d\mu)^*$ reaching the 1s state [assuming that the formation probability of $(d\mu)^*$ is c_d] and λ_{dt} being the ground-state transfer

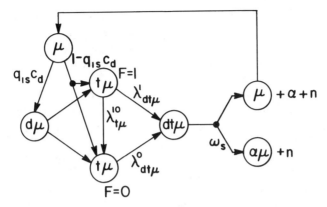

Figure 7 Simplified scheme of dtμ kinetics.

rate (Section 3.2). The second and the third terms represent the time spent in the tμ-triplet and -singlet states respectively; $\lambda_{t\mu}^{10}$ is the tμ-triplet quenching rate, and $\lambda_{dt\mu}^{1}$ and $\lambda_{dt\mu}^{0}$ are the molecular formation rates (see Table 1). The branching ratio χ is given by

$$\chi = \frac{\lambda_{t\mu}^{10} c_t}{\lambda_{t\mu}^{10} c_t + \lambda_{dt\mu}^{1} c_d}. \qquad 13.$$

Each $\lambda_{dt\mu}^{F}$ has contributions from tμ's encountering both D_2 and DT molecules.

The rate of disappearance of active muons is $\lambda_n = \lambda_0 + W\phi\lambda_c$, where λ_0 is the muon decay rate and W the total probability for muon loss (to sticking, to scavenging by $Z > 1$ impurities, etc) in each cycle. Under these steady-state conditions, the time distribution of fusions is simply

$$dN/dt = \phi\lambda_c \exp(-\lambda_n t). \qquad 14.$$

The average number of fusions per μ^- (yield) is given by $Y_n = \phi\lambda_c/\lambda_n$, so that

$$Y_n^{-1} = \lambda_0/\phi\lambda_c + W. \qquad 15.$$

4.3.2 EXPERIMENTAL RESULTS FOR STEADY-STATE CYCLING RATES
Following the theoretical prediction of large resonant dtμ formation rates in 1977 (12), experimental investigation of previously unexplored dt mixtures became the center of interest. Experimental milestones of the last decade were the confirmation of high dtμ formation rates at Dubna (62), the attainment of very large fusion yields at LAMPF (63), and the observation of unexpected transients in the fusion time distribution at PSI (64).

Further experiments have been carried out at these laboratories and at LNPI, KEK, and the Rutherford-Appleton Laboratory (RAL).

A typical setup of a high yield experiment is sketched in Figure 8. Incoming muons are registered by a counter telescope and stop in the dt mixture. The sequence of 14-MeV neutrons from dtμ fusion is observed in neutron detectors; several small detectors, each of which detect as many as four subsequent hits, are used to reduce pile-up and deadtime effects from the ~ 100 fusions catalyzed within a few microseconds. Plastic scintillators observe the electron from the final decay of the muon. Target diagnostics are essential since μCF processes depend critically on atomic and molecular target composition. Target purity better than 1 ppm is usually required. Significant quantities (grams) of tritium are present, which requires elaborate safety precautions. The LAMPF experiments

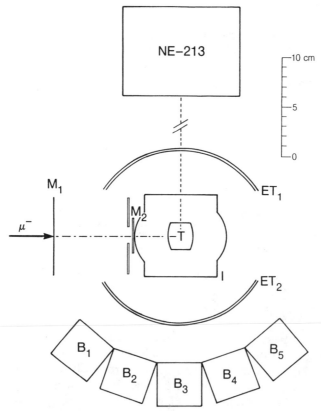

Figure 8   Experimental setup for a dtμ experiment (27): Target (T), insulation vacuum (I), μ telescope (M_1, M_2), neutron detectors (B_1–B_5, *NE-213*), electron telescopes (ET_1, ET_2).

have mainly used high pressure (one kilobar) 50-cm^3 targets over a very wide temperature range (13–800 K). The PSI collaboration, in contrast, has investigated the low temperature (12–300 K) range with much thinner, lower pressure target cells (volumes 20–1000 cm^3); this allows direct monitoring of the target (on-line mass spectrometry, detection of impurities by observation of muonic x rays).

The first dt experiment at Dubna (62) used low c_t and hence was insensitive to $\lambda_{dt\mu}$ (see Equation 12); a lower limit, $\lambda_{dt\mu} > 10^8 \, \text{s}^{-1}$, was determined, but no temperature dependence of the fusion yield was seen in the range 90–610 K. The higher c_t used at LAMPF (63) reduces the dominance of the first two terms of Equation 12, and thus reveals $\lambda_{dt\mu}$ and its temperature dependence clearly in $\lambda_c(T)$ (see Figure 9). Note that $\lambda_c(T)$ is still rising at the highest temperature reached (800 K), and does not become small at low T.

The increase of the cycling rate when the density is doubled ($\phi = 0.72$ compared to 0.36, both at $c_t = 0.5$) is very apparent in Figure 9. This unexpected effect was explored in detail in later experiments (26, 27, 65), with updated results for $\lambda_c(\phi)$ shown in Figure 10. Since the density effect is very small for low c_t points and increases with c_t, it seems clear that the effect comes from the $\lambda_{dt\mu}$ terms in Equation 12 rather than from the q_{1s}/λ_{dt} term. This is striking evidence for three-body molecular formation (see below). The PSI results extend to lower density and reveal that the extra density dependence of λ_c is more complicated than linear.

4.3.3 COMPARISON WITH THEORY   To advance our understanding, it is

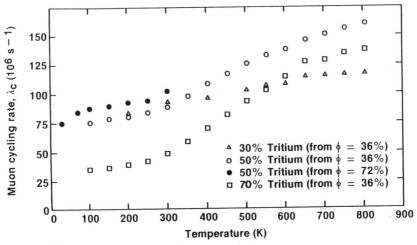

Figure 9 Cycling rate $\lambda_c(T)$ for several values of c_t (134).

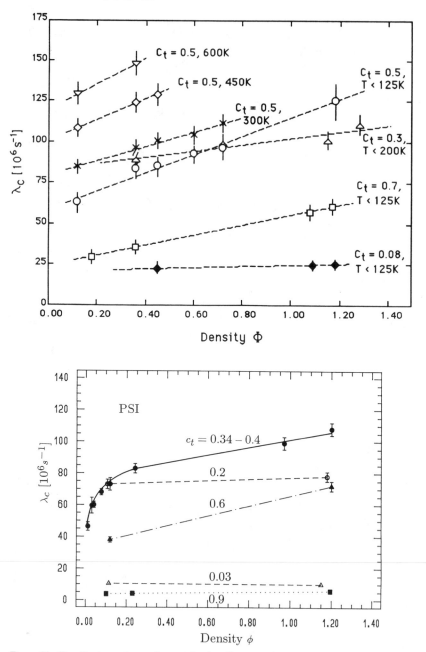

Figure 10   Density dependence of normalized cycling rates λ_c; (*top*) LAMPF (26), (*bottom*) PSI, target temperature between 20 K and 45 K (65).

essential to extract the underlying rates from the experimental λ_c's. This is difficult because four molecular formation rates (singlet- and triplet-$t\mu$'s reacting with D_2 and DT molecules), the triplet quenching rate $\lambda_{t\mu}^{10}$, and the effective d-t transfer rate λ_{dt}/q_{1s} have to be disentangled from a single observable λ_c, measured at different experimental conditions (c_t, ϕ, T). Necessarily, these analyses must rely on model-dependent expressions for λ_c, which—with our improving understanding of the dtμ cycle—have gradually become more complicated (see Equation 12). The data analysis reflects this progress toward refined models: different dtμ formation rates on D_2 and DT molecules (63), q_{1s} different from 1 (26, 27), and different molecular formation rates for the $t\mu$ hyperfine states (27, 66). At present— apart from remaining experimental problems [see, for example, discussion in (66, 67)]—the main uncertainty comes from our insufficient knowledge of q_{1s} and hyperfine effects.

Transfer reaction   Experimental and recent theoretical results for λ_{dt} are compared in Table 5. The agreement looks good, but this may be somewhat deceptive because of the model dependence mentioned above. Experiments determine λ_{dt} from λ_c measurements at low c_t where $c_d \lambda_{(dt\mu)d}^0$ is large compared with the other rates and only the first two terms in Equation 12 contribute to the cycling rate. The experimental values quoted were extracted allowing only a very weak variation of $q_{1s}(\phi, c_t)$ and simply ignoring the second term. If the second term is not ignored but instead included with the theoretical value for $\lambda_{t\mu}^{10}$, then λ_{dt} becomes $\sim 25\%$ larger (27).

Even more serious, theory predicts a much stronger dependence of q_{1s} on ϕ and c_t (see Section 3.2) than used in the experimental analyses with widely differing target conditions (Table 5). Thus, the good agreement seen in the table turns out to be rather surprising and provides experimental evidence *against* the strong variation of q_{1s} predicted by theory (24, 25). Independent confirmation of this disagreement is found in the density independence of λ_c at low tritium concentrations (Figure 10), in the analysis of pile-up events in the LNPI ionization chamber (68), and indirectly, in the study of similar charge-exchange processes $(p\pi)_n + d \rightarrow (d\pi)_n + p$ for pions (69). This failure of the calculation of q_{1s} presumably reflects our incomplete understanding of the muonic cascade, as discussed in Section 3.2.

Molecular formation and three-body effects   The resonance energies for dtμ formation clearly imply that at low temperatures, say $T \lesssim 200$ K, only the rate $\lambda_{(dt\mu)d}^0$ for molecular formation in $t\mu(F = 0) + D_2$ collisions should be large. Experimentally this certainly seems to be the case, although the

Table 5   Isotopic transfer rate λ_{dt}

Authors	λ_{dt} (10^8 s^{-1})	Experimental conditions			
		T	ϕ	c_t (%)	
Bystritsky et al (62)	2.9 ± 0.4	300 K	0.01–0.08	1–8	E[a]
Jones et al (26)	2.8 ± 0.4	< 125 K	0.4–1.2	≳ 4	E
Breunlich et al (27)	2.8 ± 0.5	23 K	1.2	≳ 4	E
Balin et al (68)[b]	2.8 ± 0.2	300 K	0.09	1.2	E
Melezhik (85)	2.7	4×10^{-2} eV			T[a]
Kobayashi et al (128)	2.6	1×10^{-2} eV			T
Kamimura (23)	2.7	1×10^{-2} eV			T

[a] E = experimental, T = theoretical.
[b] This result is less sensitive to uncertainties in q_{1s} [see (68)].

q_{1s} problem has been a severe impediment to extracting definitive molecular formation rates. We can regard the following as firmly established.

1. The molecular formation rates are large (several $\times$ 10^8 s^{-1}) and increasing up to the highest temperature reached;
2. For $T \lesssim 300$ K, $\lambda^0_{(dt\mu)d}$ is dominant and has a surprisingly flat T dependence, even down to liquid hydrogen temperatures; and
3. This same rate has the remarkable extra density dependence.

A preliminary value of $\lambda^0_{(dt\mu)d} = (3.7 \pm 0.5) \times 10^8$ s^{-1} has been extracted at $\phi = 1.2$, $T = 23$ K (27) by assuming a simple functional form of $q_{1s}(\phi, c_t)$, and $\lambda^{10}_{t\mu}$ from theory. An update of the $dt\mu$ formation rates obtained from the LAMPF experiments using a rather unconstrained q_{1s} is shown in Figure 11 (66).

The fact that, for $\lambda^0_{(dt\mu)d}$, the transitions with the largest matrix elements, $K_f = K_i + 1$ (hence $L = 0$) with $v = 2$, lie ~ 10 meV below threshold (43; Table 6) has some important consequences for both the low density, two-body rate and for the three-body contribution. The large Auger de-excitation rate for the $[(dt\mu)dee]^*$ means that the δ function in Equation 6 becomes a Lorentzian, as pointed out by Petrov (70). Hence the $K_f = K_i + 1$ transitions contribute significantly even at low ϕ (71). The $0 \to 2$ transition (hence $L = 1$) is expected to be important at low temperature, since it evidently falls just below threshold (Table 6). Since the ortho- ↔ para-transitions are rather slow for D_2, the populations of the $K_i = 0$ vs $K_i = 1$ states and hence the relative contributions of $0 \to 2$ and $1 \to 3$ transitions could depend on how the target is prepared (72).

According to Menshikov & Ponomarev (73), these below-threshold $K_f = K_i + 1$ transitions dominate the three-body molecular formation; an

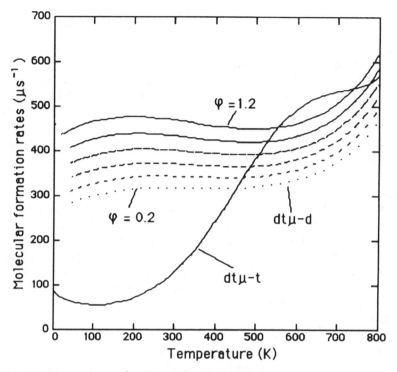

Figure 11 Extracted rates $\lambda^0_{(dt\mu)d}(T)$ and $\lambda^0_{(dt\mu)t}(T)$ from the fit to LAMPF data described by Anderson (66). The curves for $(dt\mu)d$ are for $\phi = 0.2(0.2)1.2$. (Figure provided by A. N. Anderson.)

Table 6 Transition energies[a] (in meV) for the resonant formation process $(t\mu)_{F=0} + (D_2)_{K_i,v_i=0} \rightarrow [(dt\mu)dee]^*_{K_f,v_f=2}$

	K_f				
K_i	0	1	2	3	4
0	−17.0	−12.2	−2.4	12.1	31.4
1	−24.4	−19.6	−9.8	4.7	24.0

[a] Calculated with $\varepsilon^{F=0}_{11} = -596.3$ meV (see Tables 2 and 3) and rovibrational energies for the electronic molecule from Scrinzi et al (129), including the reduced mass effects on the $(dt\mu)e$ and D binding. A value of 64.0 meV has been adopted for $\Delta\varepsilon$; when a more accurate value is available, the energies in the table should be simply shifted by the change.

adjacent spectator molecule can carry off enough kinetic energy to make these transitions possible. Starting with Petrov (70), a number of workers (71, 74–76) have calculated three-body molecular formation, making use of this below-threshold transition by including a collisional width in the Lorentzian that replaces the δ function of Equation 6. This is analogous to the impact approximation, which plays a central role in the theory of collisional line broadening in spectroscopy. However, Cohen & Leon (77) have pointed out that the basic requirements for the applicability of the impact approximation are not met for the molecular formation problem, so that a new approach is needed. Petrov & Petrov (78) suggest using many-body perturbation theory, while Leon (79) points out that con-figuration mixing of the rotational states of the $[(dt\mu)dee]^*$ molecule, arising from the torque exerted by any neighboring molecule, implies a significant amount of three-body molecular formation just by itself. Lane recently attempted to extend the theory of collisional line broadening to include the effect of the $t\mu$-momentum, but obtained explicit results only for hard-sphere molecules of rather small radius (80). This problem of the exact origin of the three-body effect in molecular formation, and of suitable methods for its calculation, is both vitally important and, for now, rather uncertain.

We also point out that, even for two-body molecular formation, the use of lowest-order perturbation theory (as outlined in Section 4.1.2) is much more suspect for the $dt\mu$ case than for $dd\mu$, since the transition matrix elements and hence the distortion of the wave functions are much greater for $dt\mu$.

4.3.4   TRANSIENT PHENOMENA   Transients in the short time interval before steady state is reached provide direct information about the rates involved in the kinetics (Figure 7). Thus far, low c_t or ϕ have been used to slow the rapid $dt\mu$ processes so that this transient interval is expanded into an experimentally accessible region.

Epithermal μCF   At $\phi = 0.01$ new features of the μCF cycle are revealed, as demonstrated by the striking transients observed at PSI (64, 81) shown in Figure 12. Though these spectra resemble the hyperfine components observed in the $dd\mu$ cycle, their T and c_t dependence is inconsistent with $t\mu$ hyperfine quenching. The current interpretation is that hot $t\mu$ atoms are initially formed, then undergo rapid molecular formation before they are thermalized by elastic collisions (82, 83).

Two questions naturally arise from these data. (*a*) Are these transients, in particular the implied high epithermal formation rates at early time, understandable in terms of today's theory? (*b*) What impact do epithermal contributions of muonic atoms have on steady-state observables?

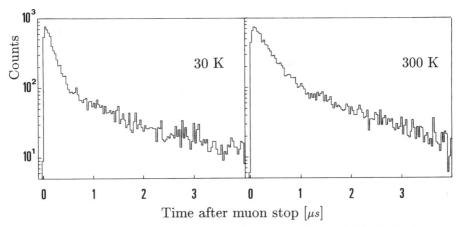

Figure 12   Time spectra of dt fusion neutrons at $\phi \approx 1\%$ and $c_t = 0.9$ (81). Initial spikes are due to high fusion yields produced by epithermal tμ atoms.

Theoretical investigations have addressed the first of these questions and found qualitative agreement with experiment (82, 83). Extremely high formation rates $\lambda^F_{dt\mu} \approx 10^{10}$ s^{-1} are due to transitions to the $\nu = 3$ level of the mesomolecular complex at tμ kinetic energies of 0.2–0.5 eV (84). The maximum of dtμ formation rates at such high energies is consistent with the increase of cycle rates continuing up to 800 K (Figure 9). Monte Carlo simulations of the thermalization process based on recent theoretical scattering cross sections (85) reproduce the main features of the c_t and T dependences of the observed transients (32, 86).

The answer to the second question above depends on the initial tμ energy distribution, as well as on the subsequent competition between molecular formation and thermalization. For high density, tμ atoms are more likely to be thermalized by elastic collisions during the cascade; if thermalization is nearly complete, then even high epithermal formation rates do not affect the steady-state rates (87). The crucial unknown is thus the energy distribution as a function of ϕ (see Section 3.2). If it were known, then a calculation of the type carried out by Menshikov et al (84) could evaluate the epithermal contribution to molecular formation in the steady state.

Hyperfine effects   As discussed in Section 4.3.3, current theory provides clear predictions for the hyperfine structure of μCF at low temperatures: Only $\lambda^0_{(dt\mu)d}$ is resonant, while $\lambda^1_{(dt\mu)d}$ has a small nonresonant value of 10^5 s^{-1} (43). Although important elements of the cycling rate analysis, the large difference between the dtμ formation rates from the two tμ hyperfine states and the theoretical value for $\lambda^{10}_{t\mu} = 1.3 \times 10^9$ s^{-1} (37) have never been verified experimentally. On the contrary, a recent experiment (88) using liquid deuterium with $c_t = 2 \times 10^{-3}$ reported the absence of the slow buildup expected from the hyperfine transition at this low c_t. Instead, a much faster transient was observed (these results were confirmed with higher statistics in 1988 at PSI). Various explanations for this effect have been considered [$\lambda^1_{(dt\mu)d}$ as high as 10^7 s^{-1}, substantially higher $\lambda^{10}_{t\mu}$, $t\mu$ spin flip on deuterium], but all disagree with theory. Clearly, the question of hyperfine effects in the $dt\mu$ cycle is still unsolved.

4.4   Nonresonant Fusion Cycles

In addition to $dd\mu$ and $dt\mu$, the other combinations ($pd\mu$, $pt\mu$, $tt\mu$) have now been studied (see Table 7). For these systems, molecules are formed

Table 7   Rates of nonresonant μCF cycles

| Process | Symbol | Rate (in 10^6 s^{-1}) | |
		Experiment[f] (Ref.)	Theory (Ref.)
Molecule formation rates			
$d\mu + H_2 \rightarrow (pd\mu)pe + e^a$	$\lambda_{pd\mu}$	5.8 ± 0.3 (130)[b]	5.9 (89)
		6.82 ± 0.25 (131)	
		5.53 ± 0.16^g (90)	
$t\mu + H_2 \rightarrow (pt\mu)pe + e$	$\lambda_{pt\mu}$	6.8 ± 0.6 (132)	6.5 (89)
$t\mu + T_2 \rightarrow (tt\mu)te + e$	$\lambda_{tt\mu}$	1.8 ± 0.6 (119)	2.96 (89)
Fusion rates			
$pd\mu \rightarrow \gamma + {}^3He\mu$	λ_γ	$0.255 \pm 0.012^{c,d}$	0.70 ± 0.17^e (100)
			0.32^h
$\rightarrow \mu + {}^3He$	λ_μ	$0.050 \pm 0.010^{c,d}$	0.11 (100)
	$\lambda_\gamma + \lambda_\mu$	0.305 ± 0.010^c (130)	
		0.287 ± 0.022^g (90)	
		0.289 ± 0.027 (133)	
$pt\mu \rightarrow \gamma + {}^4He\mu$	λ_γ	0.07 ± 0.02 (132)	
$\rightarrow \mu + {}^4He$	λ_μ		$(5 \pm 1) \times 10^{-4}$ [i]
$\rightarrow e^+e^- + {}^4He\mu$	$\lambda_{e^+e^-}$		3.7×10^{-4}[i]
$tt\mu \rightarrow 2n + {}^4He$	λ_f	15 ± 2 (119)	~ 10 (91)

[a] Experiment (60) reports different rates for H_2 and HD molecules.
[b] Complete list of older references in (11).
[c] Analyses assume fusion occurs predominantly in states with total nuclear spin $\frac{1}{2}$.
[d] Evaluation from observed $\lambda_\gamma + \lambda_\mu$ using experimental yields of gammas and conversion muons (11).
[e] Estimate based on pd scattering experiments.
[f] Liquid hydrogen temperature $T \approx 23$ K, except where indicated by Footnote g.
[g] Room temperature.
[h] Estimate based on the mirror process of thermal neutron capture on deuterium (J. D. Jackson, unpublished).
[i] Calculated for states with total nuclear spin 0 (L. N. Bogdanova and V. E. Markushin, unpublished).

by nonresonant Auger processes, since no loosely bound states are available for resonant formation (see Table 2). Apart from inherent interest, the knowledge of the rates in these branches is essential because they compete with the main $dt\mu$ cycle; thus they contribute to the total muon loss probability W (Section 5.2). In general, experimental molecular formation rates are in agreement with calculations using the two-level approximation of the three-body problem (89). The experimental values for $\lambda_{pd\mu}$ are given in some detail, since recently puzzling $pd\mu$ results have been reported by Aniol et al (60): the observed yield of γ's after $pd\mu$ fusion doubles in mixtures of $H_2 + D_2$ compared to pure HD, and both yields decrease with temperature. These results seem to imply a strong molecular and temperature dependence of $\lambda_{pd\mu}$. Both effects are surprising for a nonresonant process. Note also that earlier experiments at 30 and 300 K are in reasonable agreement with each other (Table 7).

5.   NUCLEAR FUSION AND STICKING

5.1   Nuclear Fusion

The study of fusion reactions between hydrogen nuclei bound in muonic molecules is of intrinsic interest for the physics of few-nucleon systems (91). Stimulated by the increased activity in μCF in the last few years, significant progress in this area has recently been achieved. Accurate molecular wave functions are now available for the Coulomb part of this problem. A systematic set of experimental data is being acquired for all isotopic molecules, apart from the $dt\mu$ molecule where fusion rates are too fast to be observed.

As discussed in Section 4, the $dd\mu$ and $dt\mu$ molecules are formed resonantly in their $J = 1, v = 1$ states. Because of the centrifugal barrier, fusion rates are much slower in $J = 1$ levels than in $J = 0$ levels. Nevertheless, in the symmetric molecule $dd\mu$ the $\Delta J = 1$ transition is forbidden (because it would require a change in spin state) and fusion occurs in the two states $v = 1, 0$ with $J = 1$ (50). For $dt\mu$, Auger transitions rapidly convert the $J = 1$ state to a $J = 0$ state, and fusion occurs 84% in the $J = 0, v = 1$ state and 16% in the $J = 0, v = 0$ state (17).

The fusion rate can be calculated by either of two approaches: (a) determine the molecular wave function ignoring nuclear forces and put in the nuclear information directly from the scattering cross section, or (b) determine the complex molecular wave function including nuclear forces and obtain the fusion rate from the imaginary part of the energy. In the first approach

$$\lambda_{xy\mu}^{f} = A \lim_{r_{xy} \to 0} \int |\psi_{dt\mu}|^2 \, d^3 r_\mu.$$

16a.

Here the coefficient A is simply related to the usual representation of the $x+y$ fusion cross section, in terms of the astrophysical S function (91a)

$$\sigma = \frac{S(E)}{E} \exp\left[-\frac{\sqrt{2}\pi e^2}{\hbar}\left(\frac{\mu}{E}\right)^{1/2}\right],$$

16b.

where μ is the reduced mass of $x+y$ and E is the relative (c.m. system) energy, by

$$A = \frac{\hbar}{\pi e^2 \mu} \lim_{E \to 0} S(E).$$

16c.

The above is the "traditional" approach and has been applied to all muonic molecules. In the second approach

$$\lambda_{xy\mu}^{f} = \text{Im}\,(E_{xy\mu})/\hbar.$$

16d.

In principle, this is the more accurate approach (if the nuclear forces are known) but has been applied only to dtμ. It is essential that the calculation of $\lambda_{xy\mu}^{f}$ be done consistently; e.g. it would be wrong to use the complex wave function in Equation 16a.

The effect of the nuclear interaction, dominated by the $I^{\pi} = 3/2^{+}$ resonance, on the $J = 0$ dtμ wave functions has been incorporated in three different ways: by perturbative (92) and nonperturbative (93) use of an optical potential and by the nuclear R matrix (94). The imaginary part of the optical potential or nuclear R matrix accounts for the α-n channel. The results of these calculations are in satisfactory agreement with each other, giving energy shifts in the range $\Delta E = -0.7$ to -1.0 meV and widths in the range $\Gamma = 0.80$ to 0.85 meV for the $J = 0$, $v = 0$ state. (The values for the $J = 0$, $v = 1$ state are about 15% smaller in magnitude.) The latter correspond to a fusion rate, $\Gamma/\hbar$, of $\sim 1.3 \times 10^{12}$ s^{-1}. The fusion rates for the $J = 0$ states have also been calculated in the traditional way by integrating the wave function obtained neglecting the nuclear interaction in the limit $r_{\mathrm{dt}} \to 0$; this method gave $0.71 \times 10^{12}\,\mathrm{s}^{-1}$ (95). The source of this difference is not known.

In the $J = 1$ states of dtμ, there are both s-wave and p-wave nuclear components of the wave function. The latter contribute most of the norm of the wave function, but the former are mostly responsible for fusion. Hence the scarcity of experimental data on p-wave d+t scattering is not a problem. Calculations indicate that the fusion rates in the $J = 1$ states of dtμ are about four orders of magnitude slower than in the $J = 0$ states (93, 95, 96).

In principle, there exists the possibility of a strong rearrangement of the dtμ spectrum due to the nuclear interaction (97). This could happen if the $3/2^+$ resonance happened to be nearly degenerate with a dtμ bound state, though the resonance width would have to be very small ($\lesssim 1$ keV). The above calculations do not indicate such an occurrence. Still, there is not quite a consensus that the nuclear effect is small (98).

In the ddμ and ttμ molecules, the two nuclei are prepared in pure $L = 1$ states of relative angular momentum by the molecular formation (91). Consequently, unlike nuclear scattering experiments, the fusion process in the molecule provides information about the nuclear p-wave interaction at virtually zero energy. The branching ratio Y_n / Y_p of the mirror reactions

$$\mathrm{dd}\mu \to \begin{cases} {}^3\mathrm{He} + \mathrm{n} + \mu \\ \mathrm{t} + \mathrm{p} + \mu \end{cases}$$

was measured in an ionization chamber at LNPI (53). The result, 1.39 ± 0.04, demonstrated a clear violation of charge symmetry in the p-wave that remained unexplained for several years. However, a recent calculation using a charge-independent R-matrix description of reactions in the $A = 4$ system yields $Y_n / Y_p = 1.43$ (99). A ratio of 1.3 has been obtained by the resonating group method (99a). The large deviation from unity can be attributed to the small effects of internal Coulomb isospin mixing being greatly amplified in the external Coulomb field by the proximity of broad levels having opposite isospin in the p-wave states.

Radiative fusion rates of pdμ and ptμ molecules are sensitive observables for studying the effect of meson exchange currents in the $A = 3$ and 4 systems. In addition, the rate for the conversion process pdμ (or ptμ) $\to$ ${}^3\mathrm{He}$ (or ${}^4\mathrm{He}$) $+ \mu$ yields information on exotic fusion channels not present in scattering experiments. The first microscopic calculations of this muon conversion process in pdμ have been performed (100), and both ptμ channels were measured recently (PSI, 1988). Regarding the radiative pdμ

fusion rate, we note that a recent estimate based on the mirror reaction (radiative neutron capture on deuterium) is in much better agreement with the observed λ_γ than the previous estimates, which used pd scattering data (see Table 7). Further interesting developments can be expected in the near future when the direct calculations of these rates and the analysis of the new ptμ data (including the ptμ hyperfine structure) are finished.

5.2 Sticking

As first recognized by Jackson in 1957 (5), the probability that the muon sticks to the charged fusion products (^{3}He, ^{4}He) imposes a fundamental limit on the fusion yield (see Equation 15). In practice, however, sticking has only recently been seen as the dominant constraint on μCF applications, with resonant molecular formation bringing about experimental cycling rates more than 400 times faster than muon decay. Naturally, much effort is now aimed toward a complete understanding of the crucial sticking process.

The sticking probability is determined by two steps (explained for dtμ):

$$
\begin{array}{l}
\text{dt}\mu
\begin{cases}
\xrightarrow{\;1-\omega_s^0\;} \alpha+\mu+n \\
\xrightarrow{\;\omega_s^0\;} \alpha\mu \ (3.5\ \text{MeV})+n
\end{cases} \\[2ex]
\qquad\qquad\qquad
\alpha\mu
\begin{cases}
\xrightarrow{\;R\;} \alpha+\mu \\
\xrightarrow{\;1-R\;} \alpha\mu \ (\text{thermal}).
\end{cases}
\end{array}
\tag{17.}
$$

The initial sticking ω_s^0 depends only on the intramolecular dynamics. However, since the $\alpha\mu$ is formed with an initial kinetic energy of 3.5 MeV, it can be stripped with reactivation probability R before stopping. The effective sticking is thus

$$\omega_s = \omega_s^0(1-R). \tag{18.}$$

With the original estimate (5) of $\omega_s \approx 0.8\%$ confirmed in 1981 (101, 102), experiments were to face difficult problems in measuring these comparatively small losses.

5.2.1 EXPERIMENTAL METHODS AND RESULTS

Neutron detection   The first method developed for measuring dtμ sticking ω_s is the analysis of the disappearance rate in the exponential time distribution for fusion neutrons (Equation 14):

$$\lambda_n = \lambda_0 + W\phi\lambda_c. \tag{19.}$$

In addition to simple fits with Equation 14, more sophisticated methods are applied, e.g. analyzing the time distributions of subsequent neutrons or ensuring the survival of the muon by requiring a decay electron after the analyzed time interval (26, 27, 65). All these methods determine $W\phi\lambda_c$, so that λ_c has to be known in order to extract W.

Sticking ω_s is then determined from W by subtracting all other losses. Neglecting $Z > 2$ impurities in the hydrogen mixture and diffusion of muonic atoms to the target walls, we obtain

$$\omega_s = W - \frac{c_d q_{1s}}{c_t \lambda_{dt}}(c_d 0.58\omega_d\lambda_{dd\mu} + c_p\lambda_{pd\mu} + c_{He}\lambda_{dHe})$$

$$-\frac{1}{c_d\lambda_{dt\mu}}(c_t\omega_t\lambda_{tt\mu} + c_p\lambda_{pt\mu} + c_{He}\lambda_{tHe}) - c_{He}\omega_{He}. \quad 20.$$

In this approximate relation, 0.58 is the $n + {}^3He$ fraction following dd fusion, and $\lambda_{dd\mu}$ and $\lambda_{dt\mu}$ are steady-state averages over the hyperfine contributions. The $pd\mu$ and $pt\mu$ recycling is ignored because of slow fusion rates and high sticking in these cycles, and the finite time of $tt\mu$ fusion is neglected. While He solubility is small in liquid hydrogen, ${}^3He$ builds up in gas targets by tritium decay. Thus, the μ^- is scavenged with initial capture and excited-state transfer probability ω_{He}, and by ground-state transfer with rates λ_{dHe}, λ_{tHe} (103).

To ensure understanding of the various loss terms in Equation 20, the dependence of W on T, c_{He}, and c_p was investigated (26, 65). Moreover, all competing channels have now been studied in dedicated experiments in which the relevant fusion products (neutrons, gammas, and conversion muons) were observed directly (see Tables 1 and 7). The ω_s values resulting from the subtractions are presented as functions of c_t in Figure 13 and of ϕ in Figure 14. The independence of c_t raises confidence in the procedure. Nevertheless, some model dependence will remain until the uncertainties in the $dt\mu$ kinetics (q_{1s}, $\lambda_{dt\mu}$) have been removed. As can be seen in Figure 14, at large ϕ surprisingly small sticking values of only $\sim 0.4\%$ are found, with fair agreement among the different experiments. The strong density dependence found in the LAMPF results remains controversial, since it is not present in the PSI data nor in the theoretical calculations (see below). Note, however, that the disagreement is large only at low ϕ, where systematic problems are most critical. One must know λ_n with high precision to separate the relatively small $W\phi\lambda_c$ term, and the uncertainty of $q_{1s}(\phi, c_t)$ significantly affects the corrections in Equation 20.

X-ray detection   In the initial sticking process, muons are captured in various (n, l) states of the $(He\mu)^+$ ion. Additional excitations occur during the collisional slowing down of the $He\mu$. Observation of the resulting $He\mu$

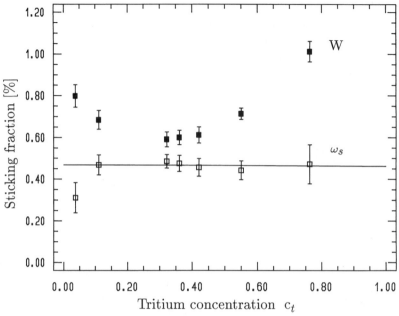

Figure 13   Sticking results as a function of c_t (27), from liquid data. At the maximum cycling rate ($c_t = 0.3$–0.4), the total loss probability W is dominated by dt sticking ω_s.

x rays demonstrates that Heμ sticking has occurred, irrespective of other muon losses.

Systematic experimental study of muonic x rays following sticking was initiated at PSI with experiments on pdμ, ddμ, and dtμ (104, 105). Investigations of dtμ sticking are being carried out at KEK (106) and RAL (107), taking advantage of the high instantaneous muon flux available at these pulsed accelerators. The dtμ case is both the most interesting and the most challenging experimentally. Apart from the extremely low yield, the enormous bremsstrahlung background from tritium β decay and the Doppler line broadening due to the high initial $\alpha\mu$ velocity are the major problems. In the PSI experiment, fusion neutrons and x rays were detected in coincidence in a liquid target of very low $c_t \approx (2$–$4) \times 10^{-4}$. A narrow coincidence width suppressed uncorrelated bremsstrahlung, and the 90° angle between the neutron and Si(Li) x-ray detector minimized Doppler broadening. The KEK experiment, in contrast, used $c_t = 0.3$ for maximum fusion yield in a small (1 cm^3) liquid target, and measured x rays delayed with respect to the intense muon pulse. Two different assumptions for the width of the K$_\alpha$ line—corresponding to detector resolution only and to maximum Doppler broadening—were used in the analysis.

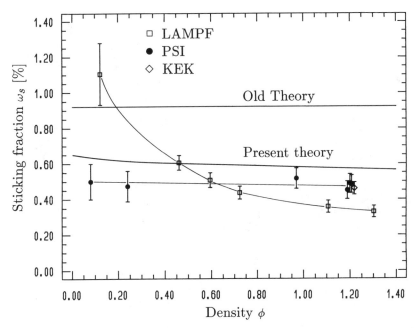

Figure 14 Experimental (neutron method) and theoretical determinations of sticking ω_s as a function of density [LAMPF (26), PSI (67, 81)—preliminary, KEK (106) using λ_c from (26, 27), old theory (102), present theory (116) for R and (114) for ω_s^0]. Curves through the experimental data points are only meant for guidance; systematic errors are included for PSI and KEK, but only statistical errors for the LAMPF points. Theory predicts 20–30% higher values than experiments at high ϕ and disagrees with the strong ϕ dependence of the LAMPF data.

The main results of these experiments are compared with recent theory in Table 8. For ddμ fusion, fair agreement between experiment and theory is achieved, with some differences remaining for the K_α yield. For dtμ, there is obvious disagreement between the experimental results. While a clear K_α peak was observed by Bossy et al (105) with a yield consistent with theory, much smaller yields were reported by Nagamine et al (106). Experiments at KEK and RAL are continuing to address this important question. With some assistance from theory, effective sticking values ω_s^X (Table 8) can be derived from the observed x-ray yields:

$$\omega_s^X = \frac{Y(K_\alpha)^{exp}}{Y(K_\alpha)^{th}} \, \omega_s^{th}, \qquad\qquad 21.$$

using recent calculations for $Y(K_\alpha)^{th}/\omega_s^{th}$ (see below). The resulting PSI value for ω_s^X corroborates the result from the neutron method.

Table 8 Heμ X-ray yield per fusion

	Reference	$Y(K_\alpha)$ (%)	$Y(K_\beta)/Y(K_\alpha)$	ω_s^{Xa}(%)	
pdμ	Bossy et al (105)	3.2 ± 0.4	0.052 ± 0.005		E[b]
	Takahashi (124)	3.10	0.055		T[b]
	Markushin (117)	2.9	0.072 ± 0.002		T
ddμ	Bossy et al (105)	1.6 ± 0.2	0.13 ± 0.02		E
	Cohen (116)	2.2 ± 0.2	0.11 ± 0.02		T
	Markushin (117)	2.5	0.12 ± 0.01		T
dtμ	Bossy et al (105)	0.19 ± 0.05	≤ 0.08	0.43 ± 0.14	E
	Nagamine et al (106)	0.049 ± 0.04^c		0.11 ± 0.09	E
	Cohen (116)	0.24 ± 0.04	0.12		T
	Markushin (117)	0.25	0.12 ± 0.01		T

[a] Using $Y(K_\alpha)/\omega_s = 0.44 \pm 0.09$ from (116).
[b] E = experimental, T = theoretical.
[c] Assuming a line width of 700 eV.

The probability of initial sticking in excited Heμ states was also determined directly from the profile of the 5.5-MeV gamma line from pdμ fusion observed with high resolution Ge detectors (104).

Detection of charged fusion products   The most direct method to measure sticking is the observation of the number N of Heμ and He fusion products. The sticking probability is then simply

$$\omega_s = \frac{N(\text{He}\mu)}{N(\text{He}) + N(\text{He}\mu)}. \qquad 22.$$

The experimental challenge lies in the unambiguous separation of the relatively rare Heμ events from the frequent He fusion products as well as from other background. In addition, experimental systematics such as detection efficiencies have to be known well.

A method for measuring ddμ sticking ω_d was developed at LNPI using a hydrogen ionization chamber as both the muon stopping target and the detector. The He ions are stopped completely in the chamber volume and detected with 100% efficiency. At the gas density used, $\phi = 0.11$, recombination losses significantly reduce the amplitudes of the observed signals. Since these losses are $\sim 50\%$ for He, but only $\sim 25\%$ for Heμ, a clear separation of Heμ from the much stronger He peak is achieved, as can be seen in Figure 15, and a precise result obtained: $\omega_d = 0.122 \pm 0.003$ (53).

This method is being pursued by an extended collaboration at PSI for dtμ sticking as well (108). The small sticking value and the high tritium activity within the chamber volume require major improvements of the

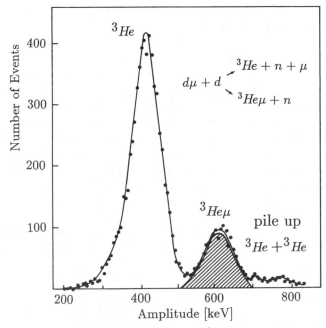

Figure 15   Measurement of dd sticking ω_d: $^3\mathrm{He}\mu$ and $^3\mathrm{He}$ fusion products are separated in the LNPI ionization chamber (53).

technique. The anode area is segmented to reduce tritium noise, and pulse shape analysis of the ionization chamber signal and coincidence with a high efficiency neutron detector array will suppress background. The proposal aims at measuring ω_s with an accuracy of 5% for target densities $\phi \approx 0.15$.

Very much lower densities ($\phi \approx 10^{-3}$) were used in a recent LAMPF experiment (109) that measured the recoiling $\alpha\mu$'s in coincidence with the d-t fusion neutron. These low densities were necessary to allow the α's and $\alpha\mu$'s sufficient range to reach the external Si surface-barrier charged-particle detector; very thin windows served to keep the tritium away from the detector. The distinction between α and $\alpha\mu$ comes from the factor of four in the ratio of the ranges. To accumulate sufficient $\alpha\mu$ events it was necessary to use a higher density than for α detection; since the effective cycling rate does not scale exactly as ϕ^2 (one factor of ϕ from the stopping rate, approximately another from the cycling rate) because of epithermal molecular formation, a systematic error in the sticking probability is introduced. The reported value (109) for initial sticking is $\omega_s^0 = (0.80 \pm 0.15 \pm$

0.12 systematic)%. To reduce systematic errors, this experiment will be pursued further at RAL (107).

5.2.2 THEORETICAL CALCULATIONS

Initial sticking   Sticking may be thought of as being due to the matching of the muon velocity in the initial state with that of possible bound final states on the recoiling daughter nucleus (102); hence sticking is small if the recoil velocity of the He nucleus is $\gg 2$ au. All existing calculations of initial sticking have been made in the sudden approximation, which simply depends on the overlap of the initial and final states, and take the final state to be just the muonic helium ion multiplied by a plane wave for its relative motion. The earliest calculations (5, 101, 102) also took a very simple approximation to the initial state, namely just the united-atom limit of the adiabatic approximation to the muonic molecule wave function, which, for example is $\mu^5 \mathrm{He}^+(1s)$ for $dt\mu$. This approximation gives $\omega_s^0(\text{adiabatic}) = 1.20\%$ for $dt\mu$ and $\omega_d^0(\text{adiabatic}) = 17.2\%$ for $dd\mu \rightarrow {}^3\mathrm{He}\mu + n$.

It is now known that the adiabatic treatment of the initial state is inadequate for the magnitude of the sticking, especially for $dt\mu$, though it gives the proportions of sticking in the various states fairly well. The first calculation made with the united-atom limit of an accurate nonadiabatic wave function used the quantum Monte Carlo method (110); this was followed by calculations using the adiabatic representation (111, 112) and the variational method (93, 95, 113, 114). At one time there was some skepticism that the variational method could determine the sticking accurately since a wave function design to minimize the energy might not be accurate at the point of coalescence of the two nucleons. This worry has been alleviated, and recent variational calculations appear to provide the most accurate values in the Coulomb-sudden approximation; the values are 0.888% for $dt\mu$ and 13.2% for $dd\mu$. These values have been averaged over the $v = 0$ and $v = 1$ states (with $J = 0$ for $dt\mu$ and $J = 1$ for $dd\mu$) for which the sticking differs very slightly. The sticking following fusion in the $J = 1$ states of $dt\mu$ is considerably smaller (115), but the cascade is apparently such that fusion does not have time to occur in them.

After adjusting for stripping (see the next subsection), the theoretical value for $dd\mu$ is in good agreement with experiment, but the value for $dt\mu$ is still 20–30% higher than experiments (see Figure 14). Since the estimated uncertainty due to the stripping calculation is only $\sim 10\%$ (116, 117), this discrepancy suggests the inadequacy of some approximation made in the calculation of initial sticking. Some assertions of a large effect due to the influence of the $I^\pi = 3/2^+$ nuclear resonance of ${}^5\mathrm{He}$ have been made (118), but preliminary estimates indicate only about a 4% effect that, in fact, *increases* the theoretical sticking rather than improving the agreement (91, 93). More rigorous calculations of the nuclear effect are in progress. The plane wave in the final state has also been questioned (G. M. Hale, unpublished), but all existing calculations use it. In a more accurate treatment, the energy dependence of the ^{5}He $3/2^+$ resonance t-matrix element would have to be taken into account. Danos et al (98) contend that an entire theoretical reformulation of the sticking problem is necessary (see Section 5.1).

For pdμ and ptμ the recoil velocity is low (γ rays usually carry the energy) and so the sticking is near unity (117). In this case, sticking in excited states is especially interesting because it arises mainly from the nonadiabatic component of the molecular wave function. Fusion of ttμ provides an interesting situation in which there are three heavy particles in the final state, $\alpha + 2$n, and the recoil of the α depends on the correlation of the neutrons. Hence the experimental determination of ttμ sticking is very desirable and was recently made by neutron detection yielding $\omega_t = 0.14 \pm 0.03$ (119). This value may be compared with the range of values of 0.05–0.18 that results from extreme assumptions about the final-state interaction of the particles (101).

Reactivation   It is now known (116, 117) that accurate calculation of R requires knowledge of the initial sticking state (1s, 2s, 2p, etc) and, especially for dt μCF, a kinetic treatment that takes into account various collisional excitation and deexcitation processes, ionization, muon transfer, and radiation; all of these rates are shown in Figure 16. Because the radiative rates are independent of target density while the collisional processes depend on density, there arises the possibility of a density dependence in R and hence a corresponding dependence in ω_s. The population of excited states of muonic helium also offers another valuable diagnostic by monitoring their x-ray emissions.

The kinetics is treated theoretically by solving the set of rate equations for $\alpha\mu$ (or ^{3}Heμ) typically with levels $n \le 10$. Higher levels can be assumed always to be ionized. The Stark mixing rates for $n \ge 3$ are fast enough that the l states can be assumed to be populated statistically, but such is not true for $n = 2$, where the metastable 2s level has a special importance (120). The stripping probability of (Heμ)$^+$ formed in the 2s state is appreciably greater than if it is formed in the 2p state; otherwise, the higher the excitation, the greater the stripping probability. Most of the sticking occurs in the 1s state: 77% for dt μCF and 71% for dd μCF. However, for ddμ most of the ^{3}Heμ atoms stripped are those initially formed in excited states, while for dtμ "ladder ionization" is important and the initial sticking state is not so critical for stripping. Because of the great importance of excited-

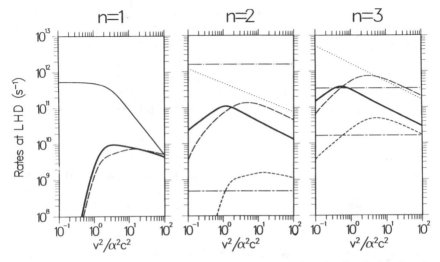

Figure 16 Some of the rates (at liquid hydrogen density) used to describe the kinetics of slowing down $\alpha\mu$ [(116), except Stark mixing rate from (120)]. The curves are energy loss (*light-solid curve*), stripping (*heavy-solid curves*), excitation (*long-dashed curves*), inelastic deexcitation (*short-dashed curves*), Auger deexcitation (*long- and short-dashed curves*), radiation (*dash-dotted curves*), and $l \neq 0 \rightarrow l = 0$ Stark transitions (*dotted curves*).

state sticking for ^{3}Heμ stripping, the factor R has a significant density dependence, rising from 0.11 at $\phi = 0.1$ to 0.18 at $\phi = 1.2$. At the lower density, radiation has time to occur and the advantage of initially sticking in an excited state is lost. For dtμ, R varies from 0.30 at $\phi = 0.1$ to 0.35 at $\phi = 1.2$. The resulting values of ω_s are shown in Figure 14.

The muonic helium x rays provide the best available probe of the distribution of initial sticking states. This determination is fairly direct for ddμ, but for dtμ it requires a major adjustment since most of the x rays follow excitation from the ground state (116, 117). The measurements of the ratios of various x-ray intensities also provide tests of the calculated state-to-state transition rates. As explained above Equation 21, this measurement can also be used to obtain the sticking. A recent theoretical calculation separating the $2p_0$ and $2p_{\pm 1}$ state populations shows that the K_α x-ray emission is not isotropic (121). Consequently, the experimental x-ray yields, which are observed only at an angle of 90° with respect to the neutron (105), need to be revised upward by about 4% (the detectors have finite size, but the solid-angle correction is relatively small since the distribution is flat at 90°). The agreement of the two PSI sticking values mentioned above (x-ray and neutron method) strongly supports the calculated value of R, for which as yet there is no direct experimental determination.

6. SUMMARY AND OPEN QUESTIONS

Great progress has been achieved in μCF in the last decade. Qualitative estimates have now given way to careful quantitative calculations and measurements.

The energies of the loosely bound states of the ddμ and dtμ molecules, which are crucial to the spectacular resonant formation rates, have been calculated with great care. The results for the nonrelativistic Coulomb binding energies can be considered final, while corrections for finite size of the nuclei, QED effects, etc are just being completed. The goal is to fix the energies to a fraction of a meV—an accuracy of some 0.1 ppm of the nearly 3-keV total binding energy. These extremely accurate calculations are required and can be tested experimentally in μCF: The ddμ binding energy has been determined within 0.3 meV from measurements of the temperature dependence of resonant ddμ formation from the upper dμ hyperfine state.

Systematic experimental studies of resonant formation in pure D_2 have been carried out at different accelerators, and the resulting ddμ formation rates are in excellent agreement with a recent *ab initio* calculation. Thus, the resonant formation mechanism is verified in great detail. Because of back decay, however, this is not a very stringent test of the magnitude of the calculated formation matrix elements; a better test can be provided by the resonant contribution to dμ hyperfine quenching, which has been observed. Here, however, there seems to be a discrepancy: The sum of the theoretical resonant and nonresonant hyperfine transition rates is somewhat too large compared with experiment.

Comparable understanding for dtμ fusion—which was essentially *terra incognita* before 1979—has encountered formidable problems. The predicted very high rates for resonant dtμ formation have been verified, but in contrast to ddμ no leveling out with increasing temperature has been seen up to the highest temperature reached (800 K). The behavior of the (normalized) rate for dtμ formation for singlet tμ colliding with D_2 is completely unexpected in two respects: First, this rate does not get small as $T \to 0$, but instead remains large and nearly constant for $T \lesssim 300$ K. Second, it exhibits a strong density dependence indicating three-body (or many-body) collision effects. The culprit for the density effect is thought to be a transition with large matrix element lying just below threshold, but going from this idea to a real calculation remains a challenging and unsolved problem. More experimental information is urgently needed, especially on the separate temperature dependence of the two- and three-body contributions to molecular formation. In addition, the strongly nonlinear density dependence of the cycling rate at low density needs to be investigated further.

Much of the difficulty in obtaining good measurements of the dtμ formation rates comes from the fact that they are not observed directly but must instead be inferred from the cycling rates, using assumptions about the initial ground-state populations of dμ and tμ. This situation stimulated new calculations of the muonic cascade to determine these populations; these theories predict large excited-state d$\mu \rightarrow$ tμ transfer probabilities and strong dependences on target density. However, these predictions apparently disagree with experiment. Clearly, there is much that is not understood about the muonic hydrogen cascade—in spite of the fact that the exotic hydrogen cascade has been studied theoretically and experimentally for at least three decades!

The sticking probabilities have now been measured rather precisely for all the fusions. Calculations have been carried out for these as well, with a significant discrepancy appearing for dtμ. Not only does this case have by far the smallest sticking probability, but for dtμ it is clear that it is the sticking that provides the ultimate limit on the yield of fusions per μ^-. Hence, sticking for dtμ has received (and indeed requires) meticulous theoretical attention. The best calculated value, 0.58%, remains 20–30% above the experimental ones (at liquid hydrogen density). This discrepancy may reflect some important physics not yet included in the calculations. On the experimental side, some inconsistencies (density effect, x-ray yield) are still unresolved. To clarify the situation, further theoretical work and more accurate experiments, e.g. using $\alpha\mu$ detection, are needed.

There are several places where the knowledge emerging from μCF impinges on other fields. One example is the possibility of measuring pure p-wave fusion of dd and tt nuclei bound in muonic molecules. For dd fusion, the surprising experimental value of 1.4 for the ratio between the n^{3}He and pt final states has recently been explained theoretically. As another example, radiative pdμ and ptμ fusion probes the effect of exchange currents in the $A = 3$ and $A = 4$ systems. We also mention that the measurement of the dμ hyperfine quenching rate is important for the interpretation of muon capture experiments for weak interaction information.

Although this review is about the basic physics of μCF, it would not be complete without at least a mention of the possibility of power production from dt μCF. In fact, the major experimental surprises, namely small sticking and the increase in dtμ formation with density, have doubled the fusion yields compared with the theoretical predictions of a decade ago. Yields of up to 125 (65) or even 150 (26) fusions per muon have been reported; thus about 2.5 GeV of energy are released. However, according to present estimates, at least 8 GeV of beam energy are needed to produce a usable μ^- with an accelerator (122); hence, direct energy production would appear to be excluded without a significant breakthrough. A hybrid system devised by Petrov (123), which is claimed to be more efficient than competing breeding schemes, combines electronuclear breeding with a μCF "reactor" to produce fuel for conventional fission reactors. Externally applied fields to enhance the molecular formation rate (124) and reactivation (125) and even μCF in dense plasmas (126) are also being explored theoretically. Finally we mention that instead of energy production, dt μCF might possibly find application as an intense source of 14-MeV neutrons.

ACKNOWLEDGMENTS

J. S. Cohen and M. Leon are grateful for support from the US Department of Energy, in large part by the Division of Advanced Energy Projects. W. H. Breunlich and P. Kammel are grateful for support by the Austrian Academy of Sciences, the Austrian Science Foundation, and PSI. Finally, each of us wants to express his gratitude to all our colleagues working in μCF, in particular to those with whom we have had the pleasure of collaborating for many years. Without their dedication, much of the progress discussed in this review would not have been possible.

Literature Cited

1. Frank, F. C., *Nature* 160: 525 (1947)
2. Sakharov, A. D., *Rep. Phys. Inst. Acad. Sci. USSR* (1948)
3. Zeldovich, Ya. B., *Dokl. Akad. Nauk SSSR* 95: 493 (1954)
4. Alvarez, L. W., et al., *Phys. Rev.* 105: 1127 (1957)
5. Jackson, J. D., *Phys. Rev.* 106: 330 (1957)
6. Zeldovich, Ya. B., Gershtein, S. S., *Usp. Fiz. Nauk* 71: 581 (1960) [*Sov. Phys. Usp.* 3: 593 (1961)]
7. Zavattini, E., in *Muon Physics*, ed. V. W. Hughes, C. S. Wu. New York: Academic (1975), 2: 219
8. Doede, J. H., *Phys. Rev.* 132: 1782 (1963); Dzhelepov, V. P., et al., *Nuovo Cimento* 33: 40 (1964); *Zh. Eksp. Teor. Fiz.* 50: 1235 (1966) [*Sov. Phys. JETP* 23: 820 (1966)]
8a. Kammel, P., et al., *Phys. Lett.* 112B: 319 (1982); *Phys. Rev.* A28: 2611 (1983)
9. Vesman, E. A., *Pis'ma Zh. Eksp. Teor. Fiz.* 5: 113 (1967) [*JETP Lett.* 5: 91 (1967)]
10. Vinitsky, S. I., et al., *Preprint JINR P4-10336*, Dubna (1976)
11. Gershtein, S. S., Ponomarev, L. I., see Ref. 7, 3: 141
12. Gershtein, S. S., Ponomarev, L. I., *Phys. Lett.* 72B: 80 (1977)
13. Bardin, G., et al., *Nucl. Phys.* A453: 591 (1988)
14. Cargnelli, M., PhD thesis, Tech. Univ. Vienna. 132 pp. (1986); Cargnelli, M., et al., in *Proc. Workshop on Fundamental Muon Physics*, ed. C. M. Hoffman, V. W. Hughes, M. Leon. Los Alamos, LA-10714-C (1986), p. 182
15. Bracci, L., Fiorentini, G., *Phys. Rep.* 86: 169 (1982); Ponomarev, L. I., *Atomkernenergie–Kerntechnik* 43: 175 (1983); Jones, S. E., *Nature* 321: 127 (1986); Rafelski, J., Jones, S. E., *Sci. Am.* 257: 84 (1987); Ponomarev, L. I., *Muon Catal. Fusion* 3: 629 (1988)
16. Markushin, V. E., *Zh. Eksp. Teor. Fiz.* 80: 35 (1981) [*Sov. Phys. JETP* 53: 16 (1981)]
17. Bogdanova, L. N., et al., *Zh. Eksp. Teor. Fiz.* 83: 1615 (1982) [*Sov. Phys. JETP* 56: 931 (1982)]
18. Cohen, J. S., *Phys. Rev.* A27: 167 (1983)
19. Anderhub, H., et al., *Phys. Lett.* 101B: 151 (1981)
20. Cohen, J. S., Martin, R. L., Wadt, W. R., *Phys. Rev.* A24: 33 (1981)
21. Korenman, G. Ya., Popov, V. P., in

354   BREUNLICH ET AL

Muon-Catalyzed Fusion, Sanibel Island, ed. S. E. Jones, J. Rafelski, H. J. Monkhorst (AIP Conf. Proc. 181). New York: Am. Inst. Phys. (1989), p. 145
22. Aniol, K. A., et al., *Phys. Rev.* A28: 2684 (1983)
23. Kamimura, M., *Muon Catal. Fusion* 3: 335 (1988)
24. Menshikov, L. I., Ponomarev, L. I., *Pis'ma Zh. Eksp. Teor. Fiz.* 39: 542 (1984) [*JETP Lett.* 39: 663 (1984)]
25. Menshikov, L. I., Ponomarev, L. I., *Pis'ma Zh. Eksp. Teor. Fiz.* 42: 12 (1985) [*JETP Lett.* 42: 13 (1985)]
26. Jones, S. E., et al., *Phys. Rev. Lett.* 56: 588 (1986)
27. Breunlich, W. H., et al., *Phys. Rev. Lett.* 58: 329 (1987)
28. Kravtsov, A. V., Mikhailov, A. I., Popov, N. P., *LNPI Preprint 1361* (1988)
29. Menshikov, L. I., *Muon Catal. Fusion* 2: 173 (1988)
30. Melezhik, V. S., Ponomarev, L. I., Faifman, M. P., *Zh. Eksp. Teor. Fiz.* 85: 434 (1983) [*Sov. Phys. JETP* 58: 254 (1983)]
31. Adamczak, A., Melezhik, V. S., *Muon Catal. Fusion* 2: 131 (1988)
32. Cohen, J. S., *Phys. Rev.* A34: 2719 (1986)
33. Breunlich, W. H., et al., *PSI Nucl. Part. Phys. Newsl.*, Vol. 44 (1988)
34. Crawford, J., et al., *Phys. Lett.* 213B: 391 (1988)
35. Ponomarev, L. I., Somov, L. N., Faifman, M. P., *Yad. Fiz.* 29: 133 (1979) [*Sov. J. Nucl. Phys.* 29: 67 (1980)]
36. Deleted in proof
37. Bracci, L., et al., *Phys. Lett.* A134: 435 (1989)
38. Puzynin, I. V., Vinitsky, S. I., *Muon Catal. Fusion* 3: 307 (1988)
39. Bhatia, A. K., Drachman, R. J., *Phys. Rev.* A30: 2138 (1984)
40. Hu, C.-Y., *Phys. Rev.* A32: 1245 (1985)
40a. Alexander, S. A., Monkhorst, H. J., *Phys. Rev.* A38: 26 (1988)
40b. Kamimura, M., *Phys. Rev.* A38: 621 (1988)
41. Bakalov, D., *Muon Catal. Fusion* 3: 321 (1988)
42. Vinitsky, S. I., et al., *Zh. Eksp. Teor. Fiz.* 74: 849 (1978) [*Sov. Phys. JETP* 47: 444 (1978)]
43. Leon, M., *Phys. Rev. Lett.* 52: 605 (1984)
44. Cohen, J. S., Martin, R. L., *Phys. Rev. Lett.* 53: 738 (1984)
45. Menshikov, L. I., Faifman, M. P., *Yad. Fiz.* 43: 650 (1986) [*Sov. J. Nucl. Phys.* 43: 414 (1986)]
46. Lane, A. M., *J. Phys.* B20: 2911 (1987)

47. Lane, A. M., *Phys. Lett.* 98A: 337 (1983)
48. Menshikov, L. I., et al., *Zh. Eksp. Teor. Fiz.* 92: 1173 (1987) [*Sov. Phys. JETP* 65: 656 (1987)]
49. Leon, M., *Phys. Rev.* A33: 4434 (1986)
50. Bogdanova, L. N., et al., *Phys. Lett.* 115B: 171 (1982), errata *Phys. Lett.* 167B: 485 (1986)
51. Faifman, M. P., *Muon Catal. Fusion* 2: 247 (1988)
52. Bystritski, V. M., et al., *Zh. Eksp. Teor. Fiz.* 76: 460 (1979) [*Sov. Phys. JETP* 49: 232 (1979)]
53. Balin, D. V., et al., *Phys. Lett.* 141B: 173 (1984); *Pis'ma Zh. Eksp. Teor. Fiz.* 40: 318 (1984) [*JETP Lett.* 40: 1112 (1984)]
54. Zmeskal, J., et al., *Muon Catal. Fusion* 1: 109 (1987)
55. Naegele, N., et al., *Nucl. Phys.* A439: 397 (1989)
56. Vorobyov, A. A., *Muon Catal. Fusion* 2: 17 (1988)
57. Bystritski, V. M., et al., see Ref. 21, p. 17
58. Doede, J., *Phys. Rev.* 132: 1782 (1963); Fetkovich, J., et al., *Phys. Rev. Lett.* 4: 570 (1960)
59. Zmeskal, J., PhD thesis. Univ. Vienna, 133 pp. (1986); Zmeskal, J., et al., in preparation
60. Aniol, K. A., et al., *Muon Catal. Fusion* 2: 63 (1988)
61. Gershtein, S. S., et al., *Zh. Eksp. Teor. Fiz.* 78: 2099 (1980) [*Sov. Phys. JETP* 51: 1053 (1980)]
62. Bystritski, V. M., et al., *Phys. Lett.* 94B: 476 (1980); *Pis'ma Zh. Eksp. Teor. Fiz.* 31: 249 (1980) [*JETP Lett.* 31: 228 (1980)]; *Zh. Eksp. Teor. Fiz.* 80: 1700 (1981) [*Sov. Phys. JETP* 53: 877 (1981)]
63. Jones, S. E., et al., *Phys. Rev. Lett.* 51: 1757 (1983)
64. Breunlich, W. H., et al., *Phys. Rev. Lett.* 53: 1137 (1984)
65. Petitjean, C., et al., *Muon Catal. Fusion* 2: 37 (1988)
66. Anderson, A. N., see Ref. 21, p. 57
67. Ackerbauer, P., et al., presented at Muon Catalyzed Fusion Workshop, Sanibel Island (1988)
68. Balin, D. V., et al., *Zh. Eksp. Teor. Fiz.* 92: 1543 (1987) [*Sov. Phys. JETP* 65: 866 (1987)]
69. Kravtsov, A. V., et al., *Muon Catal. Fusion* 2: 199 (1988)
70. Petrov, Yu. V., *Phys. Lett.* 163B: 28 (1985)
71. Petrov, Yu. V., et al., *Muon Catal. Fusion* 2: 261 (1988)
72. Leon, M., Cohen, J. S., *Phys. Rev.* A31: 2680 (1985)

73. Menshikov, L. I., Ponomarev, L. I., *Phys. Lett.* 167B: 141 (1986)
74. Leon, M., *Muon Catal. Fusion* 1: 163 (1987)
75. Menshikov, L. I., *Muon Catal. Fusion* 2: 273 (1988); *Preprint IAE-4606/2*, Moscow (1988)
76. Lane, A. M., *J. Phys.* B21: 2159 (1988)
77. Cohen, J. S., Leon, M., *Phys. Rev.* A39: 946 (1989)
78. Petrov, V. Yu., Petrov, Yu. V., *LNPI Preprint 1390* (1988)
79. Leon, M., *Phys. Rev.* A39: 5554 (1989)
80. Lane, A. M., to be published
81. Breunlich, W. H., et al., *Muon Catal. Fusion* 1: 67 (1987)
82. Cohen, J. S., Leon, M., *Phys. Rev. Lett.* 55: 52 (1985)
83. Kammel, P., *Nuovo Cimento Lett* 43: 349 (1985)
84. Menshikov, L. I., Somov, L. N., Faifman, M. P., *Zh. Eksp. Teor. Fiz.* 94: 6 (1988) [*Sov. Phys. JETP* 67: 652 (1988)]
85. Melezhik, V. S., *Muon Catal. Fusion* 2: 117 (1988)
86. Somov, L. N., *Muon Catal. Fusion* 3: 465 (1988)
87. Menshikov, L. I., Ponomarev, L. I., *Pis'ma Zh. Eksp. Teor. Fiz.* 45: 329 (1987) [*JETP Lett.* 45: 417 (1987)]
88. Kammel, P., et al., *Muon Catal. Fusion* 3: 483 (1988)
89. Ponomarev, L. I., Faifman, M. P., *Zh. Eksp. Teor. Fiz.* 71: 1689 (1976) [*Sov. Phys. JETP* 44: 886 (1976)]
90. Bystritski, V. M., et al., *Zh. Eksp. Teor. Fiz.* 71: 1680 (1976) [*Sov. Phys. JETP* 44: 881 (1976)]; *Zh. Eksp. Teor. Fiz.* 70: 1167 (1976) [*Sov. Phys. JETP* 43: 606 (1976)]
91. Bogdanova, L. N., *Muon Catal. Fusion* 3: 359 (1988)
91a. Burbidge, E. M., et al., *Rev. Mod. Phys.* 29: 547 (1957)
92. Bogdanova, L. N., Markushin, V. E., Melezhik, V. S., *Zh. Eksp. Teor. Fiz.* 81: 829 (1981) [*Sov. Phys. JETP* 54: 442 (1981)]
93. Kamimura, M., see Ref. 21, p. 330
94. Struensee, M. C., et al., *Phys. Rev.* A37: 340 (1988)
95. Hu, C.-Y., *Phys. Rev.* A34: 2536 (1986)
96. Bogdanova, L. N., et al., *Yad. Fiz.* 34: 1191 (1981) [*Sov. J. Nucl. Phys.* 34: 662 (1981)]
97. Belyaev, V. B., Gandyl, E. M., Zubarev, A. L., *Z. Phys.* A314: 107 (1983)
98. Danos, M., Biedenharn, L. C., Stahlhofen, A., see Ref. 21, p. 308
99. Hale, G. M., Dodder, D. C., unpublished
99a. Hofmann, H. M., unpublished
100. Bogdanova, L. N., et al., *Muon Catal. Fusion* 3: 377 (1988)
101. Gershtein, S. S., et al., *Zh. Eksp. Teor. Fiz.* 80: 1690 (1981) [*Sov. Phys. JETP* 53: 872 (1981)]
102. Bracci, L., Fiorentini, G., *Nucl. Phys.* A364: 383 (1981)
103. Popov, N. P., *Muon Catal. Fusion* 2: 207 (1988); Kravtsov, A. V., et al., *Muon Catal. Fusion* 2: 183 (1988)
104. Bossy, H., et al., *Phys. Rev. Lett.* 55: 1870 (1985)
105. Bossy, H., et al., *Phys. Rev. Lett.* 59: 2864 (1987)
106. Nagamine, K., et al., *Muon Catal. Fusion* 1: 137 (1987)
107. Brooks, F. D., et al., *Muon Catal. Fusion* 2: 85 (1988)
108. Ackerbauer, P., et al., *PSI proposal R-88-03.1* (1988)
109. Paciotti, M. A., et al., see Ref. 21, p. 38
110. Ceperley, D., Alder, B. J., *Phys. Rev.* A31: 1999 (1985)
111. Bogdanova, L. N., et al., *Phys. Lett.* 161B: 1 (1985)
112. Bogdanova, L. N., et al., *Nucl. Phys.* A454: 653 (1986)
113. Hu, C.-Y., Kauffmann, S. K., *Phys. Rev.* A36: 5420 (1987)
114. Haywood, S. E., Monkhorst, H. J., Szalewicz, K., *Phys. Rev.* A37: 3393 (1988); 39: 1634 (1989)
115. Hu, C.-Y., *Phys. Rev.* A36: 4135 (1987)
116. Cohen, J. S., *Phys. Rev. Lett.* 58: 1407 (1987)
117. Markushin, V. E., *Muon Catal. Fusion* 3: 395 (1988)
118. Danos, M., Müller, B., Rafelski, J., *Phys. Rev.* A34: 3642 (1986)
119. Breunlich, W. H., et al., *Muon Catal. Fusion* 1: 121 (1987)
120. Struensee, M. C., Cohen, J. S., *Phys. Rev.* A38: 44 (1988); 38: 53 (1988)
121. Cohen, J. S., Padial, N. T., *Phys. Rev.* A39: 915 (1989)
122. Petrov, Yu. V., *Muon Catal. Fusion* 3: 525 (1988); Jaendel, M., see Ref. 21, p. 394; Bertin, A., et al., see Ref. 21, p. 405
123. Petrov, Yu. V., *Nature* 285: 466 (1980); *Muon Catal. Fusion* 1: 351 (1987); *Muon Catal. Fusion* 3: 525 (1988)
124. Takahashi, H., see Ref. 21, p. 185
125. Bracci, L., Fiorentini, G., *Nature* 297: 134 (1982); Kulsrud, R. M., see Ref. 21, p. 367
126. Menshikov, L. I., Ponomarev, L. I., *Pis'ma Zh. Eksp. Teor. Fiz.* 46: 246 (1987) [*JETP Lett.* 46: 312 (1987)]
127. Ponomarev, L. I., *Muon Catal. Fusion* 3: 629 (1988)
128. Kobayashi, K., et al., *Muon Catal. Fusion* 2: 191 (1988)

129. Scrinzi, A., Szalewicz, K., Monkhorst, H. J., *Phys. Rev.* A37: 2270 (1988)
130. Bleser, E. J., *Phys. Rev.* 132: 2679 (1963)
131. Conforto, G., et al., *Nuovo Cimento* 33: 1001 (1964)
132. Hartmann, F. J., et al., *Muon Catal. Fusion* 2: 53 (1988); Ackerbauer, P., et al., *SIN Newsl.* 20: 43 (1988)
133. Bertl, W., et al., *Atomkernenergie–Kerntechnik* 43: 184 (1983)
134. Caffrey, A. J., et al., In *Proc. Muon-Catalyzed Fusion Workshop*, Jackson, WY (1984), ed. S. E. Jones, p. 53, unpublished
135. Bakalov, D., Korobov, V., *Rapid Commun. JINR–Dubna*, Vol. 2 (1989)
136. Juster, F.-P., et al., *Phys. Rev. Lett.* 55: 2261 (1985)
137. Collard, H., et al., *Phys. Rev.* 138B: 57 (1965)
138. Vinitsky, S. I., et al., *Zh. Eksp. Teor. Fiz.* 79: 698 (1980) [*Sov. Phys. JETP* 52: 353 (1980)]
139. Frolov, A. M., *Yad. Fiz.* 44: 1367 (1986) [*Sov. J. Nucl. Phys.* 44: 888 (1986)]
140. Scrinzi, A., Szalewicz, K., *Phys. Rev.* A39: 2855 (1989)
141. Scrinzi, A., Szalewicz, K., *Phys. Rev.* A39: 4983 (1989)

Annu. Rev. Nucl. Part. Sci. 1989. 39: 357–406

EXPERIMENTAL TESTS OF PERTURBATIVE QCD

G. Altarelli

Theory Division, CERN, 1211 Geneva 23, Switzerland

KEY WORDS: partons, strong interactions, asymptotic freedom, hard processes, deep inelastic processes.

CONTENTS

1. INTRODUCTION

Quantum chromodynamics (QCD) (1), the SU(3) gauge theory of colored quarks and gluons, is the current theory of strong interactions. It serves as a primary building block in the standard model of the known interactions of fundamental particles (except gravity) based on the gauge group SU(3) ⊗ SU(2) ⊗ U(1). The most striking physical properties of QCD are asymptotic freedom and confinement. The first property (2) is rigorously established and means that the effective coupling decreases logarithmically at short distances. This is the basis for perturbative QCD, which is relevant

357

0163–8998/89/1201–0357$02.00

for processes involving large momentum transfers. Color confinement means that the potential energy between colored charges increases approximately linearly at large distances, so that only color-singlet states can be produced and observed. Originally a mere conjecture (3), the basis for color confinement has become increasingly more solid, partly as a result of experiments using quarkonium spectroscopy (e.g. 4) (which are consistent with a potential that increases at long distances), and partly because of lattice simulations (e.g. 5) [for example, the calculation (6) of the potential between static color charges at large distances and the study (e.g. 7) of the deconfinement phase transition]. Although not yet proved, it is quite plausible that confinement is really implied by the theory.

Precise experimental tests of QCD are clearly as important as tests of the electroweak sector of the standard theory. Yet testing QCD is even more difficult than testing the electroweak theory. In fact, the latter interactions are so weak that perturbation theory is always reliable at present energies. Moreover, leptons as well as photons, W's, and Z's are at the same time the fields in the Lagrangian and the particles observed in our detectors. On the contrary, in QCD the perturbative approach, which is essentially the only viable method for extracting from the theory testable quantitative predictions, is only valid for processes with large momentum transfers. Even in the most favorable cases the expansion is only slowly converging because of the relatively large values of the QCD coupling, $\alpha_s(Q) \gg \alpha_{QED}$. Moreover, QCD is a theory of confined quarks and gluons (the "partons") but only hadrons are actually observed. Except for the rather limited set of completely inclusive experiments, nonperturbative effects connected with soft parton cascades and hadronization generally tend to obscure the underlying simplicity of the parton dynamics. On one hand, in spite of its limitations perturbative QCD provides a rich testing ground for the theory, as reviewed in this article. On the other hand, among nonperturbative methods, QCD on the lattice (5) is extremely promising for understanding crucial properties such as confinement, broken chiral symmetry, hadronic masses, and matrix elements. However, computer limitations impose very drastic restrictions on the lattice size, on the accuracy of the extrapolation to the continuum limit, and on the possibility of implementing the effects of quark loops on the various observables. As a consequence, lattice calculations have thus far been more useful in our understanding of the theoretical foundations of QCD than for offering any direct comparisons with experiment.

The difficulty of testing QCD is reflected in the fact that no single process or experiment by itself provides a clear-cut and precisely quantitative experimental proof of the theory, at least when practical limitations of feasible experiments are taken into account. There are no analogues in

QCD of the $g-2$ experiment for the muon or the electron, of the Lamb shift, and so on in QED. In view of the uncertainties connected with any given experiment, our confidence in QCD rests on the overall set of convergent positive indications that arise from an increasing number of different experiments with all possible beams and targets in a wide range of large energies and momentum transfers. As a result, the integrated experimental evidence in favor of QCD is now quite impressive.

In particular, QCD provides a solid field theoretical basis for the parton approach. The beautiful "naive" parton model of Bjorken, Feynman, and others (e.g. 8) has evolved into the "QCD-improved" parton model (e.g. 9). This powerful approach has become such a familiar and widely used tool for the everyday practice of high energy physics that one is led to take all its new successes for granted and, in a way, to think of them as obvious. Actually the very definite and characteristic pattern of successful predictions given by the parton model already provides quite a solid ground of experimental evidence for QCD. In addition to inheriting the beautiful experimental score of the parton approach, QCD is also tested by an increasing list of phenomena that go beyond the naive parton model and therefore provide an additional and qualitatively different experimental basis for the specific dynamical structure of the theory.

This review is devoted to a comparison of QCD with experimental data. The theory of QCD has been summarized in recent textbooks (10) and review articles (11, 12). Here only that minimum of formalism is reported which is convenient for a clear and self-contained exposition of the experimental tests of QCD.

The QCD theory is specified by the Yang-Mills (13) Lagrangian corresponding to the unbroken color gauge group SU(3), with quark matter fields in the fundamental three-dimensional representation of SU(3):

$$\mathscr{L}_{\text{QCD}} = -\frac{1}{4} \sum_{A=1}^{8} F_{\mu\nu}^{A} F^{A\mu\nu} + \sum_{j=1}^{f} \bar{q}_j(i\slashed{D} - m_j)q_j, \qquad 1.$$

with

$$F_{\mu\nu}^{A} = \partial_\mu g_\nu^A - \partial_\nu g_\mu^A - g_s c_{ABC} g_\mu^B g_\nu^C$$

$$D_\mu = \partial_\mu + ig_s \sum_{A=1}^{8} t^A g_\mu^A, \qquad 2.$$

where f is the number of active quark flavors, g_μ^A ($A = 1, \ldots, 8$) are the gluon fields, q_j ($j = 1, \ldots, f$) are the quark fields, t^A are the generators in the quark representation, normalized according to $\text{Tr}(t^A t^B) = \frac{1}{2}\delta^{AB}$. In turn the normalization of t^A fixes the coupling g_s and the structure constants c_{ABC} given by $[t^A, t^B] = ic^{ABC}t^C$. Gauge fixing and ghost terms are to be added for a correct quantization of the theory (e.g. 14), but are omitted here. Also omitted is the P-, T-, and CP-violating θ term (15) originating from instantons (16),

$$\mathscr{L}_\theta = \theta \frac{g_s^2}{32\pi^2} \varepsilon_{\mu\nu\rho\sigma} \sum_{A=1}^{8} F^{A\mu\nu} F^{A\rho\sigma}, \qquad\qquad 3.$$

whose empirical smallness ($|\theta| \lesssim 10^{-8}$) (17) is not understood and represents an important conceptual problem for the standard model (18).

The selection of SU(3) as color gauge group is unambiguous for three reasons: (a) The group must admit complex representations because it must be able to distinguish a quark from an antiquark (there are meson states made up of $q\bar{q}$ but not similar qq bound states). (b) There must be color-singlet (because we see no color replicas of known hadrons), completely antisymmetric, baryonic states made up of qqq in order to solve the statistics puzzle for the lowest-lying baryons of spin 1/2 and 3/2 in the 56 of (flavor) SU(6) (e.g. 19). (c) As discussed below, the number of colors for each kind of quark must be in agreement with the data on the total hadronic e^+e^- cross section and on the $\pi^0 \to 2\gamma$ rate and also, to some extent, on other processes such as lepton pair production and the semileptonic branching ratio of the τ lepton. Within simple groups, selection criterion (a) above restricts the choice to SU(N) with $N \geq 3$, SO($4N+2$) with $N > 1$ [SO(6) has the same algebra as SU(4)], and E6; then (b) and (c) lead unambiguously to SU(3), with each flavor of quark in a fundamental representation of the group. Too many colored quarks not only would violate criterion (c) but could also spoil asymptotic freedom (2). In particular the $\pi^0 \to 2\gamma$ rate is fixed by the Adler-Bell-Jackiw anomaly (20) to the value (21)

$$\Gamma(\pi^0 \to 2\gamma)_{\mathrm{TH}} = N_c^2 (Q_u^2 - Q_d^2)^2 \frac{\alpha^2 m_{\pi^0}^3}{32\pi^3 f_\pi^2} = (7.61 \pm 0.02) \left(\frac{N_c}{3}\right)^2 \mathrm{eV} \qquad 4.$$

obtained for $f_\pi = 131.69 \pm 0.15$ MeV (22). N_c is the number of quark color replicas, and $Q_{u,d}$ are the u, d quark charges. The theoretical error is expected to be of order m_π^2/M^2 (because of the soft-pion extrapolation from $q^2 = m_\pi^2$ down to $q^2 = 0$), where M is the scale of variation of the relevant form factor ($M \approx 1$ GeV). The experimental value (22) is

$$\Gamma(\pi^0 \to 2\gamma)_{\mathrm{exp}} \approx 7.3 \pm 0.2 \, \mathrm{eV}. \qquad\qquad 5.$$

This beautiful result is a great achievement for the theory and a remarkable proof that $N_c = 3$.

The quantity $\alpha_s = (g_s^2/4\pi)$, the analogue in QCD of the QED fine structure constant, is widely used. The precise definition of the renormalized coupling must be specified. Technically the most practical definition is in the context of dimensional regularization (23; e.g. 24) through the procedures of minimal subtraction (MS) (25) or of modified minimal subtraction ($\overline{\text{MS}}$) (26). Unless explicitly stated, we always refer to the $\overline{\text{MS}}$ definition of α_s in the following. Note that in the perturbative region the light quark masses can be neglected in most cases. Thus the definition of α_s refers to the massless theory. The renormalization group formalism (27) leads to the concept of running coupling. The running coupling $\alpha_s(Q)$ is a function of the energy scale Q. Loosely speaking $\alpha_s(Q)$ acts as an effective coupling for QCD interactions at distances of order $1/Q$. The running coupling is defined by the relation

$$t = \int_{\alpha_s(\mu)}^{\alpha_s(Q)} \frac{d\alpha}{\beta(\alpha)}, \qquad 6.$$

where $t = \ln Q^2/\mu^2$ and the initial value $\alpha_s \equiv \alpha_s(\mu)$ has been specified. The β function, obeying the relation

$$\frac{\partial}{\partial t}\alpha_s(Q) = \beta[\alpha_s(Q)], \qquad 7.$$

can be computed in perturbation theory (2, 28, 29). For f flavors, we have

$$\beta(\alpha) = -b_f\alpha^2(1+b'_f\alpha+b''_f\alpha^2+\ldots)$$

$$= -\left(11-\frac{2}{3}f\right)\frac{\alpha^2}{4\pi}\left[1+\frac{(153-19f)\,\alpha}{2(33-2f)\,\pi}\right.$$

$$\left.+\frac{3}{32(33-2f)}\left(2857-\frac{5033}{9}f+\frac{325}{27}f^2\right)\left(\frac{\alpha}{\pi}\right)^2+\ldots\right]. \qquad 8.$$

The values of b_f (2) and b'_f (28) do not depend on the renormalization prescription, while b''_f (29) and the following terms change with the definition of α_s [however, b''_f as given in Equation 8 is the same (30) in the MS and $\overline{\text{MS}}$ prescriptions]. Numerically, one obtains for $f = 5$

$$\beta(\alpha) = -0.610\alpha^2\left[1+1.261\frac{\alpha}{\pi}+1.475\left(\frac{\alpha}{\pi}\right)^2+\ldots\right]. \qquad 9.$$

When the first two terms in the expansion of $\beta(\alpha)$ are included in Equation 6 defining $\alpha_s(Q)$, one can integrate the equation to obtain

$$\frac{1}{\alpha_s(Q)} = \frac{1}{\alpha_s}+b_f t-b'_f\ln\frac{\alpha_s(Q)}{\alpha_s}. \qquad 10.$$

To the required accuracy this equation can be solved in the form

$$\alpha_s(Q) = \alpha_{0s}(Q)\{1 - b_f'\alpha_{0s}(Q)\ln\ln Q^2/\Lambda^2 + O[\alpha_{0s}^2(Q)]\}, \qquad 11.$$

where

$$\alpha_{0s}(Q) = (b_f \ln Q^2/\Lambda^2)^{-1} \qquad 12.$$

and the parameter μ appearing in $\alpha_s \equiv \alpha_s(\mu)$ has been traded for the parameter Λ:

$$\ln\frac{\mu^2}{\Lambda^2} = \frac{1}{b_f\alpha_s} + \frac{b_f'}{b_f}\ln b_f\alpha_s. \qquad 13.$$

As b_f and b_f' do not depend on the renormalization prescription, the functional form of the running coupling is universal up to terms of order $\alpha_s^2(Q)$ included. Different definitions of the coupling imply different values of $\alpha_s(\mu)$, hence of Λ. Thus one talks of Λ_{MS}, $\Lambda_{\overline{\text{MS}}}$, etc, and for us $\Lambda \equiv \Lambda_{\overline{\text{MS}}}$. Clearly the difference only appears at next-to-leading order, because the variation of α_s appears at order α_s^2. Thus only sufficiently precise measurements analyzed in terms of theoretical formulae developed up to nonleading accuracy can really be used to determine $\Lambda_{\overline{\text{MS}}}$.

Note that b_f, b_f', ... depend on the number of excited quark flavors. In fact when working at large but fixed Q, with Q much larger than the masses of some light quarks and much smaller than the masses of some heavy quarks, then according to an intuitive decoupling theorem (31), the relevant number of flavors f is that of light quarks. In fact in QCD the theory obtained by omitting heavy quarks is still renormalizable and the couplings do not grow with masses (unlike in spontaneously broken gauge theories where the couplings of Higgs bosons and of longitudinal gauge boson modes increase with masses). The behavior of amplitudes with light external particles is dictated in this range of Q by the set of diagrams in which all internal lines are also light. For example, for $Q < am_b$ with $a \approx 1$–2, the large-Q behavior of $\alpha_s(Q)$ is determined by b_4 and b_4', while b_5 and b_5' are relevant for $\alpha_s(Q)$ at $Q \gg am_b$. Of course, the physical coupling is continuous so that some matching prescription must be specified. Experiments on e^+e^- annihilation at PETRA or TRISTAN work at $Q \approx 30$–50 GeV, well above the b threshold. Their results are naturally compared with the asymptotic formula for $f = 5$. One can extrapolate their determination of $\alpha_s(Q)$ below the b threshold by setting, for example (we only consider Q values above c threshold for simplicity)

$$\alpha_s(Q) = \alpha_{s,5}(Q)\cdot\theta(Q-am_b) + \alpha_{s,4}(Q)\cdot\theta(am_b-Q), \qquad 14.$$

with

$$\alpha_{s,5}(Q) = \alpha_s(Q, 5) \qquad\qquad 15.$$

and

$$\frac{1}{\alpha_{s,4}(Q)} = \frac{1}{\alpha_s(Q, 4)} + \frac{1}{\alpha_s(am_b, 5)} - \frac{1}{\alpha_s(am_b, 4)} \qquad 16.$$

where

$$\alpha_s(Q, f) = \frac{1}{b_f \ln Q^2/\Lambda_5^2}\left[1 - \frac{b_f' \ln \ln Q^2/\Lambda_5^2}{b_f \ln Q^2/\Lambda_5^2}\right]. \qquad 17.$$

Note that the constant addition in Equation 16 does not affect the expected asymptotic behavior coming from the logarithmically growing terms in $1/\alpha_s(Q, 4)$. This constant is fixed so that

$$\alpha_{s,4}(am_b) = \alpha_{s,5}(am_b). \qquad 18.$$

On the other hand, the results of experiments on deep inelastic scattering or on Υ decays are naturally expressed in terms of the asymptotic form for $\alpha_{s,4}(Q)$:

$$\alpha_{s,4}(Q) = \frac{1}{b_4 \ln Q^2/\Lambda_4^2}\left[1 - \frac{b_4' \ln \ln Q^2/\Lambda_4^2}{b_4 \ln Q^2/\Lambda_4^2}\right]. \qquad 19.$$

In this case it is $\alpha_{s,5}(Q)$ that is defined with a constant term in complete analogy with Equation 16. In the range of Q of practical interest, the values of Λ_4 and Λ_5 are quite different and approximately related by

$$\Lambda_5 \approx 0.65\Lambda_4. \qquad 20.$$

One sees that for practically the same $\alpha_s(Q)$ the corresponding values of Λ_4 and Λ_5 are quite different. On the other hand, there is little numerical difference when the matching parameter a is moved in the interval $1 \le a \le 2$, or when other similar matching procedures are applied.

In conclusion, when a measurement of $\alpha_s(Q)$ is translated into a value of Λ or vice versa, it is essential that the exact functional form of $\alpha_s(Q)$ in terms of Λ be specified [whether one-loop or two-loop accuracy is assumed, which of the possible forms differing at $O(\alpha_s^3)$ is adopted, the relevant number of excited flavors, the procedure for making α_s continuous at thresholds].

2. EXPERIMENTAL DETERMINATIONS OF $\alpha_s(Q)$

The most direct quantitative test of QCD is to compare several measurements of α_s in different processes. As already stated, for a meaningful determination of $\alpha_s(Q)$ at a well-defined scale Q (or equivalently for a measurement of Λ) in a given process, one needs a perturbative calculation of the corresponding observable at least at next-to-leading accuracy in α_s. In fact a change of scale Q or a change of definition of α_s only affects the result at the level of nonleading terms. In the following we review and discuss the most significant determinations of α_s from the available data (for a review, see for example 32).

2.1   *Total Hadronic Cross Section in e^+e^- Annihilation*

If we neglect quark masses, we obtain (9) for $\sigma \equiv \sigma(e^+e^- \to$ hadrons)

$$\sigma = \frac{4\pi\alpha^2}{3Q^2}(1+Z)R_{e^+e^-},\tag{21.}$$

where $Q = \sqrt{s}$, Z is the effect of the weak neutral gauge boson (which can be explicitly computed in terms of $\sin^2\theta_W$ and M_Z), and $R_{e^+e^-}$ is given (33–35) by

$$R_{e^+e^-} = 3\sum_f Q_f^2 \left\{ 1 + \frac{\alpha_s(Q)}{\pi} + (1.986 - 0.115f) \right.$$

$$\times \left[\frac{\alpha_s(Q)}{\pi}\right]^2 + (70.985 - 1.200f - 0.005f^2)$$

$$\times \left[\frac{\alpha_s(Q)}{\pi}\right]^3 - \left(\sum_f Q_f\right)^2 1.679 \left[\frac{\alpha_s(Q)}{\pi}\right]^3$$

$$\underset{f=5}{=} \frac{11}{3}\left\{ 1 + \frac{\alpha_s(Q)}{\pi} + 1.411\left[\frac{\alpha_s(Q)}{\pi}\right]^2 + 64.81\left[\frac{\alpha_s(Q)}{\pi}\right]^3 \right\},\tag{22.}$$

where in the last line the small term proportional to $(\Sigma_f Q_f)^2$ has been reabsorbed for $f = 5$ into the main correction of order α_s^2. Note that $R_{e^+e^-}$ is (essentially) proportional to N_c, so that the value of this quantity is an important direct test of the existence of three (color) replicas of quarks. The correction of order α_s was computed long ago (33) and is clearly independent of the renormalization scheme, that of order α_s^2 was computed in the $\overline{\text{MS}}$ scheme by three independent collaborations (34). Finally the calculation of $R_{e^+e^-}$ at three loops in the $\overline{\text{MS}}$ scheme was recently completed by Gorishny et al (35). The precise expression of R given in Equation 22 only applies to the photon exchange term. Differences due to the axial Z couplings are neglected in Equations 21 and 22.

Two different groups have analyzed all available data in the range

$7 < Q < 56$ GeV in order to extract α_s. De Boer (36) fixes $\sin^2 \theta_W = 0.23$ (the world average value) and obtains

$$\alpha_s(Q = 34\,\text{GeV}) = 0.140 \pm 0.016. \qquad\qquad 23.$$

Marshall (37) presents a combined fit of α_s and $\sin^2 \theta_W$. Together with the fitted result $\sin^2 \theta_W = 0.242 \pm 0.017$, the corresponding value of α_s is given by

$$\alpha_s(Q = 34\,\text{GeV}) = 0.135 \pm 0.016. \qquad\qquad 24.$$

Note that the effect of the order α_s^3 correction to $R_{e^+e^-}$ is large enough to change the value of α_s by $\sim 10\%$ (the central value decreases from 0.155 to 0.140). Although this variation of α_s is within the overall error, still the size of the coefficient of $(\alpha_s/\pi)^3$ in the expansion of $R_{e^+e^-}$ is somewhat disturbing. In fact, with the value of α_s given by Equation (23) the expansion of $R_{e^+e^-}$ reads

$$R_{e^+e^-} = \frac{11}{3}\left[1 + r_1 \frac{\alpha_s}{\pi} + r_2 \left(\frac{\alpha_s}{\pi}\right)^2 + r_3 \left(\frac{\alpha_s}{\pi}\right)^3 + \cdots\right]$$

$$\simeq \frac{11}{3}[1 + 0.0446 + 0.0028 + 0.0057 + \cdots]. \qquad\qquad 25.$$

Strictly speaking, beyond one loop the coefficients of the expansion depend both on the renormalization scheme and on the choice of scale Q; however, the large ratio of r_3 with respect to r_1 and r_2 makes it hopeless to improve the situation substantially by a reasonable reparameterization. Note that as far as one can tell from the three-loop result of Equation 9, the β-function expansion in the MS or $\overline{\text{MS}}$ prescriptions is well behaved. The computation of the $O(\alpha_s^3)$ corrections to $R_{e^+e^-}$ dramatically reveals the limitations of all the so-called optimization procedures (38). For whatever choice made at the two-loop level among these procedures, the result at three loops is always about as bad as in Equation 25. This clearly shows that the optimization choices cannot pretend to reduce the theoretical error.

In conclusion, the determination of α_s through $R_{e^+e^-}$ is in principle very clear. Unfortunately this method has a limited sensitivity because the QCD correction is quite small. In spite of that, the result obtained by combining all experiments together is amazingly precise. Rounding off the errors we can summarize the results as follows:

$$\alpha_s(Q = 34\,\text{GeV}) = 0.14 \pm 0.02 \qquad\qquad 26.$$

$$\Lambda_{\overline{\text{MS}}}^{(5)} \approx 0.65 \Lambda_{\overline{\text{MS}}}^{(4)} \approx 240^{+230}_{-140}\,\text{MeV}. \qquad\qquad 27.$$

2.2   Scaling Violations in Deep Inelastic Leptoproduction

In principle, deep inelastic leptoproduction is the most solid and powerful method for testing perturbative QCD and measuring α_s. As for the total hadronic cross section in e^+e^- annihilation, the underlying theory here is very well founded (9) [for example in terms of the light-cone operator expansion (39)] and, the process being completely inclusive, there are no problems associated with the experimental definition of jets and the relation with theoretical partonic cross sections. But with respect to $\sigma(e^+e^- \to$ hadrons) there are essential advantages. First, there are many independent structure functions (e.g. 40) and all of them can be measured at different values of the Bjorken variable x for each given $Q^2 = -q^2$ (where q_μ is the virtual γ or $W^\pm$ or Z four-momentum). Thus one actually deals with a system of tests. (Typically, after a suitable binning, one can compare with theory a number of $\ln Q^2$ slopes at different values of x.) Second, for the structure functions the scaling violations are quantitatively more important than the small $O(\alpha_s)$ corrections to $\sigma(e^+e^- \to$ hadrons) because they arise from the resummation of a series of logarithmically enhanced terms.

Over the last years an imposing experimental effort has been devoted to the measurement of scaling violations in deep inelastic scattering with electron or muon, neutrino, or antineutrino beams (e.g. 41) on hydrogen, deuterium, and heavy nuclei. A new generation of high precision experiments has been completed [EMC (42), BCDMS (43), CDHSW (44)]. Many important predictions of the theory have been confirmed. The existence of scaling violations is definitely established at Q^2 values large enough to support the prediction that their asymptotic decrease is only logarithmic. The observed pattern and magnitude of the scale-breaking effects are in good agreement with the theoretical expectations. The values of Λ extracted from the data are consistent with other experimental derivations.

Upon closer inspection, however, several serious problems remain in the comparison of the data among themselves and with the theory, so that even after so much theoretical and experimental work the situation is still not clear and satisfactory. There are experimental discrepancies that are certainly beyond the declared systematic and statistical errors among different measurements of the same observable. The most disturbing of these is perhaps the difference (45) between the BCDMS and the EMC measurements of the structure function $F_2(x, Q^2)$ with muon beams on hydrogen (see Figure 1) (a priori the most accurate and comprehensive sets of hydrogen data). This discrepancy not only reflects itself in a sizeable difference in the values of the proton parton densities useful for physics at pp, p$\bar{\text{p}}$, and ep colliders (which are preferentially extracted from hydrogen data to avoid possible nuclear effects), but it also casts serious doubt on the

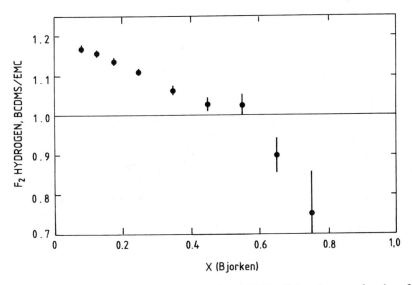

Figure 1   Ratio of F_2^p measured by the BCDMS and EMC collaborations as a function of x (45). Only statistical errors are shown.

whole analysis of scaling violations. In fact it turns out a posteriori that the level of uncontrolled systematics present in at least one of the two experiments is large enough to affect considerably the measurement of α_s and Λ (45).

A different problem has to do with experiments on an iron target. In this case (45) there is reasonable consistency among different experiments but the data show deviations from the expected behavior. In particular, complete agreement with the theory is obtained only at very large values of Q and W (the invariant mass of the produced hadronic system), where the statistical accuracy is unfortunately small.

We next summarize the status of QCD tests based on scaling violations in leptoproduction. The Q^2 dependence of structure functions is dictated by the QCD evolution equations (46). In the nonsinglet case these equations are simplest:

$$Q^2 \frac{\partial}{\partial Q^2} \ln F_a^{NS}(x, Q^2) = \frac{\alpha_s(Q)}{2\pi} \frac{1}{F_a^{NS}(x, Q^2)}$$

$$\times \int_x^1 \frac{dy}{y} F_a^{NS}(y, Q^2) \mathscr{P}_{qq}^{NS(a)}\left(\frac{x}{y}, \alpha_s(Q)\right), \qquad 28.$$

where in leading approximation the kernel $\mathscr{P}_{qq}^{NS(a)}(z, \alpha_s)$ coincides with the simple distribution (46)

$$P_{qq}(z) = \frac{4}{3}\left[\frac{1+z^2}{(1-z)_+} + \frac{3}{2}\delta(1-z)\right]. \qquad 29.$$

Note that the lowest-order kernel P_{qq} is independent of the index a, i.e. it is the same for all nonsinglet structure functions. The next-to-leading order corrections to $\mathscr{P}_{qq}^{NS(a)}(z, \alpha_s)$ have been computed (see 47).

The nonsinglet evolution equations are exactly valid for the difference of any given structure function measured on proton and neutron targets, i.e. $F^{NS} = F^p - F^n$, or for the structure function F_3, which is given by the difference of neutrino and antineutrino scattering on a given target. Thus the analysis based on F_3 (or $F^p - F^n$) is particularly clear in principle but it suffers from the relatively large errors arising from taking a difference of cross sections (in any case F_3 is not accessible to experiments with muon beams).

For a general structure function one can distinguish nonsinglet and singlet parts. For example, for $\mathscr{F}_2 = F_2/x$ one starts from

$$\mathscr{F}_2(x, Q^2) = \sum_{i=1}^{2f} c_i q_i(x, Q^2). \qquad 30.$$

where the sum runs over all active flavors of quarks and antiquarks, and where c_i are the relevant electroweak charges. Equation 30 is certainly valid to lowest order and can be taken as a redefinition of quark densities beyond leading order (48). Then

$$\mathscr{F}_2 = \mathscr{F}_2^{NS} + \mathscr{F}_2^S = \sum_{i=1}^{2f} (c_i - \langle c\rangle)q_i + \langle c\rangle\Sigma, \qquad 31.$$

where $\langle c\rangle = (\Sigma_{i=1}^{2f} c_i)/2f$ and $\Sigma = \Sigma_{i=1}^{2f} q_i$. A gluon term is also present in the singlet evolution equation (46):

$$Q^2\frac{\partial}{\partial Q^2}\mathscr{F}_2^S(x, Q^2) = \frac{\alpha_s(Q)}{2\pi}\int_x^1 \frac{dy}{y}\left\{\mathscr{F}_2^S(y, Q^2)\mathscr{P}_{qq}^{S(2)}\left[\frac{x}{y}, \alpha_s(Q)\right]\right.$$
$$\left. + 2f\langle c\rangle g(y, Q^2)\mathscr{P}_{qg}^{(2)}\left[\frac{x}{y}, \alpha_s(Q)\right]\right\}. \qquad 32.$$

The system is then closed by an analogous equation for the gluon density:

$$Q^2\frac{\partial}{\partial Q^2}g(x, Q^2) = \frac{\alpha_s(Q)}{2\pi}\int_x^1 \frac{dy}{y}\left\{\frac{\mathscr{F}_2^S(y, Q^2)}{\langle c\rangle}\mathscr{P}_{gq}^{(2)}\left[\frac{x}{y}, \alpha_s(Q)\right]\right.$$
$$\left. + g(y, Q^2)\mathscr{P}_{gg}^{(2)}\left[\frac{x}{y}, \alpha_s(Q)\right]\right\}. \qquad 33.$$

The next-to-leading order corrections to the splitting functions $\mathscr{P}$ for the singlet case have been computed (49, 50). While the lowest-order kernels are totally unambiguous, the corresponding corrections of order α_s depend on the exact definition of quark and gluon densities beyond the leading order. For example, Equation 30 provides a possible definition of quark densities to all orders (48).

An important feature of the QCD evolution equations, evident from Equations 28, 32, and 33, is that the Q^2 derivative at x of a given structure function only depends on the quark and gluon densities at $y \geqslant x$. This allows us to predict the Q^2 evolution from the values of x actually measured. In fact, at fixed Q^2, it is not possible in practice to reach too small a value of x. Furthermore it is empirically true and theoretically reasonable that glue and sea densities are negligible with respect to valence quark densities at sufficiently large x; thus the gluon term in the singlet equation can correspondingly be omitted. Since the singlet kernel $\mathscr{P}_{qq}^S$ also approaches $\mathscr{P}_{qq}^{NS}$ at large x $[|\mathscr{P}_{qq}^S - \mathscr{P}_{qq}^{NS}| \sim (1-x)^5$ near $x \to 1]$ (9), one can use the much simpler nonsinglet equation at $x > x_0$, with a suitable value of x_0. In practice, as we shall see, $x_0 \approx 0.25$–0.30 is normally adopted.

In the evolution equations there are two variables, x and Q^2. The shape in x of the structure functions at fixed $Q^2 = Q_0^2$ is not a prediction of perturbative QCD. But given the x dependence of the structure function at $Q^2 = Q_0^2$, one can predict the shape from the evolution equations at all Q^2 (in the singlet case, the shape of the gluon density is also needed at $Q^2 = Q_0^2$). Although the x and Q^2 dependences are coupled by the evolution equations, it is clear that for QCD tests what matters most is the Q^2 variation at fixed x rather than the x variation at fixed Q^2 [for a recent discussion of this point, see (50a)]. When the limited range in Q^2 and the experimental errors are taken into account, one realizes that the QCD test in the nonsinglet case essentially consists in checking that a single value of Λ can accommodate the measured logarithmic slopes $d \ln F/d \ln Q^2$ at a number of fixed values of x. It is beyond our present capabilities to measure (within a single experiment) significant deviations from a linear behavior in $\ln Q^2$. In the singlet case the gluon density, in addition to Λ, is also to be determined from the logarithmic slopes. In fact the gluons are not directly coupled and their distribution is also to be inferred from the scaling violations (or from processes other than leptoproduction, e.g. large-p_T photons in $p\bar{p}$ collisions).

After this concise theoretical summary we next review the data and the corresponding QCD analysis at next-to-leading order accuracy.

The BCDMS collaboration (43) has measured F_2 with muon beams on carbon and hydrogen. This experiment has the largest statistics at large Q^2. The carbon data has $Q^2 > 25$ GeV^2 in the range $0.275 < x < 0.75$.

The extrapolation to $x > 0.75$ does not introduce an important error because the structure functions are very small in this range. The data are analyzed in the nonsinglet approximation. The results for the logarithmic slopes are shown in Figure 2. The corresponding determination of $\Lambda_{\text{MS}}^{(4)}$ for four flavors leads to the result

$$\Lambda_{\text{MS}}^{(4)} = 230 \pm 20 \pm 60 \, \text{MeV}, \qquad 34.$$

which corresponds to

$$\alpha_{\text{s}}(Q = 10 \, \text{GeV}) = 0.160 \pm 0.003 \pm 0.010. \qquad 35.$$

The BCDMS collaboration (43) has also analyzed the data on hydrogen in the nonsinglet approximation for $x \geqslant 0.275$ with $Q^2 > 20 \, \text{GeV}^2$. The logarithmic slopes on hydrogen are shown in Figure 3, together with the QCD fit. Here one finds

$$\Lambda_{\text{MS}}^{(4)} = 205 \pm 22 \pm 60 \, \text{MeV} \qquad 36.$$

or

$$\alpha_{\text{s}}(Q = 10 \, \text{GeV}) = 0.156 \pm 0.004 \pm 0.011. \qquad 37.$$

The hydrogen fit is in perfect agreement with the results on carbon. They can be combined to give

$$\Lambda_{\text{MS}}^{(4)} \approx 1.54 \Lambda_{\text{MS}}^{(5)} = 220 \pm 15 \pm 50 \, \text{MeV}. \qquad 38.$$

The hydrogen data on F_2 from BCDMS are also available in the range $0.07 \leqslant x < 0.275$ (with $Q^2 > 8 \, \text{GeV}^2$ for $x < 0.16$ and $Q^2 > 14 \, \text{GeV}^2$ for

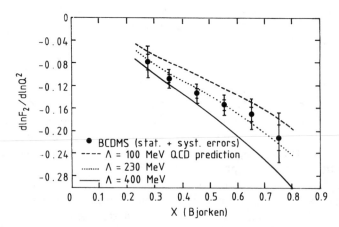

Figure 2   The logarithmic slopes of F_2 measured on carbon by BCDMS (43) compared to the next-to-leading order nonsinglet QCD evolution for the indicated values of $\Lambda \equiv \Lambda_{\text{MS}}^{(4)}$.

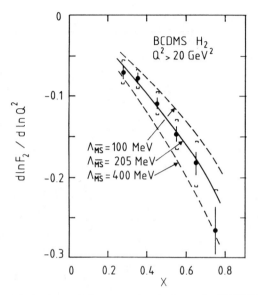

Figure 3 The logarithmic slopes of F_2 measured on hydrogen by BCDMS (43) compared to the next-to-leading order nonsinglet QCD evolution for the indicated values of $\Lambda \equiv \Lambda_{\overline{MS}}^{(4)}$.

$0.16 < x < 0.25$) so that a singlet fit can be performed at small x (Figure 4). In this region of x the logarithmic slopes require a sizeable gluon density. The resulting gluon density is shown in Figure 14. The gluon distribution is indeed concentrated at small values of x as demanded by consistency because the gluon term in the evolution equations was neglected for $x \geqslant 0.275$ in the nonsinglet analysis. In conclusion, the BCDMS analysis presents a remarkable consistency among carbon data, hydrogen data at large x, and hydrogen data at small x. It also shows a beautiful agreement with perturbative QCD predictions.

Unfortunately this idyllic picture is somewhat spoiled by the results from other experiments of a priori comparable precision. As already mentioned, there is a severe disagreement (45) (about three times larger than that allowed by the quoted systematic errors) between the BCDMS and the EMC data on F_2 for proton targets, shown in Figure 1. Previous SLAC data (51) on F_2^p (up to $Q^2 \approx 20$ GeV^2) cannot resolve this discrepancy because there is essentially no overlap in x and Q^2. It is true that the discrepancy is mainly on the normalization of F_2 at different x and not on the logarithmic slopes (Figure 5). The EMC data are indeed also consistent within errors with QCD. The nonsinglet fit to F_2^p by EMC at $x \geqslant 0.35$ and $Q^2 > 8$ GeV^2 (with $\langle Q^2 \rangle \approx 22.5$ GeV^2) leads to

$$\Lambda_{\overline{MS}}^{(4)} = 105^{+55}_{-45}\,{}^{+85}_{-45}\,\text{MeV}. \qquad\qquad 39.$$

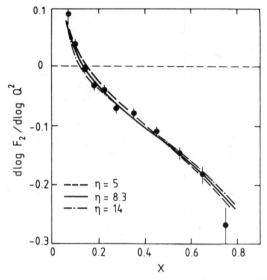

Figure 4 The logarithmic slopes of F_2 measured by BCDMS (43) on hydrogen compared to the next-to-leading order singlet QCD evolution with $\Lambda^{(4)}_{\overline{\text{MS}}} \approx 220$ MeV and gluon density $xg(x, Q^2) \approx A(1-x)^\eta$ for $Q^2 = 5$ GeV^2.

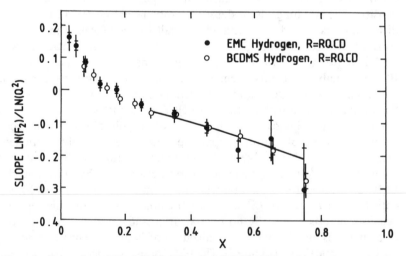

Figure 5 Comparison (45) of EMC (42) and BCDMS (43) logarithmic slopes of F_2 on hydrogen. Curve is next-to-leading order QCD fit to BCDMS data.

This value of $\Lambda_{\overline{MS}}$ is consistent with the corresponding BCDMS result in Equation 36, although the EMC central value is smaller by a factor of two. The real problem is that the discrepancy indicates an uncontrolled systematic error large enough (45) to make the agreement with QCD and the consistency of the fitted values of $\Lambda_{\overline{MS}}^{(4)}$ to some extent accidental.

The BCDMS results for $\Lambda_{\overline{MS}}$ obtained from the data on carbon are compatible with the CHARM collaboration results (52) obtained from neutrino scattering on marble ($CaCO_3$), a target not too much heavier than carbon ($A \approx 20$ vs $A \approx 12$). The CHARM result, obtained from the nonsinglet structure function F_3 at next-to-leading order accuracy for $Q^2 \approx 3$–78 GeV^2, is given by

$$\Lambda_{\overline{MS}}^{(4)} = 310 \pm 140 \pm 70 \, \text{MeV}. \qquad\qquad 40.$$

There are many high statistics experiments on the iron structure functions ($A \approx 56$). F_2^{Fe} has been measured by CCFRR (53), CDHSW (44) with v beams, and by BFP (54) and EMC (42) with muon beams. The data are in reasonable agreement within the stated uncertainties, although the EMC data are 5–10% below the other data sets. Similarly the xF_3 measurements from CCFRR and CDHSW are also consistent. For iron structure functions the logarithmic slopes in general show a steeper x dependence than expected from QCD, with values of $\Lambda_{\overline{MS}}$ compatible with Equations 38 and 39. For example, a comparison of the EMC data on F_2^{Fe} with the nonsinglet QCD fit obtained at next-to-leading order accuracy using the value of $\Lambda_{\overline{MS}}$ measured by BCDMS on carbon is shown (45) in Figure 6 (together with possible modifications induced by a model of higher twist effects and target mass corrections). A quite similar behavior is also observed in the logarithmic slopes of xF_3 recently measured by CDHSW (44). In this case agreement with QCD is restored for $Q^2 > 20$ GeV, within a more limited accuracy determined by the smaller statistics. The iron data suggest that pre-asymptotic or nonperturbative effects could be relatively more important in heavy targets.

In conclusion, the BCDMS data on H and C have the highest statistics at the largest values of Q^2. These data are in beautiful agreement with QCD and lead to the value of $\Lambda_{\overline{MS}}^{(4)}$ quoted in Equation 38. This value of $\Lambda_{\overline{MS}}^{(4)}$ is in agreement with the values of $\Lambda_{\overline{MS}}^{(4)}$ quoted by EMC on H_2 and by CHARM on $CaCO_3$. However, the agreement with EMC is somewhat illusory in view of the large discrepancies between BCDMS and EMC on F_2^p. The data on Fe are unfortunately not very conclusive as tests of QCD and measurements of α_s.

2.3 Quarkonium Decays

The rates of quarkonium (especially the Υ) provide a nominally rather precise experimental determination of α_s and $\Lambda_{\overline{MS}}$. The problem they have

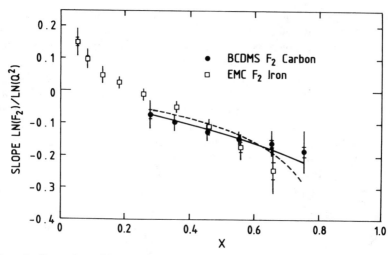

Figure 6 Comparison of data from EMC (42) on F_2 on iron and from BCDMS (43) on F_2 on carbon. The slopes on iron are apparently steeper than those on carbon. Solid curve is QCD fit to carbon data; dashed curve includes nuclear and higher twist corrections.

is the theoretical error. In the nonrelativistic approximation the decay rates are proportional to the absolute square of the wave function at the origin. Thus, within the limits of this approximation, ratios of decay rates are independent of the wave function, which is unknown. The most convenient ratios for determining α_s are $\Gamma_{\gamma gg}/\Gamma_{ggg}$ and $\Gamma_{\mu\mu}/\Gamma_{ggg}$. Next-to-leading order calculations of these ratios have been performed in the $\overline{\text{MS}}$ scheme (55). For the upsilon, the results are

$$\frac{\Gamma_{\mu\mu}}{\Gamma_{ggg}} = \frac{9\pi}{10(\pi^2-9)} \frac{\alpha^2}{\alpha_s^3(\mu)} \left\{ 1 - \frac{\alpha_s(\mu)}{\pi} \left(3\pi b_4 \ln \frac{\mu^2}{m_Q^2} + 0.43 \right) + \cdots \right\} \qquad 41.$$

$$\frac{\Gamma_{\gamma gg}}{\Gamma_{ggg}} = \frac{4}{5} \frac{\alpha}{\alpha_s(\mu)} \left\{ 1 - \frac{\alpha_s(\mu)}{\pi} \left(\pi b_4 \ln \frac{\mu^2}{m_Q^2} + 2.6 \right) + \cdots \right\}, \qquad 42.$$

where $b_4 = (33-2f)/12\pi$ ($f = 4$) and μ is an arbitrary scale of the order of the heavy quark mass $m_Q \approx (m_\Upsilon/2)$.

The experimental value of $\Gamma_{\mu\mu}/\Gamma_{ggg}$ is known for $\Upsilon(1S)$, $\Upsilon(2S)$, and $\Upsilon(3S)$. It is obtained by the relation $\Gamma_{ggg}/\Gamma_{\mu\mu} = (1-B_{\gamma gg}-B_{\pi\pi}-B_{E1})/B_{\mu\mu}-3-R$, where $R \approx 3.48$ and $B_{\pi\pi}$ and B_{E1} are absent for $\Upsilon(1S)$. Recent precise measurements (56) by CUSB and CLEO when combined lead to $B_{\mu\mu} = 2.57 \pm 0.09$, 1.38 ± 0.29, and 1.73 ± 0.16, for $\Upsilon(1S)$, $\Upsilon(2S)$, and $\Upsilon(3S)$ respectively. These values lead to $\Gamma_{ggg}/\Gamma_{\mu\mu} = 31.3 \pm 1.3$, 32.5 ± 8.3, 31.0 ± 2.7 respectively. For $\mu = m_b \approx 4.9$ GeV one obtains

1S: $\alpha_s(m_b) = 0.1743 \pm 0.0024$, $\Lambda^{(4)}_{\overline{MS}} = 159 \pm 9$ MeV   43.

2S: $\alpha_s(m_b) = 0.175 \pm 0.015$, $\Lambda^{(4)}_{\overline{MS}} = 162 \pm 54$ MeV   44.

3S: $\alpha_s(m_b) = 0.1742 \pm 0.0021$, $\Lambda^{(4)}_{\overline{MS}} = 155 \pm 18$ MeV.   45.

Similarly the world average for $\Gamma_{\gamma gg}/\Gamma_{ggg}$ measured by the CUSB, CLEO, ARGUS, and Crystal Ball collaborations is $2.78 \pm 0.15\%$ (57). From this value and Equation 42 one obtains

$$\alpha_s(m_b) = 0.181 \pm 0.009.   46.$$

The agreement among the values of α_s derived from $\Gamma_{\mu\mu}/\Gamma_{ggg}$ for $\Upsilon(1S)$, $\Upsilon(2S)$, and $\Upsilon(3S)$ and from $\Gamma_{\gamma gg}/\Gamma_{ggg}$ for $\Upsilon(1S)$ is remarkable and is an experimental check of the wave function factorization. Actually, consistent values of α_s are also obtained (though with more uncertainty) from charmonium decays. An overall fit of the available data, including a crude estimate of relativistic corrections, was performed by Kwong et al (58). They found $\alpha_s(m_c) \approx 0.278 \pm 0.014$ and $\alpha_s(m_b) = 0.185 \pm 0.006$, which correspond to $\Lambda^{(4)}_{\overline{MS}} = 199 \pm 22$ MeV.

In the stated results for α_s and $\Lambda_{\overline{MS}}$ the error shown does not clearly include the theoretical error. This is certainly the largest source of uncertainty. Corrections to the nonrelativistic approximation can still be sizeable in spite of the experimental success of factorization. The order of magnitude of v^2/c^2 is in fact ~ 0.25 for charmonium and ~ 0.10 for the Υ system. The effects of higher perturbative orders and of nonperturbative terms could be large because the energy scale is relatively small. For $\Gamma_{\gamma gg}/\Gamma_{ggg}$ I see a further problem in the fact that the observed photon spectrum is not well understood in perturbation theory. The lowest-order spectrum is definitely too hard to accommodate the data (57). Field (59) very convincingly explains the observed soft-γ spectrum as being due to an effective mass of gluon jets. While the parton gluon is massless, the physical gluon jet has a nonvanishing invariant mass. By a Monte Carlo simulation, Field (59) has shown that an average mass $\langle M \rangle \approx 1.6$ GeV should be attributed to the gluon jet in order to reproduce the data. Perturbative effects (gluon splitting into $q\bar{q}$ or gg) should indeed induce an invariant mass of order $\alpha_s M_\Upsilon$, which is not far from the inferred value of $\langle M \rangle$. However, the only existing calculation of the perturbative corrections (60) to the normalized spectrum gives a negligible improvement. This calculation could be wrong, and in fact it is somewhat obscure in many respects. It is important to clarify this point because if the spectrum is indeed dominated by nonperturbative (or higher order) effects, then the perturbative evaluation of the total width could also be to some extent affected, even if inclusive quantities are usually more protected.

In conclusion, it appears difficult to me to compress the total theoretical error below the 10–20% level. Actually, the theoretical error is relatively so small only because the different measurements on the Υ system are remarkably consistent. Thus I would tentatively conclude that

$$\alpha_s(m_b) \approx 0.175(1 \pm 15\%). \qquad 47.$$

Even with this enlarged error the resulting determination of $\Lambda_{\overline{MS}}^{(4)}$ is comparatively quite good:

$$\Lambda_{\overline{MS}}^{(4)} \approx 1.54 \Lambda_{\overline{MS}}^{(5)} = 180 \pm 80 \, \text{MeV}. \qquad 48.$$

2.4 $e^+e^- \to Jets$

All methods of measuring α_s described in the previous sections are based on totally inclusive processes. We now consider the determination of α_s from the observed properties of jets in the final state. The study of jets in e^+e^- annihilation has provided a formidable laboratory for QCD testing for about a decade. Many striking confirmations of the theory have been obtained (e.g. 61, 61a): the observation of the predicted jet structure and the expected hierarchy of two, three, four, . . . jets, an indication of gluons and their vector nature, the quantitative correspondence between the observed distributions in energy and angles, and the QCD matrix elements. The more general aspects of jet physics are discussed in the next section. We concentrate here on the measurement of α_s from jets in e^+e^- annihilation.

The principle of the method is to measure a quantity that is zero in lowest order (corresponding only to a quark-antiquark pair in the final state) and starts at order α_s (quark-antiquark-gluon). For a meaningful determination of α_s it is necessary to know the same quantity at order α_s^2. This implies computing virtual corrections to three-parton amplitudes and real four-parton matrix elements. As is well known, the contribution to the rate of virtual and real diagrams is separately divergent, while only the sum is finite for well-defined physical observables. The additional difficulty of jet physics consists in the obvious fact that the theory deals with partons and the physical observables are jets of hadrons. The relation between partons and the experimentally defined jets necessarily requires some model of nonperturbative fragmentation and hadronization. While nonperturbative effects should asymptotically become negligible, their influence on the extracted value of α_s is still sizeable at PEP/PETRA energies and is the main source of error.

For example, assume that one wants to compute some three-jet distribution $d\sigma$(three-jets). As we have seen, the perturbative calculation at order α_s^2 needs "jet-dressing" to become finite. One must add to the contribution of three partons the integral over unresolved configurations from final states with four partons:

$$d\sigma(\text{3-jets}) = d\sigma(\text{3-partons}) + \int_{NR} d\sigma(\text{4-partons}),$$ 49.

where NR indicates the nonresolved configurations, i.e. those in which any two partons are too close to be separated so that the event is observed as a three-jet event. Clearly some jet resolution criterion is needed. Typically this is a cut on the invariant mass y_{ij} of a pair i,j of partons: below a given value of y_{ij} the pair is detected as a single jet. Evidently the presence of these cuts introduces the problem of a cut-dependent result. Furthermore, the experimental jet identification criterion based on observed hadrons can only be translated into a cut on parton variables by means of a model of fragmentation and hadronization. Considerable progress has been made (62) in understanding early discrepancies from different calculations (63, 64) based on different four-parton resolution criteria. Also, to determine α_s, one now selects some global quantity [e.g. oblateness (9) or energy-energy correlations and their asymmetry (65, 66)] that is independent or less dependent on jet resolution criteria. Of course the dependence on fragmentation and hadronization effects always remains, even if care is taken to concentrate on quantities that are invariant under collinear splitting of one into two massless particles. Finally, one demands a good apparent convergence, i.e. that the resulting nonleading correction of order α_s^2 is not too large for a natural choice of the renormalization scale μ appearing in the leading term proportional to $\alpha_s(\mu)$.

At present the most common determination of α_s is based on the asymmetry of energy-energy correlations (AEEC) (65, 66). The energy-energy correlation (EEC) is defined by

$$\frac{1}{\sigma}\frac{d\Sigma}{d\cos\chi} = \frac{1}{\sigma}\sum_{i,j}\int\frac{d\sigma}{dx_i\,dx_j\,d\cos\chi}x_i x_j\,dx_i\,dx_j$$

$$\approx \frac{1}{N_{\text{events}}}\sum_{\text{events}}\sum_{i,j}x_i x_j\delta(\cos\theta_{ij}-\cos\chi),$$ 50.

where χ is the fixed angle between two calorimeter cells, $x = 2E/\sqrt{s}$. Clearly the contribution at χ not too close to 0 and π arises from noncollinear events. The energy weights make EEC infrared safe; the linearity in x_i guarantees invariance under collinear splitting of particle i. The AEEC is defined as

$$\frac{1}{\sigma}\frac{d\Sigma^{\mathrm{AEEC}}}{d\cos\chi} = \frac{1}{\sigma}\frac{d\Sigma}{d\cos\chi}(\pi-\chi) - \frac{1}{\sigma}\frac{d\Sigma}{d\cos\chi}(\chi). \qquad 51.$$

The asymmetry is different from zero because in a typical three-jet event there is a narrow jet in one hemisphere and a broad di-jet in the other one.

A purely perturbative calculation leads to

$$\frac{1}{\sigma}\frac{d\Sigma^{\mathrm{AEEC}}}{d\cos\chi} = \frac{\alpha_s(Q)}{\pi}A(\cos\chi)\left[1 + \frac{\alpha_s(Q)}{\pi}R(\cos\chi) + \cdots\right]. \qquad 52.$$

In the range $-0.95 < \cos\chi < 0.95$ the value of R (67) varies between 2.5 and 3.5, so that the expansion is apparently well behaved. For $|\cos\chi|$ near 1 the perturbative expansion should be improved by a resummation of the corresponding singularities.

Experimentally it is found that for $\chi > 30°$ the purely perturbative evaluation of the AEEC distribution at order α_s^2 provides an excellent fit to the data. The results on α_s determined from the purely perturbative fit are reproduced in Table 1 (67). By taking the average one obtains

$$\alpha_s(Q = 34\,\mathrm{GeV}) = 0.121 \pm 0.003$$
$$\text{(perturbative)}$$
$$\Lambda_{\overline{\mathrm{MS}}}^{(5)} = 100 \pm 15\,\mathrm{MeV}. \qquad 53.$$

As usual, the problem is to estimate the theoretical error, most of which is expected to arise from fragmentation and hadronization effects. Thus the most natural thing to do is to compare the purely perturbative result with those obtained by including models of jet formation. Following Ali & Barreiro (67), who discuss the idea more completely, we report the results based on two different Monte Carlo analyses including the perturbative calculations of order α_s^2 and a model of fragmentation: a model by Ali et al (68; see also 68a) and a version of the Lund model (69). The

Table 1 Values (67) of $\alpha_s(Q)$ and $\Lambda_{\overline{\mathrm{MS}}}^{(5)}$ derived from a purely perturbative treatment of the asymmetry of energy-energy correlations.

Q (GeV)	$\alpha_s(Q)$	$\Lambda_{\overline{\mathrm{MS}}}^{(5)}$ MeV	Exp.
14	0.14 ± 0.01	100^{+40}_{-35}	JADE
22	0.13 ± 0.01	100^{+55}_{-40}	JADE
34	0.115 ± 0.005	70^{+25}_{-20}	JADE
34.6	0.125 ± 0.005	125^{+35}_{-30}	PLUTO
34.8	0.125 ± 0.005	125^{+35}_{-30}	TASSO

results from these analyses are reported in Table 2 (67). The average values are

$$\alpha_s(Q = 34 \, \text{GeV}) \approx 0.128 \pm 0.003$$

$$\Lambda_{\overline{\text{MS}}}^{(5)} = 144 \pm 16 \, \text{MeV}$$

(Ali et al)   54.

$$\alpha_s(Q = 34 \, \text{GeV}) = 0.148 \pm 0.002$$

$$\Lambda_{\overline{\text{MS}}}^{(5)} = 325 \pm 20 \, \text{MeV}.$$

(Lund)   55.

These results are in agreement with some other existing measurements (61, 61a, 67) based on oblateness or the planar triple energy correlation (70). There is indeed a systematic difference in the results from the Ali et al and the Lund model. Both models lead to an increase of α_s, but in the former the change with respect to the perturbative result is much smaller than in the latter. The dispersion of the results can be taken as an indication of the theoretical error. From $e^+e^- \to$ jets one therefore concludes that

$$\alpha_s(Q = 34 \, \text{GeV}) \approx 0.135 \pm 0.015$$

$$\Lambda_{\overline{\text{MS}}}^{(5)} = 215 \pm 130 \, \text{MeV}.$$

56.

2.5 Other Processes

The main source of additional information on α_s is obtained from $\gamma\gamma$ reactions (for a recent review, see 71). The photon structure function F_2^γ measured in $\gamma\gamma$ collisions (one tagged photon of virtual squared mass $-Q^2$ on a quasi-real photon) is special because it is predicted to grow as $\ln Q^2$ (72). The logarithmic increase of F_2^γ is well supported by the data and is a nice confirmation of asymptotic freedom, which preserves the $\ln Q^2$ behavior in the presence of QCD corrections, even though the corrections

Table 2 Values (67) of $\alpha_s(Q)$ and $\Lambda_{\overline{\text{MS}}}^{(5)}$ derived from the asymmetry of energy-energy correlations by adding to the perturbative treatment a model of fragmentation and hadronization

Q (GeV)	Model	$\alpha_s(Q)$	$\Lambda_{\overline{\text{MS}}}^{(5)}$ (MeV)	Exp.
35	Ali et al (68)	0.122 ± 0.004	108^{+24}_{-21}	MARK J
44	Ali et al (68)	$0.129 \pm 0.004 \pm 0.011$	190 ± 40	TASSO
35	Ali et al (68)	0.133 ± 0.004	180 ± 30	PLUTO
35	Lund (69)	0.137 ± 0.005	217^{+48}_{-42}	MARK J
44	Lund (69)	$0.143 \pm 0.005 \pm 0.012$	340^{+70}_{-60}	TASSO
35	Lund (69)	0.142 ± 0.004	260 ± 40	PLUTO
35	Lund (69)	$0.157 \pm 0.005 \pm 0.012$	440^{+60}_{-50}	CELLO
29	Lund (69)	$0.158 \pm 0.003 \pm 0.008$	380^{+40}_{-30}	MARK II

markedly modify the shape of the structure function (making it considerably softer).

The leading pointlike component is to a large extent computable (especially at relatively large x). However, early hopes of measuring α_s, free from hadronic nonperturbative unknowns, directly from the observed values of F_s^γ (at sufficiently large Q^2 in some range of not too small x) cannot be fulfilled. It is now generally recognized that the determination of $\Lambda_{\overline{MS}}$ from F_2^γ requires data at different values of Q^2 (71). In fact it is clear that in order to measure $\Lambda_{\overline{MS}}$ complete control of the next-to-leading order terms is necessary. At that level (73) the complete separation of the computable pointlike-photon contribution from the hadronic terms becomes impossible. These terms arise from the hadronic component of the photon as visualized for example by vector-meson dominance (the virtual photon scatters on a ρ, ω, ϕ, ... in the quasi-real photon). Actually spurious singularities in the space of moments (which affect the behavior of F_2^γ at small x) are generated in the singlet sector if hadronic terms are not properly taken into account. It is therefore necessary to introduce a parametrization of the nonperturbative hadronic terms and to determine from the data the corresponding parameters together with $\Lambda_{\overline{MS}}$. This generates some ambiguity on $\Lambda_{\overline{MS}}$ that the limitations of present data cannot eliminate.

Several theoretical treatments of the hadronic component (74) have been advocated. Values of $\Lambda_{\overline{MS}}$ obtained by different experiments and procedures are listed in Table 3 (71, 75). Note that most of the listed data were analyzed in terms of formulae for $f = 3$. Thus the reported values

Table 3 Results on $\Lambda_{\overline{MS}}$ obtained by different experiments on the photon structure function

Experiment[a]	Q^2 (GeV^2)	$\Lambda_{\overline{MS}}$ (MeV)[b]	Ref.
PLUTO (a)	3–100	183^{+80}_{-54}	71
PLUTO (b)	3–100	240 ± 90	71
PLUTO (c)	3–100	160 ± 60	71
JADE	10–220	250 ± 90	71
TASSO	7–70	140^{+190}_{-60}	71
TPC/2γ (a)	0.7–22	108 ± 32	75
TPC/2γ (b)	0.7–22	232 ± 59	75

[a] Entries with the label (a), (b), ... for one given experiment differ by the assumed model for hadronic component.

[b] The quoted value of $\Lambda_{\overline{MS}}$ in most cases should read as $\Lambda_{\overline{MS}}^{(3)}$ because it is obtained from a fit to the (charm-subtracted) data done in terms of QCD formulae with f = 3.

of $\Lambda_{\overline{MS}}$ should be mainly identified with $\Lambda_{MS}^{(3)}$, which differs from $\Lambda_{MS}^{(4)}$ by about 30% (because $\Lambda_{MS}^{(3)} \approx 1.3\Lambda_{MS}^{(4)}$).

In conclusion, when translated in terms of $\Lambda_{MS}^{(4)}$ the results on the photon structure function lead to values of $\Lambda_{MS}^{(4)}$ in the range 50–300 MeV, perfectly consistent with other experiments.

There are many less precise or more ambiguous determinations of α_s from other experimental sources. For example, I can quote the determination of α_s at $\mu \approx m_\tau$ from the leptonic branching ratio of the τ lepton. This branching ratio is $B_e = 18.3 \pm 0.3\%$. The fact that it is close to 1/5 rather than to 1/3 is a proof that $N_c = 3$, because $B_e \approx 1/(2 + N_c)$. From the deviations from the value 1/5 a value of $\alpha_s(m_\tau)$ can be extracted (76) with some model dependence from the treatment of nonperturbative effects. At the other extreme, a measurement of α_s at $\mu \approx M_W$ has been attempted (77) by UA2 from W+jet production at the CERN $p\bar{p}$ collider. The results in both cases are consistent with the more precise methods already described.

2.6   Summary and Conclusion on α_s

In Section 2 we have reviewed and interpreted the large amount of experimental information on α_s. A sample of the most significant determinations of α_s is reported in Table 4. Each entry is a combination of different measurements. It is clearly remarkable that the final results on α_s from so many completely different sources are in such good agreement. This is one of the most important quantitative tests of QCD. A plot of a more extended set of data is shown in Figure 7. I do not think that it would be appropriate to combine the errors according to Gaussian statistics in order to derive an average value of Λ. In fact, the data on deep inelastic scattering are in part contradictory or not always in agreement with QCD; the error on the Υ entry is a personal estimate; the $e^+e^- \rightarrow$ jet result is obtained by a particular combination of the existing data and so on. However, I cannot

Table 4   Summary of the most significant determinations of α_s and $\Lambda_{\overline{MS}}$ (see text)

	α_s (34 GeV)	$\Lambda_{\overline{MS}}^{(4)}$ (MeV)	$\Lambda_{\overline{MS}}^{(5)}$ (MeV)
$R_{e^+e^-}$	0.14 ± 0.02	370^{+350}_{-220}	240^{+230}_{-140}
BCDMS	0.127 ± 0.006	220 ± 60	140 ± 40
Υ	0.123 ± 0.009	180 ± 80	120 ± 50
$e^+e^- \rightarrow$ jets	0.135 ± 0.015	330 ± 200	215 ± 130
γ structure function	0.120 ± 0.016	175 ± 125	115 ± 80

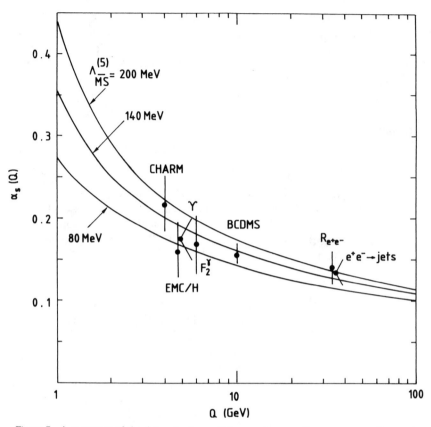

Figure 7 A summary of the determinations of the running coupling constant α_s discussed in the text. The curves for $\Lambda_{\overline{MS}}^{(5)} = 140 \pm 60$ MeV are obtained following the matching procedure at the *b* threshold explained in Equations 14–19 (with $a \approx 1$).

evade the task of providing the readers with a suggested set of values. In this spirit I propose

$$\Lambda_{\overline{MS}}^{(4)} = 220 \pm 90 \text{ MeV} \qquad\qquad 57.$$

or equivalently

$$\Lambda_{\overline{MS}}^{(5)} = 140 \pm 60 \text{ MeV}, \qquad\qquad 58.$$

where the reported error is about twice the value that would be obtained by combining the errors shown in Table 4.

A common objection is that no experiment has really detected the running of α_s. Even from the whole set of data we have discussed, once the errors are taken into account, the running of α_s cannot be clearly established in the range $Q \approx 3$–50 GeV. This fact is simply a consequence of the slow logarithmic decrease of $\alpha_s(Q)$. Actually there are experiments that claim to have observed the decrease of α_s with Q (see, for example, 78) but, in my opinion, they are not convincing.

However, the fact that the values of α_s measured at large Q are so small, as shown in Table 4, is a very strong proof of the running of α_s, because such a relatively feeble strong force could not provide hadrons with the observed tight binding. A more significant test of QCD is evident from Figure 7. A relatively loose determination of $\alpha_s(Q)$ at $Q \approx 1$ GeV leads to a very tight determination of $\alpha_s(Q)$ at large Q. For example, from the value of $\Lambda_{\overline{MS}}^{(5)}$ given in Equation 58, the prediction for α_s to be measured at LEP (and HERA) is very precise:

$$\alpha_s(Q \approx M_Z) \approx 0.11 \pm 0.01. \qquad\qquad 59.$$

Establishing that this prediction is experimentally true would be a very quantitative and accurate test of QCD, conceptually equivalent but more reasonable than trying to see the running in a given experiment.

3. THE QCD THEORY OF HARD PROCESSES

In this section we briefly discuss the experimental evidence for QCD derived from the phenomenology of hard processes. The property of asymptotic freedom provides the theoretical framework for a consistent and systematic formulation of the parton model in QCD (renormalization group, factorization theorem, etc). Thus the successes of the naive parton model are directly inherited by QCD. Clearly we are mostly interested in those predictions of the parton model in QCD that go beyond a generic and naive formulation of the parton picture and thus provide us with specific dynamical tests of the underlying QCD theory.

An enormous amount of theoretical and experimental work on hard processes in QCD has accumulated over the years. A systematic review of the phenomenology of high momentum transfer reactions would by far exceed the limits of the present relatively concise article. In the following, we mention only a few among the most significant experimental examples in support of QCD.

3.1 *Jets in* e^+e^- *Annihilation*

Experiments on e^+e^- annihilation at high energy (e.g. 61, 61a) provide a wonderful laboratory for systematically testing the distinct signatures predicted by QCD for the structure of the final state averaged over a large number of events. In the following we discuss the predictions of QCD concerning the properties of the final state.

Typical of asymptotic freedom is the hierarchy of configurations emerging from the smallness of $\alpha_s(Q)$ at high energies. Each configuration starts at a given order in $\alpha_s(Q)$ and is characterized by a specified topology. When all corrections of order $\alpha_s(Q)$ are neglected, one recovers the naive parton-model prediction for the final state: almost collinear events with two back-to-back jets with limited transverse momentum and an angular distribution as $(1+\cos^2\theta)$ with respect to the beam axis. The two-jet structure of the majority of the events and the angular distribution of jets typical of spin-$\frac{1}{2}$ quarks (scalar particles would lead to a $\sin^2\theta$ distribution) were first established at SPEAR (79) and later confirmed and extensively studied, especially at the high energy e^+e^- colliders PEP, PETRA, and more recently also at TRISTAN. For example, the angular distribution of jets in e^+e^- annihilation measured by TASSO (61, 61a) at PETRA is shown in Figure 8. At order $\alpha_s(Q)$ a tail of events is predicted to appear with large transverse momentum $p_\perp \approx Q/2$ with respect to the thrust axis (the axis that maximizes the sum of the absolute values of the longitudinal momenta). The small fraction of events with large $p_\perp$ consists mostly of three-jet events with an almost planar topology. The skeleton of a three-jet event at leading order in $\alpha_s(Q)$ is formed by three hard partons, the third

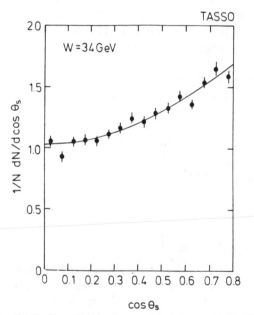

Figure 8   Angular distributions of jets in e^+e^- annihilation measured by TASSO at PETRA, compared with the expected angular distribution for spin-1/2 quarks $(1+\cos^2\theta)$.

being a gluon emitted by a quark or antiquark line. The first observation of three-jet events at PETRA gave relatively direct experimental support to gluons. At order $\alpha_s(Q)$ the transverse momentum in the event plane $\langle p_\perp \rangle_{in}$ with respect to the thrust axis is predicted to increase linearly with Q (apart from logarithms) while $\langle p_\perp \rangle_{out}$ is still fixed in this approximation. Similarly the most energetic jet, called the narrow jet, should look like a jet of a two-jet event (at somewhat scaled-down energy) and correspondingly $\langle p_\perp \rangle_{narrow}$ is fixed, while $\langle p_\perp \rangle_{broad}$ increases with Q. At order $\alpha_s^2(Q)$ a hard perturbative nonplanar component starts to build up and some small fraction of four-jet events is predicted to appear: both $\langle p_\perp \rangle_{out}$ and $\langle p_\perp \rangle_{narrow}$ start increasing.

The topological signatures just described in a qualitative way are quite well supported by the available data. For example, we reproduce in Figure 9 the data from TASSO (61, 61a) that clearly show the increase of $\langle p_\perp^2 \rangle_{in}$

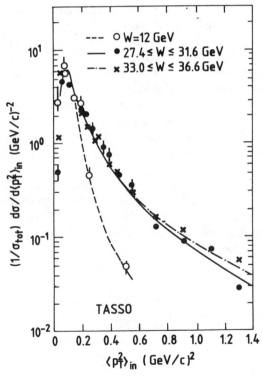

Figure 9   Increase of $\langle p_\perp^2 \rangle_{in}$ with the center-of-mass energy W (the transverse momentum squared in the event plane, with respect to the thrust axis) measured by TASSO (67) at PETRA.

with the center-of-mass energy $W = 2E_{beam} = Q$. Even more impressive is Figure 10 taken from MARK J (61, 61a) that compares the observed energy flow diagrams with the predictions of QCD and of some other (rather artificial) models. In Figure 10a F_{minor} is a measure of acoplanarity. The data support the QCD prediction, which is less spherical than phase space and more acoplanar than two-jet events. In Figure 10b and 10c the energy flow polar angle diagrams for noncollinear events are shown for two different cuts in thrust and jet angles. Finally in Figure 10d the unfolded energy flow diagram is compared with different detailed model predictions and shows very good agreement with QCD (also including a model treatment of fragmentation effects).

The precise form of the QCD matrix element for three or four partons in the final state can be confronted with experiment, although some model of the relation between computed partons and observable jets must be superimposed. The determination of α_s from the energy-energy correlation distributions is the most quantitative of these comparisons of e^+e^- annihilation data with QCD matrix elements. The determination of the gluon spin was also attempted (66, 67) from the study of three-jet distributions in continuum e^+e^- annihilation and in Υ hadronic decays ($T \rightarrow ggg, \gamma gg$). In Figure 11 we show, as an example, the results obtained by TASSO (61, 61a) on the observed distribution in $\cos\tilde{\theta}$, where $\tilde{\theta}$ is the angle between the thrust axis (roughly aligned with the most energetic jet, called jet 1) and the line of jets 2 and 3 in their own center-of-mass frame (Ellis-Karliner test) (80). The distribution for vector gluons is clearly preferred with respect to the analogous matrix element computed for scalar gluons (although this alternative does not really correspond to a sensible theory).

More recently, some more detailed aspects of the QCD predictions for the parton branching and cascade have been discussed and partially tested. An important example is the class of coherence effects (for a review, see 81, 82) and the tested difference between $\gamma q\bar{q}$ and $gq\bar{q}$ final states (83).

Figure 10   Energy flow in e^+e^- collisions at 34 GeV [Mark J (61, 61a)]. (a) The distribution $d\sigma/dF_{minor}$ in the fraction of the visible energy flow of the entire event projected along the minor axis (perpendicular to the event plane). Curves represent models with gluon emission (*solid*), without gluon emission (*short-dashed*), and phase space with a p_T cutoff (*long-dashed*). (b) Comparison of the data with QCD, and $q\bar{q}$ models, using energy flow diagrams in the thrust-major event plane for events with $O_B > 0.3$, $T_N > 0.98$, or $\theta_{minor} > 60°$. (c) Same as (b) but with $T_N < 0.98$ and $\theta_{minor} < 60°$. (d) The unfolded energy flow diagram of (c) compared with the models of QCD, $q\bar{q}$, phase space, and the $q\bar{q}$ model with $\exp(-p_T/650)$ fragmentation distribution.

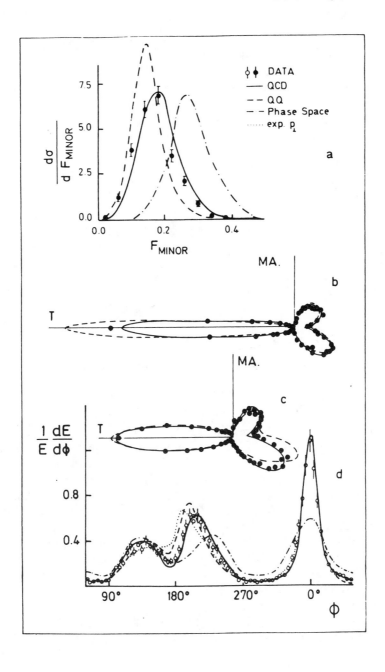

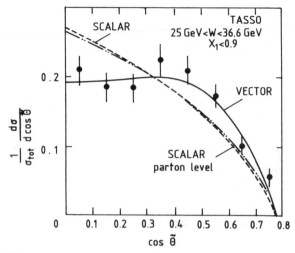

Figure 11 Ellis-Karliner test (80) on three-jet events in e^+e^- annihilation [data by TASSO (67) at PETRA]. The prediction of the QCD matrix element, with vector gluons, is strongly preferred by the data with respect to a model calculation with scalar gluons.

In conclusion the wealth of experimental results on jet physics in e^+e^- annihilation and their successful comparison with the theory have contributed much to establishing a solid observational base for QCD.

3.2   Deep Inelastic Leptoproduction and the Nucleon's Parton Densities

In Section 2.2 we summarized the determination of α_s from the observed scaling violation in deep inelastic leptoproduction. In this section we discuss the additional very important information concerning the QCD-improved parton model that can be obtained from the data on deep inelastic scattering with muon and neutrino beams. This includes tests of the parton-model predictions and the experimental determination of the quark and gluon parton densities in the proton (or the isoscalar nucleon). Once the parton densities have been measured at some Q^2, they can be evolved at all Q^2 and used to predict many other processes; some of them (Drell-Yan, W/Z production, jets in $p\bar{p}$ collisions, photons at large p_T, etc) are discussed in the following sections.

In the naive parton model, for spin-$\frac{1}{2}$ quarks, the longitudinal structure function $F_L = F_2 - 2xF_1$ is predicted to vanish asymptotically as $1/Q^2$ [Callan-Gross relation (84)]. In QCD, $F_L(x, Q^2)$ is of order $\alpha_s(Q)$ instead and therefore vanishes more slowly, as $1/\ln Q^2$. The leading QCD expression for F_L is given (85) by

$$F_{\rm L}(x, Q^2) = \frac{\alpha_{\rm s}(Q)}{2\pi} x^2 \int_x^1 \frac{{\rm d}y}{y^3} \left[\frac{8}{3} F_2(y, Q^2) \right.$$

$$\left. + 2 \sum_{i=1}^{2f} e_i^2 (y - x) g(y, Q^2) \right] + \cdots, \qquad 60.$$

where $\Sigma_{i=1}^{2f} e_i^2$ is the sum of all coefficients of q and $\bar{\rm q}$ in the naive parton-model expression of F_2/x (for $f = 4$ it is $20/9$ in electroproduction and 8 for v or $\bar{v}$ scattering from charged currents). In Figure 12 we display some recent data on $F_{\rm L}$ at large Q^2 for muon production on protons (42, 43) ($\langle Q^2 \rangle \approx$ 15–60 GeV^2) and for neutrino production on iron (44) ($\langle Q^2 \rangle \approx$ 4–69 GeV^2). One sees that the longitudinal structure function is indeed small at large Q^2, once again confirming that the charged partons have spin $\frac{1}{2}$. The data are perfectly consistent with the QCD prediction in Equation 60. The expected rise at small x of $F_{\rm L}$ due to the increasing sea and gluon contributions is indicated by the CDHSW and BCDMS data.

The QCD-corrected parton sum rules, in particular the Adler sum rule (86)

$$\int_0^1 \frac{{\rm d}x}{x} [F_2^{vn}(x, Q^2) - F_2^{vp}(x, Q^2)] = 2 \qquad 61.$$

and the Gross–Llewellyn Smith (87) sum rule

$$\int_0^1 {\rm d}x\, F_3^{vN}(x, Q^2) = 3 \left[1 - \frac{\alpha_{\rm s}(Q)}{\pi} + \cdots \right] \qquad 62.$$

(N = isoscalar nucleon target), are well supported by experiment [see, for example, Figure 13 with the data by the BEBC collaboration (88)]. Similarly the approximate prediction of a ratio of $5/18$ between F_2 for electroproduction and neutrino production on isoscalar targets is in agreement with the data (89).

The integral $\int_0^1 {\rm d}x\, F_2^{vN}(x, Q^2)$, where N is an isoscalar nucleon target, is a good measure of the total momentum fraction carried by quarks and antiquarks in the proton. It is well known (41) that this quantity is about 0.45–0.50 at $Q^2 \approx 10$ GeV^2 and is nearly constant for $Q^2 > 10$ GeV^2. The remaining fraction of momentum is attributed to gluons.

In general it is fair to say that the great wealth of accumulated data on deep inelastic leptoproduction is in very good agreement with the QCD-improved parton model. It is therefore possible to extract relatively reliable quark and antiquark distributions from the existing data. More difficult is the measurement of the gluon density because gluons are not directly coupled to electroweak currents.

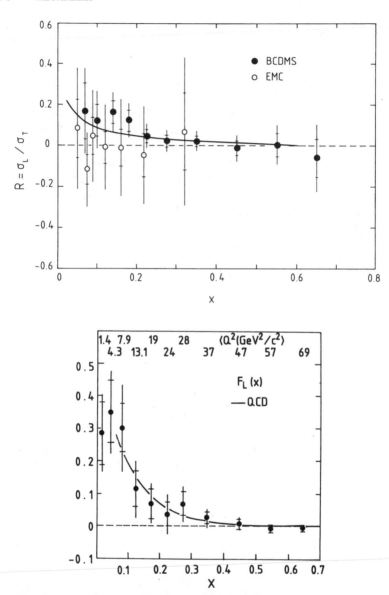

Figure 12 The longitudinal structure function for muon (*top*) and neutrino (*bottom*) deep inelastic scattering on nucleons. In the top panel the data from EMC (42) and BCDMS (43) on hydrogen are shown [$\sigma_L/\sigma_T = F_L/F_T = (F_2-2xF_1)/2xF_1$]. In the bottom panel the v data on iron from CDHSW (44) are displayed (with the Q^2 value for each point explicitly shown). The data are compared with the QCD expectation based on Equation 60 (85).

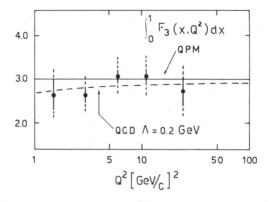

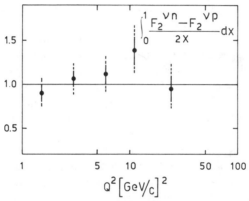

Figure 13 Tests of the Gross–Llewellyn Smith (*upper*) and Adler (*lower*) sum rules (Equations 62 and 63) by the BEBC Collaboration (88).

The sum $u_v + d_v$ of valence densities $u_v = u - \bar{u}$ and $d_v = d - \bar{d}$ is obtained from the structure function F_3 measured in neutrino scattering on isoscalar targets. The u_v and d_v distributions can be separated by using the charged-current cross sections for hydrogen and deuterium targets, which indicate that $d_v/u_v \approx 0.57(1-x)$ at $Q^2 \approx$ few GeV². Additional information on the ratio d_v/u_v is derived from the measurements of $F_2^{\mu n}/F_2^{\mu p}$ by BCDMS (43) and EMC (42), which are mutually consistent for this ratio. The sea densities can be constrained by using the data on F_2 on protons or isoscalar targets. The information on the flavor dependence of the sea densities is not very rich. One usually assumes $\bar{u} = \bar{d}$. The amount of strange sea $\bar{s}$ (presumably about equal to s) can be measured from antineutrino-induced dimuon production (explained by charm production) and its shape is consistent with that for $\bar{u}$ and $\bar{d}$ obtained from F_2. The CDHS collaboration (90) finds $\bar{s} \approx 0.4\bar{u}$ at $Q^2 \approx 5$ GeV^2. Actually the data on Drell-Yan muon pair production in pp collisions are also often used to constrain further the shape in x of the sea distributions.

Some uncertainty on the quark densities is introduced by the observed nuclear-size dependence of the structure functions per nucleon measured on complex nuclei [the so-called EMC effect (42)]. This must be taken into account when the data on heavy isoscalar targets are used to derive information on the parton densities in the proton. However, in the last few years the magnitude of the nuclear effects at small and intermediate x measured in several different experiments has settled down to a tolerable size (45). Somewhat paradoxically the main problem is now represented by the already mentioned experimental discrepancy between the BCDMS and the EMC data on $F_2^{\mu p}$ (45).

Apart from evident practical reasons, the determination of the gluon density in the proton is clearly important from the point of view of testing the theory. In fact the physical reality of quarks was first established by the study of the spectroscopy of hadrons and later confirmed by the parton-model description of leptoproduction as the quarks are directly coupled to the electroweak currents. For the gluons it is certainly more difficult to obtain a solid basis of experimental evidence. As discussed in the last section, good evidence for gluon jets has been obtained from the study of the final state in e^+e^- annihilation. We have also seen that about half of the proton momentum is carried by gluons. Here we consider the experimental information on the gluon parton density in the nucleon.

The main input on the gluon parton density in the proton is obtained from the study of scaling violations at small x in leptoproduction. Accurate analyses of the scaling violations in the singlet sector have been performed by CDHS (91), CHARM (92; M. Diemoz, private communication), EMC (42), and BCDMS (43). The value of $\Lambda_{\overline{\text{MS}}}$ and the gluon density have to be separately determined from a fit of the observed scaling violations (Figure 4). As only a few logarithmic slopes at small x are important for the fit, the trial parametrization of the gluon density can only be a very crude one. The problem is further complicated by the fact that at small x the average values of Q^2 are in general smaller and also the effects associated with the charm threshold can simulate scaling violations.

In spite of these difficulties there is a reasonable agreement on the shape of the gluon density at $Q^2 = 5$–10 GeV^2 as determined by different experiments (Figure 14). The gluon density is concentrated at small x and its effects on scaling violations are indeed found to be negligible above $x \approx 0.25$–0.3. The more recent data show a somewhat softer gluon distribution than that first obtained by CDHS (91). On the other hand, the

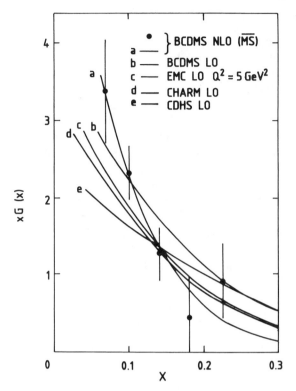

Figure 14   A collection of gluon densities at $Q^2 = 5$ GeV^2 obtained from scaling violations in deep inelastic scattering. The curves *a*, *b*, *c* are taken from Ref. (43), while I am grateful to Dr. M. Diemoz for the addition of curves *d* and *e*.

new CDHSW data (44), based on larger statistics, show marked differences at low *x* with respect to the previous structure functions. A soft gluon distribution is also supported by the data on large-p_T photons in pp and p$\bar{\text{p}}$ collisions (93, 94) and on J/ψ production (95). The CDHS gluon distribution served as an input to the widely used parametrizations of parton densities by Duke & Owens (96), Eichten et al (97), etc. More modern sets of parton densities by Diemoz et al (98) and Martin et al (95) are instead based on the now preferred soft-gluon density.

It is remarkable that the gluon density obtained from the scaling violations in deep inelastic scattering, evolved at much larger values of Q^2, is found to be a necessary contribution in interpreting ISR and p$\bar{\text{p}}$ collider data, for example on jet or photon production at large p_T or on heavy flavor hadro- or photoproduction, as discussed below.

3.3   *Drell-Yan Processes and W/Z Production*

The production of lepton pairs in hadron-hadron collisions, via virtual photon or intermediate weak boson exchange, is a process of great importance for QCD. Drell-Yan processes (9) in fact provide a crucial and quite nontrivial test of the validity of the parton approach and of its implementation in QCD through the factorization theorem (99). This theorem predicts that the Q^2-dependent quark and gluon densities measured in leptoproduction on a given hadronic target, evolved in Q^2 by the QCD evolution equations, are directly relevant to predictions of cross sections for other hard processes involving the same hadron. The key point of the factorization theorem is that the Q^2-dependent parton densities are universal, i.e. process independent. The dependence on the particular process only enters at the level of the partonic subprocesses. The resulting prediction for hadron-hadron collisions is a double convolution of the Q^2-dependent parton densities with the parton cross section, which is perturbative and can be computed as an expansion in $\alpha_s(Q)$. For an inclusive process $A + B \rightarrow X$ one has:

$$E \frac{d\sigma}{d^3 p} = \sum_{i,j} \int dx_A \, dx_B \, F_{i/A}(x_A, Q^2) F_{j/B}(x_B, Q^2)$$

$$\times E \frac{d\hat{\sigma}}{d^3 p} [x_A p_A, x_B p_B, p, \dots, \alpha_s(Q)], \qquad 63.$$

where σ ($\hat{\sigma}$) is the hadronic (partonic) cross section, $F_{i/A}$ is the density of parton "i" in hadron "A", and Q is the large energy scale in the process (typically the lepton pair mass in the Drell-Yan case). For simplicity we have here identified the factorization scale Q, which appears in the parton densities, with the renormalization scale, which enters in $\alpha_s(Q)$. Q can be chosen, with relatively wide freedom, around the natural physical scale for a given process. A change in Q is compensated by the corresponding change in $\hat{\sigma}$. Of course the compensation is exact only for the complete expression of $\hat{\sigma}$. At any fixed order in α_s a scale change produces a variation of the cross section by terms of higher order in α_s.

The class of Drell-Yan reactions (also including W/Z production) are simplest among hard hadron-hadron processes in that the final state can be totally inclusive and the observed particles (the lepton pair) are non-strongly interacting. The fact that the cross sections are quadratic in the parton densities implies that one is testing the parton model in a dynamical configuration that is far more complex than in leptoproduction. In fact the validity of the factorization theorem in Drell-Yan processes has been the subject of a long debate (100). The conclusion was in favor of the validity of the parton-model prediction (101), but the arguments are certainly not as simple and clear as for processes in which the light-cone operator expansion can be applied.

The precise measurements of muon pair production cross sections at fixed-target experiments and at the ISR have been very important for establishing the signatures of the parton-model approach to Drell-Yan processes. These characteristic properties are (a) linear A dependence for experiments on nuclei with atomic number A (a consequence of the incoherent sum of the parton contributions in the target); (b) the angular distribution of the lepton pair predominantly as $1 + \cos^2 \theta$ in their center of mass; (c) the approximate scaling of $Q^4 (d\sigma/dQ^2 \, dy)$ and other similar nondimensional quantities (in QCD the nonscaling effects are only logarithmic); and (d) intensity rules, for example, the dominance of valence-valence ($\pi^{\pm}N, K^-N, \bar{p}N$) over valence-sea cross sections (K^+N, pN). The available data at sufficiently large energies, for masses of the pair above the J/ψ and not too close to the phase-space boundary (e.g. for $\tau = Q^2/s$ not too close to 1) neatly support all the previous distinctive predictions (for a review, see 102). The study of Drell-Yan processes has also produced important information on the x behavior of the sea densities in the nucleon and on the otherwise inaccessible quark densities in pions and kaons.

The QCD-corrected parton model leads to an absolute prediction for the total cross section (9). The value of the lowest-order cross section is inversely proportional to N_c, the number of color replicas for quarks, because a given quark can only annihilate with an antiquark of the same color to produce a colorless lepton pair. The order $\alpha_s(Q)$ corrections to the cross section were computed long ago (103) and found to be large (when the parton densities are defined from the structure function F_2 measured in leptoproduction at $q^2 = -Q^2$). The ratio $\sigma_{corr}/\sigma_{LO}$ of the corrected and the lowest-order cross sections, called the K factor, is slowly varying in Q^2 and y. Recently the calculation of an important part of the two-loop K factor has also been completed (104).

What has been computed at order $\alpha_s^2(Q)$ is the contribution of the quark-antiquark annihilation channel to the partonic cross section, limited to the terms that are singular near $z = (\tau/x_A x_B) = 1$ with $\tau = Q^2/s$. [In lowest order the whole contribution arises from $z = 1$; at first order, and presumably in higher orders as well, the bulk of the correction arises from the singular terms of the form $\delta(1-z)$, $\ln(1-z)/(1-z)_+$, and $1/(1-z)_+$.] The terms of order $\alpha_s^2(Q)$ at small τ are found to be of the expected size relative to the large $O(\alpha_s)$ terms, have the same sign of the first-order correction, and somewhat exceed the estimate obtained by a simple exponentiation of the first order. For example for W/Z production at 630 GeV Matsuura et al (104) estimate that the K factor at one loop is 1.39

with $\alpha_s \equiv \alpha_s(M_W)$ and becomes 1.57 with the addition of the computed two-loop effects.

The QCD predictions for Drell-Yan processes can best be tested for W/Z production. W/Z production at CERN and Tevatron energies is ideal because $Q \approx M_{W/Z}$ is large enough that the K factor [which decreases with $\alpha_S(Q)$] is acceptable, Q and $\sqrt{s}$ are not too unbalanced ($\sqrt{\tau} = Q/\sqrt{s} = 0.13$–0.15 at $\sqrt{s} = 630$ GeV and $\sqrt{\tau} = 0.046$–0.052 at the Tevatron), and the parton densities are reasonably well known in the relevant region of $x \approx O(\sqrt{\tau})$. Note in this respect that a precise calculation of W/Z production at supercollider energies is a more difficult problem. The predicted W and Z cross sections are reported in Table 5 (105), where only the first-order K factor was taken into account, while an estimate of the higher orders assuming exponentiation was included in the stated errors. The error is asymmetric because the "central" value was computed by choosing $Q = M_W$ in the corrective terms, while the p_T distribution with $\langle p_T \rangle \ll M_W$ would perhaps suggest a lower scale, hence a larger α_s. [This suggestion is apparently confirmed by the approximate calculation of the two-loop K factor of Matsuura et al (104).] What is actually measured is the cross section times branching ratio for W → eν or Z → e$^+$e$^-$. Especially $B(W \rightarrow e\nu)$ depends on m_t: $B(W \rightarrow e\nu) \approx 0.089$ for $m_t = 40$ GeV, ≈ 0.100 for $m_t = 60$ GeV, and ≈ 0.109 for $m_t \geqslant 80$ GeV. The comparison with the data is shown in Figure 15 for $m_t = 60$ GeV (105). The agreement between theory and experiment is quite good. To appreciate this point, note that the presence of color introduces a factor of $1/N_C \approx 1/3$ in σ and a factor $\sim 5/(3 + 2N_c) \approx 5/9$ in the branching ratio $B(W \rightarrow e\nu)$ (for large m_t). The fact that the data appear to favor the upper side of the theoretical range is partly explained by the now largely computed two-loop K factor and could also be an indication for $m_t > 60$ GeV. Finally recall that if the already mentioned discrepancy between BCDMS and EMC data on F_2^p (45) is eventually resolved in favor of

Table 5 Values of W ($= W^+ + W^-$) and Z production cross sections in p$\bar{\text{p}}$ collisions for $\sin^2 \theta_W = 0.229$, $m_W = 80.8$ GeV/c^2, $m_Z = 92.0$ GeV/c^2 (105)

$\sqrt{s}$ (TeV)	σ^W (nb)	σ^Z (nb)
0.54	$4.3^{+1.3}_{-0.6}$	$1.4^{+0.4}_{-0.2}$
0.63	$5.4^{+1.6}_{-0.9}$	$1.7^{+0.5}_{-0.3}$
1.6	$17^{+4.0}_{-2.5}$	$5.1^{+1.2}_{-0.8}$
1.8	$19^{+5.0}_{-3.3}$	$5.8^{+1.6}_{-1.0}$
2.0	$21^{+6.0}_{-4.0}$	$6.4^{+1.9}_{-1.2}$

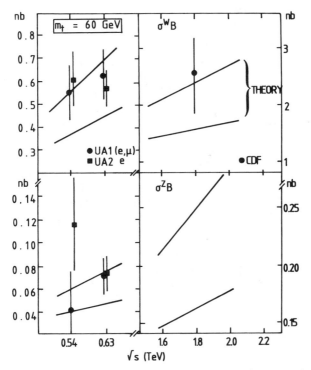

Figure 15   Theory and experiment on W/Z production cross sections, assuming $m_t = 60$ GeV (105). The solid lines define the band predicted by QCD (103–105).

BCDMS, then the quark densities would be increased by 10–15% in the relevant region of x. In conclusion, within the present uncertainties there is very good agreement between theory and experiment on the W/Z production cross sections. The agreement is a significant test of QCD, and it may soon become even more significant when the results from the recent high luminosity runs at CERN and at the Tevatron are available.

The prediction of the transverse momentum distribution of the W or Z has an even deeper dynamical significance. The purely perturbative calculations are only valid at $p_T \approx M_{W/Z}$. At smaller p_T values, in the region $\Lambda \ll p_T \ll M_{W/Z}$, where the bulk of the data on $d\sigma/dp_T$ is concentrated, the sequence of logarithmic terms of all orders arising from the soft-gluon radiation from the initial parton legs must be resummed (106; for calculations of the perturbative tail, see 106a). The corresponding Sudakov exponent is typical of vector gluons. The resulting prediction (107) of the p_T distribution with the correct perturbative limit, the soft-gluon resummation, and the exact integral under the curve to reproduce the corrected total cross section [with the $O(\alpha_s)$ K factor included] are compared with the data in Figure 16 (108; see also 108a). The observed p_T distribution is correctly reproduced in terms of a commonly adopted parametrization of parton densities (96) and with a reasonable value of $\Lambda_{\overline{MS}}$. This is very important because the p_T distribution has no analogue in the naive parton model. The average p_T for W/Z production is quite large ($\langle p_T \rangle \approx 8$ GeV at $\sqrt{s} \approx 0.63$ TeV) in comparison to all possible hadronic scales and is entirely produced by QCD radiation.

3.4   Hard Processes in pp and p̄p Collisions

In recent years the contribution of p̄p collider experiments (also including pp at the ISR) has considerably broadened the observational support of QCD. The study of the production properties of weak bosons, discussed in the previous section, is a beautiful example of the importance of collider physics for QCD. In this section we briefly consider the most prominent experimental tests of QCD at p̄p and pp colliders including jet physics, photons at large p_T, and heavy flavor production.

The study of jet production at hadron colliders opens new territory for probing the validity of the parton model. In p̄p reactions the amplitudes for $2 \to 2$ or $2 \to 3$ quarks/gluons are made accessible to experimental

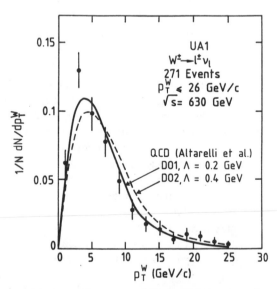

Figure 16   The UA1 data (108, 108a) on the W p_T distribution are compared with the QCD theory (107). The Duke & Owens structure functions (96) and the corresponding values of $\Lambda_{\overline{MS}}$ were used.

study. The relative simplicity of semileptonic weak interactions, as compared to the intricacies of nonleptonic weak processes, is a reminder that the strong corrections to a single electroweak current vertex are in principle much simpler than gluon exchange or radiation in presence of four or five colored legs. Finally, the scale Q of energy at $p\bar{p}$ colliders is the largest accessible to experiment at present. Thus the successful predictions by QCD (for a review, see 109) of the two-jet production rate at large p_T, of their angular distribution, of the ratio of three to two jets, and of the three-jet distributions are certainly very impressive tests of the theory. One may object that these predictions are obtained in the leading logarithmic approximation. The complex work needed for a complete calculation of jet production in next-to-leading order approximation (110, 111) is under way and about to produce results relevant for actual experiments. The uncertainties connected with the choice of the scale Q—plus the ambiguities in the value of Λ, the errors connected with our ignorance on some details of parton densities, and so on—in practice lead to predictions within a factor of two or so. While this is true, and while the experimental errors are also of about the same size, predictions within a factor of two are still extremely significant when applied to steep functions of several variables, i.e. functions that vary by many orders of magnitude in the explored domain. This is the case for the jet production ratio that varies by orders of magnitude both in $\sqrt{s}$ (from the ISR to the Sp$\bar{p}$S collider and more recently to the Tevatron) and in p_T (Figure 17) (112).

Thus the physics of large-p_T jets in $p\bar{p}$ reactions is a marvelous success of the QCD-improved parton model. For example, it is well established (109) that the agreement at relatively small p_T values (but large enough for the perturbative calculation to hold) would not be possible without gluons. Even more striking is the effect of gluons in determining the angular distribution of jets with respect to the beam axis. At large $\cos\theta$ values the dominant behavior is determined by the Rutherford singularity as $(1-\cos\theta)^{-2}$, typical of the exchange of massless vector gluons. The expected angular distribution is exactly reproduced by the data, as shown in Figure 18 (113).

The production of hard photons at large p_T has been observed in fixed-target experiments (93, 114) at the ISR (115) and by both UA1 and UA2 (116) at the CERN $p\bar{p}$ collider. In this case, the calculation of QCD predictions at the next-to-leading order accuracy has been completed (117). At all explored energies the agreement between theory and experiment is quite good. In Figure 19 the UA1 and UA2 data are compared with the QCD predictions of Aurenche et al (117). The data on large-p_T photons, especially those at relatively low energy, are also useful to obtain information on the x distribution of the gluon density in the proton.

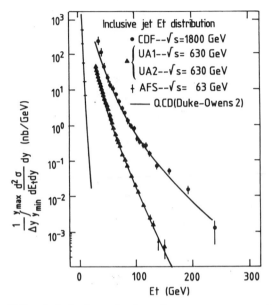

Figure 17   Data (112) on inclusive jet production at large transverse energy E_T measured at the ISR, the Sp$\bar{p}$S, and the Tevatron compared with the QCD predictions based on the parton densities of Duke & Owens (96).

Great progress in the QCD theory of heavy flavor hadro- and photo-production was recently achieved with the calculation of next-to-leading order corrections (118, 119; see also 118a). The data on the photo-production of charm, in particular the recent set of precise data by experiment E691 at FNAL (120), are in good agreement with QCD predictions from photon-gluon fusion, for a reasonable value of the effective charm mass. The hadroproduction of charm and beauty at fixed targets and at colliders is in general affected by larger theoretical errors with respect to photoproduction (121). For charm, the next-to-leading order corrections bring the effective mass of charm required by the data in closer agreement with the photoproduction data and with the mass expected from charm particle spectroscopy. For beauty, the cross section and p_T distributions observed by UA1 at the CERN p$\bar{p}$ collider (122) are in agreement with the QCD prediction (with large theoretical errors). The theoretical predictions become particularly reliable for top production at present colliders because m_t is large and the ratio $m_t/\sqrt{s}$ is not too small.

For heavy flavor production gluons are also essential. Without gluons there would be neither photon-gluon nor gluon-gluon fusion diagrams.

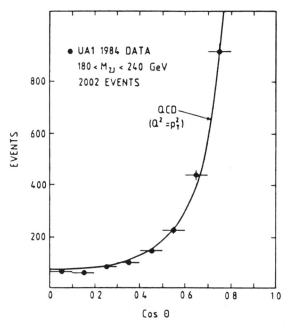

Figure 18 Angular dependence of jet-jet events in pp̄ collisions measured by UA1 (and UA2) (113) displaying the Rutherford singularity as $(1 - \cos\theta)^{-2}$ expected from gluon exchange.

For example, the predicted cross-section for beauty production at colliders would be a factor 30 or 40 smaller. Of course, one could invoke ad hoc intrinsic charm and beauty quark densities in the proton to explain the present meager data without gluons. While more data on more processes can in principle decide the issue, it remains true that QCD correctly predicts the observed amount of charm and beauty production in terms of the gluon density measured from the observed scaling violations in leptoproduction.

4. CONCLUSIONS

In this article the main results forming the experimental basis for perturbative QCD have been briefly discussed. It is a fact that a wide variety of observables related to hard processes are correctly predicted by the theory. The measured values of $\Lambda_{\overline{\text{MS}}}$ obtained from several different processes quantitatively coincide to a fairly good accuracy. To quantify the level of precision, we recall that the value of α_s at a scale close to the Z mass, relevant for QCD tests at LEP, can now be predicted with $\sim 9\%$

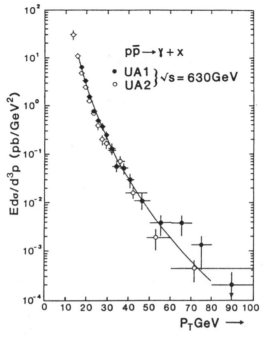

Figure 19   The inclusive cross section for direct photon production measured by UA1 and UA2 (116) compared with the QCD prediction of Aurenche et al (117).

accuracy (see Equation 59). The running coupling obtained from experiment, together with the parton densities measured in leptoproduction, computed at the relevant scale Q by the QCD evolution equations invariably produces correct predictions for all hard processes that with time become accessible to measurements. Although it is true that each individual test cannot be pushed to a high level of precision, it is also true that the data accumulated over the last 15 years or so and the related theoretical work have produced a very large number of successful and increasingly precise tests of the theory. At present, it is fair to say that the experimental support of QCD is quite solid and quantitative. The forthcoming experiments at $p\bar{p}$ colliders, at LEP, SLC, and HERA will certainly be very important with their great potential for extending the experimental investigation of the validity of QCD.

ACKNOWLEDGMENTS

I am glad to thank Drs. C. Berger, M. Diemoz, K. Ellis, G. Martinelli, P. Nason, and P. Zerwas for important contributions and clarifying discussions.

Literature Cited

1. Gell-Mann, M., *Acta Phys. Austriaca Suppl.* IX: 733 (1972); Fritzsch, H., Gell-Mann, M., in *XVIth Int. Conf. on High Energy Physics*, Batavia (1972), II: 135; Fritzsch, H., Gell-Mann, M., Leutwyler, H., *Phys. Lett.* 47B: 365 (1973)
2. Gross, D. J., Wilczek, F., *Phys. Rev. Lett.* 30: 1343 (1973); *Phys. Rev.* D8: 3633 (1973); Politzer, H. D., *Phys. Rev. Lett.* 30: 1346 (1973)
3. Weinberg, S., *Phys. Rev. Lett.* 31: 494 (1973)
4. Lee-Franzini, J., *Nucl. Phys.* B3: 139 (1988)
5. Petronzio, R., in *Proc. 24th Int. Conf. on High Energy Physics*, Munich, Berlin: Springer Verlag (1988), p. 136
6. Stack, J. D., *Phys. Rev.* D29: 1213 (1984); Barkai, B., Moriarty, K. J. M., Rebbi, C., *Phys. Rev.* D30: 1283 (1984); Hasenfratz, P., in *Proc. 23rd Int. Conf. on High Energy Physics*, Berkeley. Singapore: World Scientific (1986), p. 169
7. Karsch, F., *Z. Phys.* C38: 147 (1988); APE Collaboration, Bacilieri, P., et al., *Phys. Rev. Lett.* 61: 1545 (1988); Brown, F. R., et al., *Phys. Rev. Lett.* 61: 2058 (1988)
8. Feynman, R. P., *Photon Hadron Interactions*. New York: Benjamin (1972)
9. Altarelli, G., *Phys. Rep.* 81: 1 (1982)
10. Yndurain, F. J., *QCD: An Introduction to the Theory of Quarks and Gluons*, New York: Springer Verlag (1983); Muta, T., *Foundations of QCD*, Lecture Notes in Physics, Singapore: World Scientific (1987)
11. Jacob, M., ed., *Perturbative Quantum Chromodynamics*. Amsterdam: North Holland (1982)
12. Collins, J. C., Soper, D. E., *Annu. Rev. Nucl. Part. Sci.* 37: 383–409 (1987)
13. Yang, C. N., Mills, R. C., *Phys. Rev.* 96: 191 (1954)
14. Abers, E. S., Lee, B. W., *Phys. Rep.* 9: 11 (1973)
15. 't Hooft, G., *Phys. Rev. Lett.* 37: 8 (1976); *Phys. Rev.* D14: 3432 (1976)
16. Belavin, A. A., Polyakov, A. M., Schwartz, A. S., Tyupkin, Yu. S., *Phys. Lett.* 59B: 85 (1975)
17. Baluni, V., *Phys. Rev.* D19: 2227 (1979); Crewther, R. J., Di Vecchia, P., Veneziano, G., Witten, E., *Phys. Lett.* 88B: 123 (1979), E91B: 487 (1980)
18. Peccei, R. D., in *CP violation*, ed. C. Jarlskog. Singapore: World Scientific (1989)
19. Greenberg, O. W., *Annu. Rev. Nucl. Part. Sci.* 28: 327 (1978)

20. Adler, S. L. *Phys. Rev.* 177: 2426 (1969); Bell, J. S., Jackiw, R., *Nuovo Cimento* A51: 47 (1969)
21. Fukuda, H., Miyamoto, Y., *Prog. Theor. Phys.* 4: 347 (1949); Steinberger, J., *Phys. Rev.* 76: 1180 (1969)
22. Particle Data Group, *Phys. Lett.* B204: 1 (1988)
23. Ashmore, J. F., *Nuovo Cimento Lett.* 4: 289 (1972); Bollini, G. G., Giambiagi, J. J., *Nuovo Cimento* 12B: 20 (1974)
24. Itzykson, C., Zuber, J. B., *Introduction to Quantum Field Theory*. New York: McGraw-Hill (1980)
25. 't Hooft, G., *Nucl. Phys.* B61: 455 (1973)
26. Bardeen, W. A., Buras, A. J., Duke, D. W., Muta, T., *Phys. Rev.* D18: 3998 (1978)
27. Stückelberg, E. C. G., Peterman, A., *Helv. Phys. Acta* 26: 499 (1953); Gell-Mann, M., *Phys. Rev.* 95: 1300 (1954); Bogoliubov, N. N., Shirkov, D. V., *Dokl. Akad. Nauk USSR* 103: 206, 391 (1955); Ovsiannikov, L. V., *Dokl. Akad. Nauk USSR* 109: 1121 (1956); Callan, G. C., *Phys. Rev.* D2: 1541 (1970); Symanzik, K., *Commun. Math. Phys.* 18: 227 (1970)
28. Caswell, W., *Phys. Rev. Lett.* 33: 244 (1974); Jones, D. R. T., *Nucl. Phys.* B75: 531 (1974)
29. Tarasov, O. V., Vladimirov, A. A., Zharkov, A. Yu., *Phys. Lett.* 93B: 429 (1980)
30. Braten, E., Leveille, J. P., *Phys. Rev.* D24: 1369 (1981); Blumenfeld, A., Moshe, M., *Phys. Rev.* D26: 648 (1982)
31. Appelquist, T., Carazzone, J., *Phys. Rev.* D11: 2856 (1975).
32. Duke, D. W., Roberts, R. G., *Phys. Rep.* 120: 275 (1985)
33. Jost, R., Luttinger, J. M., *Helv. Phys. Acta* 23: 201 (1950); Appelquist, T., Georgi, H., *Phys. Rev.* D8: 4000 (1973); Zee, A., *Phys. Rev.* D8: 4038 (1973)
34. Chetyrkin, K. G., Kataev, A. L., Tkachov, F. V., *Phys. Lett.* 85B: 277 (1979); Dine, M., Sapirstein, J., *Phys. Rev. Lett.* 43: 668 (1979); Celmaster, W., Gonsalves, R. J., *Phys. Rev. Lett.* 44: 560 (1979); *Phys. Rev.* D21: 3112 (1980)
35. Gorishny, S. G., Kataev, A. L., Larin, S. A., *Phys. Lett.* 212B: 238 (1988)
36. de Boer, W., see Ref. 5, p. 905
37. Marshall, R., see Ref. 5, p. 901
38. Stevenson, P. M., *Phys. Rev.* D23: 2916 (1981); Politzer, H. D., *Nucl. Phys.* B194: 493 (1982); Stevenson, P. M., Politzer, H. D., *Nucl. Phys.* B277: 758

(1986); Grunberg, G., *Phys. Rev.* D29: 2315 (1984)

39. Wilson, K., *Phys. Rev.* 179: 1499 (1969); Brandt, R., Preparata, G., *Nucl. Phys.* 27B: 541 (1971)
40. Altarelli, G., *Riv. Nuovo Cimento* 4: 335 (1974)
41. Sloan, T., Smadja, G., Voss, R., *Phys. Rep.* 162: 45 (1988); Diemoz, M., Ferroni, F., Longo, E., *Phys. Rep.* 130: 293 (1986)
42. Aubert, J. J., et al. (EMC Collaboration), *Nucl. Phys.* B259: 189 (1985); B272: 158 (1986); B293: 740 (1987)
43. Benvenuti, A. C., et al. (BCDMS Collaboration), *Phys. Lett.* 195B: 91 (1987); 195B: 97 (1987); *CERN preprint EP/89-06* (1989) and *EP/89-07* (1989)
44. Buchholz, P. (CDHSW Collaboration), in *Proc. EPS Int. Conf. on High Energy Physics*, Bari, Laterza (1985), p. 557; Vallage, B., Thèse Univ. Paris-Sud, Orsay (1986); Perez, P., in *Proc. Int. Conf. on Neutrino Physics and Astrophysics*, Sendai. Singapore: World Scientific (1986), p. 341
45. Mount, R., see Ref. 5, p. 1007
46. Altarelli, G., Parisi, G., *Nucl. Phys.* B126: 298 (1977)
47. Floratos, E. G., Ross, D. A., Sachrajda, C. T., *Nucl. Phys.* B129: 66 (1977), EB139: 545 (1978); Curci, G., Furmanski, W., Petronzio, R., *Nucl. Phys.* B175: 27 (1980); Floratos, E. G., Lacaze, R., Kounnas, C., *Phys. Lett.* 99B: 89 (1981); Gonzales-Arroyo, A., Lopez, C., Yndurain, F. J., *Nucl. Phys.* B153: 161 (1979)
48. Altarelli, G., Ellis, R. K., Martinelli, G., *Nucl. Phys.* B157: 461 (1979)
49. Floratos, E. G., Ross, D. A., Schrajda, C. T., *Nucl. Phys.* B152 (1979) 493; Furmanski, W., Petronzio, R., *Phys. Lett.* 97B: 438 (1980); Floratos, E. G., Lacaze, R., Kounnas, C., *Phys. Lett.* 98B: 285 (1981); Herrod, R. T., Wada, S., *Phys. Lett.* 96B: 195 (1981); *Z. Phys.* C9: 351 (1981)
50. Herrod, R. T., Wada, S., *Phys. Lett.* 96B: 195 (1981), *Z. Phys.* C9: 351 (1981)
50a. Ouraou, A., Virchaux, M., theses, Orsay (1988)
51. Bodek A., et al. (SLAC-MIT Collaboration), *Phys. Rev.* D20: 1471 (1979)
52. Bergsma, F., et al. (CHARM Collaboration), *Phys. Lett.* 153B: 111 (1985)
53. McFarlane, D. B., et al. (CCFRR Collaboration), *Z. Phys.* C26: 1 (1984)
54. Meyers, P. D., et al. (BFP Col-

laboration), *Phys. Rev.* D34: 1265 (1986)
55. Mackenzie, B. P., Lepage, G. P., *Phys. Rev. Lett.* 47: 1244 (1981); Barbieri, R., Gatto, R., Kögerler, R., Kunszt, Z., *Phys. Lett.* 57B: 455 (1975); Celmaster, W., *Phys. Rev.* D19: 1517 (1979); Barbieri, R., Caffo, M., Gatto, R., Remiddi, E., *Phys. Lett.* 95B: 93 (1980)
56. Franzini, P., in *Proc. Les Rencontres de Physique de la Vallée d'Aoste*, ed. M. Greco, Editions Frontières (1989), to be published
57. Bizzeti, A., see Ref. 5, p. 909
58. Kwong, W., Mackenzie, P., Rosenfeld, R., Rosner, J. L., *Phys. Rev.* D37: 3210 (1988)
59. Field, R. D., *Phys. Lett.* 133B: 248 (1983)
60. Photiadis, D. M., *Phys. Lett.* 164B: 160 (1985)
61. Kramer, G., *Springer Tracts in Mod. Phys.*, Vol. 102 (1984); Wu, S. L., *Phys. Rep.* 107: 59 (1984); Adeva, B. et al., *Phys. Rep.* 109: 131 (1984); Naroska, B., *Phys. Rep.* 148: 67 (1987)
61a. Ali, A., Söding, P., eds., *Adv. Ser. on Directions in High Energy Physics*. Singapore: World Scientific, Vol. 1 (1988)
62. Gottschalk, T. D., *Phys. Lett.* B109: 331 (1982); Gottschalk, T. D., Shatz, M. P., *Phys. Lett.* B150: 451 (1985); Sjöstrand, T., *Z. Phys.* C26: 93 (1984)
63. Ellis, R. K., Ross, D. A., Terrano, A. E., *Phys. Rev. Lett.* 45: 1226 (1980); *Nucl. Phys.* B178: 321 (1981); Vermaseren, J. A. M., Gaemers, K. J. F., Oldham, S. J., *Nucl. Phys.* B187: 301 (1981)
64. Fabricius, K., Kramer, G., Schierholz, G., Schmitt, I., *Z. Phys.* C11: 315 (1982); Gutbrod, F., Kramer, G., Schierholz, G., *Z. Phys.* C21: 235 (1984)
65. Basham, C., Brown, L., Ellis, S., Love, S., *Phys. Rev. Lett.* 41: 1585 (1978); *Phys. Rev.* D19: 2018 (1979); D24: 2382 (1981)
66. Ali, A., Barreiro, F., *Phys. Lett.* 118B: 155 (1982); *Nucl. Phys.* B236: 269 (1984); Richards, D. G., Stirling, W. J., Ellis, S. D., *Phys. Lett.* 119B: 193 (1982); *Nucl. Phys.* B229: 317 (1983)
67. Ali, A., Barreiro, F., see Ref. 61a, p. 612.
68. Ali, A., Pietarinen, E., Kramer, G., Willrodt, J., *Phys. Lett.* 93B: 155 (1980)
68a. Hoyer, P., et al., *Nucl. Phys.* B161: 349 (1979)
69. Anderson, B., Gustafson, G., Ingelman, F., Sjöstrand, T., *Phys. Rep.* 97: 31 (1983)

70. Csikor, F., et al., *Phys. Rev.* D31: 1025 (1985); D34: 129 (1986)
71. Berger, C., Wagner, W., *Phys. Rep.* 146: 1 (1987); Kolanoski, H., Zerwas, P., see Ref. 61a, p. 697
72. Witten, E., *Nucl. Phys.* B120: 189 (1977)
73. Bardeen, W. A., Buras, A. J., *Phys. Rev.* D20: 166 (1979)
74. Antoniadis, I., Grunberg, G., *Nucl. Phys.* B213: 445 (1983); Antoniadis, I., Marleau, L., *Phys. Lett.* 161B: 163 (1985); Field, J. H., Kapusta, F., Poggioli, L., *Phys. Lett.* 181B: 362 (1986); *Z. Phys.* C36: 121 (1987); Glück, M., Reya, E., *Dortmund Univ. preprint* (1988)
75. Maxfield, S. J., see Ref. 5, p. 661
76. Braaten, E., *Phys. Rev. Lett.* 60: 1606 (1988); Narison, S., Pich, A., *Phys. Lett.* B211: 183 (1988)
77. Meier, K., see Ref. 5, p. 729
78. JADE Collaboration, *DESY preprint DESY88–105* (1988); AMY Collaboration, *KEK preprint KEK 88-113* (1989)
79. Schwitters, R. F., et al., *Phys. Rev. Lett.* 35: 1320 (1975); Hanson, G. G., et al., *Phys. Rev. Lett.* 35: 1609 (1975)
80. Ellis, J., Karliner, I., *Nucl. Phys.* B148: 141 (1979)
81. Dokshitzer, Yu. L., Khoze, V. A., Troyan, S. I., Mueller, A. H., *Rev. Mod. Phys.* 60: 373 (1988)
82. Bartel, W., et al. (JADE Collaboration), *Phys. Lett.* 101B: 129 (1981); 134B: 275 (1984); *DESY Report 88–015* (1988); Aihara, H., et al. (TPC Collaboration), *Z. Phys.* C28: 31 (1985)
83. Sheldon, P., et al. (MARK II Collaboration), *Phys. Rev. Lett.* 57: 1398 (1986); Aihara, M., et al. (TPC Collaboration), *Phys. Rev. Lett.* 57: 945 (1986)
84. Callan, C., Gross, D. J., *Phys. Rev. Lett.* 22: 156 (1969)
85. Altarelli, G., Martinelli, G., *Phys. Lett.* 76B: 89 (1978)
86. Adler, S., *Phys. Rev.* 143: 1144 (1965)
87. Gross, D. J., Llewellyn Smith, C. H., *Nucl. Phys.* B14: 337 (1969)
88. Allasia, D., et al., *Z. Phys.* 28: 321 (1985)
89. Voss, R., in *Proc. 1987 Int. Symp. Lepton and Photon Interactions at High Energies*, Hamburg, ed., W. Bartel, R. Rückl. Amsterdam: North-Holland (1988), p. 581.
90. Abramowicz, H., et al. (CDHS Collaboration), *Z. Phys.* C15: 19 (1982)
91. Abramowicz, H., et al. (CDHS Collaboration), *Z. Phys.* C12: 289 (1982); C17: 237 (1983)
92. Bergsma, F., et al. (CHARM Collaboration), *Phys. Lett.* 123B: 269 (1983)
93. Bonesini, M., et al. (WA70 Collaboration), *Z. Phys.* C38: 371 (1988); Bernasconi, A., et al. (UA6 Collaboration), *Phys. Lett.* 206B: 163 (1988)
94. Aurenche, P., et al., *LPTHE Orsay preprint 88/38* (1988)
95. Martin, A. D., Roberts, R. G., Stirling, W. J., *Phys. Rev.* D37: 1161 (1988)
96. Duke, D. W., Owens, J. F., *Phys. Rev.* D30: 49 (1984)
97. Eichten, E., Hinchliffe, I., Lane, K., Quigg, C., *Rev. Mod. Phys.* 56: 579 (1984); E58: 1065 (1986)
98. Diemoz, M., Ferroni, F., Longo, E., Martinelli, G., *Z. Phys.* C39: 21 (1988)
99. Radyushkin, A. V., *Phys. Lett.* 69B: 461 (1978); Politzer, H. D., *Nucl. Phys.* B129: 301 (1977); Amati, D., Petronzio, R., Veneziano, G., *Nucl. Phys.* B140: 54 (1978); B146: 29 (1978); Ellis, R. K., Georgi, H., Machacek, M., Politzer, H. D., Ross, G. G., *Nucl. Phys.* B152: 285 (1979); Libby, S., Sterman, G., *Phys. Rev.* D18: 3252, 4737 (1978)
100. Collins, J. C., Sterman, G., *Nucl. Phys.* B185: 172 (1981); Bodwin, G., Brodsky, S. J., Lepage, G. P., *Phys. Rev. Lett.* 47: 1799 (1981)
101. Collins, J. C., Soper, D. E., Sterman, G., *Nucl. Phys.* B250: 199 (1985); Bodwin, C. T., *Phys. Rev.* D31: 2616 (1982)
102. Rutherfoord, J. P., in *Proc. 1985 Int. Symp. on Lepton and Photon Interactions at High Energies*, Kyoto, ed. M. Konuma, K. Takahashi (1985)
103. Altarelli, G., Ellis, R. K., Martinelli, G., *Nucl. Phys.* B143: 521 (1978); EB146: 544 (1978); B157: 461 (1979); Kubar-André, J., Paige, F., *Phys. Rev.* D19: 221 (1979); Kubar-André, J., Le Bellac, M., Meunier, J. L., Plaut, G., *Nucl. Phys.* B157: 251 (1980)
104. Matsuura, T., van der Marck, S. C., van Neerven, W. L., *Phys. Lett.* 211B: 171 (1988)
105. Altarelli, G., Di Lella, L., *Adv. Ser. on Directions in High Energy Physics.* Singapore: World Scientific (1989), 4: 177
106. Dokshitzer, Yu. L., Dyakonov, D. I., Troyan, S. I. *Phys. Lett.* 78B: 290 (1978); *Phys. Rep.* 58: 269 (1980); Parisi, G., Petronzio, R., *Nucl. Phys.* B154: 427 (1979); Curci, G., Greco, M., Srivastava, Y., *Phys. Rev. Lett.* 43: 434 (1979); *Nucl. Phys.* B159: 451 (1979); Collins, J., Soper, D. E., *Nucl. Phys.* B139: 381 (1981); B194: 445 (1982); B197: 446 (1982); Collins, J., Soper, D.

E., Sterman, G., *CERN preprint TH.3923/84* (1984); Kodaira, J., Trentadue, L., *Phys. Lett.* 112B: 66 (1982); 123B: 335 (1983); Davies, C. T., Stirling, W. J., *Nucl Phys.* B244: 337 (1984)

106a. Altarelli, G., Parisi, G., Petronzio, R., *Phys. Lett.* 76B: 351, 356 (1978); Ellis, R. K., Martinelli, G., Petronzio, R., *Nucl. Phys.* B211: 106 (1983); Bawa, A., Stirling, J. K., *Phys. Lett.* 203B: 172 (1988); Arnold, P. B., Hall-Reno, M., *Fermilab-pub-88/168-T*(1988)

107. Altarelli, G., Ellis, R. K., Greco, M., Martinelli, G., *Nucl. Phys.* B246: 12 (1984); Altarelli, G., Ellis, R. K., Martinelli, G., *Z. Phys.* C27: 617 (1985), *Phys. Lett.* 151B: 457 (1985)

108. Albajar, C., et al., *Z. Phys.* in press (1989)

108a. Stubenrauch, C., *thesis, Note CEA-N-2532*, Saclay (1987)

109. Ellis, R. K., Scott, W. G., see Ref. 105, 4: 131

110. Ellis, R. K., Sexton, J. C., *Nucl. Phys.* B269: 445 (1986)

111. Ellis, S. D., Kunszt, Z., Soper, D. E., *Phys. Rev. Lett.* 62: 726 (1989) and *ETH-PT/89-1* (1989); Aversa, F., Chiappetta, P., Greco, M., Guillet, J. Ph., *Phys. Lett.* 210B: 225 (1988); 211B. 465 (1988); *preprint CPT-88/P.2186* (1988)

112. Schochet, M. T., see Ref. 5, p. 18.

113. Arnison, G., et al. (UA1 Collaboration), *Phys. Lett.* 158B: 494 (1985); 177B: 244 (1986); Bagnaia, P., et al. (UA2 Collaboration), *Phys. Lett.* 144B: 283 (1984)

114. Badier, J., et al. (NA3 Collaboration), *Z. Phys.* C31: 341 (1986); De Marzo, C., et al. (NA24 Collaboration), *Phys. Rev.* D36: 8 (1987)

115. Akesson, T., et al. (AFS Collaboration), *Phys. Lett.* 158B: 282 (1985)

116. Albajar, C., et al. (UA1 Collaboration), *Phys. Lett.* 209B: 385 (1988); Appel, J. A., et al. (UA2 Collaboration), *Phys. Lett.* 176B: 239 (1986)

117. Aurenche, P., et al., *Phys. Lett.* 140B: 87 (1984); *Nucl. Phys.* B297: 661 (1988)

118. Nason, P., Dawson, S., Ellis, R. K., *Nucl. Phys.* B303: 607 (1988)

118a. Beenakker, W., Kuijf, H., van Neerven, W. L., Smith, J., *Leiden preprint* (1988)

119. Ellis, R. K., Nason, P., *Fermilab preprint FERMILAB-PUB* 88/54-T (1988)

120. Anjos, J. C., et al. (E691 Collaboration), *Fermilab preprint, FERMILAB-PUB-88/125-EP*(1988)

121. Altarelli, G., Diemoz, M., Martinelli, G., Nason, P., *Nucl. Phys.* B306: 724 (1988); Berger, E., see Ref. 5, p. 987

122. Kenyon, I. R. (UA1 Collaboration), see Ref. 5, p. 981

Annu. Rev. Nucl. Part. Sci. 1989. 39: 407–65

EXOTIC LIGHT NUCLEI[1]

Claude Détraz

Grand Accélérateur National d'Ions Lourds, 14021 Caen Cedex, France

David J. Vieira

Isotope and Nuclear Chemistry Division, Los Alamos National Laboratory, Los Alamos, New Mexico 87545, USA

KEY WORDS: atomic masses, decay properties of nuclei far from stability, radioactive ion beams.

CONTENTS

[1] The US Government has the right to retain a nonexclusive, royalty-free license in and to any copyright covering this paper.

1. INTRODUCTION

With advances in accelerators and experimental techniques, the study of nuclei lying far from β-stability continues to open the nuclear frontier, providing us with valuable insight into the properties of nuclei with extreme neutron-to-proton composition. The ability to produce and separate (or identify) such exotic nuclei allows researchers to use the nucleus as a microscopic laboratory in which to investigate interesting structural effects that occur in nuclei. One example is the rapid onset of prolate deformation found in the ^{31}Na-^{32}Mg region, where one would normally expect to find a spherical shape for the ground state due to the completion of the $N = 20$ closed shell (1–3). The fact that $N = 20$ is no longer "magic" for such neutron-rich nuclei came as a complete surprise—this deformation could not have been predicted from what was known about other nearby nuclei lying closer to β-stability. Because both strong and electroweak forces play important roles in the binding and decay of these nuclei, fundamental information about the very nature of these forces can be obtained. These investigations are typified by the continuing study of superallowed Fermi β-decay transitions, which have been characterized for a variety of nuclei ranging from ^{14}O to ^{54}Co (4, 5). The constancy of these transition rates represents an important verification of the conserved-vector-current hypothesis. Not only does the study of exotic nuclei broaden our understanding of nuclear structure and test our knowledge of fundamental interactions taking place within nuclei, but such work also has important ramifications for various astrophysical processes and the production of energy via nuclear chain reactions.

In this review we concentrate on studies of the exotic light nuclei (see Figure 1), where recent developments have enabled the production and investigation of a wide variety of nuclei lying near or at the neutron and proton drip lines. Such drip-line nuclei are particularly interesting because their neutron-to-proton composition is pushed to the very limits imposed by nuclear binding. Predicting the properties of these nuclei provides a demanding test of nuclear theory. Identifying the nucleon stability (or instability) of these nuclei is often the first piece of information with which to test qualitatively our understanding of such nuclei. Mass measurements of these isotopes usually follow, providing a quantitative test of current mass models. In particular, systematic mass measurements have proven valuable in defining regions in which the nuclear structure of these nuclei is changing, such as the onset of nuclear deformation or supporting shell closures. Moreover, because the masses of nuclei increase rapidly with neutron (or proton) excess and thus make their Q_β decay energies large, a variety of beta-delayed nucleon and beta-delayed multinucleon decay

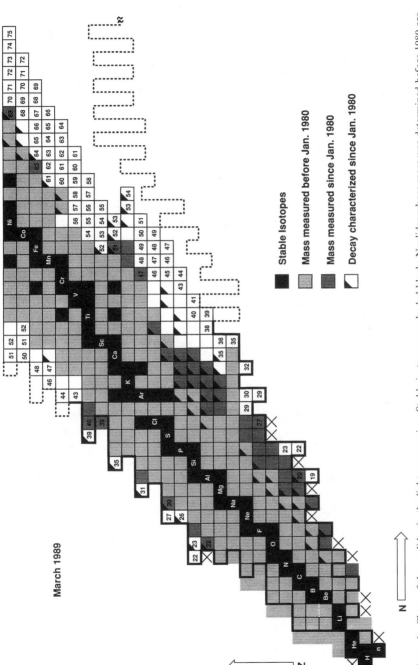

Figure 1 Chart of the nuclides in the light mass region. Stable isotopes are shaded black. Nuclides whose mass was determined before 1980 are light gray; those measured since 1980 are dark gray. Isotopes discovered to be stable with respect to nucleon emission since 1980 are labeled by their respective mass number; those nucleon unbound are indicated by an "X" if their mass is unmeasured or by an unbordered shaded box if their mass has been determined. Heavy solid lines indicate the extent of the known neutron and proton drip lines; dashed lines are the predicted (6) drip lines. Black triangles in the upper left corner of selected boxes indicate isotopes for which initial decay information has been reported since 1980.

channels become energetically possible. Along the drip line itself, the search for direct nucleon or dinucleon emission is presently under intense study. Such decay processes are of intrinsic interest and many provide nuclear structure information that cannot be obtained otherwise. Thus the study of exotic nuclei explores a frontier where new phenomena are observed and new insights can be gained.

From these technical developments in producing and separating exotic nuclei, emerged radioactive ion beams. Although still quite new, such beams of β-unstable nuclei are of high interest as an isospin and mass selective probe for studying nuclear reactions. Moreover, such radioactive ion beams can be used to characterize various reactions of astrophysical interest that cannot be investigated by any other means. In Section 5, we highlight some work in this new field and examine its future possibilities.

For additional information about the properties of nuclei far from stability, the reader is referred to review articles by Hansen (7) and Hamilton (8) and to chapters in books edited by Poenaru & Ivascu (9) and by Bromley (10). More popular accounts of the advances in the field of exotic nuclei are contained in reports by Cerny & Poskanzer (11) and Hardy (12). Recent conferences detailing the development of this field have been held in Helsingør in 1981 (13), Darmstadt-Seeheim in 1984 (14), and Rosseau Lake in 1987 (15).

2. TOWARD THE DRIP LINES

2.1 Production of Exotic Nuclei

The production of exotic nuclei has benefited from several breakthroughs, ranging from the discovery of important new reaction mechanisms to the development of high energy and/or high intensity accelerators. Although the ingenuity of nuclear scientists often provides ad hoc ways of producing a rare and sought-after nuclear species, several production mechanisms have proven themselves particularly valuable for studying exotic nuclei. For a long time, fission induced by thermal neutron capture has been one of the most effective sources of exotic neutron-rich isotopes in the $A = 70$-160 region (16). This method remains highly productive, as is evident by the recent discovery of several neutron-rich isotopes of Ni and Cu (17). In the early 1970s, deep inelastic heavy ion reactions provided a new way of producing exotic nuclei; tens of new neutron-rich isotopes were observed by Artukh et al (18) and Volkov (19). Years later the same mechanism was used again to produce another set of even more neutron-rich nuclei (20–22). Multinucleon transfer reactions using low energy, heavy ion beams have also proven to be a viable method of producing exotic nuclei. Compound nucleus formation followed by the evaporation of a few nucleons, usually neutrons, has long been a favorite way of producing exotic nuclei, most notably neutron-deficient nuclei. If a suitable projectile, target, and bombarding energy are chosen, such fusion-evaporation reactions have a reasonable degree of selectivity. Because both multinucleon transfer and fusion-evaporation reactions are well suited to the use of advanced spectroscopic methods, these production methods are likely to continue to play an important role in the study of exotic nuclei (23). Finally, medium energy, proton-induced fragmentation, fission, and spallation reactions have proven to be another rich source of exotic nuclei.

This last mode of production and that of neutron-induced fission retain several distinct advantages: (a) the production cross sections are large; (b) thick targets can be used because of the small rate of energy loss and the long range of the primary beam; and (c) the range of the reaction products of interest is short. The first two points make it possible to obtain high production rates; the latter point ensures that a high percentage of the reaction products are retained in the target, which makes them amenable to powerful target-ion source mass separator and/or mass spectrometer techniques (see Section 2.2).

A few years ago another method of producing exotic nuclei was advanced. This method takes advantage of projectile fragmentation reactions that are potentially as general as those described above. Related to target fragmentation, this reaction mechanism results in high yields of exotic nuclei. Because of the large momentum of the incident projectile, the fragments are concentrated in a small opening angle cone around zero degrees. This strong kinematic "focusing" of all reaction products at forward angles is well suited to the use of magnetic spectrometers. The pioneering work done at the Bevalac used ^{40}Ar and ^{48}Ca beams of 200 MeV/u to discover over 15 exotic isotopes (24, 25).

The availability of new heavy ion accelerators, with energies reaching 100 MeV/u and intensities roughly four orders of magnitude larger than those of the Bevalac, has opened up further possibilities for this production method. A wealth of new results have been obtained (26–28). Projectile fragmentation at an energy of 50 MeV/u is certainly not the dominant process that makes up nearly all of the reaction cross section, as is the case at higher energies. The momentum distribution of the fragments, which to first order result from the Fermi momentum of the participant nucleons, is relatively broader at these lower energies; hence, the collection efficiencies of fragments within a certain angular acceptance is lower. The angular distribution of projectile fragments is further broadened by dissipation effects that are reminiscent of low energy mechanisms that persist even at energies of 20–50 MeV/u (29). Yet the fact that the beam intensities for the projectiles of interest reach 5×10^{11} pps and are expected to increase over the coming years offers an unmatched opportunity to produce new exotic isotopes.

The projectile fragmentation-like process, at energies of 30–100 MeV/u, presents features that are both favorable and detrimental to the production of exotic nuclei. On the positive side is the possibility of additional nucleon transfer, either by direct or dissipative reaction mechanisms, with sizable cross sections. Evidence that this occurs is found in the work of Pougheon et al (30), in which new neutron-deficient Cu ($Z = 29$) isotopes were observed in reactions using 55 MeV/u ^{58}Ni ($Z = 28$) projectiles. However, detracting from this positive feature is the fact that the exotic fragments produced with the highest intensities had Z values that were 2 to 3 units smaller than those of the projectile. This results from a substantial mass loss in the fragmentation process, which leads to abraded nuclei that have very different shapes than those of the ground state and hence are left highly excited with large surface energies. This excitation is rapidly dissipated by the emission of nucleons, which—being primarily a statistical process—drives the neutron-to-proton composition of the fragment back toward the valley of stability and thus reduces the production rates for the most exotic species.

The projectile fragmentation process described above suggests that the next step in enhancing the production of exotic nuclei will come from beams of several hundreds of MeV/u (or even several GeV/u) with higher intensities. The forthcoming availability of such beams at the Schwerionen Synchrotron (SIS) accelerator now under construction in Darmstadt should open new possibilities. Yet one feature of projectile fragmentation reactions at intermediate energies should retain an important advantage: stopping these recoils so that their β decay (or other slow decay processes) can be characterized. Short of implementing a sophisticated and rapid deceleration process, this stopping mechanism must be accomplished in some type of degrader-stopping material. At intermediate energies, only a few percent of the nuclei brought to rest in a silicon detector telescope undergo nuclear reactions that change their nature. On the other hand, at relativistic energies, as many as 80% of the exotic fragments produced could be lost to such processes. This effect is compounded by the need for low-background measuring conditions, which suggests the need for increased shielding (more prohibitive with increasing energy). Thus we expect intermediate energy reactions will continue to be one of the most valuable sources of exotic nuclei for years to come.

2.2 Advances in Separation and Identification Techniques

In addition to the many advances made in producing exotic nuclei, there has been significant progress in developing techniques for separating

and/or identifying exotic species from a host of other, potentially interfering and more intense reaction products. Attempts to perform these
separations (identifications) rapidly, with high efficiencies, and with or
without chemical (Z) selectivity have led to a variety of methods—each
with its own distinct advantages and disadvantages. Most notable among
these methods are the use of on-line isotope separators/mass spectrometers
and the development of recoil spectrometers/separators.

On-line isotope separators and mass spectrometers (referred to col-
lectively as ISOL systems) have been the workhorse instruments of the
field (7, 31). Most of these systems use ion sources and relatively thick
targets, which, as mentioned above, can result in high intensity radioactive
beams for those elements that can be readily ionized. For example, at
ISOLDE-2, where targets of ≤ 100 g/cm^2 are bombarded by ~ 2 μA of
600-MeV protons, secondary beam intensities in excess of 10^{10} pps at the
peak of the isotopic yield distribution can be obtained for approximately
30 elements with a reasonable degree of chemical selectivity (32, 33). Ion
source developments continue to expand the capabilities of such systems
by enhancing the chemical selectivity and/or ionization efficiency, or by
decreasing the holdup times in the ion source (34). New developments,
such as those embodied in the design of ISOLDE-3 now being completed,
promise to make substantial improvements in separation (enhancement)
factors and in mass resolving powers, for which the direct separation of
isobaric members, particularly for the most exotic member(s), appears
possible. All of these factors are important in obtaining good yields of
exotic species that are well separated from other interfering reaction products. For an overview of present ISOL systems and the status of their
technical development, see the review article by Ravn & Allardyce (35)
and consult the proceedings of the latest electromagnetic isotope separator
conference (36).

The development of recoil separators and recoil spectrometers has
played an ever increasing role in the study of exotic nuclei, most notably
in the light mass region. Such systems have several advantages over ISOL
systems. Among these are (a) the separation occurs rapidly (typically
flight times are on the order of 1 μs); (b) the separation efficiencies are
independent of atomic number; and (c) these systems are well suited to
angle concentrated production mechanisms, such as those of heavy ion
fusion-evaporation and projectile fragmentation reactions. These advantages make it possible to use such recoil systems for both in-beam prompt
experiments and out-of-beam delayed coincidence studies. Therefore,
important information about the reaction mechanism itself can be
obtained using recoil spectrometers. Finally, the speed and universality of
these recoil devices enable short-lived (i.e. $T_{1/2} \gtrsim 1$ μs) exotic isotopes to

be characterized simultaneously with other known and unknown nuclei. This feature is of particular value to systematic or survey-type experiments. Many of these advantages are well exemplified by the investigations performed on the selector for heavy ion reaction products (SHIP) velocity separator, where experiments ranging from subbarrier fusion and radiative capture reaction mechanisms studies to the discovery of ground-state proton radioactivity and the sequential alpha-decay characterization of elements up to $Z = 109$ have been performed (37). For additional information about recoil spectrometers, refer to the review articles by Enge (38) and Cormier (39).

Three recoil spectrometers not covered in the above articles are worthy of special note here because of their recent impact in the field of exotic light nuclei. These are the time-of-flight isochronous (TOFI), the energy loss spectrometer (SPEG), and ligne d'ions super-epluchés (LISE) recoil spectrometers.

A layout of the TOFI recoil spectrometer is shown in Figure 2. It was designed (40) to perform systematic mass measurements of neutron-rich fragmentation and fission products as produced by the 1-mA 800-MeV proton beam of the Los Alamos Meson Physics Facility (LAMPF). Reaction products that recoil out of the target (typically 1 mg/cm^2 Th) are captured in a transport line located $\sim 90°$ to the primary beam. A crude

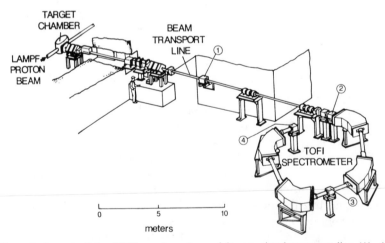

Figure 2 Layout of the TOFI spectrometer and its associated transport line (41, 42). Collimators at positions (1) and (2) crudely limit the A/Q and p/Q of the ions before they enter the spectrometer; the collimator at position (3) defines the p/Q acceptance of the spectrometer. Fast-timing, transmission detectors are located at positions (1), (2), and (4); a total energy and Z identification counter system is also placed at position (4)

mass-to-charge (A/Q) prefilter built into the transport line prevents the high yield, uninteresting ions from entering the spectrometer (41). The spectrometer itself is made up of four $81°$ integrated function dipoles arranged to be focusing in x, y, and time for a given set of A/Q ions (42). The system is also momentum-nondispersive overall (i.e. achromatic) so that a reasonably large acceptance is made possible ($\delta p/p = 4\%$, $\Omega = 2.2$ msr). The unique time focusing or isochronous feature of this system is accomplished by sending ions of different velocities on different path lengths through the system so that all ions of a given A/Q take the same amount of time. In such a system, the recoil's transit time provides a precise measure of its A/Q ratio. From separate measurements of the ion's stopping power, velocity, and energy, it is possible to determine both the atomic number and charge state of the recoil. Thus, given sufficient counting statistics, direct mass measurements of several isotopes can be determined simultaneously (see Section 3.1).

Equally impressive are the mass measurements made using SPEG, a general purpose, high resolution, energy loss spectrometer (43), at GANIL. Direct mass measurements are possible with SPEG because the production target (typically 350 mg/cm^2 Ta) is moved well upstream of the spectrometer and the intervening transport section is retuned for maximum transmission of the exotic projectile fragments of interest (see Figure 3). The α-shaped momentum analyzing spectrometer located immediately downstream of the production target eliminates the primary beam and selects which species are to be measured. The cyclotron radiofrequency (RF) or a fast-timing detector located after the momentum analyzer and a fast-timing detector at the focal plane of SPEG measures the ion's time of flight over a flight path of ~ 80 m. By combining this high precision velocity measurement with the magnetic rigidity measurement of the recoil using SPEG, it was possible to determine the mass of several exotic nuclei (see Section 3.1).

The LISE spectrometer (44), also located at GANIL, is shown in Figure 4. This system is composed of two $45°$ dipoles and associated quadrupole focusing elements. Projectile fragments are separated from the primary beam on the basis of their different magnetic rigidities in the first dipole. The recoils of interest are focused onto an intermediate focal plane where a collimator defines the momentum acceptance (e.g. $\delta p/p = 5\%$, $\Omega = 1$ msr) and a thin stripper foil (typically 5 mg/cm^2 Al) is often placed to reduce contamination from incompletely stripped primary beam ions. The second half of the spectrometer serves to cancel the dispersion produced in the first half in such a way that a double achromatic focus [i.e. $(x/\delta p) = (x'/\delta p) = 0$] is achieved. At this final position a multi-element solid-state detector telescope is usually placed to determine the Z and total

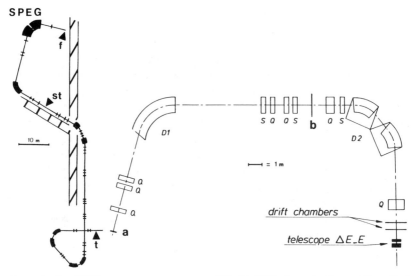

Figure 3   The SPEG mass measurement setup (98). (*Left*) The beam is shown at the exit of the second GANIL cyclotron; *t* stands for the target, *st* the stripping foil, and *f* the focal plane where a $\Delta E - E$ telescope is set up behind two drift chambers. (*Right*) The SPEG spectrometer: *Q* stands for the quadrupoles, *S* the sextupoles, *D*1 the analyzing dipole and *D*2 the spectrometer dipole. Position *a* marks the location of the stripper foil; a position sensitive parallel plate counter is placed at position *b*.

energy of the fragment. As a consequence of the achromatic condition mentioned above, the overall path length is independent of the fragment's initial position and angle. Thus, by combining the overall flight time with the known momentum, one can determine the fragment's A/Q, which at these velocities ($\beta = 0.3$) is effectively A/Z up to $Z \approx 30$. Redundant Z and A (or Q verification) determinations are obtained when multiple energy loss and total energy measurements are combined with the information above. In this way LISE has proven remarkably versatile and successful in identifying ~ 70 previously unknown isotopes (see Sections 2.3 and 2.4) and in performing a variety of decay spectroscopy studies (see Section 4).

Of further interest here is the use of LISE as a momentum-loss recoil separator. By using a suitably shaped, wedge-like degrader placed at the intermediate focal plane and by lowering the magnetic rigidity of the second dipole according to the desired energy loss in the degrader, one can effect a physical separation based on the fragment's A and Z (44, 45). Although this separation method does not yield unique A and Z values at GANIL energies (typically a set of isotones can be selected), it greatly

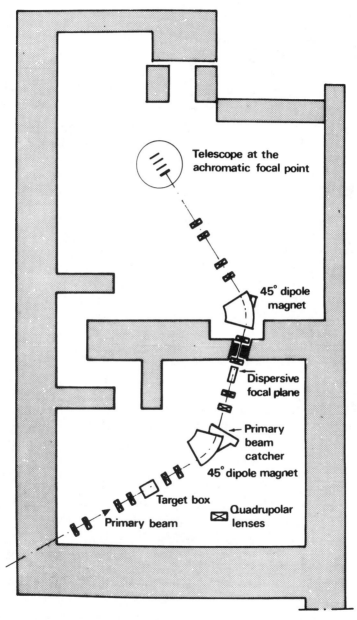

Figure 4   Schematic layout of the LISE spectrometer (44, 45).

reduces background problems associated with the high yield recoils and makes production of secondary radioactive beams a reality (see Section 5).

2.3   *Defining the Neutron Drip Line Up to* $Z = 10$

Although the effect of the Coulomb force strongly limits the number of bound neutron-deficient isotopes, all theoretical estimates of nuclear binding energies predict the occurrence of literally thousands of neutron-rich nuclei (46). This is seen in part in Figure 1, where the predicted drip line for neutrons is further from stability than for neutrons. It is much more difficult to reach the neutron drip line than the proton drip line because a more exotic combination of neutrons and protons is required. From a practical point of view, it is difficult to foresee ever approaching the neutron drip line in any but the lightest elements.

The proof of stability or instability for nucleon emission is generally shown in a two-dimensional plot in which the axes' parameters are related to the Z and A values of each species. A good example is shown in Figure 5, where the energy loss as measured in a transmission counter ($\Delta E \propto Z^2$) is plotted versus the time of flight (TOF) measured through the LISE spectrometer (TOF $\propto A/Z$). The grouping of events (or the lack thereof) around discrete A and Z values proves the particle stability (or instability) of each species. Naturally, such particle stability searches are enhanced by improved production yields, large collection/detection efficiencies, good A and Z resolutions, and zero or nearly zero backgrounds made possible through redundant measurements and/or high quality separations.

Until recently, the neutron drip line had been established only through $Z = 4$ with the surprising discovery of ^{14}Be (47), which was predicted to be neutron unbound at the time. In 1984 the last predicted neutron bound isotope of boron, ^{19}B, was found and ^{18}B was shown to be unbound (48). Since then, experiments using the LISE spectrometer have confirmed these observations and gone on to (*a*) identify ^{22}C, ^{23}N, and $^{29-30}$Ne (49, 50); (*b*) confirm the nucleon stability of ^{20}C and ^{27}F (51); and (*c*) prove the instability of ^{21}C and ^{25}O. Moreover, recent GANIL experiments were successful in observing ^{29}F (52) and ^{32}Ne and indicate that ^{26}O is unbound (53). Comparing these results with the recently updated mass calculations (6) that use the Garvey-Kelson mass relationships, it appears that all neutron-rich nuclei predicted to be bound with respect to neutron or two-neutron emission have been observed through $Z = 10$.

Two useful pieces of information are derived from this work. First, the odd-even neutron binding energy effect observed in these neutron-rich nuclei (i.e. where isotones with even neutron numbers are bound and those with odd numbers are unbound) indicates the continued presence of large neutron pairing energies. This contradicts the observation made by Vogel

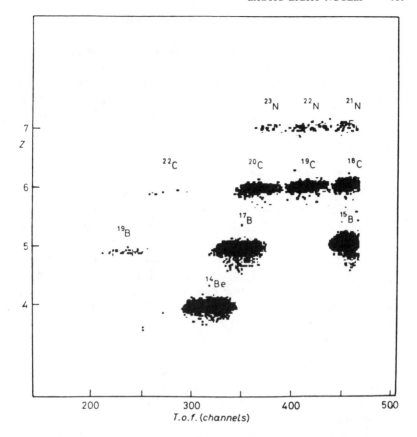

Figure 5   Atomic number (Z) vs time-of-flight ($\propto A/Z$) two-dimensional plot of particle-identified ions collected on the LISE spectrometer as produced in ^{40}Ar (44 MeV/u)+Ta fragmentation reactions. One can see the nucleon stability of ^{22}C and ^{23}N and the nucleon instability of ^{21}C (49).

et al (54) and Jensen et al (55) for heavier nuclei, which indicate that pairing energies decrease quadratically with increasing neutron excess. In fact, if the empirically determined pairing energy formula of References 54 and 55 is taken at face value, it would predict little (if not negative?) neutron pairing in these nuclei, in clear disagreement with nucleon stability observations mentioned above. Second, the bound or unbound characterization of these exotic nuclei serves to constrain atomic mass models. An interesting example, if only preliminary at this time, is the observation that ^{26}O is unbound, whereas ^{32}Ne is bound. Neither result is predicted by most mass models/relationships (46).

Recent research has shown that many new isotopes exist above $Z = 10$, (17, 52, 56–58) (see Figure 1). Although these nuclei are not sufficiently close to the predicted neutron drip line to allow any insight into their nature, these forefront experiments illustrate the significant technical advances in producing and identifying (separating) such nuclei, and they herald the day when spectroscopic information about these exotic isotopes will be available.

2.4   *Extending the Proton Drip Line Up to Z = 23*

As a result of the increased Coulomb energy that arises with the addition of each proton to a nucleus, the proton drip line lies much closer to the nearest β-stable isotope of each element than does the neutron drip line. This proximity to stable nuclei, in general, makes it experimentally easier to reach the proton drip line than the neutron drip line, but the fact that proton (or two-proton) emission can be significantly delayed by the Coulomb barrier makes the true establishment of the proton drip line more problematical. This fact is illustrated in Figure 6, in which we have plotted the estimated $T_{1/2} \approx 100$ ns to $T_{1/2} \approx 100$ ms partial half-life windows for proton and two-proton emission from different Z parents vs the available decay energy. These calculations are based on penetrability considerations only [i.e. the full single-particle Wigner limit (59) has been assumed]. Thus, for example, a $Z = 21$ neutron-deficient isotope that is unbound with respect to proton (two-proton) emission by 330 keV (950 keV) would be expected to have a half-life as short as ~ 100 ns and thus could probably be observed in an on-line experiment. Therefore, such particle identification (or beta-decay related) experiments must be qualified by uncertainties of this order (note that these uncertainties increase with increasing atomic number). Mass measurements and direct proton (two-proton) radioactivity experiments can be more definitive.

By 1980, the proton drip line was established up through $Z = 11$ following the mass measurement of ^{19}Na, which indicated that this isotope was proton unbound by ~ 320 keV (60, 61). It might have been said that all neutron-deficient isotopes up to $Z = 12$, including $Z = 15, 17$–19, and 21 that were predicted (62) to be proton (two-proton) bound had been observed. Since then, some 20 other neutron-deficient isotopes near or at the proton drip line have been observed through $Z = 29$. ^{22}Al, ^{26}P, and ^{35}Ca were first produced in compound nuclear reactions and identified by means of their unique β-delayed two-proton decay process (63, 64). Mass measurements using pion double charge exchange reactions proved that both ^{28}S and ^{40}Ti were particle bound (65). Several LISE particle identi-

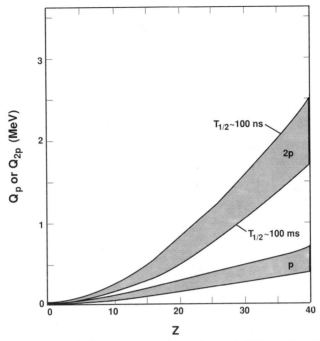

Figure 6 A plot of the calculated proton (two-proton) decay half-life windows ($T_{1/2} = 100$ ns to 100 ms) as a function of available decay energy (Q_p or Q_{2p}) and atomic number (Z) of the parent; based on Coulomb barrier penetrability considerations only.

fication experiments led to the observation of (*a*) ^{23}Si, ^{27}S, and ^{31}Ar (66); (*b*) the unbound nature of ^{21}Al and the discovery of the $T_z = -3$ nucleus, ^{22}Si (67) (see Figure 7); and (*c*) a host of other isotopes above $Z = 22$, which include ^{43}V, ^{44}Cr, ^{46,47}Mn, ^{48}Fe, $^{50-52}$Co, ^{51,52}Ni, and ^{56,57}Cu (30). Mass measurements of ^{39}Sc proved that this isotope is proton unbound by ~ 600 keV (68, 69). And finally, evidence for the existence of ^{39}Ti has recently been found through its characterized β-delayed proton decay (70).

According to a comparison with updated Garvey-Kelson mass predictions (6), all proton or two-proton bound nuclei have been found through $Z = 23$ and for $Z = 25$, 27, and 29. [For completeness, we also mention the known ground-state proton radioactivities at $Z = 53$, 55, 69, and 71 (71).] Of further interest is the fact that both the Garvey-Kelson and the isobaric multiplet mass equation calculations (72) indicate that ^{22}Si, ^{26}P, ^{39}Ti, and ^{43}V are either proton (odd-Z) or two-proton (even-Z) unbound. Because of the uncertainties related to Coulomb barrier hindrance mentioned above, none of the present experiments are inconsis-

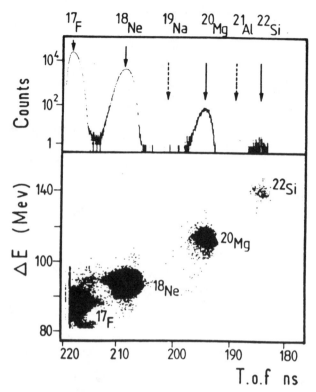

Figure 7 The discovery of ^{22}Si and the identification of neighboring nuclei as produced in ^{36}Ar (85 MeV/u)+Ni fragmentation reactions (67). (*Lower panel*) The two-dimensional display of fragments identified in the focal plane of the LISE spectrometer. (*Upper panel*) The corresponding projected time-of-flight (A/Z) spectrum.

tent with these predictions; therefore, these nuclei represent tantalizing candidates for future mass measurements and/or proton (two-proton) radioactivity search experiments.

3. MASS MEASUREMENTS

After initial experiments to produce a previously unobserved isotope and to prove (or disprove) its nucleon stability, the ground-state mass is often the next measurement to be pursued. As a fundamental property, the mass of a nucleus provides valuable quantitative information to test our understanding of nuclear binding and nuclear structure. This information may range from the exploration of shell structure effects, such as those found near closed shells, or the changes in binding energy resulting from changes in nuclear shape, to the testing of charge symmetry as are inherently assumed in various mass relationships such as the isobaric mass multiplet equation (IMME) and the Garvey-Kelson (GK) mass relationships. Thus the mass measurements of exotic nuclei often provide our first glimpse into the interesting nuclear structure features found in nuclei of extreme neutron-to-proton composition.

In what follows, comments are restricted to those nuclei whose masses have been measured for the first time since January 1980.

3.1 *Measurements of Neutron-Rich Nuclei*

For the very lightest nuclei, researchers attempt the mass determinations of nuclei with large neutron-to-proton ratios [e.g. ^{9}He ($N/Z = 3.5$), compared to ^{34}Na ($N/Z = 2.1$) and ^{252}Cf ($N/Z = 1.6$)] to simulate the behavior and improve our understanding of neutron-like matter as well as to explore the stability and nuclear structure of few-nucleon systems. For the neutron-rich isotopes of helium, it is interesting to note that ^{8}He is more bound than ^{6}He with respect to both one- and two-neutron emission and, similarly, ^{7}He is more one-neutron bound than ^{5}He. This behavior is not found in any other isotopic series and has led to the speculation that ^{9}He and ^{10}He may be more bound than previously expected. In response to this conjecture, Seth et al (73) recently measured the mass of ^{9}He by using the pion double charge exchange (DCX) reaction ^{9}Be$(\pi^-,\pi^+)^9$He. Resting on a mixed multibody breakup phase space, the observed ground state and three excited states had reasonably narrow line widths ($\Gamma = 420$–550 keV, FWHM) that were largely limited by the experimental energy resolution (Figure 8). The resulting mass proved that the ground state was unstable with respect to single-neutron emission by 1.1 MeV. This is 1.3 to 2.7 MeV more bound than predicted by either the GK mass relationships (62) or the modified shell model formalism (74), respectively. Of special interest here are the $(0+1)\hbar\omega$ basis space, no "inert core," shell model calculations of Poppelier et al (75), which proved remarkably successful in predicting both the ground-state mass and the energies of the observed excited states.

Applying the mass of ^{9}He to the local GK transverse relationship, Seth et al (73) predict that ^{10}He will be one-neutron and two-neutron unbound by 0.3 MeV and 1.4 MeV, respectively. Being significantly more bound than previous calculations, this new estimate further strengthens the argument that ^{10}He may be much more bound than was thought. In an attempt to address this question, a new search for the nucleon stability of ^{10}He was undertaken using ^{18}O projectile fragmentation reactions as the production source (76). During this experiment, no ^{10}He events were detected where one would have expected to have observed ~ 1000 events if ^{10}He had a

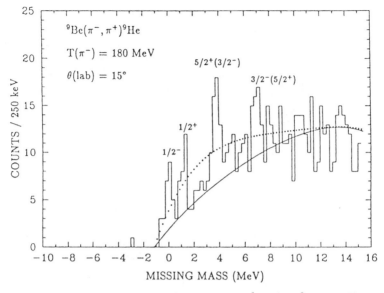

Figure 8 The missing mass spectrum for the reaction $^9Be(\pi^+,\pi^-)^9He$ (73). The curves represent various fits to the multibody phase space.

half-life greater than 100 ns. This strongly suggests that ^{10}He is nucleon unbound. A definitive answer to the neutron binding of ^{10}He, however, now awaits a Q-value reaction measurement such as a heavy-ion-induced DCX reaction on a radioactive ^{10}Be target.

Of further interest in the light mass region are the recent measurements of ^6H, ^{13}Be, and ^{14}Be; ^6H was investigated by Aleksandrov et al (77) and Belozyorov et al (78), using the $^7Li(^7Li,^8B)^6H$ and $^9Be(^{11}B,^{14}O)^6H$ reactions, respectively. Although their results are complicated by background from pile-up events, reactions taking place on target impurities, $^3H + 3n$ breakup phase space, and low counting statistics, the authors point to an enhancement observed in their spectra with a width of $\Gamma = 1.8$ MeV and attribute this to the ground state of ^6H. If this is the case, the resulting mass of ^6H would be three-neutron unbound by ~ 2.7 MeV, which is 3 to 6 MeV more bound than that predicted by the shell model (75). In an attempt to verify these results, a measurement of the $^6Li(\pi^-,\pi^+)^6H$ reaction was attempted (79). The findings of this experiment yielded no enhancement above phase space, and an upper limit on the cross section of 1.6 nb/sr was set, 20 times smaller than that measured for ^9He(g.s.) (73). This disagreement now clouds the mass determination of ^6H and another measurement with improved statistics and cleaner experimental conditions is needed.

The mass of ^{13}Be was measured in the ^{14}C(^{7}Li,^{8}B)^{13}Be reaction (80). Although severely hampered by many of the same problems that plagued the experiments mentioned above, this work extracted a ^{13}Be mass that is neutron unbound by 1.8 (± 0.5) MeV—in good agreement with most mass predictions. Nonetheless, a cleaner and statistically more accurate measurement would be highly desirable.

The mass of ^{14}Be was first measured by means of the ^{14}C(π^-,π^+)^{14}Be reaction (81). The resulting mass was 0.6 to 1.0 MeV more bound than predicted by the GK mass relationships (62) and the modified shell model (74), but it is in reasonable agreement with the shell model calculations of Poppelier et al (75).

In summarizing this work in this light mass region, we have found that the measured masses are consistently 1–2 MeV more bound than those predicted by either the GK mass relationships or the modified shell model, but in good agreement with the "no inert" core shell model calculations of Poppelier et al (75). Therefore, this work represents another excellent verification of the shell model—in this case for few-body neutron-rich systems. For additional information about such nuclei, see the review article by Ogloblin & Penionzhkevich (82).

Beyond this light mass region, heavy ion mutinucleon reactions have also been widely used to determine the masses of neutron-rich nuclei ranging from ^{33}Al to ^{69}Ni. Most of these reactions are three-nucleon transfer reactions such as (^{14}C,^{15}O) (two-proton pickup/one-neutron stripping), (^{36}S,^{33}Al) (three-proton stripping), or (^{14}C,^{11}C) (three-neutron stripping) reactions on isotopically enriched, rare isotope targets. Difficulties arise because (*a*) the production of the ground state is generally not favored over the excited states, and this complicates level assignments; (*b*) the production cross sections are small (ground-state cross sections are typically on the order of 1 μb/sr at $\theta_{lab} = 10°$) and fall off rapidly with increasing angle, which makes it difficult to check kinematic shifts; and (*c*) reactions on target impurities can often cause serious interferences in the energy spectrum of the ejectile.

Target impurities were a problem in the mass measurement of ^{35}Si (83, 84) and ^{36}P (83, 85), in which an 80% enriched ^{36}S target in the form of Ag$_2$S on a carbon backing was used. Careful measurements on targets enriched in ^{32,34}S and ^{12}C were needed to unravel the complicated spectra that were obtained. Similar problems plagued the measurements of ^{47}Ar (86) and ^{51}Ca (86, 87). For ^{47}Ar, additional data was collected at $\theta_{lab} = 21°$ to kinematically shift away interfering reactions on oxygen contaminants present in the ^{48}Ca target, whereas in the ^{51}Ca measurements the presence of ^{40}Ca impurities proved to be a serious source of background. A 1.1-MeV discrepancy between these two measurements was found, which, as suggested by Benenson et al (86), could be accounted for by the misassignment of the ^{51}Ca(g.s.) to ^{40}Ca impurity background events observed in the experiment of Brauner et al (87).

This problem of interfering reactions on target contaminants can be solved, in part, by direct detection of the exotic species of interest as the reaction ejectile. This method was illustrated in the mass measurements of ^{33}Al by means of the ^{48}Ca(^{36}S,^{33}Al)^{51}V reaction (88) and of ^{37}P through the ^{48}Ca(^{36}S,^{37}P)^{47}Sc reaction (89). In these cases, reactions taking place on target impurities that produce ^{33}Al or ^{37}P have more negative Q values than those resulting from the ^{48}Ca target, and therefore they do not cause an interference in the region of the ground state. As exemplified in the measurement of ^{37}P during which misidentified ^{36}S ions partially contaminated the spectrum, detecting of the exotic species as the ejectile requires the clean identification of all reaction products. Thus, this method of reducing target contaminant interference is applicable only in the light mass region where the nucleon-bound ejectile of interest can be cleanly identified.

The science issues addressed by these measurements have three main themes. The first set of measurements (^{33}Al,^{35}Si,^{36,37}P) dealt with the question of deformation observed at $N = 20$ for ^{31}Na (1, 2) and ^{32}Mg (3) and how far this deformed region extends into neighboring nuclei. No enhanced binding above that expected from systematic trends was observed, which suggests a normal spherical shell closure at $N = 20$ occurs for $Z \geq 13$ species (see more detailed discussion below). The measurement of ^{51}Ca was motivated by the conjecture (90) that ^{52}Ca is doubly magic. However, no definitive statement about enhanced binding at $N = 32$ can be made from this work; further measurements are needed. The mass measurement of ^{69}Ni (91) was undertaken, in part, to determine if the subshell closure at $N = 40$ is reinforced by the $Z = 28$ closed shell. Recent work (92) related to the first excited state of ^{68}Ni tends to support this idea [in analogy to what is observed in $^{90}_{40}$Zr], yet no strong indication of this reinforcement is evident in the two-neutron binding energy trend. In general, all of these mass measurements for $Z \geq 13$ neutron-rich isotopes were in good agreement with a variety of mass predictions.

Several masses in this light region were determined by the β-endpoint method. These include measurements of ^{31,32}Mg (93), ^{52}Ca–^{52}Sc (94), ^{49,50}K (95), and ^{40,42}Cl (96). Such work requires a detailed knowledge of the β-decay scheme and a relatively pure source. The above radioactivities were produced at either the on-line mass spectrometer located on the CERN PS or at ISOLDE-2 by the selective ionization of either an alkali element (e.g. Na or K) or a halogen element (e.g. Cl as a negative ion) with intensities of approximately 10^3 ions per second. The decay electrons were detected in a ΔE-E plastic scintillator telescope in which the spectrum was gated by either gamma rays or, as demonstrated in the work of Miéhe et al (95), by delayed neutrons. Because of the β-feeding of high-lying excited states, which may de-excite through one of the gamma rays used to gate the β-spectra, only the upper portion ($\sim$ 3 MeV) of the gated β-spectra are generally fit to extract Q_β values. Thus, a reasonably large set of data is needed to obtain reliable results with accuracies of 300 keV or better. Of the measurements mentioned above, most yielded masses that are in good agreement with theoretical expectations, given their uncertainties. However, serious discrepancies (where the experimental Q_β values are low by as much as 1.5 MeV) were found for ^{31}Mg, ^{52}Ca, and ^{52}Sc. In the case of ^{31}Mg, this measurement was severely hampered by low counting statistics; subsequent measurements (mentioned below) have since rectified this situation. The ^{52}Ca and ^{52}Sc measurements remain somewhat suspect, and additional measurements are warranted.

A new development in the area of exotic nuclei is the advancement of recoil spectrometers to perform direct mass measurements. As described in Section 2, several mass measurements have been made with the TOFI spectrometer located at LAMPF and the SPEG spectrometer at GANIL. In the TOFI experiments, 800-MeV proton-induced fragmentation and fission reactions on Th targets are used to produce a wide variety of exotic neutron-rich nuclei. A small fraction of these recoils are captured by a secondary beam line and introduced into TOFI, where their mass is obtained from a single, high precision measurement of the ions' time of flight as they pass through this isochronous recoil spectrometer. Typical mass-to-charge ($A/Q \propto \mathrm{TOF}$) spectra are shown in Figure 9. In the direct mass measurement method, the centroid of each mass line determines the mass of that species (isotopes with known masses are collected simultaneously with the unknowns and serve as calibrations). Through the collection of sufficient statistics, meaningful mass measurements are made.

A similar direct mass measurement method is employed at GANIL using the SPEG spectrometer. Here the exotic recoils are produced through 40–60-MeV/u projectile fragmentation reactions and a two-parameter, high resolution measurement of the ion's velocity and momentum-to-charge ratio, from which the mass is determined. The benefits of high beam intensities available at LAMPF are offset at GANIL by the ability to (a) use thicker targets and (b) capture a larger fraction of the exotic recoils. The latter is due to the kinematic concentration of all reaction products at forward angles resulting from the projectile fragmentation process. The two methods yield comparable results, and both groups have reported

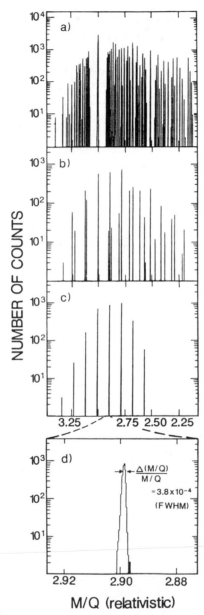

Figure 9   Mass-to-charge spectra obtained with the TOFI spectrometer for proton-induced Th fragmentation products with (*a*) $Z = 6$–15; (*b*) $Z = 11$; (*c*) $Z = 11$, $Q = 9$; and (*d*) an enlarged view of the $^{26}\mathrm{Na}^{9+}$ mass line (97).

significant improvements with each new set of measurements. At present, the mass resolutions are on the order of 4×10^{-4} (FWHM), with systematic errors ranging from 4 to 10 ppm.

Together these two research groups have determined the masses of 22 previously unmeasured neutron-rich nuclei ranging from ^{17}B to ^{38}P and have measured another 11 isotopes with improved (or comparable) error bars (97–100). In general, there is a good agreement among these measurements. The exceptions are the measurements of ^{27}Ne and ^{30}Na by Vieira et al (97), in which an isobaric contamination from a neighboring higher-Z element appears to have biased measurements in such a way as to produce a more bound mass. Isobaric cross contamination and the existence of unknown isomeric states that could lead to a mixed mass measurement are potential problems when this fast-recoil direct-mass measurement method is used. However, in the work of Gillibert et al (99) where a high geometry γ array was placed around the recoil stopping detector, no evidence for an isomeric activity with a half-life longer than 1 μs was observed, with the exception of ^{32}Al, which was only weakly affected.

A plot of the two-neutron separation energies (S_{2n}) resulting from these measurements (combined with appropriate Q-value reaction measurements mentioned above) is shown in Figure 10. Such a plot is a useful way of removing the odd-even neutron pairing effects, so that other nuclear structure effects stand out more clearly. Two things are of particular note in Figure 10: (a) the sudden decrease in S_{2n} values occurring after $N = 15$ for the neutron-rich isotopes of O, F, and Ne; and (b) the sudden upturn in S_{2n} observed at ^{31,32}Na, which is not matched by the S_{2n} trends observed in the neighboring Mg and Al isotopes.

Concentrating first on the downward break in the S_{2n} trend observed at $N = 15$ one finds that nearly all mass theories and relationships have underpredicted the binding of neutron-rich nuclei in this region (46). In particular, for ^{24}O the two-neutron separation energies are underestimated by as little as ~ 1 MeV in the Garvey-Kelson mass relationship predictions of Jänecke & Masson (6) to values as high as 2.8 MeV in the microscopic-macroscopic model predictions of Tachibana et al (102). However, as seen at the bottom of Figure 11, this $N = 14$–16 S_{2n} trend is well reproduced by the shell model calculations of Wildenthal et al (103).

From a careful investigation of the shell model calculations, it was found that the observed break in the S_{2n} trend at $N = 15$ can be explained in terms of two-body interaction energies (100, 105). The single-particle energy spacing between the $0d_{5/2}$ and the $1s_{1/2}$ levels is fairly small (~ 0.9 MeV), so that the strong S_{2n} decrease observed in going from $N = 15$ to $N = 16$ results primarily from the interplay of (a) the strongly attractive $[0d_{5/2}\text{-}0d_{5/2}]_{J=0}$ two-body interaction and (b) the approximate cancellation

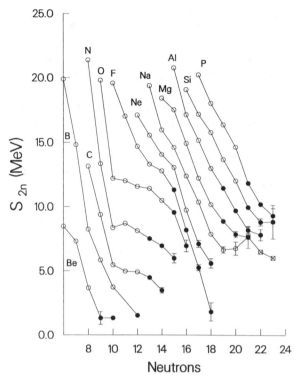

Figure 10 Two-neutron separation energy vs neutron number for isotopes of boron to phosphorus. Open circles indicate previously measured nuclei (masses taken from Ref. 101 and supplemented by recent measurements); solid circles represent nuclei measured for the first time (the mass represents the weighted average of values discussed in the text). Cross circles indicate Na isotopes from the work of Thibault et al (1) that have not yet been remeasured. (To reduce confusion, the large error bars on these latter points have been excluded.)

of the attractive $(1s_{1/2}\text{-}1s_{1/2})$ and $[0d_{5/2}\text{-}1s_{1/2}]_{J=2}$ two-body interactions by the repulsive $[0d_{5/2}\text{-}1s_{1/2}]_{J=3}$ two-body interaction. The net effect is a small decrease in the S_{2n} value from $N = 14$ to $N = 15$, followed by a larger S_{2n} decrease between $N = 15$ and $N = 16$. Considering the relatively pure configurations of the ground states and the strong dependence of these particular two-body interaction energies on the binding of these exotic nuclei, this work represents a powerful test and, in this case, verification of the shell model.

The second feature prominent in Figure 10 is the dramatic upturn in the S_{2n} trend occurring at $N = 20$ in the neutron-rich Na isotopes. Both the measurement of these sodium isotopes (1) and the explanation (2) of this change in binding energy in terms of the onset of prolate deformation

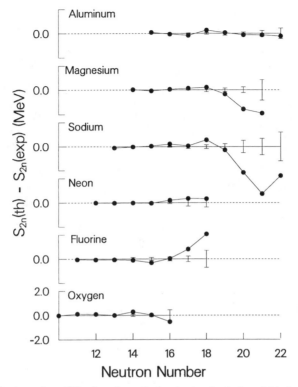

Figure 11 A comparison of the S_{2n} values calculated using the shell model (taken from Ref. 103 for $N \leq 20$ and Ref. 104 for $N = 21$–22) minus the experimental S_{2n} values shown in Figure 10 vs neutron number for oxygen to aluminum isotopes. The experimental uncertainties are indicated by error bars.

have been known for some time. However, only recently has one been able to measure the masses of their neighboring $N = 20$–22 isotones in Mg and Al to explore the extent of this deformation.

As advanced by the Hartree-Fock calculations of Campi et al (2), the explanation of how this deformation occurs involves the inversion of the lowest $0f_{7/2}$-1/2[330] neutron Nilsson level with the highest $0d_{3/2}$-3/2[202] Nilsson level at high prolate deformation (see Figure 12). The binding of the nucleus is thus increased as a result of this deformation (with the related destruction of the $N = 20$ closed shell); a similar increase in S_{2n} values is observed at the onset of deformation in the rare earth region. Another way of stating this effect is to say that a deformed intruder state has dropped below the energy of the normal spherical state, thus leading to a more bound ground state (107).

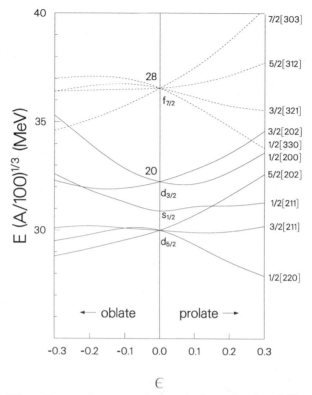

Figure 12   Nilsson diagram of neutron and proton levels as a function of ellipsoidal deformation, defined by the eccentricity coordinate ε. (extracted from Ref. 106).

As is seen in Figure 11, the sd basis shell model calculations of Wildenthal et al (103) and the sdpf basis, $0\hbar\omega$ calculations of Becker et al (104) dramatically fail in the region of $^{31-33}$Na. This is the result of 2 particle–2 hole (and to a lesser extent 4p–4h) excitations into the $f_{7/2}$ shell that are explicitly excluded from these calculations. The latter was shown to be true by the extended calculations of Watt et al (108) and Poves & Retamosa (109). Looking to higher Z, $N = 20$–22 isotones, we observe that the discrepancies are reduced to half of those observed in the Na isotopes for ^{32,33}Mg and that good agreement with the shell model calculations is fully restored for $^{33-35}$Al.

A qualitative understanding of this effect can be obtained by referring to the Nilsson diagram of Figure 12 and considering the levels occupied by the protons. In Na nuclei, the odd proton resides in the $0d_{5/2}$-3/2[211] Nilsson level, in which the energy is largely independent of deformation.

However, in Al the odd proton resides in the $0d_{5/2}$-5/2[202] Nilsson level, in which the energy increases rapidly with increasing prolate deformation. Thus the net gain in binding energy, made possible in the Na isotopes by the neutron filling the $0f_{7/2}$-1/2[330] Nilsson level instead of the $0d_{3/2}$-3/2[202] level, is countered in the Al isotopes by the increased energy that is required in placing protons in the $0d_{5/2}$-5/2[202] Nilsson level at such large deformations. As a consequence, the deformation observed in the Na isotopes appears to be very localized; the Mg isotopes are less deformed than those of Na and the Al isotopes are spherical. Carrying this argument one step further, one might expect ^{29}F to show an even more pronounced S_{2n} deformation effect than that of $^{31-32}$Na, but this remains to be determined. Further improvements in the fast-recoil, direct-mass measurement technique and additional measurements in this region and for adjacent heavier mass regions appear likely in the near future.

3.2   Neutron-Deficient Nuclei

Far fewer mass measurements have been performed in the light mass region on the neutron-deficient side of stability than on the neutron-rich side (see Figure 1). This is due in part to the rapid increase in the Coulomb energy that occurs with each additional proton in the nucleus. Thus, the Q values for reactions producing neutron-deficient nuclei are, in general, more negative, and their cross sections are correspondingly smaller than those of neutron-rich nuclei that are equally far removed from their nearest stable isotope. Fast-recoil direct-mass measurement techniques have not yet been applied to neutron-deficient nuclei, but this is likely to change in the future.

As with neutron-rich nuclei, the use of pion-induced double charge exchange (DCX) reactions has led to the first-time mass measurements of several neutron-deficient nuclei, including ^{28}S, ^{32}Ar, and ^{40}Ti, by means of (π^+, π^-) reactions on $N = Z$ targets (65, 110). The mass of ^{24}Si was first measured by the four-neutron pickup reaction ^{28}Si$(\alpha, ^{8}$He$)^{24}$Si (111), but a pion-induced DCX reaction measurement (110) soon confirmed these results. Although the pion-induced DCX reaction cross section was determined to be ~ 20 times larger than that measured for the $(\alpha, ^{8}$He) reaction, the much larger ^{4}He beam intensity provided a more precise mass determination. The ground-state cross sections for both of these reactions decrease linearly, if not faster, with increasing mass number (65, 112). This fact has, for the most part, limited the use of these reactions to studies of lighter mass nuclei; a notable exception is the mass measurement of ^{58}Zn through the ^{58}Ni(π^+, π^-) reaction (113).

By combining these ground-state mass measurements with the masses of analog $T = 2$ isospin multiplet members [many $T_z = -1$ members have recently been determined through the beta-delayed proton decay studies (see Section 4.1)], it was possible to confirm further the validity of the quadratic form of the isobaric multiplet mass equation (IMME) (114). Of the twenty-seven $T = 3/2$ quartets that have been measured up to $A = 41$ and the nine $T = 2$ quintets, in which at least four members of the multiplet have been measured through $A = 36$ (115–117), all cases except $A = 9$, $16(2_1^+)$, and 20 agree with the quadratic form of the IMME. Deviations from the quadratic form of the IMME can come about only through second-order (or higher) corrections involving charge-symmetry-breaking interactions that are estimated to be fairly small [e.g. calculations of the cubic "d" coefficient are 1 to 3 keV (118)]. Thus, relatively precise mass measurements for all members of the multiplet are required to provide a meaningful test of the quadratic relationship. Of the three discrepancies mentioned above, ^{9}C gives the strongest evidence of a charge-symmetry-breaking force by yielding a nonzero "d" coefficient equal to 5.8 (± 1.5) keV (115). The other two cases are suggestive but will require higher precision measurements before they are convincing.

Given the success of the quadratic form of the IMME, one can reliably predict the mass of the unmeasured multiplet members from the mass of three (or more) other members. Generalizing this approach, Pape & Antony (72) have used the systematic variation of the linear "b" coefficient with isospin and mass number to predict the masses of several neutron-deficient nuclei ranging up to $A = 117$.

Also related to the IMME studies is the work of Ormand & Brown (120), who used the systematic variation of the IMME "b" and "c" coefficients to extract an empirical isospin-nonconserving (INC) Hamiltonian for use in shell model calculations. The resulting Hamiltonian contains a small charge-symmetry-breaking interaction in which the neutron-neutron two-body interaction is 1.8% more attractive than the proton-proton interaction, and a charge-dependent interaction in which the proton-neutron interaction is 2.4% stronger than the average of the p-p and n-n interactions. Using this INC Hamiltonian, these authors calculated both the isospin-mixing corrections to the Fermi matrix element in superallowed Fermi β-decay transitions (5, 121) and the spectroscopic amplitudes for the isospin-forbidden nucleon decay of $T = 3/2$ states to $T = 0$ states (122) with a reasonable degree of success. Further applications of this empirical INC Hamiltonian are under way in investigations of other isospin-nonconserving processes and for predictions of isotopic mass shifts.

Related to the search for nuclei that undergo ground-state proton radio-activity, recent measurements of the mass of ^{39}Sc employed the reactions ^{40}Ca(^{14}N, ^{15}C) (68) and ^{40}Ca(^{7}Li, ^{8}He) (69). These results, which agree well with one another, show that ^{39}Sc is proton unbound by ~ 600 keV, as predicted by both the Garvey-Kelson transverse mass relationship and the INC formalism discussed above. Based on Coulomb barrier penetrability and spectroscopic factor calculations, ^{39}Sc is estimated to undergo proton decay with a partial half-life on the order of 10^{-14} seconds. This half-life is too fast for most experimental decay methods. Such studies exemplify the unique advantage of using two-body reactions to measure the masses of short-lived nucleon-unbound isotopes.

Using a comparison of nuclei with known masses to those whose masses were predicted by a general charge-symmetric Garvey-Kelson mass relationship (125), Comay, Kelson & Zidon (126) argue that many neutron-deficient nuclei that are proton (two-proton) unbound by 1 MeV or more exhibit a significant increase in binding energy (400 keV to 1 MeV), which results from a reduced Coulomb energy associated with the unbound proton (two-protons). This Coulomb perturbation effect is known as the Thomas-Ehrman shift (127); if verified in heavier mass nuclei, such a mass shift could have important consequences for the search for proton (two-proton) radioactivities (see Section 4.4).

Of further interest in the neutron-deficient region are the $T_z = -1/2$ nuclei ^{57}Cu and ^{59}Zn, which lie just above the region shown in Figure 1. The masses of ^{57}Cu and ^{59}Zn have been measured using the reactions ^{58}Ni(p,2n)^{57}Cu $\rightarrow \beta$ endpoint (128), ^{58}Ni(^{7}Li, ^{8}He)^{57}Cu (129), ^{58}Ni(^{14}N, ^{15}C)^{57}Cu (130) and ^{58}Ni(^{3}H,2n)^{59}Zn $\rightarrow \beta$ endpoint (131), ^{58}Ni(p,π^-)^{59}Zn (132), respectively. Coulomb displacement energies (CDE) determined from these measurements and the masses of their mirror nuclei further extend our knowledge of the CDE anomaly known as the Nolen-Schiffer anomaly (133). In particular, the experimental CDE are found to be 5 to 11% larger than those calculated from the Coulomb force and the known rms charge radii. As seen in Figure 13, the size of this anomaly is nuclear-structure dependent: single-particle $2p_{3/2}$ and $1f_{7/2}$ states exhibit a larger discrepancy than those of $(1d_{3/2})^{-1}$ hole states. Core polarization resulting from the influence of the valence nucleon(s) would appear to explain, at least partially, this particle-hole orbital dependence (129), but the underlying reason for the anomaly itself remains uncertain (118, 134, 135). The role of charge-symmetry-breaking forces that could explain this anomaly warrants further investigation.

3.3 Pairing Energies and Mass Model Improvements

Looking more globally at the nuclear mass surface, we note that the surface is actually split into four subsurfaces, composed of even-Z–even-N; odd-N; odd-Z; and odd-Z–odd-N nuclei, due in large part to the presence of neutron and proton pairing. A reexamination (54, 55) of the neutron (and proton) pairing energies as deduced from third-order mass differences

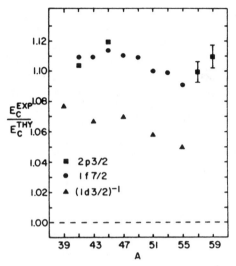

Figure 13   The ratio of experimental Coulomb displacement energies to those calculated using the spherical Hartree-Fock model (129).

has revealed a quadratic dependence on neutron excess $(N-Z)/A$ (see Figure 14). Of further interest, the difference between the neutron and proton pairing energies (55) smoothly oscillates with mass number—an artifact that is believed to be shell structure related.

In the Hartree-Fock-Bogoliabov calculations of Jensen & Miranda (136), the neutron excess dependence of these pairing energies could not be reproduced using the normal Skyrme plus pairing effective interactions. However, by including an explicit neutron-proton pairing interaction, a reasonable agreement with the observed dependence was obtained. Additional work on pairing energies has been performed by Madland & Nix (137). Their derivations, based on the BCS approximation applied to systems with dense, equally spaced levels, led to an exponential dependence of the pairing energies on neutron excess. Their deduced formulas agreed well with the average pairing energy trends, although systematic $N(Z)$ fluctuations in the experimental neutron (and to a lesser extent, proton) pairing energies—reminiscent of shell effects—are clearly evident. In an attempt to deal with the latter problem, Möller & Nix (138) have incorporated a microscopic pairing correction into their mass model, which uses a calculated set of levels as obtained from a folded-Yukawa single-particle potential instead of a uniform level distribution.

As pointed out in Section 2.3, where the establishment of the neutron drip line for light mass nuclei shows the continued presence of strong

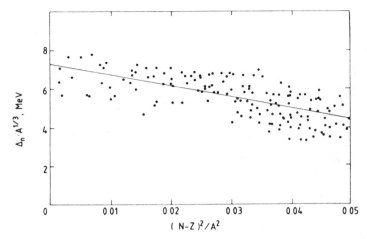

Figure 14   The neutron pairing energy Δ_n multiplied by $A^{1/3}$ as a function of the neutron excess squared $[(N-Z)/A]^2$. These data include all masses for even-even nuclei with $A > 40$, except those involving magic N or Z nuclei. The line represents a least squares fit to the data (55).

neutron pairing, in general contradiction with the above formulations, it is clear that these new pairing energy calculations warrant careful examination. In particular, the systematic mass measurements for a set of adjoining, far-ranging isotopic series within a limited Z region(s) is needed.

Valuable insight into the systematic deficiencies of mass models and/or relationships can be gained from a careful comparison of theory and experiment. This is illustrated in Figure 15, where the mass differences between the predictions of the macroscopic-microscopic mass model of Möller & Nix (138) and the known mass surface are shown. Localized regions where the predictions lead to too much binding are found interspersed by regions of too little binding. One notes further that the degree of localization (and the maximum extent of these deficiencies) and the frequency of these regions increase with decreasing N and Z. The largest discrepancies are observed for transitional nuclei, i.e. those nuclei lying between regions of spherical and deformed shaped nuclei.

Based on the nature of these deviations, the authors speculate that these differences could represent either a natural collapse of the macroscopic-microscopic method for light mass systems or deficiencies in the single-particle model used to calculate the microscopic shell correction energies. Given that parameters for the folded-Yukawa potential used in their single-particle calculations were deduced from experimental data in the rare earth and actinide regions and held constant throughout the periodic

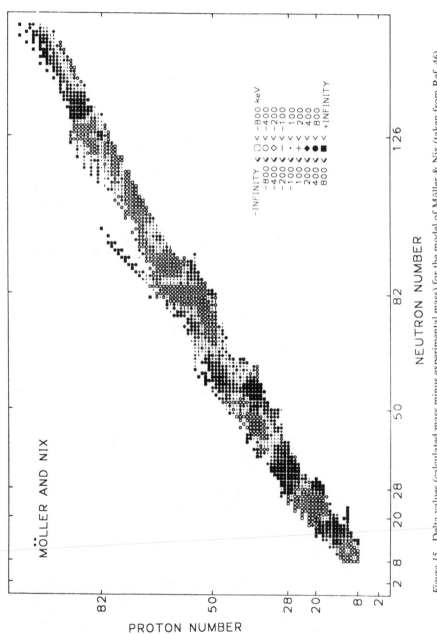

Figure 15   Delta values (calculated mass minus experimental mass) for the model of Möller & Nix (taken from Ref. 46).

system, one suspects increasing deviations to result the farther one moves away from these "optimized" regions and for nuclei with much different shapes. Clearly, the validity of their single-particle model in predicting the level structure of nuclei in other mass regions should be carefully investigated. The incorporation of octupole (ε_3) and dodecapole (ε_6) shape degrees of freedom into their calculations is also expected to prove valuable, particularly for those regions where octupole deformed ground states have been observed (139). Another important advantage of such global approaches, exemplified by macroscopic-microscopic models and Hartree-Fock calculations, is their ability to calculate many experimental observables with a relatively small number of parameters. Thus, such theories provide valuable physical insight into the nature of both exotic and common nuclei.

Examining the predictive quality of several mass models and/or relationships, Haustein (140) points out that those approaches using more adjustable parameters in general have a larger ratio of rms deviation for "new" masses (e.g. those not used to constrain the model parameters) as compared to the rms deviation obtained from the original data set. Thus caution is advised in accepting the rms deviation quoted accuracies for many of these predictions when a large number of adjustable parameters are involved and the isotope of interest is far removed from the known mass surface.

As an interesting example, consider the predictions based on GK mass relationships where the deviations between theory and experiment increase proportional to T_z^3 (see Figure 16). Although such higher order isospin terms have been empirically included in the recent predictions of Masson & Jänecke (142) with some improvement in fitting the most exotic species, a fundamental understanding of their source is still lacking. Along the same line, the successes resulting from the inclusion of phenomenological neutron-proton interactions into some models and mass systematics (102, 143) warrant further investigations, particularly as they relate to understanding the true nature of these empirical interactions.

4.   EXOTIC DECAY MODES

The β-decay of nuclei far from stability is characterized by large Q_β values, which allow for a wide variety of β-delayed particle emissions. The interest in such processes lies mainly in the fact that they are directly related to the structural properties of well-identified nuclear states. For example, although uncertainties in the details of reaction mechanisms make it difficult to extract information on two-proton correlations in nuclei from two-proton transfer, the proton pairs observed through β-delayed two-proton emission provide a direct measure of two-proton correlations inside a

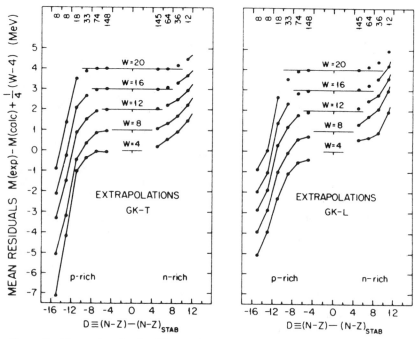

Figure 16 Mean values for the difference between experimental and predicted masses as a function of the distance from the line of β-stability for the Garvey-Kelson transverse (GK-T) and Garvey-Kelson longitudinal (GK-L) mass relationships. The horizontal bars (displayed upward) represent the increasing widths (*w*) of the data base used in these predictions (141).

well-identified nuclear state in a way that is free from strong-interaction distortions. Also, the partial decay width of β-fed isobaric analog states through isospin-forbidden one-proton or two-proton emission provides valuable information on isospin mixing. Another incentive for studying very neutron-rich isotopes is to understand what role they play in astrophysical nucleosynthesis. In recent years there has been an emphasis on the study of β-delayed particle emission, which rapidly sets in as the dominant decay mode for nuclei lying far from the valley of β-stability.

4.1 β-Delayed Neutron and Proton Emission

Excited nuclear states populated in the daughter nucleus by β-decay can be particle unbound. They then undergo a fast strong-interaction decay by emission of the unbound proton or neutron. These two processes, β-n and β-p, are charge symmetric, yet they provide rather different information about exotic neutron-rich and neutron-deficient nuclei. Because unmoderated neutron detection methods have limited efficiencies, neutron energy spectra are not always within experimental reach. Often, only the integrated β-delayed neutron-emission probability, P_n, is measured. This information is essential to evaluate the level of competition taking place between neutron capture and β-decay for those neutron-rich nuclei of importance to the astrophysical r-process. Conversely, exact measurement of the proton energies in β-p decays allows precise level-by-level spectroscopy in neutron-deficient nuclei.

The review by Hardy & Hagberg (144) gives examples of the unique information that can be extracted from studies of β-p processes, in particular on the quenching of the Gamow-Teller transition in nuclei. The study of β-p decay for ^{28}S (117) is one of the most recent tests of the shell model description of exotic nuclei lying far from the stable or nearly stable isotopes on which it was founded.

The β-decay of neutron-rich nuclei was recently reviewed (146). The most important results obtained since then are the measurements of even more exotic nuclei of interest to stellar nucleosyntheis. Kratz et al (147) showed how sensitive the half-life of ^{44}S is to constrain the flux and duration of neutron exposures that account for the observed ^{46}Ca/^{48}Ca solar abundance ratio. When the half-life of ^{44}S was measured (148) from the fragmentation of ^{48}Ca projectiles, the value obtained, $T_{1/2} \approx 200$ ms, was shorter than predicted values. According to Kratz et al (147), this value would result in higher neutron fluxes to account for the solar ^{46}Ca/^{48}Ca ratio. Other nuclei in this region, such as ^{46}Cl, whose radioactive properties are still unknown, are also produced through ^{48}Ca fragmentation. The $T_{1/2}$ and P_n values of these nuclei would further constrain the network calculations in such a way as to check the consistency of the nucleosynthesis process, which accounts for both the ^{46}Ca/^{48}Ca ratio and meteoritic Ca-Ti-Cr isotopic abundances.

More generally, the fragmentation of ^{48}Ca beams carries the opportunity of measuring the radioactive properties of many new neutron-rich isotopes. This measurement would provide a stringent test for theoretical calculations that often predict widely different $T_{1/2}$ and P_n values. For instance, recent results by Lewitowicz et al (148) were compared to the predictions of (a) the "old" gross theory of β-decay by Takahashi (149); (b) a recently improved version of the theory by Tachibana et al (150); (c) the microscopic model of Klapdor et al (151); and (d) the quasi-particle RPA calculations by Staudt & Klapdor (152). Clearly experimental results can be reproduced by theoretical calculations only if microscopic effects are thoroughly included, thus showing the limited ability of gross theories to account for the properties of neutron-rich nuclei of astrophysical interest.

4.2   β-Delayed Two-Proton Emission

A recent review by Moltz & Cerny (153) on β-2p emission reports in detail the existing results and their relevance to nuclear structure. Beyond the three emitters ^{22}Al, ^{26}P, and ^{35}Ca, discussed in this review, several new cases of this radioactive process have recently been observed. The $T_z = -5/2$ series of isotopes observed a few years ago (66) appears to be the first set of nuclei that systematically undergoes this radioactive decay, as exemplified by the earlier observation of ^{35}Ca (154). The study of ^{31}Ar decay has also been reported recently (155). In this case, the isobaric analog state in ^{31}Cl is expected to decay both by one-proton and two-proton emission. There is no evidence for any proton peak with energy around 12 MeV or 9.8 MeV. This absence is due to the combined effects of a low branching ratio and the strong decrease in detection efficiency for high energy protons in the thin ΔE detectors where ^{31}Ar ions are implanted. In contrast, a group of peaks near 7.5 MeV, decaying with the ^{31}Ar half-life, appears in the summed energy spectrum (see Figure 17). This group may

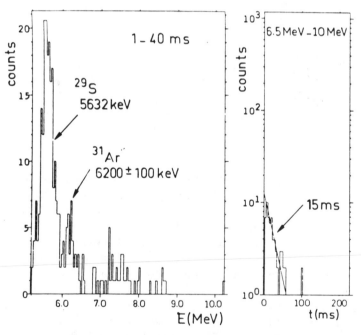

Figure 17   High-energy protons from ^{31}Ar decay (155). (*Left*) The summed energy spectra for protons. (*Right*) The events above 6.5 MeV are assigned to ^{31}Ar according to their common decay time characteristics.

represent the summed energy of the two protons emitted by the decay of the β-fed $T = 5/2$ analog state in ^{31}Cl to the ground state of ^{29}P, for which a 2p energy of 7.8 MeV is calculated. Although the origin of the charged-particle activity in this study is unambiguously assigned to ^{31}Ar, there is no direct evidence that it corresponds to the emission of two protons. However, a recent report by Reiff et al (156) eliminates this ambiguity. Although their work cannot rule out the possibility that another nucleus other than ^{31}Ar could be the source of their activity, it does unambiguously show that their activity results from two-proton emission. Taken together, these two independent and complementary results provide evidence for a fourth case of β-2p emission.

Although this process is now established, the question of its mechanism remains somewhat uncertain. In principle, the emission of two protons can proceed by means of several processes: (a) decay through an $L = 0$ final-state interaction between the two protons, sometimes referred to as ^{2}He emission; (b) uncoupled simultaneous emission; and (c) sequential emission. In the first case, a strong angular correlation at small relative opening angles is expected, whereas in the latter, successive proton emission through an intermediate state restricts the individual energy distribution of the two protons. Only in the second case are both the angle and energy of the protons uncorrelated. The existence of the sequential process has been established for ^{22}Al and ^{26}P (63). However, a small component of ^{2}He emission, which is allowed in ^{22}Al on the basis of angular momentum conservation, could not be ruled out (157). Further insight into β-2p process is expected from additional studies of ^{31}Ar now in progress.

These questions are clearly related to fundamental aspects of the coupling of nucleons inside nuclei. Such efforts to extract clearcut information from these experimental studies should prompt theoretical investigation of the mechanism of β-2p processes. Spectroscopic factors and phase space configurations provide the essential ingredients for advancing this work. In particular, it would be useful to examine whether one can reproduce a trend appearing in current experimental results: that branching ratios for β-p and β-2p processes are of the same order of magnitude when both channels are open and effectively unhindered by the Coulomb barrier.

4.3   New β-Delayed Processes

The catalogue of radioactive decay processes continues to expand as increasing Q values for nuclei lying farther and farther from the stable nuclei open up new decay channels. One well-documented case is that of ^{11}Li. It might appear to be an exceptional isolated phenomenon, but ^{11}Li decays through many different β-delayed radioactivities: one, two, and three neutrons (158) with α and ^{3}He, and also by the very unusual emission of tritons (159). However, the availability of new separation techniques and enhanced production yields make it possible to observe such rate decay modes in many other exotic isotopes.

The β-t process observed in ^{11}Li decay is energetically possible for many other neutron-rich isotopes. Another example has been reported (160) in a study of ^{8}He decay (see Figure 18) with a sizable branching ratio via ^{3}He + ^{3}H + n (or a + t + n) of 0.9%. This raises the possibility of measuring unusual spectroscopic factors in a rather direct way, with a minimum of model-dependent assumptions. Although a considerable number of nuclear reaction measurements have been made for one-proton and one-neutron (and in some cases two-nucleon) spectroscopic factors to determine nuclear structure, very little information exists on the amplitude of more complex substructures. Most of the reaction mechanisms that might allow one to measure such information [for example through (^{3}He,^{6}Li) reactions] are difficult to analyze because of uncertainties related to nuclear distortions and the interference between several multinucleon configurations that are likely to contribute. Processes such as β-t are largely free from these difficulties; therefore, the study of such processes offers a

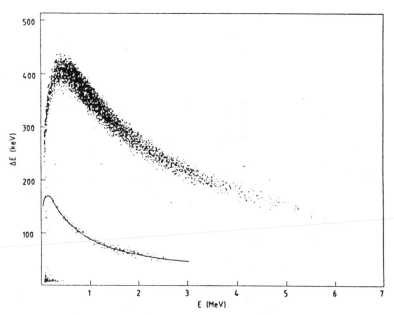

Figure 18 A two-dimensional ΔE-E spectrum of charged particles associated with the β-decay of ^{8}He (160). The calculated losses of energy for alphas and tritons (continuous line) are in agreement with the experimental distributions shown here.

remarkable opportunity to explore multinucleon amplitudes. Yet, to date very little has been done in this direction.

A recent study (161) of the β-delayed neutron emission of ^{14}Be, ^{17}B, and ^{20}C, produced by fragmentation of 60 MeV/u ^{22}Ne in a carbon target, revealed several new cases of multineutron emission. In this work, the detector telescope in which the recoils were stopped was surrounded by a neutron ball filled with 500 liters of liquid gadolinium-doped scintillator. Because there were spurious β-n coincidences that result from residual background neutrons and gamma rays that mimic neutron events, it was necessary to make careful measurements of the background so that their contribution could be subtracted out. The results are presented in Table 1. As mentioned above for β-p and β-2p, these results show relatively large probabilities for such β-delayed multineutron processes and give evidence that even the β-4n process occurs in such neutron-rich light nuclei. Further work on angular and energy correlations of the emitted neutrons should prove valuable in understanding the decay mechanism and observing the two-neutron substructures of the emitting nuclear states.

4.4   Search for Direct Two-Proton Emission

As outlined in the preceding sections, a growing number of β-delayed radioactive modes have been identified and studied. In addition, the variety of direct emissions of hadronic fragments dramatically expands the process of radioactivity beyond just single-nucleon emission. The main developments in this area have been the discovery of (a) proton radioactivity, as recently reviewed by Hofmann (71), a well-documented phenomenon occurring for many nuclei at the proton drip line; and (b) heavy fragment radioactivities, such as the emission of ^{14}C, ^{24}Ne, or ^{28}Mg from the heavy actinide elements (162, 163).

As mentioned before, the observation of the emission of more than one nucleon provides information on the formation of such clusters in nuclei. After the effect of barrier penetrability is calculated, the emission probability is directly related to the spectroscopic amplitude of the emitted nuclear fragment inside the nucleus. One process, the two-proton radioactivity, will provide valuable information about nucleon-nucleon cor-

Table 1   Probabilities (in percentage) of β-delayed multineutron emissions (from 161)

^{A}Z	$T_{1/2}$ (ms)	P_{0n}	P_{1n}	P_{2n}	P_{3n}	P_{4n}
^{14}Be	4.35 (17)	0.14 (3)	0.81 (4)	0.05 (2)	—	—
^{17}B	5.08 (5)	0.21 (2)	0.63 (1)	0.11 (7)	0.035 (7)	0.004 (3)
^{19}C	49 (4)	0.46 (3)	0.47 (3)	0.07 (3)	—	—

relations in nuclei. This predicted 2p radioactivity is expected to result from the fact that pairing effects make the binding energy of many even-Z neutron-deficient nuclei stable with respect to one-proton emission, but two-proton unbound. At the proton drip line, a nucleus might be bound with respect to one-proton emission but able to decay by the emission of two protons. This process was discussed by Goldanskii (164) and Jänecke (165). The probability of emission is governed by barrier penetration, which itself is strongly dependent upon the 2p energetics. Calculations indicate that the most favored configuration with respect to barrier penetrability is for two protons coupled to $L = 0$ and equally sharing the available decay energy.

Several exotic light nuclei are good candidates for exhibiting 2p radio-activity. To make this process easier to observe, the available two-proton decay energy should not be too large (e.g. ~ 0.8–1.0 MeV at $Z = 20$) so that the partial 2p half-life will be long enough for characterization; on the other hand, the 2p decay energy should not be so small (e.g. ~ 0.3–0.4 MeV at $Z = 20$) that the 2p partial half-life is long compared to the β-decay partial half-life (i.e. in order to have a reasonable branching ratio). Thus, a narrow energy window exists for potential 2p radioactivity candidates with Z values as shown in Figure 6.

Two nuclei, ^{31}Ar and ^{39}Ti, are within reach of current experimental capabilities and are considered good candidates for 2p radioactivity. The decay of ^{31}Ar was recently investigated (155) and determined to have a half-life of ~ 15 ms. This time scale appears consistent with that expected for β decay, and so the possibility of a significant 2p branch seems marginal. This result is in line with the predicted (6) 230-keV energy available for the two protons, a value too low to affect a 2p partial half-life that is competitive with β-decay. The other candidate, ^{39}Ti, was sought in the fragmentation of ^{58}Ni projectiles at 65 MeV/u using the same experimental equipment employed in the study of ^{31}Ar. The fragments analyzed by the LISE spectrometer were implanted in an identifying telescope, and low energy, charged-particle activities were examined during a short beam-off period (10–200 ms) directly following the detection of the recoil. Nearly 80 ^{39}Ti events were observed. A correlated charged-particle activity with a half-life of ~ 25 ms was observed (166). One can surmise, as for ^{31}Ar, that 2p direct emission does not successfully compete with β-decay.

The latter result is in itself surprising, because charge symmetry formulae predict that ^{39}Ti is unbound by ~ 0.8 MeV with respect to 2p emission. Mass predictions for neutron-deficient nuclei based on local mass relationships fall typically within 200 keV of the actual mass values. The two methods of Goldanskii (167) and Kelson & Garvey (168) have been shown to be equivalent (166). Their success rests on the effectiveness of charge symmetry in nuclei. The apparent inability of ^{39}Ti to decay rapidly through 2p emission could be linked to a systematic deviation of its mass from the predicted value resulting from a decrease in Coulomb repulsion when the last proton(s) are only weakly bound, or even unbound. This effect, known as a Thomas-Ehrman shift (127), was discussed in an analysis of binding energy trends for neutron-deficient nuclei lying near the proton drip line (126). The experimental values of mass excesses are increasingly smaller than those predicted when the available energy for p (or 2p) emission becomes larger (see Figure 19). The more proton unbound the nucleus is, the more predictions tend to overestimate the proton (two-proton) decay energy. From the systematics of Comay et al (126), the two-proton decay energy for ^{39}Ti might thus be reduced by 0.4 to 0.8 MeV, which would account for a very small 2p branching ratio that effectively prevented its observation. Accordingly, one might expect that the best candidates for 2p radioactivity in the same mass region would have predicted two-proton decay energies in the $\sim$ 1.0–1.2-MeV range. Such nuclei as ^{45}Fe or ^{48}Ni would then be good candidates for 2p radioactivity. Pending energy and

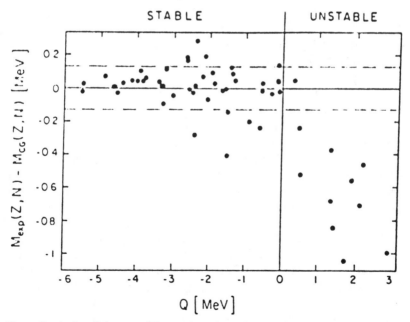

Figure 19 A plot of the mass difference (experimental mass minus calculated mass using the generalized charge symmetric mass relationship of Garvey & Kelson) vs the Q value for proton (or two-proton) emission for all neutron-deficient nuclei with $Z \geq N+2$ (126). The one-standard-deviation limits are shown for the particle-stable population (*dashed line*).

intensity accelerator upgrades at GANIL (169) and the SIS/ESR facility now under construction at Darmstadt (170) are expected to facilitate further searches for this predicted decay mode.

5.   REACTIONS INDUCED BY RADIOACTIVE NUCLEI

The possibility of using radioactive nuclei as projectiles to induce new reactions is a long-standing dream in nuclear physics and has fueled some proposals for ad hoc (and usually costly) facilities. The RIKEN group working at the Bevalac has shown that such experiments could be efficiently performed at existing facilities. Nuclear reactions using radioactive ion beams are expected to exhibit novel features. For instance, close to the neutron drip line, the occurrence of a pair of loosely bound neutrons may result in larger than expected matter radii. The calculations of Hansen & Jonson (171) and the results obtained at both the Bevalac and GANIL seem to support the neutronization of the nuclear surface in the form of a neutral halo around the nucleus that extends out to several times the nuclear radius. These calculations led to the conjecture that such loosely bound structures might exhibit high cross sections for Coulomb-dissociation that would vary in an inversely proportional way to the two-neutron binding energy (171).

5.1   *Matter Radii Measurements*

Charge radii for nuclei far from stability have been measured with great success through the advancement of collinear laser spectroscopy techniques (172). The corresponding matter radii are often difficult to measure because of technical limitations involving nuclear reactions (173) and because exotic nuclei are not readily available as projectiles or targets. Yet the measurement of total reaction cross sections requires only a small number of incident or target nuclei; the limitations are more a matter of systematic errors than statistical uncertainties. In the search for exotic nuclei, using projectile fragmentation reactions, the isotopes of interest are usually selected by a magnetic spectrometer. It is not as good a secondary beam as physicists would like for performing specific nuclear reactions because both its emittance and its energy width are much broader than acceptable for that purpose. However, especially at higher energy, these parameters are sufficiently restricted to allow for the measurement of global parameters, such as the total reaction cross section.

Tanihata, Sugimoto, and their coworkers (174, 175) first made use of this remarkable opportunity to measure total reaction cross sections for neutron-rich fragments of He, Li, and Be that are produced at the Bevalac using projectile fragmentation of an 800-MeV/u ^{11}B beam. The produced fragments were separated according to their magnetic rigidity. The typical intensity of the secondary beams thus obtained was limited to $\sim 10^{3}$ per second for ions such as ^{8}He and ^{9}Li. The interaction cross section, defined as the total cross section for processes in which at least one nucleon is removed, was extracted from transmission measurement of the secondary beam through a target. The cross section was taken as

$$\sigma = \frac{1}{N} \log{(\gamma_0/\gamma)}, \qquad \qquad 1.$$

where N is the number of target nuclei per cm^{2}, and γ (γ_0) is the ratio of the number of noninteracting projectiles to the number of incident projectiles in a run with (without) a target.

The interpretation of the cross section in terms of matter radii is not straightforward (for a review, see 176). A simple assumption can be made at high incident energy where free nucleon-nucleon interactions dominate, so that the cross section simply reflects the geometry of the interaction:

$$\sigma = \pi(r_{\mathrm{p}} + r_{\mathrm{t}})^{2}. \qquad \qquad 2.$$

The subscripts "p" and "t" refer to the projectile and target nuclei, respectively. Around the Fermi energy, care must be taken to allow for collective effects and the Coulomb barrier. Kox et al (177) derived a very efficient parametrization

$$\sigma = \pi r_0^2 \left(A_{\mathrm{p}}^{1/3} + A_{\mathrm{t}}^{1/3} + a \frac{A_{\mathrm{p}}^{1/3} A_{\mathrm{t}}^{1/3}}{A_{\mathrm{p}}^{1/3} + A_{\mathrm{t}}^{1/3}} - c \right)^2 (1 - B_{\mathrm{C}}/E_{\mathrm{cm}}). \qquad 3.$$

The major result obtained in this work was that the matter radii of the light neutron-rich isotopes studied did not always vary as $A^{1/3}$. In particular, ^{11}Li exhibited a striking increase of its matter radius over the expected value (see Figure 20).

An alternative to the transmission method described above rests on the assumption that in each nuclear reaction at least one photon is emitted. This has been proven true for collisions between stable nuclei, yet for nuclei with very low nucleon separation energies, nucleons might be removed without exciting the residual nuclear system. Thus, significant differences might exist between the transmission method and the γ-ray method because they may respond differently to different reaction processes of high probability—for example inelastic excitations, nucleon removal from the target without transfer to the projectile, and transitions without γ emission.

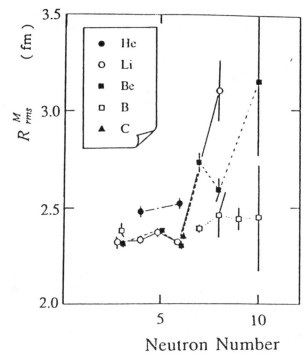

Figure 20   The effective root-mean-square matter radii (R_{RMS}^{M}) determined by the interaction cross-section measurements of Tanihata et al (180). Lines in the figure are only for guiding the eye. Note the large radius of ^{11}Li.

To verify the interesting and somewhat surprising results shown in Figure 21, two groups at GANIL used the γ-ray deexcitation technique to measure total reaction cross sections (178, 179). Both used Equation 3 to extract the value of r_0, the reduced strong absorption radius, which is less model dependent than the effective matter radii determined by Tanihata et al (180). The measurements performed by Mittig et al (178) used a thick target, actually a Si detector, in the course of the mass measurements described in Section 3. A set of NaI detectors, covering 87% of 4π, surrounded the identifying telescope. Although careful corrections were made for random coincidence, ambiguous identification, and efficiency, several assumptions were made. The variation of cross section had to be parameterized as a function of energy, whereas the effective target thickness at each energy was deduced from tables of specific energy loss. Furthermore, because only one target, ^{nat}Si, was available in this method, no check could be made on the consistency of the r_0^2 values obtained for different mass

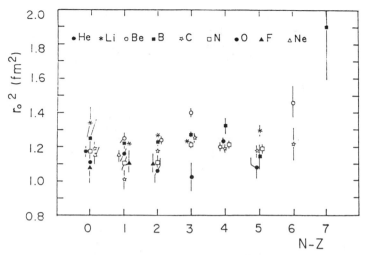

Figure 21 The reduced strong absorption radius squared (r_0^2) as a function of $N-Z$ for all the isotopes measured in the work of Saint-Laurent et al (179).

targets. The striking result showed a strong increase in r_0^2 with increasing neutron excess of the identified recoil independent of its atomic number (178). In spite of the great care taken to avoid systematic errors, this rather unexpected result called for an independent check.

The method used by Saint-Laurent and coworkers (179) avoided two limitations inherent in the Mittig studies. First, thin targets were used so that the results were free of the energy-dependent parameterizations (the method was applied for each isotope at several energies as a consistency check). Second, measurements with several different target materials were performed. The effect observed by Mittig et al was not reproduced, as is seen in Figure 21. On the other hand, general agreement was observed with the results of Tanihata et al (175) (see Figure 22). Because these sets of data were obtained by different techniques and with very different beam energies, the general agreement in the absolute values of r_0^2 is remarkable. Also, the trends in the N dependence (odd-even staggering) are similar, although somewhat different in amplitude. It is not obvious in what way this systematic discrepancy could be related to the different experimental conditions. For the Li isotopes, the results of Saint-Laurent et al suggest much smaller variations in r_0^2 than the data of Tanihata et al, whereas for Be and B the situation is the opposite. Note the smaller r_0^2 values reported for ^{11}Li and the larger r_0^2 values given for ^{14}Be and ^{17}B by Saint-Laurent et al. A check of these interesting results is certainly needed because, as explained above, one cannot exclude the possibility that reactions involving

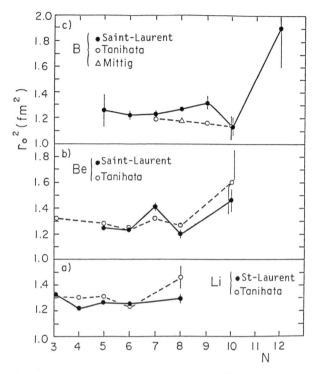

Figure 22   The reduced strong absorption radius squared (r_0^2) for (*a*) Li, (*b*) Be, and (*c*) B isotopes. Full black dots are data points obtained in the work of Saint-Laurent et al (179); open circles are from Tanihata et al (175); and the triangle is from Mittig et al (178). The lines are only guides for the eye.

weakly bound nuclei might occur without γ deexcitation, which would systematically bias the results for such nuclei. Therefore, all conclusions concerning a dramatic increase of matter radii for exotic nuclei close to the neutron drip line should be drawn with some caution at this time.

The variation of radii with isospin was compared by Tanihata et al (180) with Hartree-Fock calculations. Good agreement was obtained for a Skyrme SIII potential including a strong density-dependent interaction. On the other hand, the droplet model (181), using parameters determined from the data of heavier nuclei, systematically predicted radii about 0.3 fm larger than the experimental values and did not reproduce the larger radii observed for large neutron excess.

The marked increase of the matter radius for the weakly bound isotopes close to the neutron drip line has been related to the occurrence of an extended dineutron halo (171). This would be the case for ^{11}Li, ^{14}Be, and

^{17}B because recent mass measurements (see Section 3.1) yield two-neutron separation energies of 0.25, 1.34, and 1.49 MeV, respectively.

Two interesting questions are raised by the possibility of a halo structure for these isotopes. First, the possibility of halos should lead to particularly large cross sections for some reaction channels, as was suggested by Hansen & Jonson (171). Second, the corresponding increase of r_0^2, apparently supported by experimental results, is in sharp disagreement with recent shell model calculations (182) of $^{6-11}$Li radii, which predict a much smoother dependence of r_0^2 on the neutron number than has been suggested by experimental data. In particular, the increase of r_0^2 becomes smaller as one approaches the magic neutron number $N = 8$ for ^{11}Li—a trend that is in agreement with shell effects.

The availability of high quality data on matter radii for long chains of isotopes far from stability is certainly one of the major recent developments in the field of nuclear science. The development of versatile secondary beams is expected to open new possibilities in this direction. The most promising short-term prospect comes from the completion of the SIS/ESR facility in Darmstadt, which will provide cooled secondary beams with energies ranging from 3 to 2000 MeV/u (170).

5.2 Momentum Distributions of Projectile Fragments

Fragmentation of radioactive projectiles is another type of reaction that is now within reach of experiments. This reaction has been extensively studied using beams of stable nuclei (183, 184). The regularity of the observed momentum distribution for projectile fragments has been successfully explained by the effects of internal momentum distribution of abraded nucleons (185), even at energies barely above the Fermi energy (184).

The momentum distribution of fragments emitted in the interaction of ^{11}Li projectiles on a carbon target was measured at 790 MeV/u by Tanihata et al (176). The transverse momentum distribution (see Figure 23) obtained for ^{11}Li fragments exhibits a two-component structure in which a narrow peak lies on top of a much wider one. Their half-widths are 23 (± 5) and 95 (± 12) MeV/c, respectively. The latter is consistent with those values generally observed for fragmentation. On the other hand, the extremely small momentum fluctuation corresponds, in accordance with the Heisenberg uncertainty principle, to an extension of several fermis for the space function of the abraded nucleons, i.e. to a long tail in the neutron density distribution of the ^{11}Li projectile.

It is most interesting to note that preliminary data on the fragmentation of ^{14}Be (176) also shows the two-Gaussian structure observed for ^{11}Li.

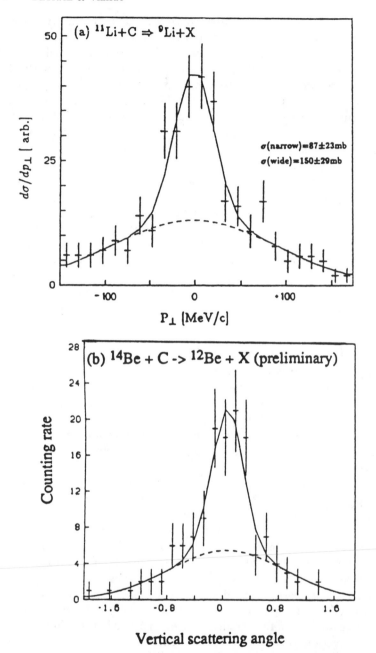

Figure 23   The transverse momentum spectrum obtained from (*top*) ^{11}Li and (*bottom*) ^{14}Be collisions with carbon (176).

This remarkable fact can be related to the very low two-neutron separation energy and the large matter radius that ^{14}Be has in common with ^{11}Li, as discussed in the preceding section.

5.3   Coulomb Dissociation

As indicated at the beginning of Section 5, enhanced Coulomb dissociation might be a special feature of nuclei close to the neutron drip line. This phenomenon is expected to occur when the electromagnetic field of a high-Z target nucleus interacts, through a virtual photon, with the projectile at impact parameters larger than the range of nuclear forces.

There should be a strong dependence of the cross section on the atomic number of the target. This was tested recently with ^{8}He, ^{11}Li, and ^{14}Be beams of 790 MeV/u on various targets (176, 186). Figure 24 shows a comparison between experimental cross sections and predictions of the geometrical model known to be generally valid at such high incident energy. Clearly there is a strong enhancement of the interaction cross section for high Z values and for one of the fragmentation channels (i.e. for two-neutron removal) as well. The excess of 2n-removal cross section over the geometrical model prediction is as large as 0.9 b. The full Coulomb dissociation cross section, which appears in Figure 24, reaches 2.4 ± 0.6 b for a Pb target. This is compatible with the effect predicted by Hansen & Jonson (171) as a consequence of the neutron halo of ^{11}Li. The uncertainty is the result of the limited accuracy for the binding energy of the two neutrons, as it is currently known. The dependence of the dissociation cross section on the Z value of the target also appears in Figure 24. It is in overall agreement with Hansen & Jonson's model, although the Z^2 variation appearing in their calculation might be stronger than is experimentally observed.

This topic will be actively studied in the near future because it is clearly a case in which exotic nuclei provide us with structural features differing markedly from those encountered near the valley of β-stability. For instance, proposals have been presented at the new SIS facility in Darmstadt (187) to search more specifically for the neutrons emitted in the ^{11}Li $\rightarrow$ ^{9}Li + 2n Coulomb dissociation reaction, and an experiment is under way at ISOLDE (188) to search for a signature of the independent β decay of the loosely bound two neutrons of ^{11}Li into free deuteron.

5.4   Radioactive Beams for Astrophysics

The fourth area in which reactions induced by radioactive nuclei provide unique information is that of nuclear astrophysics. For a review of research related to the use of radioactive beams with an emphasis on their use for astrophysics, refer to the workshop edited by Buchmann & D'Auria

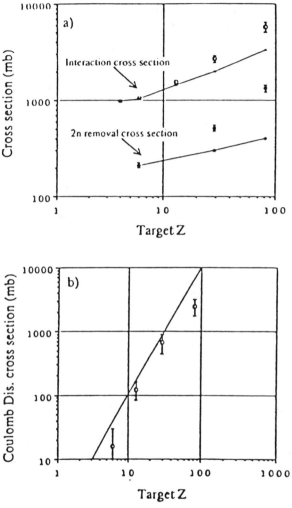

Figure 24 (*a*) Target Z dependence of ^{11}Li interaction cross sections and ^{11}Li $\rightarrow$ ^{9}Li two-neutron removal cross sections. Solid lines show geometrical estimates normalized for the Be target. (*b*) Coulomb dissociation part of the interaction cross sections for ^{11}Li. The solid line is a model prediction of the Coulomb dissociation (176).

(189). In astrophysical processes, many nuclear reactions that play a role in the evolution of stellar systems involve radioactive nuclei. There are exciting possibilities using fragmentation processes to produce radioactive nuclei and to analyze them with recoil separators such as LISE in order to form beams of sizable intensity. Table 2 presents some typical secondary beam intensities that have been obtained recently (190). However, the beams directly formed in this way are not always of practical use. First, as exemplified in Figure 25, they are not pure; the level of contamination by other species depends strongly on the selection process. Furthermore, their velocities are too high to allow for two-body reaction studies, and their emittances are very poor.

All these defects can be suppressed if such beams can be cooled in a ring and decelerated, as planned in the forthcoming SIS/ESR facility at Darmstadt (170). However, one should note that these beam handling

Table 2 Some examples of intensities obtained for secondary beams as compiled by Bimbot et al (190)

		Secondary beam		
Production mode	^AZ	Energy (MeV/u)	I/I_0	I (pps)
^{40}Ar (44 MeV/u)	^{41}K	32	5×10^{-5}	1.5×10^7
+	^{39}Ar	35	3×10^{-4}	10^8
Be (99 mg/cm^2)	^{38}Ar	34	10^{-4}	3×10^7
	^{39}Cl	34	10^{-4}	3×10^7
	^{38}S	36	6×10^{-6}	2×10^6
^{22}Ne (45 MeV/u)	^{18}C	36	3×10^{-10}	3×10^2
+	^{11}Li	36	2.5×10^{-10}	2.5×10^2
Ta (417 mg/cm^2)				
^{18}O (45 MeV/u)	^{17}N	39	10^{-5}	10^7
+	^{18}N	39	5×10^{-7}	5×10^5
Be (187 mg/cm^2)	^{14}C	39	10^{-5}	10^7
	^{16}C	38	5×10^{-7}	5×10^5
	^{17}C	38	2×10^{-8}	2×10^4
	^{18}C	38	3×10^{-10}	3×10^2
	^{13}B	39	2×10^{-7}	2×10^5
	^{14}B	39	10^{-8}	10^4
	^{15}B	39	2×10^{-9}	2×10^3
	^{11}Be	39	2×10^{-8}	2×10^4
	^{12}Be	39	10^{-8}	10^4
	^{8}Li	38	8×10^{-8}	8×10^4
	^{9}Li	38	2×10^{-8}	2×10^4
^{18}O (65 MeV/u)	^{17}N	46	5×10^{-5}	5×10^7
+	^{18}N	47	2×10^{-6}	2×10^6
Be (567 mg/cm^2)	^{14}C	46	5×10^{-5}	5×10^7
	^{16}C	50	2×10^{-6}	2×10^6
	^{13}B	47	2×10^{-6}	2×10^6
	^{14}B	47	2×10^{-8}	2×10^4

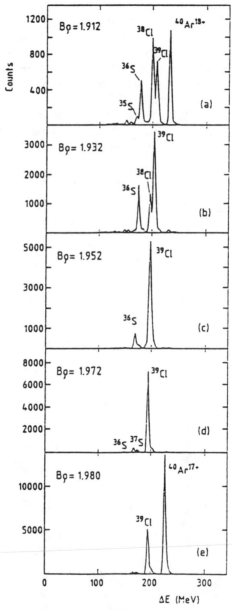

Figure 25 Effect of a *Bρ* variation on the ^{39}Cl secondary-beam purity. The beam composition (Δ*E* spectra) is shown for five different *Bρ* values (in Tesla-meter). The most abundant isotopes in the secondary beam are labeled (190).

techniques are not fast and therefore cannot be applied to beams of radioactive beams with half-lives significantly shorter than one second. Nevertheless, this facility should open up the long expected possibility of studying nuclear reactions induced by radioactive nuclei. Improved beam quality is also expected for radioactive beams already developed at several laboratories. For instance, a velocity filter will be added to the LISE facility to purify the beam further and to facilitate the use of even higher primary beam intensities now under development at GANIL (169).

Another way of producing pure radioactive beams is to post-accelerate radioactive species that have been selected by an ISOL system. Several projects or proposals based on this method were recently presented at a Tokyo symposium (191). One example is a new radioactive beam facility being built at Louvain-la-Neuve (192). It uses one smaller cyclotron as a radioisotope factory to produce ^{13}N atoms that are fed into an ECR ion source and then accelerated by the larger cyclotron to induce the $H(^{13}N, \gamma)^{14}O$ reaction of high astrophysical interest. Beams of up to 8.5 nA of ^{13}N are expected.

In conclusion, it should be stressed that reactions induced by radioactive nuclei do not just bring a mere addition to the long list of studied nuclear reactions. As discussed above, they will allow us to observe novel nuclear structure features, as well as specific reaction processes that are at variance with those observed in nuclei lying close to stability. In addition, they present, at last, the opportunity to measure cross sections essential to the understanding of stellar evolution. Recent progress in this direction, as well as prospects for the near future, amounts to a clear breakthrough in the fields of nuclear science and astrophysics. For additional information about radioactive ion beams, see the recent review by Tanihata (193) and the popular article by Nitschke (194).

6. SUMMARY AND OUTLOOK

Significant developments in production and separation (identification) techniques have heralded a broad advance on the exotic nuclei frontier. Nowhere is the progress more evident than in the light mass region, where the use of projectile fragmentation reactions and the development of fast recoil spectrometers and separators have led to the discovery of ~ 100 isotopes—many of which were not known five years ago. No longer is the research on nuclei far from stability restricted to arduous investigations of isolated isotopes, but rather these new techniques lend themselves to the study of several nuclei simultaneously. With improved methods, a large part of the light mass region on the chart of the nuclides has now been explored. Combined with theoretical advances, this research sheds new light on the nuclear structure, binding, and decay characteristics of exotic nuclei.

In particular, the establishment of both the neutron and proton drip lines has been extended by approximately a factor of two. Wide-ranging and systematic mass measurements have provided a wealth of new data with which to test and refine our understanding of the nuclear mass surface. With the increase in available β-decay energy, a wide range of new decay modes, such as β-2p, β-2n, β-3n, β-4n, and β-t, has been observed. Soon further studies along the proton drip line are expected to yield additional examples of proton and, perhaps, two-proton radioactivity. Taken together, such decay studies provide valuable spectroscopic information and details of multiparticle decay mechanisms that in many cases could not be obtained in other ways.

These developments have also led to the advent of radioactive ion beams. Only a few experiments using such beams have been published (such as the measurements of matter radii for several light neutron-rich isotopes that suggest the existence of dineutron halos) and much work remains to be done. The development of high intensity and high purity radioactive ion beams presents an entirely new vista of research opportunities for the fields of nuclear science and nuclear astrophysics.

We are heartened by the exciting progress made in the area of exotic nuclei during the last decade, especially in light of planned improvements to existing facilities, the operation of other facilities just coming on-line, and the current construction of (or plans for) other major facilities. The prospects for future research are excellent.

ACKNOWLEDGMENTS

The authors are grateful to a number of colleagues for helpful discussions and suggestions during the writing of this review and for supplying us with preprints and reprints of their publications. Special thanks are extended to Jody Heiken, Carla Lowe, Alex Mueller, Jan Wouters, and Zhou Zong-Yuan for assisting us in preparing and improving this manuscript. This work was performed in part under the auspices of the US Department of Energy.

Literature Cited

1. Thibault, C., Klapisch, R., Rigaud, C., Poskanzer, A. M., Prieels, R., et al., *Phys. Rev.* C12: 644–57 (1975)
2. Campi, X., Flocard, H., Kerman, A. K., Koonin, S., *Nucl. Phys.* A251: 193–205 (1975)
3. Détraz, C., see Ref. 13, pp. 361–63 and references therein
4. Towner, I. S., Hardy, J. C., see Ref. 14, pp. 564–71; Towner, I. S., Hardy, J. C., Harvey, M., *Nucl. Phys.* A284: 269–81 (1977)

5. Ormand, W. E., Brown, B. A., *Phys. Rev. Lett.* 62: 866–69 (1989) and references therein

6. Jänecke, J., Masson, P. J., *At. Data Nucl. Data Tables* 39: 265–71 (1988)

7. Hansen, P. G., *Annu. Rev. Nucl. Part Sci.* 29: 69–119 (1979)

8. Hamilton, J. H., *Prog. Part. Nucl. Phys.* 15: 107–34 (1985); Hamilton, J. H., Hansen, P. G., Zganjar, E. F., *Rep. Prog. Phys.* 48: 631–708 (1985)

9. Poenaru, D. N., Ivascu, M. S., eds., *Particle Emission from Nuclei*, Vols. 1-3. Boca Raton: CRC (1989)

10. Bromley, D. A., ed., *Treatise on Heavy-Ion Science*, Vol. 8. New York: Plenum (1989)

11. Cerny, J., Poskanzer, A. M., *Sci. Am.* 238(6): 60–72 (1978)

12. Hardy, J. C., *Science* 227: 993–99 (1985)

13. *Proc. 4th Int. Conf. on Nuclei Far From Stability*, Helsingør, 1981. *CERN Rep. 81-09* (1981)

14. Klepper, O., ed., *Proc. 7th Int. Conf. on Atomic Masses and Fundamental Constants*, Darmstadt-Seeheim, 1984. Darmstadt: THD Schriftenreike Wissenschaft Tecknik 26 (1984)

15. Towner, I. S., ed., *Proc. 5th Int. Conf. on Nuclei Far From Stability*, Rosseau Lake, 1987. *AIP Conf. Proc.* 164 (1988)

16. Forsling, W., Herrlander, C. J., Ryde, H., eds., *Conf. on Nuclides Far Off the Stability Line*, Lysekil, 1966. Stockholm: Almqvist & Miksell (1967); also published in *Ark. Fys.* 36: 1–686 (1967)

17. Armbruster, P., Bernas, M., Bocquet, J. P., Brissot, R., Faust, H. R., et al., *Europhys. Lett.* 4: 793–97 (1988)

18. Artukh, A. G., Avdeichikov, V. V., Gridnev, G. F., Mikheev, V. L., Volkov, V. V., et al., *Nucl. Phys.* A176: 284–88 (1971)

19. Volkov, V. V., in *Proc. Int. Conf. on Nuclear Physics*, Munich, 1973, ed. J. de Boer, H. J. Mang. Amsterdam: North-Holland (1972) 2: 279–304

20. Auger, P., Chiang, T. H., Galin, J., Gatty, B., Guerreau, D., et al., *Z. Phys.* A289: 255–59 (1979)

21. Guerreau, D., Galin, J., Gatty, B., Tarrago, X., Girard, J., et al., *Z. Phys.* A295: 105–6 (1980)

22. Breuer, H., Wolf, K. L., Glagola, B. G., Kwiatkowski, K. K., Mignerey, A. C., et al., *Phys. Rev.* C22: 2454–56 (1980)

23. Roeckl, E., *Nucl. Phys.* A488: 95c–112c (1988)

24. Symons, T. J. M., Viyogi, Y. P., Westfall, G. D., Doll, P., Greiner, D. E., et al., *Phys. Rev. Lett.* 42: 40–43 (1979)

25. Westfall, G. D., Symons, T. J. M., Greiner, D. E., Heckman, H. H., Lindstrom, P. J., et al., *Phys. Rev. Lett.* 43: 1859–62 (1979)

26. Détraz, C., in *Proc. Int. Nucl. Phys. Conf.*, ed. J. L. Durell, J. M. Irvine, G. C. Morrison. *I.O.P. Conf. Ser.* 2(86): 495–506 (1986)

27. Détraz, C., in *Proc. 5th Int. Conf. on Clustering Aspects*, Kyoto, 1988, ed. K. Ikeda, K. Katori, Y. Suzuki, suppl. to *J. Phys. Soc. Jpn.* 58: 219–31 (1989)

28. Mikolas, D., Brown, B. A., Benenson, W., Chen, Y., Curtin, M. S., et al., see Ref. 15, *AIP Conf. Proc.* 164: 708–17 (1988)

29. Guerreau, D., *Nucl. Phys.* A447: 37c–66c (1985)

30. Pougheon, F., Jacmart, J. C., Quiniou, E., Anne, R., Bazin, D., et al., *Z. Phys.* A327: 17–24 (1987)

31. Klapisch, R., *Annu. Rev. Nucl. Sci.* 19: 33–60 (1969)

32. Ravn, H. L., *Phys. Rep.* 54: 201–59 (1979)

33. Kluge, H. J., ed., ISOLDE Users' Guide. *CERN Rep. 86-05* (1986)

34. Kirchner, R., *Nucl. Instrum. Methods* B26: 204–212 (1987); 186: 275–93 (1981)

35. Ravn, H. L., Allardyce, B. W., see Ref. 10, pp. 363–439. *CERN Rep. 87-105* (1987)

36. Talbert, W. L., ed., *Proc. 11th Int. Conf. on Electromagnetic Isotope Separators and Techniques Related to their Applications*, Los Alamos , 1986. *Nucl. Instrum. Methods* B26: 1–500 (1987)

37. Münzenberg, G., Armbruster, P., Berthes, G., Hessberger, F. P., Hofmann, S., et al., *Nucl. Instrum. Methods* B26: 294–300 (1987)

38. Enge, H. A., in *Treatise on Heavy-Ion Science*, ed. D. A. Bromely. New York: Plenum (1985), 7: 403–28

39. Cormier, T. M., *Annu. Rev. Nucl. Part. Sci.* 37: 537–65 (1987)

40. Wouters, J. M., Vieira, D. J., Wollnik, H., Enge, H. A., Kowalski, S., et al., *Nucl. Instrum. Methods* A240: 77–90 (1985)

41. Vaziri, K., Wohn, F. K., Vieira, D. J., Wollnik, H., Wouters, J. M., *Nucl. Instrum. Methods* B26: 280–85 (1987)

42. Wouters, J. M., Vieira, D. J., Wollnik, H., Butler, G. W., Kraus, R. H., et al., *Nucl. Instrum. Methods* B26: 286–93 (1987)

43. Bianchi, L., Fernandez, B., Gastebois, J., Gillibert, A., Mittig, W., et al. *Nucl. Instrum. Methods* A276: 509–20 (1989)

462 DÉTRAZ & VIEIRA

44. Anne, R., Bazin, D., Mueller, A. C., Jacmart, J. C., Langevin, M., *Nucl. Instrum. Methods* A257: 215–32 (1987)
45. Dufour, J. P., Del Moral, R., Emmermann, H., Hubert, F., Jean, D., et al., *Nucl. Instrum. Methods* A248: 267–81 (1986)
46. Haustein, P. E., ed., *At. Data Nucl. Data Tables* 39: 185–393 (1988)
47. Bowman, J. D., Poskanzer, A. M., Korteling, R. G., Butler, G. W., *Phys. Rev. Lett.* 31: 614–16 (1973)
48. Musser, J. A., Stevenson, J. D., *Phys. Rev. Lett.* 53: 2544–46 (1984)
49. Langevin, M., Quiniou, E., Bernas, M., Galin, J., Jacmart, J. C., et al., *Phys. Lett.* 150B: 71–74 (1985)
50. Pougheon, F., Guillemaud-Mueller, D., Quiniou, E., Saint-Laurent, M. G., Anne, R., et al., *Europhys. Lett.* 2: 505–09 (1986)
51. Stevenson, J. D., Price, P. B., *Phys. Rev.* C24: 2102–05 (1981)
52. Guillemaud-Mueller, D., Penionzhkevich, Y. E., Anne, R., Artukh, A. G., Bazin, D., et al., *Z. Phys.* A332: 189–93 (1989)
53. Guillemaud-Mueller, D., Penionzhkevich, Y. E., et al., private communication and to be published (1989)
54. Vogel, P., Jonson, B., Hansen, P. G., *Phys. Lett.* 139B: 227–30 (1984)
55. Jensen, A. S., Hansen, P. G., Jonson, B., *Nucl. Phys.* A431: 393–418 (1984)
56. Langevin, M., Détraz, C., Guillemaud-Mueller, D., Mueller, A. C., Thibault, C., et al., *Phys. Lett.* 130B: 251–53 (1983)
57. Guillemaud-Mueller, D., Mueller, A. C., Guerreau, D., Pougheon, F., Anne, R., et al., *Z. Phys.* A322: 415–18 (1985)
58. Zhan, W., Audi, G., Bianchi, L., Cunsolo, A., Dumont, H., et al., *Nouvelles du GANIL* 25: 22–27 (1988) and to be published
59. Marion, J. B., Young, F. C., *Nuclear Reaction Analysis Graphs and Tables.* Amsterdam: North-Holland (1968), pp. 84–91
60. Cerny, J., Mendelson, R. A., Wozniak, G. J., Esterl, J. E., Hardy, J. C., *Phys. Rev. Lett.* 22: 612–15 (1969)
61. Benenson, W., Guichard, A., Kashy, E., Mueller, D., Nann, H., et al., *Phys. Lett.* 58B: 46–48 (1975)
62. Jänecke, J., *At. Data Nucl. Data Tables* 17: 455–62 (1976)
63. Cable, M. D., Honkanen, J., Schloemer, E. C., Ahmed, M., Reiff, J. E., et al., *Phys. Rev.* C30: 1276–85 (1984) and references therein
64. Äystö, J., Moltz, D. M., Xu, X. J., Reiff, J. E., Cerny, J., *Phys. Rev. Lett.* 55: 1384–87 (1985)
65. Morris, C. L., Fortune, H. T., Bland, L. C., Gilman, R., Greene, S. J., et al., *Phys. Rev.* C25: 3218–20 (1982)
66. Langevin, M., Mueller, A. C., Guillemaud-Mueller, D., Saint-Laurent, M. G., Anne, R., et al., *Nucl. Phys.* A455: 149–57 (1986)
67. Saint-Laurent, M. G., Dufour, J. P., Anne, R., Bazin, D., Borrel, V., et al., *Phys. Rev. Lett.* 59: 33–35 (1987)
68. Woods, C. L., Catford, W. N., Fifield, L. K., Orr, N. A., *Nucl. Phys.* A484: 145–54 (1988)
69. Mohar, M. F., Adamides, E., Benenson, W., Bloch, C., Brown, B. A., et al., *Phys. Rev.* C38: 737–40 (1988)
70. Lewitowicz, M., Détraz, C., Anne, R., et al., presented at 27th Winter Meet. on Nucl. Phys., Bormio, Italy (January 1989). *GANIL Rep. P. 89-09* and to be published
71. Hofmann, S., see Ref. 9, 2: 25–72
72. Pape, A., Antony, M. S., *At. Data Nucl. Data Tables* 39: 201–3 (1988) and references therein
73. Seth, K. K., Artuso, M., Barlow, D., Iversen, S., Kaletka, M., et al., *Phys. Rev. Lett.* 58: 1930–33 (1987); Seth, K. K., see Ref. 13, pp. 655–63
74. Jelley, N. A., Cerny, J., Stahel, D. P., Wilcox, K. H., *Phys. Rev.* C11: 2049–55 (1975)
75. Poppelier, N. A. F. M., Wood, L. D., Glaudemans, P. W. M., *Phys. Lett.* 157B: 120–22 (1985) and references therein
76. Stevenson, J., Brown, B. A., Chen, Y., Clayton, J., Kashy, E., et al., *Phys. Rev.* C37: 2220–23 (1988)
77. Aleksandrov, D. V., Ganza, E. A., Glukhov, Y. A., Novatskii, B. G., Ogloblin, A. A., et al., *Sov. J. Nucl. Phys.* 39: 323–25 (1984)
78. Belozyorov, A. V., Borcea, C., Dlouky, Z., Kalinin, A. M., Kalpakchieva, R., et al., *Nucl. Phys.* A460: 352–60 (1986)
79. Seth, K. K., see Ref. 15, pp. 324–33
80. Aleksandrov, D. V., Ganza, E. A., Glukhov, Y. A., Dukhanov, V. I., Mazurov, I. B., et al., *Sov. J. Nucl. Phys.* 37: 474–75 (1983)
81. Gilman, R., Fortune, H. T., Bland, L. C., Kiziah, R. R., Moore, C. F., et al., *Phys. Rev.* C30: 958–61 (1984)
82. Ogloblin, A. A., Penionzhkevich, Y. E., see Ref. 10, pp. 261–360
83. Mayer, W. A., Henning, W., Holzwarth, R., Körner, H. J., et al., *Z. Phys.* A319: 287–93 (1984)
84. Fifield, L. K., Woods, C. L., Catford, W. N., Bark, R. A., Drumm, P. V., et al., *Nucl. Phys.* A453: 497–504 (1986)

85. Drumm, P. V., Fifield, L. K., Bark, R. A., Hotchkis, M. A. C., Woods, C. L., et al., *Nucl. Phys.* A441: 95–108 (1985)

86. Benenson, W., Beard, K., Bloch, C., Sherrill, B., Brown, B. A., et al., *Phys. Lett.* 162B: 87–91 (1985)

87. Brauner, M., Rychel, D., Gyufko, R., Wiedner, C. A., Thornton, S. T., *Phys. Lett.* 150B: 75–78 (1985)

88. Woods, P. J., Chapman, R., Durell, J. L., Mo, J. N., Smith, R. J., et al., *Phys. Lett.* 182B: 297–300 (1986)

89. Fifield, L. K., Chapman, R., Durell, J. L., Mo, J. N., Smith, R. J., et al., *Nucl. Phys.* A484: 117–24 (1988)

90. Tondeur, F., see Ref. 13, pp. 81–89

91. Dessagne, P., Bernas, M., Langevin, M., Morrison, G. C., Payet, J., et al., *Nucl. Phys.* A426: 399–412 (1984)

92. Bernas, M., Dessagne, P., Langevin, M., Pougheon, F., Russel, P., et al., see Ref. 14, pp. 95–101 and references therein

93. Détraz, C., Langevin, M., Goffri-Kouassi, M. C., Guillemaud, D., Epherre, M., et al., *Nucl. Phys.* A394: 378–86 (1983)

94. Huck, A., Klotz, G., Knipper, A., Miehé, C., Richard-Serre, C., et al., *Phys. Rev.* C31: 2226–37 (1985)

95. Miehé, C., Dessagne, P., Baumann, P., Huck, A., Klotz, G., et al., *Phys. Rev.* C33: 1736–39 (1986)

96. Miehé, C., Dessagne, P., Baumann, P., Huck, A., Klotz, G., et al., *Phys. Rev.* C39: 992–96 (1989)

97. Vieira, D. J., Wouters, J. M., Vaziri, K., Kraus, R. H., Wollnik, H., et al., *Phys. Rev. Lett.* 57: 3253–56 (1986)

98. Gillibert, A., Bianchi, L., Cunsolo, A., Fernandez, B., Foti, A., et al., *Phys. Lett.* B176: 317–21 (1986)

99. Gillibert, A., Mittig, W., Bianchi, L., Cunsolo, A., Fernandez, B., et al., *Phys. Lett.* B192: 39–43 (1987)

100. Wouters, J. M., Kraus, R. H., Vieira, D. J., Butler, G. W., Löbner, K. E. G., *Z. Phys.* A331: 229–33 (1988)

101. Wapstra, A. H., Audi, G., Hoekstra, R., *At. Data Nucl. Data Tables* 39: 281–87 (1988)

102. Tachibana, T., Uno, M., Yamada, M. Yamada, S., *At. Data Nucl. Data Tables* 39: 251–58 (1988)

103. Wildenthal, B. H., Curtin, M. S., Brown, B. A., *Phys. Rev.* C28: 1343 (1983); Wildenthal, B. H., *Prog. Part. Nucl. Phys.* 11: 5–51 (1984)

104. Becker, J. A., Warburton, E. K., Brown, B. A., *Bull. Am. Phys. Soc.* 33: 1563 (1988); private communication and to be published

105. Wildenthal, B. H., private communication and to be published

106. Wapstra, A. H., Nijgh, G. J., van Lieshout, R., *Nuclear Spectroscopy Tables*. Amsterdam: North-Holland (1959)

107. Wood, J. L., see Ref. 15, *AIP Conf. Proc.* 164: 245–54 (1988)

108. Watt, A., Singhal, R. P., Storm, M. H., Whitehead, R. R., *J. Phys.* G7: L145–48 (1981); Storm, M. H., Watt, A., Whitehead, R. R., *J. Phys.* G9: L165–68 (1983)

109. Poves, A., Retamosa, J., *Phys. Lett.* 184B: 311–15 (1987)

110. Burleson, G. R:, Blanpied, G. S., Daw, G. H., Viescas, A. J., Morris, C. L., et al., *Phys. Rev.* C22: 1180–83 (1980)

111. Tribble, R. E., Tanner, D. M., Zeller, A. F., *Phys. Rev.* C22: 17–20 (1980)

112. Tribble, R. E., Cossairt, J. D., Kenefick, R. A., *Phys. Rev.* C15: 2028–31 (1977)

113. Seth, K. K., Iversen, S., Kaletka, M., Barlow, D., Saha, A., et al., *Phys. Lett.* 173B: 397–99 (1986)

114. Wigner, E. P., in *Proc. Robert A. Welch Conf. on Chemical Research*. Houston: Welch Foundation (1957), 1: 67–91

115. Benenson, W., Kashy, E., *Rev. Mod. Phys.* 51: 527–40 (1979)

116. Antony, M. S., Britz, J., Bueb, J. B., Pape, A., *At. Data Nucl. Data Tables* 33: 447–78 (1985)

117. Pougheon, F., Borrel, V., Jacmart, J. C., Anne, R., Détraz, C., et al., *Nucl. Phys.* (1989), in press; *IPN Rep. IPNO DRE 89-02* (1989)

118. Auerbach, N., *Phys. Rep.* 98: 273–341 (1983) and references therein

119. Deleted in proof

120. Ormand, W. E., Brown, B. A., *Nucl. Phys.* A491: 1–23 (1989)

121. See Ref. 5; Ormand, W. E., Brown, B. A., *Nucl. Phys.* A440: 274–300 (1985)

122. Ormand, W. E., Brown, B. A., *Phys. Lett.* B174: 128–32 (1986)

123. Deleted in proof

124. Deleted in proof

125. Garvey, G. T., et al., *Rev. Mod. Phys.* 41: S1–80 (1969)

126. Comay, E., Kelson, I., Zidon, A., *Phys. Lett.* 210B: 31–34 (1988)

127. Thomas, R. G., *Phys. Rev.* 80: 136–36 (1950); 88: 1109–25 (1952); Ehrman, J. B., *Phys. Rev.* 81: 412–16 (1951); Lane, A. M., Thomas, R. G:, *Rev. Mod. Phys.* 30: 257–353 (1958)

128. Shinozuka, T., Fujioka, M., Miyatake, H., Yoshii, M., Hama, H., et al., *Phys. Rev.* C30: 2111–14 (1984)

129. Sherrill, B., Beard, K., Benenson, W., Block, C., Brown, B. A., et al., *Phys. Rev.* C31: 875–78 (1985)

130. Stiliaris, E., Bohlen, H. G., Chen, X. S.,

Gebauer, B., Miczaika, A., et al., *Z. Phys.* A326: 139–46 (1987)

131. Arai, Y., Fujioka, M., Tanaka, E., Shinozuka, T., Miyatake, H., et al., *Phys. Lett.* 104B: 186–88 (1981)

132. Sherrill, B., Beard, K., Benenson, W., Brown, B. A., Kashy, E., et al., *Phys. Rev.* C28: 1712–17 (1983)

133. Nolen, J. A., Schiffer, J. P., *Annu. Rev. Nucl. Sci.* 19: 471–526 (1969)

134. Shlomo, S., *Rep. Prog. Phys.* 41: 957–1026 (1978)

135. Koch, V., Miller, G. A., *Phys. Rev.* C31: 602–12 (1985); C32: 1106–7 (1985)

136. Jensen, A. S., Miranda, A., *Nucl. Phys.* A449: 331–53 (1986)

137. Madland, D. G., Nix, J. R., *Nucl. Phys.* A476: 1–38 (1988)

138. Möller, P., Nix, J. R., *At. Data Nucl. Data Tables* 39: 213–23 (1988)

139. Leander, G. A., Chen, Y. S., *Phys. Rev.* C37: 2744–78 (1988) and references therein

140. Haustein, P., see Ref. 14, pp. 413–19

141. Jänecke, J., see Ref. 14, pp. 420–27

142. Masson, P. J., Jänecke, J., *At. Data Nucl. Data Tables* 39: 273–80 (1988)

143. Haustein, P., Brenner, D. S., Casten, R. F., see Ref. 15, *AIP Conf. Proc.* 164: 84–88 (1988)

144. Hardy, J. C., Hagberg, E., see Ref. 9, 3: 100–25

145. Deleted in proof

146. Hansen, P. G., Jonson, B., see Ref. 9, 3: 157–201

147. Kratz, K. L., Möller, P., Hillebrandt, W., Ziegert, W., Harms, V., et al., see Ref. 15, pp. 558–67

148. Lewitowicz, M., et al., *Nucl. Phys.* A496: 477–84 (1989)

149. Takahashi, K., *Prog. Theor. Phys.* 47: 1500–16 (1972)

150. Tachibana, T., Ohsugi, S., Yamada, M., see Ref. 15, pp. 614–23

151. Klapdor, H. V., Metzinger, J., Oda, T., *At. Data Nucl. Data Tables* 31: 81–111 (1984)

152. Staudt, A., Klapdor, H. V. (1990), to be published

153. Moltz, D. M., Cerny, J., see Ref. 9, 3: 133–56

154. Äystö, J., Moltz, D. M., Xu, X. J., Reiff, J. E., Cerny, J., *Phys. Rev. Lett.* 55: 1384–87 (1985)

155. Borrel, V., Jacmart, J. C., Pougheon, F., Richard, A., Anne, R., et al., *Nucl. Phys.* A473: 331–41 (1987)

156. Reiff, J. E., Hotchkis, M. A. C., Moltz, D. M., Lang, T. F., Lang, J. D., et al., *Nucl. Instrum. Methods* A276: 228–32 (1989)

157. Jahn, R., McGrath, R. L., Moltz, D.

M., Reiff, J. E., Xu, X. J., et al., *Phys. Rev.* C31: 1576–78 (1985)

158. Azuma, R. E., Björnstad, T., Gustafsson, H. A., Hansen, P. G., Jonson, B., et al., *Phys. Lett.* 96B: 31–34 (1980)

159. Langevin, M., Détraz, C., Epherre, M., Guillemaud-Mueller, D., Jonson, B., et al., *Phys. Lett.* 146B: 176–78 (1984)

160. Borge, M. J. G., Epherre-Rey-Campagnolle, M., Guillemaud-Mueller, D., Jonson, B., Langevin, M., et al., *Nucl. Phys.* A460: 373–80 (1986)

161. Dufour, J. P., Del Moral, R., Hubert, F., Jean, D., Pravikoff, M. S., et al., *Phys. Lett.* 206B: 195–98 (1988)

162. Rose, H. J., Jones, G. A., *Nature* 307: 245–47 (1984)

163. Price, P. B., *Annu. Rev. Nucl. Part. Sci.* 39: 19–42 (1989)

164. Goldanskii, V. I., *Nucl. Phys.* 27: 648–664 (1961); *Sov. Phys. Usp.* 8: 770–79 (1966); *Phys. Lett.* 212B: 11–12 (1988)

165. Jänecke, J., *Nucl. Phys.* 61: 326–41 (1964)

166. Détraz, C., et al., to be published

167. Goldanskii, V. I., *Nucl. Phys.* 19: 482–95 (1960)

168. Kelson, I., Garvey, G. J., *Phys. Lett.* 23: 689–92 (1966)

169. Détraz, C., see Ref. 191, pp. 151–70

170. Armbruster, P., Clerc, H. G., Dufour, J. P., Franczak, B., Geissel, H., et al., see Ref. 15, *AIP Conf. Proc.* 164: 839–44 (1988) and references therein

171. Hansen, P. G., Jonson, B., *Europhys. Lett.* 4: 409–14 (1987)

172. Otten, E. W., see Ref. 10, pp. 517–638

173. Bruandet, J. F., in *Proc. Int. Conf. on Heavy Ion Nuclear Collisions in the Fermi Energy Domain*, Caen 1986. *J. Phys. Colloq.* 47 (C4): 125–39 (1986)

174. Tanihata, I., Hamagaki, H., Hashimoto, O., Nagamiya, S., Shida, Y., et al., *Phys. Lett.* 160B: 380–84 (1985)

175. Tanihata, I., Hamagaki, H., Hashimoto, O., Shida, Y., Toshikawa, N., et al., *Phys. Rev. Lett.* 55: 2676–79 (1985)

176. Tanihata, I., *Nucl. Phys.* A488: 113c–126c (1988) and references therein

177. Kox, S., Gamp, A., Perrin, C., Arvieux, J., et al., *Phys. Rev.* C35: 1678–91 (1987)

178. Mittig, W., Chouvel, J. M., Long, Z. W., Bianchi, L., Cunsolo, A., et al., *Phys. Rev. Lett.* 59: 1889–91 (1987)

179. Saint-Laurent, M. G., Anne, R., Bazin, D., Guillemaud-Mueller, D., Jahnke, U., et al. *Z. Phys.* A332: 457–65 (1989)

180. Tanihata, I., Kobayashi, T., Yamakawa, O., Shimoura, S., Ekuni, K., et al., *Phys. Lett.* 206B: 592–96 (1988)

181. Mayers, W. D., Schmidt, K. H., *Nucl. Phys.* A410: 61–73 (1983)

182. Poppelier, N. A. F. M., de Vries, J. H., Wolters, A. A., Glaudemans, P. W. M., see Ref. 15, pp. 334–43
183. Goldhaber, A. S., Heckman, H. H., *Annu. Rev. Nucl. Sci.* 28: 161–205 (1978)
184. Dayras, R., see Ref. 173, pp. 13–28 and references therein
185. Hüfner, J., Nemes, M. C., *Phys. Rev.* C23: 2538–47 (1981)
186. Kobayashi, T., et al., *KEK Rep. 89-27*; submitted to *Phys. Lett.* B (1989)
187. Cronqvist, M., Hansen, P. G., Johannsen, L., Jonson, B., Nyman, G., et al., *SIS/ESR Letter of Intent* (1988)
188. Hansen, P. G., ISOLDE (CERN) Exp. IS.210 (1989) and private communication
189. Buchmann, L., D'Auria, J. M., eds., *Proc. of the Accelerated Radioactive*

Beams Workshop, Parksville, 1985. *TRIUMF Rep. TRI-85-1* (1985)
190. Bimbot, R., Della Negra, S., Aguer, P., Bastin, G., Anne, R., et al., *Z. Phys.* A322: 443–56 (1985); Bimbot, R., *I.P.N. Orsay Rep. IPNO-DRE-87-35* (1987)
191. Kubono, S., Ishihara, M., and Nomura, T., eds., *Proc. Int. Symp. on Heavy Ion Physics and Nuclear Astrophysical Problems*, Tokyo, 1988. Singapore: World Scientific (1989)
192. Arnould, M., Baeten, F., Darquennes, D., Delbar, Th., Dom, C., et al., *Belgian Inter-University Rep. RIB 1988-01* (1988)
193. Tanihata, I., see Ref. 10, pp. 443–514 (1989)
194. Nitschke, M., *New Scientist* 121(1653): 55–59 (1989)

NOTE ADDED IN PROOF

Please note the additional mass measurement of ^{57}Cu using the reaction ^{58}Ni(^{7}Li, ^{8}He) by Gagliardi et al (195), which had been overlooked in the text. This new result is considerably more precise than the previous work (128–130); all values are found in good agreement. An additional determination of ^{51}Ca using the reaction ^{48}Ca(^{18}O, ^{15}O) has recently been reported by Catford et al (196). Although the masses of two excited states would appear to agree well with one of the previous measurements (86), the ground-state masses are in disagreement by 1 MeV. Of further interest is the first mass determination of ^{39}S as measured in the ^{40}Ar(^{13}C, ^{14}O) reaction (197).

195. Gagliardi, C. A., Semon, D. R., Tribble, R. E., Van Ausdeln, L. A., *Phys. Rev.* C34: 1663–66 (1986)
196. Catford, W. N., Fifield, L. K., Ophel, T. R., Orr, N. A., Weisser, D. C., et al., *Nucl. Phys.* A489: 347–67 (1988)
197. Drumm, P. V., Fifield, L. K., Bark, R. A., Hotchkis, M. A. C., Woods, C. L., *Nucl. Phys.* A496: 530–38 (1989)

Annu. Rev. Nucl. Part. Sci. 1989. 39: 467–506

SOLAR NEUTRINOS

Raymond Davis, Jr.

Department of Astronomy, University of Pennsylvania, Philadelphia, Pennsylvania 19104

Alfred K. Mann

Department of Physics, University of Pennsylvania, Philadelphia, Pennsylvania 19104

Lincoln Wolfenstein

Department of Physics, Carnegie Mellon University, Pittsburgh, Pennsylvania 15213

KEY WORDS: neutrino oscillations, sun, underground experiments, neutrino detectors.

CONTENTS

0163–8998/89/1201–0467$02.00

1.   INTRODUCTION

It is generally believed that a realistic model of the Sun and other main sequence stars has been constructed based on well-established physical principles. Since the solar interior is poorly understood, a number of reasonable assumptions are introduced. The Sun is assumed to be in hydrostatic equilibrium, with energy generated near the center by nuclear fusion based on the p-p chains and CNO cycle and transported to the surface by photons and convection. The standard solar model (SSM), based on these principles and utilizing various input data, yields the solar density, temperature, and composition as a function of radius. The initial conditions are chosen to reproduce the present mass, luminosity, surface temperature, composition, and radius of the Sun.

Neutrinos produced in the nuclear reactions near the center of the Sun interact only weakly with the matter of the Sun and emerge directly from their source without significant loss of intensity or energy. Accordingly, neutrinos provide the most direct probe of the central region of the Sun and may serve to test the standard solar model predictions of the energy production in that region. Solar neutrinos are also a unique source for studying the intrinsic properties of neutrinos. If neutrinos have mass, there exists the possibility that electron-type neutrinos v_e are converted, i.e. may oscillate, into other types (v_μ, v_τ). As a consequence of the great distance neutrinos travel from the Sun to the Earth and the high density they traverse in emerging from the core to the surface of the Sun, it is possible to explore a large range of oscillation parameters (masses and mixings) that cannot be studied in laboratory experiments. Still another possibility is that the v_e may have a nonzero magnetic dipole moment (nonzero mass is a prerequisite) that, when acted upon by the solar magnetic field, might lead to a modification of their expected intensity as observed on Earth.

These possibilities for understanding the physics of the central region of a star and the properties and interaction of neutrinos are the primary motivations for studying solar neutrinos. In addition, there are questions raised by comparison of the limited existing solar neutrino data with predictions of the SSM. Until recently the only data were those of the ^{37}Cl radiochemical detector, sensitive primarily to the neutrinos from ^{8}B decay. These data yield a neutrino flux averaged over the period 1970–1985 that is approximately one quarter of the flux predicted by the SSM. More recent data collected by the same detector during 1987 and the first half of 1988 give a larger value, about one half the SSM prediction. During this period a second, independent, electronic detector, Kamiokande-II, was operating and measured a rate of about one half the SSM prediction for the high energy (≥ 9 MeV) region of the ^{8}B spectrum. Furthermore, the electronic detector provided directional information demonstrating that the signal came from the Sun.

The data accumulated so far might be viewed as qualitatively confirming the SSM and as providing the first direct evidence of nuclear reactions inside the Sun. On the other hand, the difference between the data and the prediction of the SSM, commonly called the solar neutrino puzzle, has been taken as a sign that either the SSM must be modified or that heretofore undetected neutrino properties are present. Furthermore, the totality of data from 1970 to 1988 suggests an anticorrelation of the observed neutrino flux with the number of sunspots. Such a correlation between neutrinos produced in the central region of the Sun with sunspots originating in surface phenomena would be difficult to explain.

It is clear that an important beginning has been made but more observations are needed. Here we review the present observations and the outlook for the future. In what follows, the calculations that yield the predicted solar neutrino flux are described in Section 2, and the effects of nonzero neutrino mass on the observed flux are discussed in Section 3. The interactions of neutrinos with matter that are the basis of different detection techniques are discussed in Section 4. Sections 5 and 6 describe radiochemical and real-time electronic detectors, respectively, including detectors in operation, under construction, and proposed for the future. The results of the ^{37}Cl and Kamiokande-II detectors are compared in Section 7. The present status and prospects of the field are summarized in Section 8.

2. SOLAR NEUTRINO FLUX

The source of energy in the Sun is a set of nuclear reactions that convert hydrogen into helium. The main chain of reactions begins with the weak processes

$$p+p \to {}^2H+e^+ +v_e \qquad\qquad \text{1a.}$$

or

$$p+e^- +p \to {}^2H+v_e. \qquad\qquad \text{1b.}$$

The resultant deuterium reacts quickly through the reaction $p+{}^2H \to {}^3He+\gamma$, with two possibilities for ^{3}He: (a) it reacts with another ^{3}He to form ^{4}He and two protons, or (*b*) it reacts about 15% of the time with ^{4}He present in the Sun through the process

$$^3\text{He} + {}^4\text{He} \rightarrow {}^7\text{Be} + \gamma. \hspace{2cm} 2.$$

Most of the ^{7}Be is converted to ^{7}Li by electron capture,

$$e^- + {}^7\text{Be} \rightarrow {}^7\text{Li} + \nu_e, \hspace{2cm} 3a.$$

rapidly followed by

$$^7\text{Li} + p \rightarrow 2\,{}^4\text{He},$$

but about one ^{7}Be nucleus in a thousand undergoes the nuclear reaction

$$p + {}^7\text{Be} \rightarrow {}^8\text{B} + \gamma. \hspace{2cm} 3b.$$

These ^{8}B beta-decay into unstable ^{8}Be*:

$$^8\text{B} \rightarrow {}^8\text{Be*} + e^+ + \nu_e \hspace{2cm} 4.$$
$$\hspace{0.5cm} \hookrightarrow 2\,{}^4\text{He}.$$

There are therefore three main sources of ν_e expected from the Sun:

1. By far the most copious source, called p-p neutrinos, is Reaction 1a, which produces a continuous spectrum with an endpoint energy of 420 keV.
2. Reaction 3a produces a line source of ν_e with an energy of 862 keV and a secondary line at 383 keV. The integrated flux is thirteen times smaller than for Reaction 1a.
3. The ^{8}B decay (Reaction 4) produces a continuous spectrum with an endpoint energy of approximately 15 MeV but a total flux 10^{-4} times that of p-p neutrinos.

There are also less important contributions from the p-e-p reaction (Reaction 1b) and from ^{13}N, ^{15}O, and ^{17}F as a consequence of the CNO cycle (see Table 1). The highest energy neutrinos could emerge from the reaction $^3\text{He} + p \rightarrow {}^4\text{He} + e^+ + \nu_e$ with a continuous spectrum and endpoint energy of 18 MeV, but with an integrated flux about 10^{-7} times that of p-p neutrinos. Detailed calculations by Bahcall & Ulrich (1) result in the neutrino fluxes shown in Figure 1 and Table 1.

These fluxes are derived from the standard solar model (SSM). The most important assumptions made in the model are as follows:

1. The interior of the Sun is in a state of hydrostatic equilibrium.
2. Energy is transported from the interior by radiative transfer except in the outer convective zone.

Table 1 Solar neutrino fluxes and capture rates

Neutrino source	Neutrino flux at Earth per cm² per sec		Neutrino capture rate in SNU = 10^{-36} sec^{-1} atom^{-1} [a]				
	Bahcall & Ulrich[a]	Turck-Chièze et al[b]	^{37}Cl	^{71}Ga	^{7}Li	^{81}Br	^{115}In
pp → De$^+\nu_e$	$6.0\pm0.12 \times 10^{10}$	$5.98\pm0.18 \times 10^{10}$	0	70.8	0	0	468
pe$^-$p → Dν_e	$1.4\pm0.07 \times 10^{8}$	1.30×10^{7}	0.2	3.0	9.2	1.1	8.1
^{3}Hep → ^{4}He e$^+\nu$	7.6×10^{3}	—	0.03	0.06	0.06	0.07	0.05
^{7}Be → ^{7}Li ν_e	$4.7\pm0.7 \times 10^{9}$	4.18×10^{9}	1.1	34.3	4.5	8.6	116
^{8}B → ^{8}Be* e$^+\nu_e$	$5.8\pm2.1 \times 10^{6}$	$3.8\pm1.1 \times 10^{6}$	6.1	14.0	22.5	15.3	14.4
^{13}N → ^{13}C e$^+\nu$	$6.1\pm3.1 \times 10^{8}$	—	0.1	3.8	2.6	0.9	13.6
^{15}O → ^{15}N e$^+\nu$	$5.2\pm3.0 \times 10^{8}$	—	0.3	6.1	12.8	1.9	18.5
^{17}F → ^{17}O e$^+\nu$	$5.2\pm2.4 \times 10^{6}$	—	0.004	0.06	0.1	0.02	0.2
			7.9 ± 2.6	132 ± 20	51.8 ± 16	27.8 ± 17	639 ± 640

[a] Authors' 3σ error (from 1).
[b] 1σ errors, Total Rate = 5.8±1.3 and 125±5 SNU for ^{37}Cl and ^{71}Ga respectively.

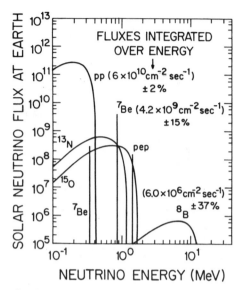

Figure 1   Solar neutrino fluxes predicted by the standard solar model (from 1). The energy thresholds for the gallium, chlorine, and Kamiokande-II neutrino detectors are 0.233, 0.814, and 9.3 MeV, respectively.

3. There is spherical symmetry and no significant rotation.
4. Chemical homogeneity exists at formation and is changed only locally as a result of nuclear reactions.
5. No unknown physics plays any significant role in the Sun.

These assumptions are preserved throughout the lifetime of the Sun. The calculations use standard equations (2)—beginning with a homogeneous zero-age main sequence model 4.6 × 10⁹ years ago to the present—and are iterated until the present observed properties (luminosity and radius) of the Sun are reproduced.

The major physical input parameters for the calculations are nuclear reaction cross sections, chemical abundances, and calculated radiative opacities. Most of the nuclear reaction cross sections (3) have been measured in the laboratory, but the cross sections must be extrapolated to the low energies relevant in the Sun using barrier penetration formulas. Reactions 1a and 1b must be calculated (4) using theoretical nuclear matrix elements together with the empirical value of (g_A/g_V) from neutron beta decay. The greatest uncertainty for the fundamental p-p reaction (Reaction 1a) may come from meson-exchange corrections (5). The chemical abundances of elements heavier than helium are deduced primarily from analysis of solar surface abundances (6). The helium abundance is allowed to vary in the iterated calculations and is eventually fixed by the fit to the solar luminosity. However, the result that is found (27 to 28%) is reasonably consistent (7, 8) with estimates of the primordial helium abundance. The calculation of the opacity as a function of temperature, density, and composition requires a detailed knowledge of the photon absorption cross section of atoms, ions, and in some cases molecules. These calculations have been carried out over many years at Livermore and Los Alamos (9); they are often used "almost as a black box since detailed checks are practically impossible" (7).

To estimate the uncertainty in the calculated flux, Bahcall & Ulrich (1) compound the uncertainties in the input parameters. For example, they give a 3σ deviation of the ^{8}B flux from its central value of 37%. In fact, one thousand calculations were carried out with random Gaussian variations of the input parameters and none gave a ^{8}B flux below the central value by more than 30%. This result depends entirely on the estimate of the uncertainties in the input parameters, which is partly subjective.

A completely independent calculation based on the same SSM has recently been carried out by a French-Belgian group (10), who compared their own numerical results with those of other calculations based on the SSM (see Table 1). Their final solar model is similar to that of Bahcall & Ulrich but uses different nuclear and other parameters. They obtain exactly the same flux of p-p neutrinos. However, as seen in Table 1, they calculate a flux of ^{8}B neutrinos lower by 35% and a flux of ^{7}Be neutrinos lower by 11%. The lower value for the ^{8}B flux has two main sources: (a) they choose a cross section for Reaction 3b that is 15% lower than that given in all other analyses, and (b) they use a different opacity. Taking these differences into account the two calculations agree to within better than 10%.

It is difficult to quantify the uncertainty that arises from the assumptions used in the standard model calculation. The best one can do is to look at a few nonstandard models for which calculations exist. If one assumes that the Sun has a much lower initial abundance of heavy elements, thus decreasing its opacity, the result is a lowering of the central temperature results with a consequent decrease in the ^{8}B neutrino flux and to a lesser extent that from ^{7}Be (1). However, at the same time, a much lower abundance of helium is required, contrary to expectations based on primordial helium abundance. The possibility that the Sun was formed with twice its present mass and then lost mass during its first billion years as a result of pulsation and rotation has been proposed by Willson and collaborators (11). The consequence is a higher present central temperature and an increase in the ^{8}B flux by as much as a factor of two. Schatzman (12) has argued that there could be turbulent mixing, which would increase the amount of hydrogen toward the center of the Sun and allow for a lower central temperature and a reduced ^{8}B neutrino flux. It is difficult to find a mechanism to produce the large mixing required (13). A number of other nonstandard models specifically designed to reduce the ^{8}B neutrino flux are reviewed by Newman and by Rood (14).

The data from helioseismology provide information concerning the solar interior and reveal small, possibly significant, deviations from the standard solar model predictions. These were reviewed by Bahcall & Ulrich (1). They find one ad hoc model based on inhomogeneous initial composition that could explain the data; this model yields a small increase in the neutrino flux.

An exotic possibility that has been proposed (15) is the existence of relic weakly interacting massive particles (WIMPs) concentrated near the center of the Sun. For a narrow range of masses and cross sections such particles would have evaded detection until now and would not have drastically altered stellar evolution (16). On the other hand the WIMPs would transfer energy from the solar interior, thus lowering the central temperature and the ^{8}B neutrino flux. It also claimed that this model gives a better fit to helioseismological data. From the point of view of elementary particle physics, WIMPs with the necessary properties do not occur in reasonable extensions of the standard electroweak theory.

3. NEUTRINO MASS EFFECTS ON DETECTED FLUX

3.1 *Neutrino Mass and Oscillations*

Neutrinos exist in three varieties, sometimes called flavors, v_e, v_μ, and v_τ, together with their antiparticles $\bar{v}_e$, $\bar{v}_\mu$, and $\bar{v}_\tau$. All are distinguished by the way they interact. The neutrinos emerging from the Sun are v_e emitted together with e^+ or from electron capture. A characteristic interaction used in their detection is inverse beta decay,

$$v_e + (n) \rightarrow (p) + e^-, \qquad\qquad 5.$$

where (n) and (p) indicate a neutron or proton bound in a nucleus. The known interactions conserve lepton number, where v_e and e^- have lepton number $+1$, $\bar{v}_e$ and e^+ have lepton number -1. The other neutrinos, v_μ and v_τ, are related to the charged leptons μ^- and τ^- in the same way as v_e is related to e^-. Thus the interaction of v_μ is

$$v_\mu + (n) \rightarrow (p) + \mu^-. \qquad\qquad 6.$$

The first detection of neutrinos from an accelerator revealed this reaction and demonstrated that ν_μ, resulting from pion decay, was different from ν_e. Various tests of decay processes and ν_μ interactions have so far shown that lepton flavor conservation is a valid concept.

There exists no compelling experimental evidence or theoretical argument that any of the neutrinos has a mass. Nevertheless the possibility of neutrino mass is of great importance for elementary particle physics and cosmology, as well as for understanding solar neutrino experiments. The standard electroweak theory has massless neutrinos, but zero mass does not result from any essential feature of the theory. Most extensions of the standard theory do yield nonzero masses, and the exploration of very small neutrino masses is considered a possible way to uncover fundamental new physics. The theory and phenomenology of neutrino mass are discussed in a number of recent reviews (17).

Direct kinematic experiments provide the following mass limits (18)

$$m(\nu_e) < 20 \text{ eV}$$

$$m(\nu_\mu) < 300 \text{ keV}$$

$$m(\nu_\tau) < 35 \text{ MeV}. \tag{7.}$$

In addition, the arrival times of $\bar{\nu}_e$ from supernova SN1987A provide a limit of around 20 eV for $m(\nu_e)$ (19). None of these direct limits is likely to be improved by more than a factor of 4 or so. It is quite natural to expect that if neutrinos have mass at all, then

$$m(\nu_\tau) \gg m(\nu_\mu) \gg m(\nu_e). \tag{8.}$$

This expectation, combined with the cosmological upper limit (20) of 80 eV for stable neutrinos, suggests that all neutrino masses may be far below directly measurable limits.

A promising way to study very small neutrino masses is indirectly through neutrino oscillations. The fundamental idea is that the neutrino "flavor eigenstates" such as ν_e emitted in decay processes are mixtures of mass eigenstates ν_i. For illustration, considering only two neutrino types, we write

$$\nu_e = \nu_1 \cos\theta_v + \nu_2 \sin\theta_v \tag{9a.}$$

$$\nu_\mu = -\nu_1 \sin\theta_v + \nu_2 \cos\theta_v. \tag{9b.}$$

Because of their different masses, ν_1 and ν_2 acquire different phase factors as a function of time so that the state which is ν_e at time $t = 0$ becomes a mixture of ν_e and ν_μ. It is easy to show that

$$P_{e\mu} \equiv |\langle v_\mu | v(x) \rangle|^2 = \sin^2 2\theta_v \sin^2 \frac{\pi x}{l_v}$$ 10.

with

$$l_v = 4\pi p/(m_2^2 - m_1^2) = 2.5 \text{ meters} \times p(\text{MeV})/\Delta m^2 \text{ (eV}^2\text{)},$$ 11.

where $v(x)$ is the state beginning at $x = 0$ as v_e; m_2 and m_1 are the mass eigenvalues; and p is the neutrino momentum. Experiments at reactors and accelerators (21, 22) have resulted in correlated limits on Δm^2 and θ_v; for example, reactor experiments give a limit of $\Delta m^2 < 10^{-1}$ eV^2 provided $\sin^2 2\theta_v > 0.1$.

These reactor experiments have looked for oscillations over distances of up to 65 meters. In contrast, solar neutrinos, which have comparable energy, travel over 10^{11} m. Thus for all values of Δm^2 down to 10^{-10} eV^2 there might be oscillations of v_e into v_μ on the way from the Sun to Earth. Since the ^{37}Cl detector cannot detect low energy v_μ at all, and the Kamiokande-II detector is relatively insensitive to low energy v_μ, there would be a deficiency in the detected flux. In addition, since l_v depends on neutrino energy, any deficiency would oscillate with energy. In short, solar neutrinos provide a method of probing very small values of Δm^2 (down to 10^{-11} eV^2 for p-p neutrinos) but only for quite large mixing angles.

Many models, particularly grand unified theories (23), suggest that θ_v may be similar in magnitude to the Cabibbo angle θ_c, which describes the mixing of d and s quarks. In this case $\sin^2 2\theta_v$ would be of the order of 0.1, so that on the average the deficiency due to oscillations would be less than 10%, too small to be detected in solar neutrino experiments. These models also yield the mass hierarchy of Equation 8 with $m(v_\mu)$ much less than 1 eV. Rather remarkably, for a large range of parameters suggested by the models one finds large transformation probabilities of v_e to v_μ or v_τ in the passage of neutrinos from the core of the Sun to the surface. This possibility occurs because oscillations within a material medium can be very different than in vacuum (24).

3.2   MSW Effect

Since neutrino oscillations depend on the change of phase of different components of the neutrino state vector, it is necessary for neutrinos propagating in a medium to take into account index of refraction effects. This yields the factor

$$\exp[ip(n-1)x].$$

The index of refraction n is given by the optical theorem

$$p(n-1) = 2\pi N f(0)/p, \qquad 12.$$

where N is the number of scatterers per unit volume and $f(0)$ is the forward scattering amplitude. Within the framework of the standard model there is one type of scattering that is not the same for all types of neutrinos: the charged-current scattering of v_e (or $\bar{v}_e$) from electrons. Since we are not interested in an overall phase factor, we consider only this scattering, which gives for v_e

$$p(n-1) = -\sqrt{2}GN_e, \qquad 13.$$

where N_e is the number of electrons per unit volume and G is the Fermi constant.

The oscillations for two neutrino flavors are now described by the equation

$$i\frac{d}{dx}\begin{pmatrix} v_e \\ v_\mu \end{pmatrix} = \begin{pmatrix} -\dfrac{\Delta m^2}{4p}\cos 2\theta_v + \dfrac{GN_e}{\sqrt{2}} & -\dfrac{\Delta m^2}{4p}\sin 2\theta_v \\ -\dfrac{\Delta m^2}{4p}\sin 2\theta_v & \dfrac{\Delta m^2}{4p}\cos 2\theta_v - \dfrac{GN_e}{\sqrt{2}} \end{pmatrix}\begin{pmatrix} v_e \\ v_\mu \end{pmatrix}. \qquad 14.$$

When $N_e = 0$ the eigenstates are just v_1 and v_2, as given by Equation 9, and the solution as a function of distance gives Equation 10. For fixed N_e the mixing angle and oscillation length in matter differ from those in vacuum. In particular, the mixing angle in matter θ_m is given by

$$\tan 2\theta_m = \sin 2\theta_v/(\cos 2\theta_v - l_v/l_0) \qquad 15.$$

$$l_0 = 2\pi/\sqrt{2}GN_e = 1.6 \times 10^7 \text{ meters}/\rho_e, \qquad 16.$$

where ρ_e is the electron number density divided by Avogadro's number. Even for small θ_v we can have $\theta_m = 45°$, provided that

$$l_v = l_0 \cos 2\theta_v = \frac{4\pi p}{m_2^2 - m_1^2}. \qquad 17.$$

The importance of this resonant amplification of neutrino oscillations was first pointed out by Mikheyev & Smirnov (25) and is often called the MSW effect. It is important to note that the electron neutrino must be the lighter neutrino ($\Delta m^2 > 0$) in order to be converted in this way. If the electron neutrino is not the lighter neutrino, the MSW effect would convert $\bar{v}_e$ into $\bar{v}_\mu$ or $\bar{v}_\tau$. As N_e varies from the interior of the Sun to its surface, there exists a large range of values of Δm^2 for which Equation 17 can be satisfied; for $\cos 2\theta_v$ close to unity this corresponds to

$$\Delta m^2(\text{eV}^2) = 1.5 \times 10^{-7}E_v(\text{MeV})\rho_e. \qquad 18.$$

Numerical solutions (26–28) and approximate analytical solutions (29) for the transition probability of v_e to v_μ in going through the Sun are given in many papers (see 25). Detailed results have also been given for transitions involving all three types of neutrinos (29a). A clear paper by Bethe (29b) stimulated great interest in the subject.

Several features of the solutions are important to note:

1. The transition probabilities are energy dependent. Some typical curves (27) of transition probability vs energy, corresponding to values of Δm^2 and $\sin^2 2\theta$ that significantly depress the ^8B v_e flux, are shown in Figure 2. It is seen that the spectrum of ^8B neutrinos can be significantly distorted. Furthermore, the overall effect on the ^8B neutrinos (mainly

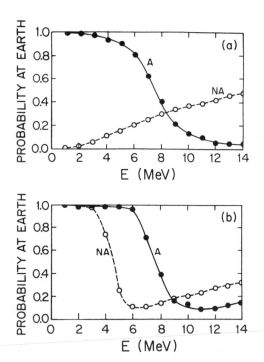

Figure 2 Probability (as a function of neutrino energy) for mixing angles θ and mass differences for a v_e originating near the center of the Sun to arrive at Earth as a v_e. (*a*) $\sin^2 2\theta = 10^{-1.5}$. The curve labeled *A* is for $\Delta m^2 = 10^{-4}$, *NA* is for $\Delta m^2 = 1.1 \times 10^{-6}$. (*b*) $\sin^2 2\theta = 10^{-3}$. The *A* curve is for $\Delta m^2 = 10^{-4}$, the *NA* is for $\Delta m^2 = 6 \times 10^{-5}$. All the curves give the same total suppression factor for the ^{37}Cl detector. *A* stands for the "adiabatic" and *NA* for the "nonadiabatic" solutions of Equation 14 (from 27).

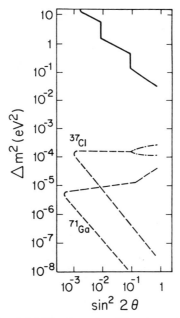

Figure 3 Values of Δm^2 and $\sin^2 2\theta$ to be explored by solar neutrino experiments. To the right of and below the dashed curves there is a greater than 50% reduction of the ν_e flux: The ^{37}Cl curve is for detectors mainly sensitive to ^{8}B neutrinos, and the ^{71}Ga curve is for detectors mainly sensitive to p-p neutrinos. Solid and dash-dot curves indicate present limits for ν_μ-ν_e oscillations.

in the range 4 to 12 MeV) can be very different from that on the low energy p-p neutrinos (energy below 0.4 MeV). Thus the two curves in Figure 2a correspond to the same overall suppression of ^{8}B neutrinos, but solution *A* gives practically no suppression for p-p neutrinos and solution *NA* gives almost total suppression.

2. Solar neutrinos provide the possibility of exploring a large range of neutrinos masses and mixings that cannot be studied in any other way. This is illustrated in Figure 3, which shows the $\Delta m^2 - \sin^2 2\theta$ plane with the regions indicated that can be explored by detectors sensitive to ^{8}B neutrinos and by those sensitive to p-p neutrinos. It should be emphasized that these hold equally well for ν_e-ν_τ oscillations. Also shown are the current limits from reactor and accelerator experiments for ν_e-ν_μ oscillations.

3. For a small range of parameters there can be a significant transformation of ν_μ back into ν_e in traversing the Earth. In this range there

can be a difference in flux between day and night and also between different seasons (30).

3.3 Other Effects of Neutrino Mass: Decay and Magnetic Moment

There are other possible effects of neutrino mass on the solar neutrino flux. If v_e is not the lightest of the neutrinos, it is conceivable that it could decay into three lighter neutrinos or into a lighter neutrino and a boson. The observation of $\bar{v}_e$ arriving from supernova SN1987A at a distance of 50 kpc largely rules out this simple possibility.

If the neutrino has a mass, it may also have an electric or magnetic dipole moment. Cisneros (31) pointed out that the normal left-handed v_e might precess in the solar core magnetic field into a right-handed v_e that does not have the usual weak interactions and that would not be detected. This idea has been revived by Voloshin et al (32) and applied to the convective zone of the Sun to explain a possible anticorrelation of the ^{37}Cl data with solar activity. This explanation requires that the neutrino have a magnetic moment of 10^{-10} to 10^{-11} Bohr magnetons, with precession taking place over a distance of order 10^{10} cm in the convective zone of the Sun with transverse magnetic fields of a few thousand gauss. It is suspected that solar magnetic fields are toroidal in nature and have strengths of this magnitude. The fields increase as the solar cycle develops and decrease as the solar cycle declines, which could explain a solar cycle variation. It was further suggested that a half-year solar latitude effect could arise in the observed solar neutrino flux. The solar rotational axis is inclined by 7.25 degrees with respect to the plane of the ecliptic, and the magnetic fields exist only at higher solar latitudes, the solar equator being essentially free of magnetic fields. It follows that the neutrino flux would exhibit a biennial sinusoidal variation at the Earth, particularly for the ^8B and ^7Be neutrinos, which are produced near the center of the Sun.

Such a large magnetic moment is not anticipated in most models of massive neutrinos, given the upper limit on the mass of v_e (33). For example, in the standard model, if the v_e is given a Dirac mass, the bound from Equation 7 requires that the magnetic moment be less than 10^{-17} Bohr magneton. A special class of models that might yield a neutrino magnetic moment as large as 10^{-11} was discussed by Voloshin (34). Upper limits of less than 10^{-12} for the magnetic moment of v_e have been deduced (35) from the observation of neutrinos from SN1987A, but these may not hold if one considers transition moments for Majorana neutrinos (36) or if neutrinos have new types of interactions. There is as well the possibility of resonant spin-flavor precession of solar and supernova neutrinos, $v_e \rightarrow v_\mu$ (see 36).

4. NEUTRINO INTERACTIONS FOR DETECTION PURPOSES

The interactions of neutrinos with matter appear to be well described by the standard electroweak model (37). These interactions occur either through the charged current ($W^{\pm}$ exchange) or neutral current (Z^0 exchange). The characteristic charged-current reaction for v_e is the inverse beta decay (Reaction 5). The corresponding charged-current reaction for v_μ (Reaction 6) has a threshold of 105 MeV and cannot be used to detect low energy v_μ from the Sun. Neutral-current interactions are the same for v_e, v_μ and v_τ and lead to elastic scattering from a neutron, proton, electron, or nucleus without change of flavor. Scattering from electrons for v_e involves both the charged and neutral currents, but for v_μ and v_τ it occurs only via neutral currents.

Reaction 5 always involves bound neutrons, for which the simplest example is

$$v_e + d \rightarrow p + p + e^-. \qquad\qquad 19.$$

For the low energy (< 15 MeV) v_e of interest, the final $p + p$ are expected to be in the $^1S_0$ state. Thus the transition from the initial $^3S_1$ deuteron is almost a pure Gamow-Teller transition. From the measured value of g_A, the axial-vector coupling from neutron beta decay, the rate of Reaction 19 may be calculated (38) using wave functions for the deuteron and the $^1S_0$ state. The uncertainty of 5 to 10% in the result is partly due to the uncertainty in the matrix element and partly due to meson-exchange corrections.

In most applications of inverse beta decay, the initial neutron and the final proton are in bound nuclei. To calculate the absorption cross section of the detector it is necessary to determine the matrix element of the weak interaction between initial and final nuclear states. If the final state decays back to the initial state, this can be done directly using the measured ft value of the decay. This is true in ^{71}Ga for low energy neutrinos (those from the p-p reaction) for which the only final state is the ground state of ^{71}Ge. On the other hand, for higher energy neutrinos such as those from ^{8}B decay, usually many excited states of the final nucleus can be reached. In ^{37}Cl, 60% of the transitions are expected to go to one excited state, the isobaric analog state. The matrix element for excited transitions in ^{37}Cl can be deduced from data on the decay of ^{37}Ca to ^{37}K, which is an isobaric mirror transition. The uncertainty in the ^{37}Cl cross section averaged over the ^{8}B spectrum is estimated to be less than 10%.

The greatest uncertainty in absorption cross sections arises when the matrix element cannot be related to a measured beta decay. For these cases, a promising method is the use of (p,n) reactions (39). In the distorted-wave Born approximation (DWBA) the forward cross section of (p,n) reactions is directly related to the Gamow-Teller matrix element. The quantitative uncertainty of this approximation is a subject of disagreement. In ^{71}Ga the ^{8}B neutrinos nearly all lead to transitions to excited states. These are expected in the SSM to give 10 to 15% of the rate in the gallium experiment. Grotz et al (40) estimate that the cross section to these states is uncertain by about 25%, whereas Bahcall & Ulrich (1) allow for a factor-of-two uncertainty. The uncertainty in the total gallium counting rate as a result of cross-sectional uncertainties is probably less than 10%.

The cross section for the elastic reaction $v_e + e^- \rightarrow v_e + e^-$ can be calculated theoretically with little uncertainty (41). For v_e from ^{8}B beta decay, the scattering of v_e from electrons has a cross section one or two orders of magnitude smaller than typical inverse beta-decay cross sections. On the other hand, the recoiling electron from $v_e + e^- \rightarrow v_e + e^-$ preserves the direction of the incident v_e and therefore allows a directional correlation to be made between the observed final-state electron and the Sun. Because of the possibility of oscillations, it is of interest to try to detect v_μ or v_τ, which requires exploiting pure neutral-current interactions. The elastic scattering on electrons can be used for this but the cross section from the neutral current alone gives a cross section 6 to 7 times smaller than for v_e. Thus it is especially interesting to use the neutral-current inelastic scattering from nuclei, which is the same for all types of neutrinos. Analogous to Reaction 19 one has

$$v + d \rightarrow v + p + n. \qquad 20.$$

It is also possible to consider the excitation of a nucleus through weak neutral-current scattering of the neutrino and then through detection of the deexcitation γ ray emitted in the transition to the ground state. An example (42) is the excitation of three states in ^{11}B, where it is possible to estimate the excitation cross section from analysis of electromagnetic excitation. It might also be possible to compare the excitation of levels in the mirror nucleus ^{11}C by the charged-current reaction. Finally, one can consider the coherent elastic scattering from the nucleus as a whole (43). Because of the coherence, the cross section increases as A^2, where A is the atomic mass number. However, the only signal is a recoiling nucleus.

5.   RADIOCHEMICAL EXPERIMENTS

5.1   *General*

A radiochemical neutrino detector is based on the inverse beta-decay process. Radiochemical neutrino telescopes employ hundreds of tons of target material, extract the radioactive product by a simple chemical procedure, and observe the decay of the product nucleus in a small counter. This method of neutrino detection can be designed to have high sensitivity, low energy threshold, efficiency, and a low background rate from cosmic rays, neutrons, and gamma radiation. The high sensitivity is achieved by choosing an abundant target isotope with a favorable inverse beta transition, allowed or superallowed, to yield a product nucleus with a convenient lifetime and decay scheme that will permit a clear separation of its radiations from background radiations. The efficiency of chemical isolation of the product can be high, in excess of 90%. The recovery yield is easily measured by using an enriched isotopic carrier. It is essential that the chemical process be simple, well understood, and tested.

A radiochemical detector does not have any means of identifying the neutrino source, nor does any other detector based on an inverse beta process. In the case of solar neutrinos one relies on the presumption that the Sun provides the highest flux of neutrinos. However, background processes can give a false signal that might be attributed to neutrinos. The major background arises from cosmic-ray muons. Energetic muons, usually in the range of hundreds of GeV, interact with nuclei by electromagnetic processes to create secondary particles, pions, and evaporation products. These may produce the desired radioactive product from the target isotope by a (p,n) reaction. To reduce such background to acceptable levels requires that the detector be located deep underground. Auxiliary experiments are also needed to evaluate the remaining background from cosmic rays. There are also backgrounds from fast neutrons that are produced in the surrounding rock by spontaneous fission of ^{238}U and (α,n) reactions. Fast neutrons produce the product isotope by (n,p) followed by (p,n) reactions. This secondary process is usually small, and fast neutrons can be eliminated easily by a water or paraffin shield.

An insidious background arises from internal radioactive contaminants, usually uranium, thorium, and their decay products that can give rise to the product nucleus by alpha-induced reactions. These contaminants can, however, be reduced to an insignificant level by careful monitoring of the target material, the walls of the containment vessel, and any reagents used in the chemical processing.

At the present stage of development of experimental neutrino detection, radiochemical detectors are the only immediate means of observing the very low energy neutrino sources in the Sun, particularly those from the p-p reaction, ^{7}Be, and CNO cycle products. Here we discuss in some detail the ^{37}Cl and ^{71}Ga radiochemical detectors that are operating and will be operating in the very near future. In the course of the last twenty years, there have also been a number of excellent suggestions for other radiochemical detectors. A brief discussion is given of these suggestions and the progress in developing them. For coverage in greater depth the reader is referred to earlier reviews on solar neutrino research (44–49) and to three conferences devoted specifically to that topic (50–52).

5.2   The Chlorine-37 Experiment

The chlorine solar neutrino experiment was built in the Homestake Gold Mine at Lead, South Dakota, USA, by Brookhaven National Laboratory in the period 1965 to 1967. The construction of the detector was stimulated by the realization that the ^{3}He(^{4}He,γ)^{7}Be and the ^{7}Be(p,γ)^{8}B reactions were of importance in the p-p chain (53–55). Prior to 1958 it was generally believed that the p-p chain ended with the ^{3}He + ^{3}He → ^{4}He + 2H reaction (56). Furthermore, it was thought that the CNO cycle produced less than 2% of the Sun's energy, a conclusion that still holds. Therefore the principal source of neutrinos from the Sun was believed to be the low energy neutrinos from the p-p reaction. The energy of these neutrinos was well below the 0.814-MeV threshold of the chlorine radiochemical detector. When it was realized that the Sun could be the source of more energetic neutrinos, particularly from ^{7}Be and ^{8}B decay, there was great interest in measuring these components of the neutrino flux.

Pontecorvo first pointed out (in 1948) that the reactions ^{37}Cl + ν_e → ^{37}Ar + e$^-$, producing the 35-day ^{37}Ar electron-capturing isotope, could be used for detecting neutrinos by the radiochemical method (57). Alvarez (in 1949) amplified this suggestion in the form of a proposal to carry out an experiment at a nuclear reactor (58). It is interesting to note that Alvarez's proposal contained a calculation of the neutrino capture cross section, attributed to Leonard Schiff, and he pointed out the importance of including captures that produce ^{37}Ar in excited states in the cross-section calculations. The method was developed (59) and applied in two experiments at the Savannah River reactors (60, 61) to test lepton number conservation (1956–1962). A pilot solar neutrino experiment (62) was carried out in a mine in Ohio (1960–1962).

The Homestake experiment is located at a depth of 4200 ± 100 hectograms cm^{-2} standard rock (4850 feet) in a cavity designed to house the facility and provide containment and a water shield. The detector contains 615 metric tons of liquid perchloroethylene (CCl$_2$=CCl$_2$) corresponding to 2.16 × 10^{30} atoms of ^{37}Cl. The apparatus consists of a single horizontal steel tank 14.6 m long and 6.1 m in diameter filled with liquid perchloroethylene and helium gas. Argon is extracted with a helium gas purging system driven by two liquid circulation pumps that force the helium simultaneously through the entire liquid volume by means of a double set of 20 eductor nozzles. To extract the argon, the helium is circulated in series through a condenser, through a molecular sieve adsorber to remove perchloroethylene vapors, and finally through liquid-nitrogen-cooled charcoal to collect the argon. The helium is then returned to the tank. The extraction efficiency is determined by the volume of helium circulated through the trapping system, and it is also measured with isotopically separated ^{36}Ar or ^{38}Ar introduced in each experiment. The detailed procedures followed and various tests performed are described in several articles and reports (47, 50–52, 63).

The entire sample of argon collected is purified and placed in a small proportional counter to observe ^{37}Ar decays. Initially only the pulse height and the time distribution were recorded. An upper limit of 3 SNU was given in 1968 for the neutrino capture rate (63). Here SNU stands for a solar neutrino unit, defined as 10^{-36} captures per second per target atom. In 1970 a pulse rise-time system was introduced that greatly increased the selectivity of ^{37}Ar decay events and thereby increased the sensitivity of the chlorine detector (64). The time, pulse height, and pulse rise-time are continuously recorded. The events with the correct pulse height and rise-time are selected, and the time distribution of these ^{37}Ar-like events is resolved into a decaying component with a 35-day half-life and a constant background component through the use of a maximum likelihood method (65).

The chlorine experiment has operated continuously since 1967 except for an 18-month period from May 1985 to October 1986 when the liquid circulation pumps suffered successive electrical failures (October 1984 and May 1985, respectively). The new pumps have a 10–20% lower pumping speed and require a correspondingly longer time to collect argon. The Homestake facility was operated by Brookhaven National Laboratory until 1986, and subsequently by the University of Pennsylvania. During the long running period, improvements were continually made in the ^{37}Ar detection sensitivity. Furthermore, the long record allowed a search for possible short-term and long-term changes in signal rate. Over the period from March 1970 to March 1988 the average ^{37}Ar production rate in the 615-ton detector was 0.518 ± 0.037 ^{37}Ar per day (66), see Figure 4. There is a cosmic-ray background of 0.08 ± 0.03 atoms per day (67–70) that must be subtracted. The total ^{37}Ar production rate ascribed to solar neutrinos is 0.438 ± 0.050 atoms per day or 2.33 ± 0.25 SNU. It is this result that should be compared to solar model calculations.

When the solar fluxes from the standard solar model (Table 1 and Figure 1) are combined with the calculated absorption cross sections of ^{37}Cl (71) discussed in Section 4, Bahcall & Ulrich (1) predict a total neutrino capture rate in ^{37}Cl of 7.9 ± 1.1 SNU (here we use an approximate 1σ error; the authors elect to use 3σ errors, 3.3 SNU). As seen in Table 1 most of the

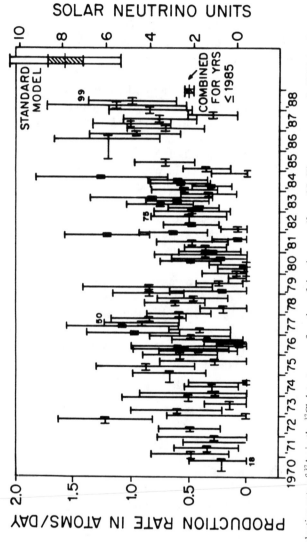

Figure 4   Production rate of ^{37}Ar in the ^{37}Cl detector as a function of time in years (from 66). The average rate in SNU for the period 1970–1985 is shown as 2.1 ± 0.3. Inclusion of the data for 1987–1988 raises the average to 2.3 ± 0.3. The prediction of the SSM from the calculation of Bahcall & Ulrich (1) is also shown, with the 3σ error quoted by them.

neutrinos originate from ^8B decay. The production of ^8B in the Sun is extremely sensitive to the central temperature and therefore to the details of the solar model calculation. The calculation of Turck-Chièze et al (10) yields 5.8 ± 1.2 SNU (1σ).

5.3 Possible Time Variation of the Observed ^8B Neutrino Flux

The long observational record obtained by the chlorine experiment, 1970 to the present time, allows one to search for both long- and short-period changes in the ^8B neutrino flux. The data have been studied in a number of analyses (72–84, and references therein). We do not review them in detail here but point out several features of interest.

Increases in the observed neutrino flux might conceivably result from collapsing stars, neutron star quakes, solar flares, intense gamma-ray bursts, etc. The highest single-rate observations might be considered as candidates for events of this nature. There were three data points corresponding to an ^{37}Ar production rate of 1.25 ^{37}Ar atoms per day. Two of these coincided with intense solar flares, and one coincided with the most intense gamma-ray burst observed by the Solar Maximum Mission satellite (78, 81, 83). A Monte Carlo simulation of the chlorine data assuming random errors on a flat distribution projects that one run of this magnitude should have been observed, an indication that three high runs are not significant evidence for such events. Searches in another detector with much higher energy threshold did not reveal an increased rate during these periods (85, 86).

The Sun's activity, studied for hundreds of years, reveals a periodic 11-year sunspot cycle and a 22-year magnetic cycle. The physical processes causing these phenomena are poorly understood. It is informative to examine the data from the chlorine experiment for a possible relation to solar activity. By combining the data in yearly averages or by taking a running average, one can smooth the data to reveal long-period trends. Figure 5 shows a plot of a five-point running average of the data compared to sunspot numbers in solar cycles 20, 21, and the beginning of 22, which suggests that the ^{37}Ar production rate anticorrelates with the sunspot numbers (66). The most significant change in the ^{37}Ar production rate occurs at a time when the solar activity cycle begins. There is a 5% probability that the data in Figure 5 reflect a statistical fluctuation (80). If, however, solar cycle 22, now rapidly developing, will cause a drop from the present high levels, the case for a solar cycle variation will be strengthened.

The data in Figure 5 show that in the 1987–1988 run the average ^{37}Ar production rate above muon background is 0.79 ± 13 atoms per day or

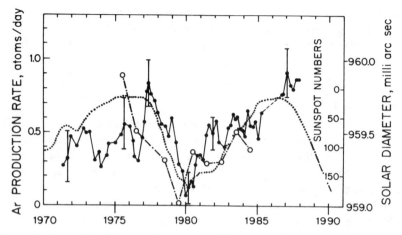

Figure 5 Plots of five-point running average of ^{37}Ar production and smoothed sunspot numbers against time in years (from 130). Solid circles, ^{37}Ar production; dotted curve, sunspot numbers; open circles, solar diameter.

4.2 ± 0.7 SNU, about three standard deviations above the previous long-term average. This result is compared in Section 7 with the Kamiokande-II result in the same time period.

5.4   The Two Gallium Experiments

The gallium solar neutrino detector was suggested by Kuzmin in 1964, who pointed out that the neutrino capture reaction $v_e + {}^{71}Ga \rightarrow {}^{71}Ge + e^-$ had a low enough threshold to permit observation of the neutrinos from the p-p reaction in the Sun. Moreover, the 11.4-day decay of ^{71}Ge has an unusually low *ft* value (4.3 seconds, spin change from 3/2 to 1/2), and therefore this reaction would have a favorable cross section (87). Approximately 20 tons of gallium are needed to obtain one neutrino capture per day, at that time an impossibly large amount of gallium. However, in the early 1970s, the electronics industry developed a need for gallium to produce light-emitting diodes, and the world production of gallium increased enormously. The price of the metal was (and still is) high, approximately $0.50 per gram.

Work on a radiochemical gallium detector began in the summer of 1974 at Brookhaven National Laboratory (BNL). By the end of the summer two methods were devised for removing trace quantities of germanium from gallium target material. One was an extraction of germanium from gallium metal (melting point 30°C), the usual industrial product, by an acid solution of hydrogen peroxide. The other method was based upon removing volatile germanium chloride from a concentrated, aqueous, gallium chloride solution containing hydrochloric acid. [The metal method was invented by John Evans (BNL) and the gallium chloride method was tried following the suggestion of A. Pomanskii (INR, Moscow).] Subsequently, these techniques were refined, and background processes from natural alpha processes and high energy muons were studied. These studies showed that a gallium detector could be made using either of the two methods (89, 90). A Soviet group at the Institute of Nuclear Research (INR) in Moscow independently developed and later adopted the gallium metal process. Since there are no deep mines in USSR, it was necessary for them to build an underground facility to house their solar neutrino experiments (91, 92).

A collaboration was formed in 1978 between Brookhaven National Laboratory and the Max-Planck-Institut für Kernphysik, Heidelberg, to perform the full-scale experiment. They chose the gallium chloride solution method and prepared a pilot experiment with a volume of 2700 liters of solution containing 1.3 tons of gallium. The solution method is easier to carry out and uses a chemical procedure that is well understood. The acidity of the solution is chosen to optimize the evolution of germanium chloride. The procedure is simply to pass gas (usually air) through the solution; the germanium chloride vapors are recovered from the gas stream by a water-spray scrubber. Germanium can be transferred to the scrubber with 100 percent efficiency. The recovered germanium chloride is purified and concentrated further by solvent extraction to a volume of 50–100 cm^3. Finally, the germanium chloride is reduced to germane (GeH_4) gas, a suitable gas for counting. The ^{71}Ge decays are recorded in a small proportional counter as in the ^{37}Cl detector. These developments are described in a proposal to carry out a full-scale experiment (93, 94). The full-scale experiment was not approved for funding, and the collaboration was dissolved. Subsequently, the Max-Planck-Institut in Heidelberg formed a new collaboration and obtained funding for a 30-metric-ton gallium detector. Their detector (GALLEX) is located in the Gran Sasso Laboratory in Italy. The detector is expected to have its full amount of gallium chloride solution in late 1989 (95).

Soviet scientists built a 7-ton gallium metal pilot experiment and demonstrated that germanium can be recovered from this quantity of metallic gallium with an efficiency of over 90%. Various background processes were studied, and appropriate counting techniques were developed (96). The metal process uses large volumes of hydrochloric acid. To avoid possibly contaminating their detector with germanium from the environment, the scientists developed a recovery system that allowed the acid to be reused. An underground laboratory in Baksan valley in the North

Caucasus mountains is now completed, and a 60-ton gallium metal detector is being installed. At the time of this writing one half of the detector is operational, and measurements with the full detector are expected during 1989. At a late stage, collaborators from the University of Pennsylvania and Los Alamos National Laboratory joined the effort.

It was expected that the gallium detectors would be sensitive primarily to the low energy neutrinos from the p-p reaction. The next most intense neutrino source, ^{7}Be, would populate two low-lying states at 0.175 and 0.500 MeV in ^{71}Ge, and ^{8}B decay neutrinos would feed many excited states. A series of (p,n) reaction experiments at 120 and 200 MeV indicated a large formation of highly excited states (98). Cross sections derived from shell model studies of the ^{71}Ga$(v,e^-)^{71}$Ge reaction are in essential agreement with the (p,n) reaction values (99). The analog state in ^{71}Ge decays by particle emission and therefore does not contribute to the production of ^{71}Ge (97). The predictions of Bahcall & Ulrich based on the SSM are shown in Table 1. It is seen that 54% of the rate comes from the p-p reaction and only 10% from ^{8}B decay. Bahcall & Ulrich predict a total neutrino capture rate in ^{71}Ga of 132 ± 6 SNU, while Turck-Chièze et al predict 125 ± 5 SNU, where these 1σ errors include both solar flux and nuclear cross-section uncertainties.

It is interesting that the isobaric analog state of ^{69}Ge decays primarily ($>95\%$) by gamma emission (100), and therefore the cross section of ^{69}Ga would be enhanced for ^{8}B neutrinos. Neutrino capture in ^{69}Ga has an energy threshold of 2.22 MeV, and hence this isotope (60% abundant) will serve only as a detector of ^{8}B decay neutrinos. A gallium detector may therefore be used to measure independently the lower energy solar neutrino sources and ^{8}B by measuring ^{71}Ge, and only ^{8}B decay neutrinos by measuring ^{69}Ge. The two radioactive decays can easily be separated by their radiation and lifetime. However, with only 60 tons of gallium it will be difficult to observe the low production rate of ^{69}Ge. The gallium chloride method is amenable to rapid recovery of germanium, and it would be possible to carry out experiments directed toward measuring the 39-hour ^{69}Ge production rate.

5.5   A Geochemical Experiment

It has often been noted that the Sun's energy production may vary over periods of millions of years or more. This thought has brought forward suggestions to measure the solar neutrino flux in the past. Cowan and Haxton noted that the neutrino capture cross sections of ^{97}Mo and ^{98}Mo to form the long-lived technetium isotopes ^{97}Tc ($t_{1/2} = 2.6$ million years) and ^{98}Tc ($t_{1/2} = 4.2$ million years) are useful for monitoring the neutrino flux during the past few millions of years (114). In this case one takes advantage of the fact that these products would be contained in molybdenum ores (MoS_2) being mined at great depth in the Henderson Mine in Colorado. In addition, the processing of the minerals by flotation not only concentrates the technetium, but also reduces the background production of technetium isotopes by uranium and other isotopes. The result is that the molybdenum experiment appears to be capable of measuring the [8]B decay solar neutrino flux in the past few million years. Another interesting outcome of the experiment may be information on the past Galactic neutrino flux from collapsing stars (116).

An experiment is now in progress at Los Alamos National Laboratory (115), where workers devised a method of collecting the relatively volatile oxides of technetium and rhenium from the ore-roasting stage of the process. These products will be purified and analyzed by surface ionization mass spectroscopy. A result from this interesting experiment is expected soon.

The only means currently available for determining the neutrino capture cross sections for the molybdenum isotopes are nuclear systematics and (p,n) reaction studies. Studies have been made with 200-MeV protons using the (p,n) method and enriched [98]Mo. These studies give a value of 3×10^{-42} cm^2 for the [8]B neutrino cross section (117). Bahcall (private communication) argues that the uncertainties in the cross sections do not allow a significant test for long-period solar neutrino variations. It is interesting to note that the production of [99]Tc ($t_{1/2} = 2.1 \times 10^5$ years) by energetic muons through electromagnetic interactions on [100]Mo may be observed in this experiment. This conclusion is based upon the studies of energetic muons underground by Fireman and his associates (69). A proper analysis of the rate may allow one to determine the average depth of the molybdenum during the last few hundred million years.

5.6 Proposed Radiochemical Experiments

A number of possible solar neutrino detectors have been considered over the last 20 years. Many of these were suggested as means of studying solar neutrinos by radiochemical detectors with different sensitivities to the principal components of the spectrum, the p-p reaction, [7]Be, and [8]B neutrinos.

Much effort has been devoted to a few promising cases, but at the present time none of these is sufficiently well developed to build a full-scale detector. We review here some of those cases in the hope of stimulating further work in this field. Table 1 summarizes the expected solar neutrino capture rate from the standard solar model for the radiochemical detectors that are discussed.

THE LITHIUM DETECTOR   The lithium detector is based on the neutrino capture reaction $^7Li(\nu_e,e^-)^7Be$ with a low threshold energy (0.861 MeV). ^7Li and ^7Be are mirror nuclei, and therefore the neutrino capture cross section to produce ^7Be in the ground state and its first excited state at 0.480 MeV have high values (71). As a result, the neutrinos from the p-e-p reaction and from ^7Be, barely over threshold, contribute a significant rate, as shown in Table 1. Lithium is a relatively inexpensive target element, the ^7Li isotope is 93% abundant, and the half-life of ^7Be is 53 days. Because of these considerations, a lithium detector was regarded as the most promising to follow the chlorine experiment.

At Brookhaven National Laboratory a chemical solvent extraction procedure was developed for removing ^7Be from concentrated aqueous solutions of lithium chloride (101). There are two not especially serious background problems associated with this approach. A Soviet group at the Institute for Nuclear Research has developed a method of extracting ^7Be from metallic lithium that avoids even these background problems (102). Their technique involves filtering lithium metal through a stainless steel gauze that apparently efficiently collects beryllium on the surface of the gauze! This method requires modest temperatures to melt the lithium, over 180°C, but using metal greatly increases the cost of the target material.

A major problem with the lithium detector is the absence of an efficient method of counting and characterizing ^7Be decays. One can observe the 480-keV gamma ray that occurs in 10% of the decays with a low background level germanium crystal, but this method is inefficient. Other detection methods have been suggested but have not been developed. Some of these are to observe the 57-eV recoiling ^7Li ion from ^7Be decay electrostatically or by thermal evaporation (101, 103), to count the recoiling ion with a cryogenic solid-state counter (M. Lowry, Princeton University, private communication), to excite the resulting lithium atom with laser light (105), and to detect ^7Be by accelerator mass spectroscopy (106).

With a solution to the counting problem, the lithium radiochemical detector could be a useful solar neutrino detector. It has the largest rate per unit target weight of all radiochemical solar neutrino detectors, 0.38 ^7Be atoms per day per ton of lithium. The major contribution (43%) to this rate comes from ^8B.

THE BROMINE DETECTOR   A second promising detector is based on the $^{81}Br(\nu,e^-)^{81}Kr^*$ reaction, producing $^{81}Kr^*$ in the 190-keV isomeric state with a 13-second half-life that decays to ^{81}Kr with a half-life of 2×10^5 years. The effective neutrino threshold energy is 0.522 MeV. Because of the long lifetime of the product, the bromine method was originally suggested as a geochemical experiment in which ^{81}Kr would be extracted from a salt deposit (107). This approach is impractical because salt deposite are impure and contain low concentrations of bromine (108). Hurst suggested that a real-time ^{81}Br experiment could be made practical by single-atom counting of ^{81}Kr via a laser–mass spectroscopic method (109). This method was developed at Oak Ridge National Laboratory and has the capability of measuring a few hundred atoms of ^{81}Kr (110). A feasible detector would include 400,000 liters of a suitable bromine liquid compound and would use the same extraction procedure as in the chlorine experiment (111).

Initially, the bromine detector was thought to be mainly sensitive to the neutrinos from $e^- + {}^7Be \rightarrow {}^7Li + v_e$. The estimates (111) of the neutrino capture cross sections for ^{81}Br have since been reevaluated. The decay of the 13-second isomeric state of ^{81}Kr* has been measured by two independent groups (112), and (p,n) reaction studies have been carried out (113). These new results have been incorporated into the standard model of Bahcall & Ulrich (1) to yield a total predicted rate of 28 ± 17 SNU (3σ error); of this rate, 35% and 55%, respectively, result from ^7Be and ^8B decay neutrinos. Accordingly, interest in the bromine experiment has for now subsided. It would be revived if the ^{37}Cl detector, which could easily be converted, were to become available.

THE IODINE DETECTOR   Suggested by Haxton (118), the iodine detector takes advantage of a halogen-containing target material transformed to a rare gas product by neutrino absorption. Iodine is mono-isotopic (^{127}I), and neutrino capture produces the ^{127}Xe with a lifetime of 36 days. The ground-state transition is forbidden, but the 125-keV state should be a favorable transition. This capture reaction has a threshold (0.789) low enough to observe ^7Be decay neutrinos. Haxton makes a rough estimate of the cross section for ^8B neutrinos by a simple extrapolation of (p,n) data and nuclear modeling. The result of these estimates indicates that an iodine detector would have a very favorable neutrino capture rate for ^7Be and ^8B decay neutrinos. A detector using 400,000 liters of methylene iodide (CH_2I_2) may have a total neutrino capture rate 15–20 times the rate of the present chlorine detector. It is, of course, important to verify these estimates by (p,n) reaction studies.

6.   ELECTRONIC DETECTORS

6.1   *General*

There is not yet any universal solar neutrino detector. Depending largely on individual inspiration and technical feasibility, each solar neutrino detector now in operation, in construction, or in proposal form is designed to emphasize the detection of a certain region of the solar neutrino spectrum. Furthermore, the large mass of high purity target material required and the need for appreciable shielding of the detector have restricted the opportunities for detector development. Accordingly, the serious experimental challenge presented by a neutrino spectrum extending from very low energy to 18 MeV and including several relatively intense monoenergetic lines, has until now been met only by one radiochemical detector and one electronic detector.

Electronic detectors are designed to record the time of occurrence and the energy of a solar neutrino interaction, and in some cases, the direction of the recoiling electron. That interaction may be purely leptonic, e.g. $v_e e \rightarrow v_e e$, which is the basis of detection in the currently operating electronic detector, the imaging water Čerenkov counter (119), Kamiokande-II. Alternatively, a detector material may be employed that is responsive to a given region of the energy spectrum through a specific semileptonic neutrino interaction with the target nuclei. Detectors employing large quantities of deuterium (120), boron, and indium (122) have been proposed for that purpose.

If their design requires them to be especially massive, the detectors of semileptonic neutrino reactions may also observe the leptonic reaction $v_e e \rightarrow v_e e$ with its smaller cross section, and certain targets, ^{2}H and ^{11}B for example, may allow the detection of weak neutral-current interactions initiated by v_μ and v_τ, as well as by v_e. The neutral-current reactions induced by neutrino flavors other than v_e would, if unambiguously identified, provide prima facie evidence of neutrino oscillations (see Section 3).

6.2 Proposed Detectors

Table 2 lists the electronic detectors directed toward studies of the solar neutrino spectrum and some of their salient properties. It is also useful to be aware of the cosmic-ray muon rate in the underground laboratories in which these detectors are or might be located, particularly because of the scarcity of adequate laboratories for this purpose. This is shown in Figure 6. Almost all of the detectors listed as proposed have been extensively described by their proponents. Here, we simply touch on their principal goals and methods.

Five of the detectors are capable of observing $v_e e \rightarrow v_e e$, initiated principally by v_e from ^{8}B decay. One of those proposed detectors, an imaging water Čerenkov detector using heavy water as target material (120), can also detect efficiently the electrons from $v_e d \rightarrow e^- pp$, since the binding energy of the deuteron is only 2.2 MeV. Similarly, the proposed ^{11}B detector (42) is a boron-loaded liquid scintillation counter designed to observe electrons from the reaction ^{11}B$(v_e, e^-)^{11}$C induced by v_e from ^{8}B decay. Because of the specific properties of ^{2}H and ^{11}B, both of those detectors may be capable of observing weak neutral-current reactions produced by

Table 2 Electronic detectors for the study of the solar neutrino spectrum

Target	Threshold (MeV)	Source(s)	Reaction(s)	Location	Detector group	Status[a]	Remarks
Light water	5–10	^{8}B	$\nu_e e \to \nu_e e^b$	Japan	Kamiokande-II	O	Imaging water Čerenkov detector; 20% photocathode coverage; 2.14-kton total mass
Light water	5–10	^{8}B	$\nu_e e \to \nu_e e^b$	Japan	SuperKamiokande	P	Imaging water Čerenkov detector; 40% photocathode coverage; 32-kton total mass
Heavy water	5–10	^{8}B	$\nu_e e \to \nu_e e^b$ $\nu_e d \to e^- pp$ $\nu_\mu(\nu_\tau)d \to \nu_e(\nu_\mu)(\nu_\tau)pn$	Canada	Sudbury Neutrino Observatory	P	Imaging water Čerenkov detector; 40% photocathode coverage; 1-kton useful mass submerged in light water shield
^{11}B	≤2	^{8}B	$\nu_e + {}^{11}B \to e^- + {}^{11}C$ $\nu_e(\nu_\mu)(\nu_\tau) + {}^{11}B \to \nu_e(\nu_\mu)(\nu_\tau) + {}^{11}B^*$	Italy	ATT Bell Labs Drexel	P	Boron-loaded liquid scintillator
^{40}Ar	≤3	^{8}B	$\nu_e e \to \nu_e e^b$	Italy	Icarus/ Gran Sasso	P	Liquid argon drift chambers; (1–3-kton useful mass)
Liquid scintillator $(CH_2)n$	≥0.25	$e^- + {}^7Be \to {}^7Li + \nu_e$	$\nu_e e \to \nu_e e$	?	?	P	50–100 tons of liquid scintillator sufficient but required to be submerged in a multikiloton water shield
Indium loaded liquid scintillator	≥0.128	$e^- + {}^7Be \to {}^7Li + \nu_e$	$\nu_e + {}^{115}In \to {}^{115}Sn^* + e$	France	Indium Collab./ Fréjus	P	11 tons of indium in cellular structure of 145 tons of liquid scintillator

[a] O = operational, P = proposed.   [b] Neutrino elastic scattering reaction on electrons with larger cross section than those for other neutrino types.

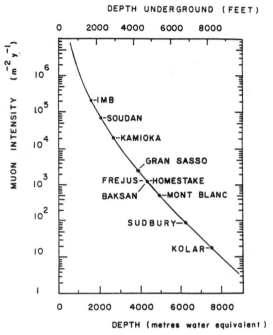

Figure 6   The cosmic-ray muon intensity as a function of the underground depth of various laboratories.

v_μ and v_τ, provided that backgrounds from natural radioactivity in the detector materials and containment structures can be sufficiently reduced and adequate energy resolution is achieved (121). Neither of these detectors can detect a directional correlation with the Sun when observing semi-leptonic reactions. The proposed SuperKamiokande is ten times larger in volume than Kamiokande-II (123). It is a light water, imaging Čerenkov detector following and improving on the design of the existing Kamiokande-II. SuperKamiokande will be at the same depth as Kamiokande-II, which is described in detail below.

The proposed Icarus detector uses liquid argon as the target material (124). The energy and direction of electrons produced in the reaction $v_e e \rightarrow v_e e$, initiated by v_e from ^8B decay, will be detected in a series of drift chambers for which the argon would constitute the drift liquid. Low energy electron tracks in argon of sufficient purity have been successfully detected in small prototypes. It remains to extend this accomplishment first to a detector roughly a ton in mass, and then to a full-size detector of kilotons.

Systems directed primarily toward detecting the monoenergetic 861-keV v_e from the reaction $e^- + {}^7Be \rightarrow {}^7Li + v_e$ have also been suggested. The motivation for measurement of this neutrino flux is its close relationship to the flux of ${}^8B$ neutrinos; the former proceeds by electron capture in ${}^7Be$, while the latter is the result of proton capture in ${}^7Be$. The ratio of the neutrino fluxes from electron capture and from proton capture is independent of the number density of ${}^7Be$ in the Sun's core; it is dependent only on the ratio of the approximately equal number densities of electrons and protons, and on the temperature T_c in the solar central region through the Gamow-Coulomb barrier penetration factor. It can be shown directly that a measurement of the ratio of those fluxes with an accuracy of $\pm 20\%$, for example, would yield T_c with an uncertainty of approximately $\pm 1.5\%$. This measured value of T_c might then be compared with the value predicted by the SSM, which is thought to have a similar uncertainty (e.g. 10). A significant disagreement would signal either a departure from the SSM or an energy-dependent oscillation effect, or both.

One such suggestion has been for a detector consisting of a liquid scintillator (126). The advantage of a liquid scintillator as a detector of the 861-keV neutrino flux is the high event rate (approximately 0.35 event per day per ton) for $v_e e \rightarrow v_e e$. The principal disadvantages are the need for extremely small contamination of natural radioactive elements in the scintillator and the lack of a distinctive event signal that would ensure discrimination against the backgrounds from residual radioactivity in the detector itself and from the outside.

An attractive alternative for observing v_e from $e^- + {}^7Be \rightarrow {}^7Li + v_e$ is a detector utilizing ${}^{115}In$. It was suggested as early as 1976 that an electronic detector with ${}^{115}In$ as its target material should be considered seriously for the detection of v_e from the $p + p \rightarrow d + e + v_e$ and $e^- + {}^7Be \rightarrow {}^7Li + v_e$ reactions because of the low threshold (128 keV) for v_e in ${}^{115}In$. [A solar neutrino detector based on indium was initially suggested by Raghavan (127).] The natural radioactivity of ${}^{115}In$, which beta-decays with a half-life of 5×10^{14} year and which has an endpoint energy of 495 keV, mitigates against ${}^{115}In$ as a detector of the neutrinos from the $p + p$ reaction up to 420 keV. However, a detector constructed of cells of indium-loaded liquid scintillator appears to be technically feasible and may be capable of measuring the flux of 861-keV v_e from the electron capture reaction in ${}^7Be$.

Finally, attempts to develop a detector that would be sensitive over a wide range of neutrino energies and to all neutrino flavors without discrimination have been in progress for several years. In general, these involve superconducting materials in which low energy neutral-current neutrino interactions would excite microscopic grains in a solid or phonons in a liquid. These attempts to extend the energy range of neutrino detection to extremely low energies are important primarily in the promise they hold for the future (see 128).

6.3 The Kamiokande-II Solar Neutrino Detector

The Kamiokande-II detector is an imaging water Čerenkov detector of total useful mass 2140 metric tons, of which the central 680 tons comprise the fiducial mass for observation of the ^{8}B solar neutrinos (119). Solar neutrinos are detected through the reaction $v_e + e^- \rightarrow v_e + e^-$ and by measuring the initial position and vector momentum of the recoiling electron. It is difficult to detect electrons with energy less than about 6 MeV in such large water Čerenkov detectors, and consequently the energy threshold for solar neutrino detection is at least that value. On the other hand, the kinematics of the elastic scattering reaction imposes an angular constraint on the recoiling electron, $\theta_e^2 \leq 2m_e/E_e$, which implies that the recoiling electron direction is closely aligned with the incident neutrino direction. This, combined with the imaging property of the detector, makes it possible to project the incident neutrinos back to the Sun.

The properties of the Kamiokande-II detector relevant to the detection of solar neutrinos are given in Table 3. The resolutions in energy, position, and angle in Table 3 are limited by the low intensity of Čerenkov radiation, by scattering of the Čerenkov light in the detector water, and principally by multiple scattering of the low energy recoiling electrons in the water. The energy calibration is achieved through measurement of photons from a radioactive source inserted in the detector, through measurement of the energy spectra of electrons from the decays of muon-induced spallation products in the detector water, and through measurement of the spectrum of electrons from the decays of muons that stop in the detector.

For the low energy electrons produced by solar neutrino interactions, there are backgrounds arising from natural radioactivity, principally radon

Table 3 Properties of the Kamiokande-II detector at low energies

Trigger threshold (at present)[a]	7.5 MeV
Energy resolution (10 MeV)[b]	±22%
Vertex position resolution (10 MeV)[b]	±1.7 meters
Angular resolution (10 MeV)[b,c]	±28°
Energy calibration uncertainty[c]	≤5%

[a] Detection efficiency is ∼52% at 7.5 MeV, ∼78% at 9.0 MeV, and ∼97% at 12 MeV. Trigger requires ≥20 photomultiplier tubes to give signal within 100 nanoseconds.
[b] Evaluated by Monte Carlo calculations.
[c] Measured with source(s).

dissolved in the detector water, from gamma rays emitted by radioactive elements in the rock of the cavity housing the detector, and from the beta decays of the muon-induced spallation nuclei in the water.

After elimination of electrons from the spallation products, the ratio of the observed integrated (mostly background) event rate above 9 MeV to the corresponding event rate predicted by the standard solar model is roughly 20 to 1. The criterion to extract the ^{8}B solar neutrino signal from the background is the angular correlation of the neutrino signal with the direction of the Sun. Figure 7 shows event rate versus cos θ_{sun}, where θ_{sun} is the measured angle of the trajectory of each observed electron with respect to the direction of the Sun, averaged over 450 live days of data-taking. The positive signal near cos $\theta_{sun} = 1$ is evident. The shape of the signal is consistent with the measured angular resolution of the detector; it is independent of reasonable changes in the binning of events, and not especially sensitive to the electron energy threshold in the interval 9 to 12 MeV. Furthermore, the signal can be made to vanish by purposely inserting

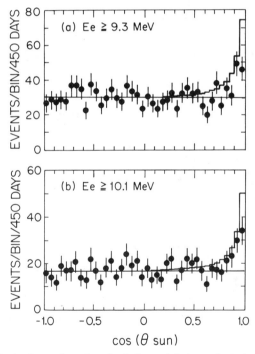

Figure 7   Distribution in cos θ_{sun}, where θ_{sun} is the angle between the trajectory of the recoil electron and the direction of the Sun at a given time from the Kamiokande-II detector. The data are restricted to the 680-ton fiducial region (of 2140 tons total); (*top*) $E_e \geq 9.3$ MeV and (*bottom*) $E_e \geq 10.1$ MeV (from 129).

an incorrect temporal or spatial relationship between the coordinates of the detector and the coordinates of the Sun.

The energy distribution of the events comprising the solar neutrino signal in the Kamiokande-II detector is also obtained by subtraction of background below the peak. It is shown in Figure 8, where it is compared with the corresponding spectrum predicted by the SSM. Given the low statistics, the agreement on the spectrum shape between the data for $E_e \geq 9.3$ MeV and the solar model is satisfactory.

The ^{8}B solar neutrino flux for $E_e \geq 9.3$ MeV recorded (129) by the Kamiokande-II detector in the time period 1 January 1987 to 1 June 1988 (450 live detector days) is compared to the SSM,

$$\frac{\text{KAM-II Data}}{\text{SSM}} = 0.46 \pm 0.13(\text{stat}) \pm 0.08(\text{syst}). \qquad 21.$$

Here SSM represents the value predicted by a Monte Carlo calculation based on the standard solar model (1) subject to the same event criteria and experimental resolutions as the data. Only a relative number is given because the flux below $E_e \leq 9.3$ MeV is not determined in this observation.

7. SUMMARY OF THE PRESENT EXPERIMENTAL SITUATION

The result of Kamiokande-II may be compared with that obtained by the ^{37}Cl detector in essentially the same time interval (130). Any discrepancy

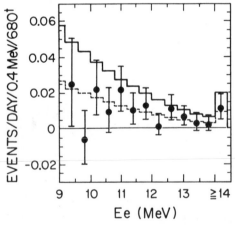

Figure 8 Differential energy distribution of the solar neutrino signal in the Kamiokande-II detector with background subtracted. The histograms are the distributions predicted by 1.0 and 0.46, respectively, times the standard solar model.

between the ^{37}Cl data and the SSM may be due entirely to a loss of ^{8}B flux, or, in the other limit, all contributions (see Table 1) to the total flux observed by and predicted for the ^{37}Cl detector may be proportionally reduced. One finds for those limiting cases the following result for the ^{8}B flux alone:

$$0.39 \pm 0.1 \leq \frac{^{37}\text{Cl Data}}{\text{SSM}} = 0.53 \pm 0.1, \qquad 22.$$

which is in agreement with the KAM-II value within the statistical errors.

The agreement between the two detectors during appreciably overlapping time periods confirms the accuracy of the two very different experimental methods and, equally important in view of the directional observation of Kamiokande-II, confirms the assumption that the ^{37}Cl detector observed solar neutrinos at least during the time period stated above.

Note that if the magnitude of the ^{8}B signal observed by the ^{37}Cl and Kamiokande-II detectors and the shape of the electron energy spectrum in Figure 8 remain the same as their statistical significance improves, then certain MSW solutions (for example, solution A in Figure 2a) will be disfavored. The average value of the neutrino rate observed in the ^{37}Cl detector in the fifteen years prior to 1985 (the detector was off during 1986 because of equipment failure) is, as we have seen, 2.1 $\pm$ 0.3 SNU, or approximately 0.27 times the value predicted by the SSM (see Figure 5). This is also lower than the combined result of the two solar neutrino detectors in 1987–1988. In this connection it is instructive to recall Figure 6 in which there is a suggestion of an anticorrelation of the ^{37}Ar production in the ^{37}Cl detector with the number of sunspots, i.e. with the solar magnetic cycle; this suggestion is reinforced by the 1987–1988 data.

While the results in Figure 6 are stimulating, it is too early to accept them as conclusive. If, however, the anticorrelation were confirmed, it would be difficult to explain by any reasonable modification of the standard solar model, and it consequently would require at least in part a particle physics explanation, i.e. as yet unobserved properties of neutrinos.

The agreement of the two detectors on a higher value of the ^{8}B solar neutrino flux in 1987–1988, approximately 0.5 the value predicted by the standard solar model, when taken in conjunction with the suggested anticorrelation of the solar neutrino flux with sunspot number, may be an indication of a more elaborate solar neutrino puzzle than has been recognized heretofore. At the onset of solar cycle 22, which is expected to reach a higher than usual maximum of sunspot numbers in the year 1990 $\pm$ 1, we can expect further data from the ^{37}Cl and Kamiokande-II detectors, as well as from the forthcoming gallium detectors, of special relevance to this puzzle.

8.   OUTLOOK FOR THE FUTURE

The study of solar neutrinos, which has been kept alive for twenty years largely by the radiochemical ^{37}Cl detector results of R. Davis, Jr. and his collaborators and by the theoretical work of John Bahcall and others, has in the past few years been infused with new vitality.

First, the demonstrated ability of the Kamiokande-II detector to observe ^{8}B solar neutrinos introduced a new experimental method to the field. The Kamiokande-II and the ^{37}Cl detector data recorded during the period 1987–1988 are in agreement on the relative magnitude of the ^{8}B flux, approximately 0.5 times the value predicted by the SSM. The average neutrino capture rate in the ^{37}Cl detector since 1970 is 2.3 ± 0.3 SNU, about 30% of the rate predicted by the SSM. In addition, the directional capability of Kamiokande-II yields strong evidence that the observed signal originates in the Sun, and it also determines the energy spectrum (above a given threshold) of the electrons that constitute the signal. Note that the ^{8}B flux is extremely sensitive to the central solar temperature as well as to certain nuclear reaction cross sections. Consequently the prediction of the SSM may be changed by relatively modest changes in the input parameters of the SSM, although no single particularly plausible change is indicated at present. The uncertainties in the predictions of the various neutrino sources are estimated to be (1): p-p and p-e-p reactions, 1%; ^{7}Be, 5%; ^{8}B, 12%; ^{13}N, 17%; and ^{15}O, 20% (where all are 1σ values). There is also a suggestion in the data of an anticorrelation of the ^{8}B flux with solar activity; this, if it persists, is difficult to explain solely by modification of the SSM.

Second, two gallium detectors will soon be operational. In observing neutrinos from the fundamental p-p reaction, they will open a new portion of the solar neutrino spectrum to exploration. The p-p reaction supplies 85% of all solar neutrinos, and the flux is relatively insensitive to uncertainties in the SSM. Any significant deficiency in the signal observed in the gallium detector will be important, but only about half of the total signal is expected to come from the p-p neutrinos. In this connection, the construction of two major laboratories dedicated to physics underground— Baksan in the Soviet Union and Gran Sasso in Italy—indicates a growing recognition of the potential importance of neutrino astronomy, as well as the foresight of the scientists whose efforts brought these laboratories into being. Data should be forthcoming soon from these detectors, and comparison of those results with the SSM will add a new chapter to the solar neutrino story.

Third, the carefully worked out proposals for new solar neutrino detectors, radiochemical and electronic, and the serious consideration given to them by the scientific community may signify the onset of a new era in solar neutrino astronomy. One of the goals is to search for possible neutrino oscillations corresponding to a range of masses and mixing angles inaccessible by any other means. If neutrino oscillations do occur then some of the neutrino flux arrives as v_μ or v_τ. The possibility exists for detecting the total flux including v_μ and v_τ by neutral-current reactions on nuclear targets since these reactions have the same cross section for all types of neutrinos. The proposed SNO detector using deuterium as a target would be sensitive to the neutrino disintegration of the deuteron. Another possibility is the search for nuclear excitation in ^{11}B. Also, those electronic detectors with relatively high event rates for ^8B neutrino-initiated inverse beta-decay reactions will provide precise ^8B flux data and a refined final-state electron energy spectrum with threshold perhaps as low as 6 MeV. Any problem with the SSM might change the total flux of ^8B neutrinos but cannot change the observed electron spectrum. On the other hand, neutrino oscillations in general are energy dependent and thus can distort the energy spectrum, as illustrated in Figure 2. An observed energy spectrum of greater statistical significance than that in Figure 8 will be invaluable in helping to select among possible hypotheses.

In summary, there appears to have developed in the last few years a new determination to confront the SSM with empirical data beyond that obtained from the ^{37}Cl detector. This determination stems from several sources: (a) better understanding of the limitations of accelerator-based neutrino experiments in searching for as yet undetected neutrino properties; (b) recognition that the solar interior provides an absolutely unique environment for the study of neutrinos with Earth-based detectors; and (c) realization that the exploitation of the Sun as a neutrino source for the study of intrinsic neutrino properties can be accomplished only by simultaneously testing the SSM with measurements of the total neutrino spectrum to provide unambiguous, precise quantitative conclusions regarding the validity of its predictions. Given the time scale set by the nature of solar neutrino detectors and the rate of observed solar neutrino interactions, this determination will be the herald of a decade of new results on and new understanding of solar neutrinos.

ACKNOWLEDGMENTS

This work was supported under US Department of Energy contracts DE-AC02-76-ER-3071 and National Science Foundation grants AST88-12418 and AST86-11924.

Literature Cited

1. Bahcall, J. N., Ulrich, R. K., *Rev. Mod. Phys.* 60: 297 (1988); Bahcall, J. N., *Neutrino Astrophysics.* Cambridge Univ. Press (1989)
2. Clayton, D. D., *Principles of Stellar Evolution and Nucleosynthesis.* New York: McGraw Hill (1968)
3. Filippone, B. W., *Annu. Rev. Nucl. Part. Sci.* 36: 717 (1986); Parker, P. D., see Ref. 9, 1: 17; Kajino, T., et al., *Astrophys. J.* 319: 531 (1987)
4. Bahcall, J. N., May, R. N., *Astrophys. J.* 155: 501 (1969)
5. Gari, M., Huffman, A. H., *Phys. Rev.* C7: 994 (1973); Bargholtz, C., *Astrophys. J. Lett.* 233: L161 (1979)
6. Grevasse, N., *Phys. Scr.* T8: 49 (1984); Aller, L. H., in *Spectroscopy of Astrophysical Plasmas*, ed. A. Dalgarno, D. Layzer. Cambridge Univ. Press (1986), p. 89
7. Casse, M., Cahen, S., Doom, C., in *Neutrinos and the Present Day Universe*, ed. T. Montmerle, M. Spiro, CEA, CEN Saclay (1985), p. 49
8. Iben, I. Jr., *Ann. Phys.* 54: 164 (1969)
9. Huebner, W. F., in *Physics of the Sun*, ed. P. A. Sturrock, et al. Dordrecht: Reidel (1986), 1: 33
10. Turck-Chièze, S., et al., *Astrophys. J.* 335: 415 (1988)
11. Willson, L. A., et al., *Comments Astrophys.* 12: 17 (1987); Guzik, J. A., et al., *Astrophys. J.* 319: 957 (1987)
12. Schatzman, E., Maeder, A., *Astron. Astrophys.* 96: 1 (1981); *Nature* 290: 683 (1981); Schatzman, E., in *I.A.U. Symp. No. 105, Observational Tests of Stellar Evolution Theory*, ed. A. Maeder, A. Renzini. Dordrecht: Reidel (1984), p. 491
13. Spruit, H. C., *Hydromagnetics of the Sun.* European Space Agency (1984); in *The Internal Solar Angular Velocity*, ed. B. R. Durney, S. Sofia. Dordrecht: Reidel (1984), p. 185
14. Newman, M. J., see Ref. 9, 3: 33; Rood, R. T., see Ref. 51
15. Faulkner, J., Gilliland, R. L., *Astrophys. J.* 299: 994 (1985); Spergel, D. N., Press, W., *Astrophys. J.* 294: 663 (1985); Spergel, D. N., Faulker, J., *Astrophys. J. Lett.* 331: L21 (1988); Boyd, R. N., et al., *Phys. Rev. Lett.* 51: 609 (1983)
16. Rood, R. T., in *Third ESO/CERN Symp., Astronomy, Cosmology, and Fundamental Physics*, ed. M. Caffo, et al. Dordrecht: Kluwer (1988)
17. Bilenky, S. M., Petcov, S. T., *Rev. Mod. Phys.* 59: 671 (1987); Langacker, P., in *Neutrino Physics*, ed. H. V. Klapdor. Berlin: Springer (1988), pp. 71–116
18. Robertson, R. G. H., Knapp, D. A., *Annu. Rev. Nucl. Part. Sci.* 38: 185 (1988)
19. Spergel, D. N., Bahcall, J. N., *Phys. Lett.* B200: 366 (1988)
20. Steigman, G., *Annu. Rev. Nucl. Part. Sci.* 29: 313 (1979); Bernstein, J., Feinberg, G., *Phys. Lett.* B101: 39 (1981)
21. Boehm, F., Vogel, P., *Physics of Massive Neutrinos*, Cambridge Univ. Press (1987), pp. 83–115; *Annu. Rev. Nucl. Part. Sci.* 34: 125 (1984)
22. Kitigaki, T., Yuta, H., eds., *Neutrino '86.* Singapore: World Scientific (1986), pp. 135–86
23. Langacker, P., *Phys. Rep.* 72: 185 (1981)
24. Wolfenstein, L., *Phys. Rev.* D17: 2369 (1978)
25. Mikheyev, S. P., Smirnov, A. Yu., *Nuovo Cimento* 9C: 17 (1986); *Sov. Phys. Usp.* 30: 759 (1987)
26. Bouchez, J., et al., *Z. Phys.* C32: 499 (1986)
27. Rosen, S. P., Gelb, J., *Phys. Rev.* D34: 969 (1986)
28. Parke, S. J., Walker, T. P., *Phys. Rev. Lett.* 57: 2322 (1986)
29. Petcov, S. T., *Phys. Lett.* 200B: 373 (1988); Haxton, W. C., *Phys. Rev.* D35: 2352 (1987)
29a. Kuo, T. K., Pantaleone, J., *Phys. Rev.* D35: 3432 (1987); *Rev. Mod. Phys.* (1989), in press
29b. Bethe, H., *Phys. Rev. Lett.* 56: 1305 (1986)
30. Cribier, M., et al., *Phys. Lett.* B182: 89 (1986); Baltz, A. J., Weneser, J., *Phys. Rev.* D35: 528 (1987)
31. Cisneros, A., *Astrophys. Space Sci.* 10: 87 (1971)
32. Voloshin, M. B., Vysotsky, M. I., *Yad. Fiz.* 44: 845 (1986) (*Sov. J. Nucl. Phys.* 44: 440); Voloshin, M. B., Vysotsky, M. I., Okun, L. B., *Zh. Eksp. Teor. Fiz.* 91: 754 (1986) (*Sov. Phys. JETP* 64: 446)
33. Liu, J., *Phys. Rev.* D35: 3447 (1987)
34. Voloshin, M. B., *ITEP Preprint*; Barbieri, R., Mohapatra, R. N., *Phys. Lett.* B218: 225 (1989)
35. Barbieri, R., Mohapatra, R. N., *Phys. Rev. Lett.* 61: 27 (1988); Lattimer, J., Cooperstein, J. *Phys. Rev. Lett.* 61: 23 (1988)
36. Leurer, M., Liu, J., *Phys. Lett.* B219: 304 (1989); Lim, C.-S., Marciano, W. J., *Phys. Rev.* D37: 1368 (1988);

Akhmedov, E. Kh., *Phys. Lett.* 213: 64 (1988)

37. Marciano, W., Parsa, Z., *Annu. Rev. Nucl. Part. Sci.* 36: 171 (1986)

38. Nozawa, S., et al., *J. Phys. Soc. Jpn.* 55: 2636 (1986); Bahcall, J. N., et al., *Phys. Rev.* D38: 1030 (1988)

39. Goodman, C. D., in *Solar Neutrinos and Neutrino Astronomy*, ed. M. L. Cherry, et al. New York: Am. Inst. Phys. (1985), p. 109

40. Grotz, K., et al., *Astron. Astrophys.* 154: L1 (1986); Krofcheck, D., et al., *Phys. Rev. Lett.* 55: 1051 (1985); Knofcheck, D., PhD thesis, Ohio State Univ. (1987)

41. Bahcall, J. N., *Rev. Mod. Phys.* 59: 505 (1987)

42. Raghavan, R. S., et al., *Phys. Rev. Lett.* 57: 1801 (1986)

43. Drukier, A., Stodolsky, L., *Phys. Rev.* D30: 2295 (1984); Cabrera, B., et al., *Phys. Rev. Lett.* 55: 25 (1985)

44. Bahcall, J. N., Sears, R., *Annu. Rev. Astron. Astrophys.* 10: 25 (1972)

45. Kuckowitz, B., *Rep. Prog. Phys.* 39: 21 (1976)

46. Bahcall, J. N., *Space Sci. Rev.* 24: 227 (1979); *Prog. Nucl. Part. Phys.* 6: 111 (1980)

47. Bahcall, J. N., Davis, R., *Science* 191: 264 (1976)

48. Weneser, J., Friedlander, G. F., *Science* 235: 755, 760 (1987)

49. Baltz, A., Weneser, J., *Comments Nucl. Part. Phys.* 18(5): 227 (1988)

50. Reines, F., Trimble, A., eds., *Proc. Solar Neutrino Conference*, Univ. California, Irvine (1972), Reports Summary in *Rev. Mod. Phys.* 45: 1 (1973)

51. Friedlander, G., ed., *Proc. Informal Conf. on the Status and Future of Solar Neutrino Research*, 2 Vols., BNL 50879 (1978), Summary in *Comments Astrophys.* 8(2): 47–54 (1978)

52. Cherry, M. L., Fowler, W. A., Lande, K., eds., *Proc. Am. Inst. Phys. Conf.*, No. 126. New York: Am. Inst. Phys. (1985)

53. Bahcall, J. N., Davis, R., in *Essays in Nuclear Astrophysics*, ed. C. A. Barnes, D. D. Clayton, D. N. Schramm. Cambridge Univ. Press (1982), p. 243

54. Pinch, T., *Confronting Nature, The Sociology of Solar Neutrino Detection*, Dordrecht: Reidel (1986)

55. Davis, R., "Fred Reines and Solar Neutrinos," in *Proc. Neutrino '88 Conf.*, Boston, ed. J. Schnepps. Singapore: World Scientific (1989), in press

56. Burbidge, E. M., Burbidge, G. R.,

Fowler, W. A., Hoyle, F., *Rev. Mod. Phys.* 29: 547 (1957)

57. Pontecorvo, B., *Chalk River Lab. Rep.*, PD-205 (1948)

58. Alvarez, L., *Univ. Calif. Radiat. Lab. Rep.*, UCRL-328 (1949)

59. Davis, R., *Phys. Rev.* 97: 766 (1955)

60. Davis, R., in *Radioisotopes in Scientific Research, Proc. 1st UNESCO Conf.*, Paris, ed. R. C. Extermann. New York: Pergamon (1957), 1: 728

61. Davis, R., Harmer, D. S., *Bull. Am. Phys. Soc.* 4: 217 (1959)

62. Davis, R., *Phys. Rev. Lett.* 12: 303 (1964)

63. Davis, R., Harmer, D. S., Hoffman, K. C., *Phys. Rev. Lett.* 14: 20 (1968)

64. Davis, R., Evans, J. C., Rogers, L. C., Radeka, V., *Neutrino '72 Conf.*, ed. A. Frenkel, G. Marx (1972), 1: 5

65. Cleveland, B. T., *Nucl. Instrum. Methods* 214: 451 (1983)

66. Davis, R., Lande, K., Cleveland, B. T., Ullman, J., Rowley, J. K., see Ref. 55

67. Wolfendale, A. W., Young, E. C. M., Davis, R., *Nature Phys. Sci.* 238: 1301 (1972)

68. Cassidy, G. L., in *Proc. 13th Int. Conf. on Cosmic Rays* (1973), 3: 1958

69. Fireman, E. L., see Ref. 52, p. 22

70. Zatsepin, G. T., Kopylov, A. V., Shirokova, E. K., *Sov. J. Nucl. Phys.* 33: 200 (1981)

71. Bahcall, J. N., *Rev. Mod. Phys.* 50: 881 (1978); *Phys. Rev.* B135: 137 (1964)

72. Sakurai, K., *Nature* 278: 146 (1979); *Solar Phys.* 74: 35 (1981); *Space Sci. Rev.* 38: 243 (1984)

73. Lanzerotti, L. J., Raghavan, R. S., *Nature* 293: 122 (1981)

74. Ehrlich, R., *Phys. Rev.* 22: 2282 (1982)

75. Raychaudhuri, P., *Solar Phys.* 104: 415 (1986)

76. Gavrin, V. N., Kopysev, Yu. S., Makeev, N. T., *JETP Lett.* 35: 608 (1982)

77. Subramanian, A., *Astron. Nachr.* 308: 127 (1987)

78. Bazilevskaya, G. A., et al., *Sov. J. Nucl. Phys.* 39: 543 (1984)

79. Haubold, H. J., Gerth, E., *Astron. Nachr.* 306: 203 (1985)

80. Bahcall, J. N., Field, G., Press, W., *Astrophys. J.* 320: L69 (1987); Bahcall, J. N., *Nature* 330: 318 (1987); Bahcall, J. N., *Phys. Rev. Lett.* 61: 2650 (1988)

81. Rowley, J. K., Cleveland, B. T., Davis, R., see Ref. 52, p. 1

82. Basu, D., *Solar Phys.* 81: 363 (1982)

83. Davis, R., in *ICOBAN Conf.*, Toyama, Japan, ed. J. Anfune, p. 237; Davis, R., Cleveland, B. T., Rowley, J. K., in *Int.*

Cosmic Ray Conf., Moscow, Vol. 4, 328 (1987); in *Underground Sci. Conf.*, Baksan Valley, USSR, ed. G. V. Domogatski, et al. (1988), p. 1

84. Davis, R., Evans, J. C., see Ref. 68, 3: 2001; Evans, J. C., Davis, R., Bahcall, J. N., *Nature* 251: 486 (1974)
85. Alexeyev, E. N., Alexeyeva, L. N., Chudakov, A. E., Krivosheina, I. V., in *20th Int. Cosmic Ray Conf.*, Moscow (1987), 4: 351
86. Hirata, K. S., et al., *Phys. Rev. Lett.* 61: 2653 (1988)
87. Kuzmin, V. A., *Sov. Phys. JETP* 22: 1051 (1966)
88. Deleted in proof
89. Dostrovsky, I., see Ref. 51, p. 231
90. Bahcall, J. N., et al., *Phys. Rev. Lett.* 40: 1351 (1978)
91. Markow, M. A., in *Neutrino '77 Conf.*, Baksan Valley, Moscow: Publishing Office NAUK (1977)
92. Zatsepin, G. T., *Proc. 8th Int. Workshop on Weak Interactions and Neutrinos*, Javea, Spain, ed. A. Morales (1982), p. 754
93. Internal proposal to the US Department of Energy, Chemistry Department, Brookhaven Natl. Lab., Upton, NY (1 April 1981)
94. Hampel, W., *Am. Inst. Phys. Conf. Proc.*, No. 96, ed. M. M. Neito, et al. (1983), p. 88; see Ref. 55
95. Hampel, W., in *Proc. Workshop Neutrino Physics*, Heidelberg, ed. V. Klapdor. Heidelberg: Springer (1988)
96. Barabanov, I. R., et al., see Ref. 52, p. 175; Gavrin, V. N., see Ref. 55
97. Champagne, A. E., et al., *Phys. Rev.* C38: 900 (1988)
98. Krofcheck, D., et al., *Phys. Rev. Lett.* 55: 1051 (1985)
99. Mathews, G. J., et al., *Phys. Rev.* C32: 796 (1985)
100. Champagne, A. E., et al., *Phys. Rev.* C38: 2430 (1988)
101. Rowley, J. K., see Ref. 51, p. 267
102. Veretenkin, E. P., Gavrin, V. N., Yanovich, E. A., *Sov. J. Atomic Energy* 58: 65 (1985)
103. Davis, R., *Phys. Rev.* 86: 976 (1952)
104. Deleted in proof
105. Kramer, S., in *Conf. Resonance Ionization Spectroscopy*; Hurst, G. S., Payne, M. G., Kramer, S. D., Young, J. P., *Rev. Mod. Phys.* 51: 767 (1979)
106. Litherland, T., Fireman, E. L., Rowley, J. K., *Nucl. Instrum. Methods* B29 (1987)

107. Scott, R., *Nature* 264: 729 (1976)
108. Rowley, J. K., Cleveland, B. T., Davis, R., Hampel, W., Kirsten, T., in *Proc. of the Conf. on the Ancient Sun*, ed. R. O. Pepin, J. A. Eddy, R. B. Merrill; *Geochim. Cosmochim. Acta*, Suppl. 13, p. 45 (1980)
109. Hurst, G. S., et al., *Phys. Today* 33(9): 24 (1980)
110. Chen, C. H., Kramer, S. D., Allman, S. L., Hurst, G. S. *Appl. Phys. Lett.* 44: 640 (1983)
111. Hurst, G. S., et al., *Phys. Rev. Lett.* 53: 1116 (1984)
112. Davids, C. N., Wang, T. F., Almad, I., Holzmann, R., Janssens, R. V. F., *Phys. Rev.* C35: 1114 (1987); Lowry, M. M., Kouzes, R. T., Loeser, F., McDonald, A. B., Naumann, R. A., *Phys. Rev.* C35: 1950 (1987)
113. Krofcheck, D., et al., *Phys. Lett.* B189: 299 (1987)
114. Cowan, G. A., Haxton, W. C., *Science* 216: 51 (1982)
115. Wolfsberg, K., et al., Summary in *Comments Astrophys.* 8(2): 196 (1985)
116. Haxton, W. C., Johnson, C. W., *Nature* 333: 325 (1988)
117. Rapaport, J., et al., *Phys. Rev. Lett.* 54: 2325 (1985)
118. Haxton, W. C., *Phys. Rev. Lett.* 60: 768 (1988)
119. Hirata, K. S., et al. (Kamiokande-II Collaboration), *Phys. Rev.* D38: 448 (1988), and references therein
120. Sinclair, D., in *Neutrino Physics*, ed. H. V. Klapdor, B. Povh. Berlin: Springer-Verlag (1988), p. 239
121. Chen, H. H., *Phys. Rev. Lett.* 55: 1534 (1985)
122. Booth, N. E., Salmon, G. L., Hukin, D. A., see Ref. 52, p. 216
123. Koshiba, M., *Phys. Today* Dec., p. 38 (1987)
124. Bahcall, J. N., et al., *Phys. Lett.* B178: 324 (1986)
125. Deleted in proof
126. Mann, A. K., in *Proc. 1986 Snowmass Summer Study*, ed. R. Donaldson, J. Marx (1986), p. 701
127. Raghavan, R. S., *Phys. Rev. Lett.* 37: 259 (1976)
128. Booth, N. E., in *Proc. Workshop on Superconductive Particle Detectors*, Torino, Italy, Oct. 1987, and references therein
129. Hirata, K. S., et al., *Phys. Rev. Lett.* 63: 16 (1989)
130. Davis, R. Jr., see Ref. 55

CUMULATIVE INDEXES

CONTRIBUTING AUTHORS, VOLUMES 30–39

A

Adelberger, E. G., 35:501–58
Alcock, C., 38:161–84
Allison, W. W. M., 30:253–98
Altarelli, G., 39:357–406
Amsel, G., 34:435–60
Andersen, J. U., 33:453–504
Arianer, J., 31:19–51
Arima, A., 31:75–105
Armbruster, P., 35:135–94
Arnold, J. R., 33:505–37
Ashery, D., 36:207–52

B

Bagnaia, P., 38:659–703
Benczer-Koller, N., 30:53–84
Berg, U. E. P., 37:33–69
Berger, E. L., 37:463–91
Berko, S., 30:543–81
Berry, H. G., 32:1–34
Bethe, H., 38:1–28
Birkelund, J. R., 33:265–322
Blok, H. P., 33:569–609
Bloom, E. D., 33:143–97
Boal, D. H., 37:1–31
Boehm, F., 34:125–53
Bohigas, O., 38:421–53
Bollinger, L. M., 36:475–503
Bonderup, E., 33:453–504
Braun-Munzinger, P., 37:97–131
Breunlich, W. H., 39:311–56
Brown, B. A., 38:29–66
Bucksbaum, P. H., 30:1–52
Bugg, D. V., 35:295–320
Busza, W., 38:119–59

C

Cahill, T. A., 30:211–52
Carr, J. A., 36:29–81
Cassiday, G. L., 35:321–49
Castaldi, R., 35:351–95
Chanowitz, M. S., 38:323–420
Charpak, G., 34:285–349
Chrien, R. E., 39:113–50
Cline, D., 36:683–716
Cobb, J. H., 30:253–98
Coester, F., 37:463–91
Cohen, J. S., 39:311–56

Cole, F. T., 31:295–335
Collins, J. C., 37:383–409
Colson, W. B., 35:25–54
Commins, E. D., 30:1–52
Cooper, S., 38:705–49
Cormier, T. M., 32:271–308; 37:537–65

D

Darriulat, P., 30:159–210
Davis, R. Jr., 39:467–506
DeTar, C. E., 33:235–64
Détraz, C., 39:407–65
DeVolpi, A., 36:83–114
Diamond, R. M., 30:85–157
Dieperink, A. E. L., 35:77–105
DiLella, L., 35:107–34
Donoghue, J. F., 33:235–64; 39:1–17
Douglas, D. R., 38:455–96
Dover, C. B., 39:113–50
Dragt, A. J., 38:455–96
Drees, J., 33:385–452

E

Edwards, H. T., 35:605–60
Ellis, D. V., 37:213–41
Ellis, J., 32:443–97
Ellis, S. D., 38:659–703
Engfer, R., 36:327–59
Ericson, T. E. O., 35:271–94

F

Fabjan, C. W., 32:335–89
Fernow, R. C., 31:107–44
Fick, D., 31:53–74
Filippone, B. W., 36:717–43
Fisk, H. E., 32:499–573
Franzini, P., 33:1–29
French, J. B., 32:35–64
Friar, J. L., 34:403–33
Frois, B., 37:133–76

G

Gaillard, J.-M., 34:351–402
Gaillard, M. K., 32:443–97
Gaines, I., 37:177–212
Gaisser, T. K., 30:475–542; 38:609–57

Garrett, J. D., 36:419–73
Geller, R., 31:19–51
Gibbs, W. R., 37:411–61
Gibson, B. F., 34:403–33; 37:411–61
Girardi, G., 32:443–97
Goeke, K., 32:65–115
Goldhaber, G., 30:337–81
Grenacs, L., 35:455–99
Grosso-Pilcher, C., 36:1–28

H

Hagemann, G. B., 36:419–73
Hasenfratz, A., 35:559–604
Hasenfratz, P., 35:559–604
Hass, M., 30:53–84; 32:1–34
Haxton, W. C., 35:501–58
Healy, L. M., 38:455–96
Heisenberg, J., 33:569–609
Herrmann, G., 32:117–47
Herskind, B., 36:419–73
Hinchliffe, I., 36:505–43
Hitlin, D. G., 38:497–532
Hofmann, W., 38:279–322
Holmes, S. D., 35:397–454
Hughes, V. W., 33:611–44
Huizenga, J. R., 33:265–322
Hung, P. Q., 31:375–438

I

Iachello, F., 31:75–105

J

Jackson, A. D., 33:105–41
Jones, W. V., 37:71–95

K

Kahana, S. H., 39:231–58
Kajantie, K., 37:293–323
Kammel, P., 39:311–56
Keefe, D., 32:391–441
Kienle, P., 36:605–48
Knapp, D. A., 38:185–215
Kneissl, U., 37:33–69
Kohaupt, R. D., 33:67–104
Kolb, E. W., 33:645–96
Kota, V. K. B., 32:35–64

507

CHAPTER TITLES, VOLUMES 30–39

Annual Reviews Inc.

A NONPROFIT SCIENTIFIC PUBLISHER

4139 El Camino Way
P.O. Box 10139
Palo Alto, CA 94303-0897 • USA

ORDER FORM

ORDER TOLL FREE
1-800-523-8635
(except California)

Telex: 910-290-0275

Annual Reviews Inc. publications may be ordered directly from our office by mail, Telex, or use our Toll Free Telephone line (for orders paid by credit card or purchase order*, and customer service calls only); through booksellers and subscription agents, worldwide; and through participating professional societies. **Prices subject to change without notice.** ARI Federal I.D. #94-1156476

- **Individuals:** Prepayment required on new accounts by check or money order (in U.S. dollars, check drawn on U.S. bank) or charge to credit card—American Express, VISA, MasterCard.
- **Institutional buyers:** Please include purchase order number.
- **Students:** $10.00 discount from retail price, per volume. Prepayment required. Proof of student status must be provided (photocopy of student I.D. or signature of department secretary is acceptable). Students must send orders direct to Annual Reviews. Orders received through bookstores and institutions requesting student rates will be returned. You may order at the Student Rate for a maximum of 3 years.
- **Professional Society Members:** Members of professional societies that have a contractual arrangement with Annual Reviews may order books through their society at a reduced rate. Check with your society for information.
- **Toll Free Telephone orders:** Call 1-800-523-8635 (except from California) for orders paid by credit card or purchase order and customer service calls only. California customers and all other business calls use 415-493-4400 (not toll free). Hours: 8:00 AM to 4:00 PM, Monday-Friday, Pacific Time. *Written confirmation is required on purchase orders from universities before shipment.
- **Telex: 910-290-0275**

Regular orders: Please list the volumes you wish to order by volume number.
Standing orders: New volume in the series will be sent to you automatically each year upon publication. Cancellation may be made at any time. Please indicate volume number to begin standing order.
Prepublication orders: Volumes not yet published will be shipped in month and year indicated.
California orders: Add applicable sales tax.
Postage paid (4th class bookrate/surface mail) **by Annual Reviews Inc.** Airmail postage or UPS, extra.

ANNUAL REVIEWS SERIES		Prices Postpaid per volume USA & Canada/elsewhere	Regular Order Please send:	Standing Order Begin with:
Annual Review of **ANTHROPOLOGY**			Vol. number	Vol. number
Vols. 1-14	(1972-1985)...............	$27.00/$30.00		
Vols. 15-16	(1986-1987)...............	$31.00/$34.00		
Vol. 17	(1988)	$35.00/$39.00		
Vol. 18	(avail. Oct. 1989)	$35.00/$39.00	Vol(s). _____	Vol. _____
Annual Review of **ASTRONOMY AND ASTROPHYSICS**				
Vols. 1, 4-14, 16-20	(1963, 1966-1976, 1978-1982) ..	$27.00/$30.00		
Vols. 21-25	(1983-1987)...............	$44.00/$47.00		
Vol. 26	(1988)	$47.00/$51.00		
Vol. 27	(avail. Sept. 1989)	$47.00/$51.00	Vol(s). _____	Vol. _____
Annual Review of **BIOCHEMISTRY**				
Vols. 30-34, 36-54	(1961-1965, 1967-1985).......	$29.00/$32.00		
Vols. 55-56	(1986-1987)...............	$33.00/$36.00		
Vol. 57	(1988)	$35.00/$39.00		
Vol. 58	(avail. July 1989)............	$35.00/$39.00	Vol(s). _____	Vol. _____
Annual Review of **BIOPHYSICS AND BIOPHYSICAL CHEMISTRY**				
Vols. 1-11	(1972-1982)...............	$27.00/$30.00		
Vols. 12-16	(1983-1987)...............	$47.00/$50.00		
Vol. 17	(1988)	$49.00/$53.00		
Vol. 18	(avail. June 1989)	$49.00/$53.00	Vol(s). _____	Vol. _____
Annual Review of **CELL BIOLOGY**				
Vol. 1	(1985)	$27.00/$30.00		
Vols. 2-3	(1986-1987)...............	$31.00/$34.00		
Vol. 4	(1988)	$35.00/$39.00		
Vol. 5	(avail. Nov. 1989)............	$35.00/$39.00	Vol(s). _____	Vol. _____

ANNUAL REVIEWS SERIES		Prices Postpaid per volume USA & Canada/elsewhere	Regular Order Please send:	Standing Order Begin with:

Annual Review of **COMPUTER SCIENCE**

			Vol. number	Vol. number
Vols. 1-2	(1986-1987)................	$39.00/$42.00		
Vol. 3	(1988)	$45.00/$49.00		
Vol. 4	(avail. Nov. 1989)...........	$45.00/$49.00	Vol(s). _____	Vol. _____

Annual Review of **EARTH AND PLANETARY SCIENCES**

Vols. 1-10	(1973-1982)................	$27.00/$30.00		
Vols. 11-15	(1983-1987)................	$44.00/$47.00		
Vol. 16	(1988)	$49.00/$53.00		
Vol. 17	(avail. May 1989)...........	$49.00/$53.00	Vol(s). _____	Vol. _____

Annual Review of **ECOLOGY AND SYSTEMATICS**

Vols. 2-16	(1971-1985)................	$27.00/$30.00		
Vols. 17-18	(1986-1987)................	$31.00/$34.00		
Vol. 19	(1988)	$34.00/$38.00		
Vol. 20	(avail. Nov. 1989)...........	$34.00/$38.00	Vol(s). _____	Vol. _____

Annual Review of **ENERGY**

Vols. 1-7	(1976-1982)................	$27.00/$30.00		
Vols. 8-12	(1983-1987)................	$56.00/$59.00		
Vol. 13	(1988)	$58.00/$62.00		
Vol. 14	(avail. Oct. 1989)...........	$58.00/$62.00	Vol(s). _____	Vol. _____

Annual Review of **ENTOMOLOGY**

Vols. 10-16, 18	(1965-1971, 1973)			
20-30	(1975-1985)................	$27.00/$30.00		
Vols. 31-32	(1986-1987)................	$31.00/$34.00		
Vol. 33	(1988)	$34.00/$38.00		
Vol. 34	(avail. Jan. 1989)...........	$34.00/$38.00	Vol(s). _____	Vol. _____

Annual Review of **FLUID MECHANICS**

Vols. 1-4, 7-17	(1969-1972, 1975-1985).......	$28.00/$31.00		
Vols. 18-19	(1986-1987)................	$32.00/$35.00		
Vol. 20	(1988)	$34.00/$38.00		
Vol. 21	(avail. Jan. 1989)...........	$34.00/$38.00	Vol(s). _____	Vol. _____

Annual Review of **GENETICS**

Vols. 1-19	(1967-1985)................	$27.00/$30.00		
Vols. 20-21	(1986-1987)................	$31.00/$34.00		
Vol. 22	(1988)	$34.00/$38.00		
Vol. 23	(avail. Dec. 1989)...........	$34.00/$38.00	Vol(s). _____	Vol. _____

Annual Review of **IMMUNOLOGY**

Vols. 1-3	(1983-1985)................	$27.00/$30.00		
Vols. 4-5	(1986-1987)................	$31.00/$34.00		
Vol. 6	(1988)	$34.00/$38.00		
Vol. 7	(avail. April 1989)	$34.00/$38.00	Vol(s). _____	Vol. _____

Annual Review of **MATERIALS SCIENCE**

Vols. 1, 3-12	(1971, 1973-1982)...........	$27.00/$30.00		
Vols. 13-17	(1983-1987)................	$64.00/$67.00		
Vol. 18	(1988)	$66.00/$70.00		
Vol. 19	(avail. Aug. 1989)...........	$66.00/$70.00	Vol(s). _____	Vol. _____

Annual Review of **MEDICINE**

Vols. 9, 11-15	(1958, 1960-1964)			
17-36	(1966-1985)................	$27.00/$30.00		
Vols. 37-38	(1986-1987)................	$31.00/$34.00		
Vol. 39	(1988)	$34.00/$38.00		
Vol. 40	(avail. April 1989)	$34.00/$38.00	Vol(s). _____	Vol. _____